Lecture Notes in Electrical Engineering 157

For further volumes:
http://www.springer.com/series/7818

Ford Lumban Gaol (Ed.)

Recent Progress in Data Engineering and Internet Technology

Volume 2

Springer

Editor
Ford Lumban Gaol
Bina Nusantara University
Perumahan Menteng
Jakarta
Indonesia

ISSN 1876-1100 e-ISSN 1876-1119
ISBN 978-3-642-28797-8 e-ISBN 978-3-642-28798-5
DOI 10.1007/978-3-642-28798-5
Springer Heidelberg New York Dordrecht London

Library of Congress Control Number: 2012933326

Printed on acid-free paper

Springer is part of Springer Science+Business Media (www.springer.com)

Foreword

In the recent years we are being faced with flooding of data whose handling has promoted rapid advancements in internet technology. This book is a collection of selected papers, all of which were presented at the International Conference on Data Engineering and Internet Technology (DEIT 2011).

This conference, which took place in Bali Dynasty Resort, Bali, Indonesia on 15th–17th March 2011, brought together researchers and scientists from academia, industry, and government laboratories to share their results and identify future research directions in data engineering and internet technology. Topics of interest include, among others: computational algorithms and tools, database management and database technologies, intelligent information systems, data engineering applications, internet security, internet data management, web search, data grids, cloud computing, as well as web-based application.

The book makes fascinating reading and will be not only of great interest to researchers involved in all aspects of Data engineering and Internet Technology, but also it will surely attract more practitioners and academics towards these research areas.

Dr. Ford Lumban Gaol

Contents

Applying Moving Boundary and Nested Grid to Compute the Accretion, Erosion at the Estuary

Duong Thi Thuy Nga and Nguyen Ky Phung

Abstract. To evaluate the accretion, erosion at the estuary, we use a current model based on a 2D finite-difference grid, an alluvium transport model and a bed sediment transport model. The first one gives the velocity distribution on the surface of water body and in the case of transient analysis, the velocity distribution is computed at each computational time step. This velocity distribution will be taken as the input for the other models. Those models often run very slowly. There are two improvements in this research: moving boundary to get more correct results and nested grid to help the models run fast. The computational results in study area are agreement to our experimental results and the real data in Ca Mau coastal zone. It shows that those models are more accuracy and faster than the old ones.

Keywords: current model, moving boundary, nested grid, sediment transport model.

1 Introduction

Accretion and erosion at estuaries affect much to our economy, ecological environment. Erosion is a disaster which damages our lives and the surrounding environment. Although accretion can create good alluvium, it makes the ability

Duong Thi Thuy Nga
University of Science, VNU HCM,
Ho Chi Minh City, Vietnam
e-mail: ngadtt@hcmus.edu.vn

Nguyen Ky Phung
SIHYMETE,
Ho Chi Minh City, Vietnam
e-mail: kyphungng@gmail.com

F.L. Gaol (Ed.): Recent Progress in DEIT, Vol. 2, LNEE 157, pp. 1–10.
springerlink.com

of flood drainage decreases. Therefore, computing and predicting those phenomena are very important in protecting our environment.

In the world, there are many research groups implemented this project. Most of them used Defant and Hansen methods [1, 7] at Atlantic, Pacific. Some of them used ADI method to solve the problem [2, 3, 4, 6, 9, 10]. The strength of those models is that the software is easy to use because they simplified the equations by removing many parameters. Unfortunately it is also their limit. The accuracy is low. In addition, their models usually run slowly. Therefore, two main objectives for the development of these models are:

- Improving the accuracy of the available models by applying moving boundary, and
- Using nested grid to help the models run fast.

The remainder of this paper is following, section 2 presents detailed expositions for the models. The experiment results are undertaken in section 3. Section 4 focuses on future developments of the model. Finally, section 5 gives conclusions of the study.

2 Methodology

2.1 Current Model

The model solves the depth averaged 2D shallow water equations. They are equations of floods, ocean tides and storm surges. They are derived by using the hypotheses of vertically uniform horizontal velocity and negligible vertical acceleration (i.e. hydrostatic pressure distribution).

$$\frac{\partial u}{\partial t}+u\frac{\partial u}{\partial x}+v\frac{\partial u}{\partial y}-fv+g\frac{\partial \varepsilon}{\partial x}+gu\frac{\sqrt{u^2+v^2}}{(h+\varepsilon)C^2}-\frac{\tau_x}{h+\varepsilon}=f_x(t)$$

$$\frac{\partial v}{\partial t}+u\frac{\partial v}{\partial x}+v\frac{\partial v}{\partial y}+fu+g\frac{\partial \varepsilon}{\partial y}+gv\frac{\sqrt{u^2+v^2}}{(h+\varepsilon)C^2}-\frac{\tau_y}{h+\varepsilon}=f_y(t)$$

$$\frac{\partial \varepsilon}{\partial t}+\frac{\partial}{\partial x}(h+\varepsilon)u+\frac{\partial}{\partial y}(h+\varepsilon)v=0$$

where u, v: depth averaged velocity components in X and Y directions [m/s]
ε : water surface elevation [m]
h+ ε : depth of water [m]
t: time [s]
x, y: distance in X and Y directions [m]
C: Chezy coefficient
f: Coriolis force coefficient
τ_x, τ_y: horizontal diffusion of momentum coefficient in X and Y directions
$f_x(t)$, $f_y(t)$: sum of components of external forces in X and Y directions

2.2 *Alluvium Transport Model*

$$\frac{\partial C}{\partial t}+u\frac{\partial C}{\partial x}+v\frac{\partial C}{\partial y}=\frac{1}{H}\frac{\partial}{\partial x}\left(HK_x\frac{\partial C}{\partial x}\right)+\frac{1}{H}\frac{\partial}{\partial y}\left(HK_y\frac{\partial C}{\partial y}\right)+\frac{S}{H}$$

*where:*C: depth averaged concentration [kg/m^3].
u, v: depth averaged velocity components in X and Y directions [m/s]
K_x, K_y: diffusion coefficients in the X and Y directions [m^2/s]
H: relative depth [m], $H = h + \zeta$
S: source of sediment particles [g/m^2.s]

To compute K_x, K_y, we use Elder formula:

$$K_x = 5.93\sqrt{g}h|u|C^{-1}$$

$$K_y = 5.93\sqrt{g}h|v|C^{-1}$$

where C: Chezy coefficient [m$^{0.5}$/s]

2.3 *Bed Sediment Transport Model*

$$\frac{\partial h}{\partial t}=\frac{1}{1-\varepsilon_p}\left[S+\frac{\partial}{\partial x}\left(HK_x\frac{\partial C}{\partial x}\right)+\frac{\partial}{\partial y}\left(HK_y\frac{\partial C}{\partial y}\right)+\frac{\partial q_{bx}}{\partial x}+\frac{\partial q_{by}}{\partial y}\right]$$

$$q_b = 0.053((S-1)g)^{0.5} d_m^{1.5} T^{2.1} D_*^{-0.3} \frac{(u,v)}{\sqrt{u^2+v^2}}$$

where S: source function.
$S = E$ *when* $\tau_b > \tau_e$; $S = -D$ *when* $\tau_b < \tau_d$; $S = 0$ *when* $\tau_d \le \tau_b \le \tau_e$

2.4 *Algorithm*

All of the models use an alternating direction implicit (ADI) finite different method to solve the problems. The method involves two stages. In each stage, a tri-diagonal matrix for the computational domain is built to solve the values:

$$\begin{bmatrix} B_1 & C_1 & & & & & \\ A_2 & B_2 & C_2 & & & & \\ & A_3 & B_3 & C_3 & & & \\ & & \cdots & & & & \\ & & A_i & B_i & C_i & & \\ & & & \cdots & & & \\ & & & A_{l-1} & B_{l-1} & C_{l-1} \\ & & & & A_l & B_l \end{bmatrix} * \begin{bmatrix} C_1^{t+1/2} \\ C_2^{t+1/2} \\ C_3^{t+1/2} \\ \vdots \\ C_i^{t+1/2} \\ \vdots \\ C_{l-1}^{t+1/2} \\ C_l^{t+1/2} \end{bmatrix} = \begin{bmatrix} D_1 \\ D_2 \\ D_3 \\ \vdots \\ D_i \\ \vdots \\ D_{l-1} \\ D_l \end{bmatrix}$$

- **Initial conditions**

At time t = 0: u = 0, v = 0, $\varepsilon = 0$

$$C(x,y,0) = C_0(x,y) \text{ or } C(x,y,0) = constant$$

- **Boundary conditions**
  - *For the current model*

- Compute tidal components $\varsigma = \sum_{i=1}^{N} A_i \sin(\omega_i t + \varphi_i)$, or

- Compute Q = U * W

Then, compute the velocities at the boundaries.

- *For the alluvium transport model*

- Solid boundary: $\dfrac{\partial C}{\partial n} = 0$

- Liquid boundary:
  - Water flow runs from outside to the domain: $C = C_b(t)$
  - Water flow runs out of the domain: $\dfrac{\partial^2 C}{\partial x^2} = 0$

2.5 Moving Boundary

In tidal fall, the surface area decreases, the shallow area appears and not be inundated. Then, we will not calculate the tidal features in these areas. Limit calculation will be moved by water withdrawal of tide.

In tidal rise, these shallow areas will be gradually submerged in water. Limit calculation will be gradually enlarged as the tidal water level increases.

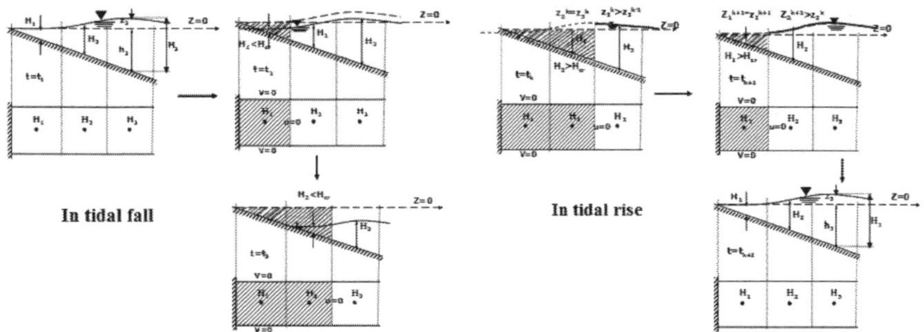

Fig. 1 Boundary in tidal fall

Fig. 2 Boundary in tidal rise

2.6 Nested Grid

The models often run slowly when being applied in a large domain. Hence, we use nested grid technique to help them run faster.

A computational grid is overlapped into the domain. The models will compute values such as u, v, ε, h. Locations of those values are described as follow:

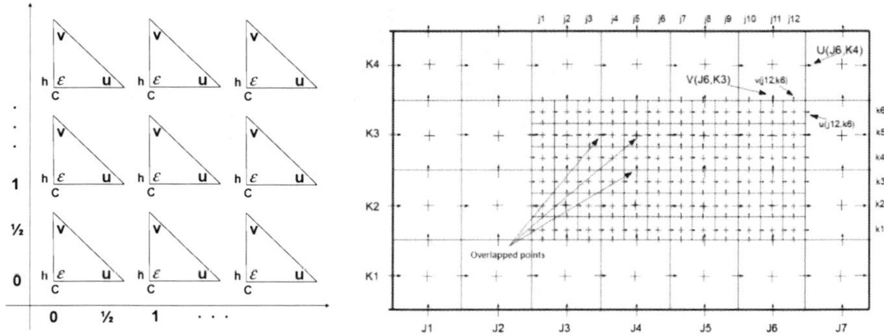

Fig. 3 Locations of u, v, ε, h

Fig. 4 Location of a closed grid on a sparse grid

The models relies on a nested grid to run fast. It uses a sparse grid for all the domain but closed grids for interesting areas.

- **Linking grids**

Values in close grids (lower case characters) are interpolated by values in the sparse grid (upper case characters). Then, the values in closed grids are used to edit the ones in the sparse grid.

$$u(j12,k6) = \frac{\alpha+1}{2\alpha}U(J6,K3) + \frac{\alpha-1}{2\alpha}U(J6,K4)$$

$$v(j12,k6) = \frac{\alpha+1}{2\alpha}V(J6,K3) + \frac{\alpha-1}{2\alpha}V(J7,K3)$$

$$\varsigma(j12,k6) = \left(\frac{\alpha+1}{2\alpha}Z(J6,K3) + \frac{\alpha-1}{2\alpha}Z(J6,K4)\right)\frac{2}{\alpha+1} + \frac{\alpha-1}{2\alpha}\varsigma(j12,k6)$$

3 Experiment Results

3.1 Applying the Models in Ca Mau Coastal Zone

The study area (W=61 km; L=88 km) is computed under the following conditions:

- Boundary condition for the current model

On the open boundary, the water levels are given by computing tidal components as shown in table 1 and 2. The parameters of these tidal components are from [11, 12, 13].

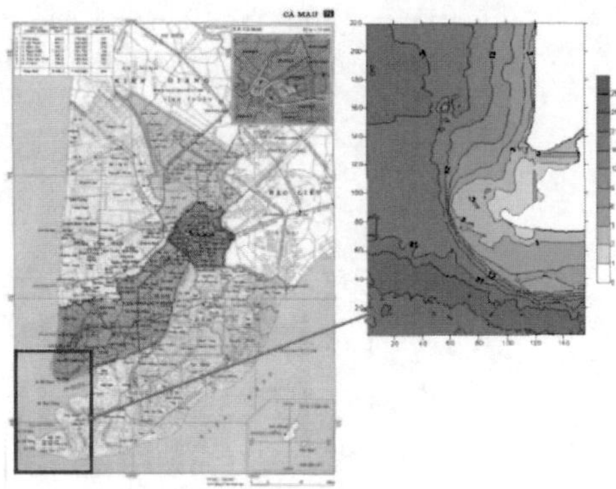

Fig. 5 Map of the study area

Table 1 Tidal characteristics at eastern sea

No.	Name of tidal components	Amplitude (m)	Phase (rad)
1	M_2	0.72	0.59
2	N_2	0.15	0.08
3	S_2	0.3	1.3
4	K_2	0.08	1.3
5	K_1	0.59	5.4
6	O_1	0.42	4.6
7	P_1	0.19	5.4
8	Q_1	0.01	4.2
9	M_4	0.01	4.8
10	MS_4	0.01	5.8
11	M_6	0.004	2.6

Table 2 Tidal characteristics at Thai Lan bay

No.	Name of tidal components	Amplitude (m)	Phase (rad)
1	M_2	0.15	1.35
2	N_2	0.15	0.08
3	S_2	0.12	1.35
4	K_2	0.08	1.3
5	K_1	0.38	0.18
6	O_1	0.25	1.8
7	P_1	0.49	5.4
8	Q_1	0.07	4.2
9	M_1	0.08	3.5

- Wind: Southern West direction, 4.2m/s.
- Alluvium concentration at boundaries: 0.0001g/ml

The models were tested in two different configurations with PC Core Duo, 2GB Ram and gave two notable results:

$\Delta x = \Delta y = 200m$, $\Delta t = 15s$: *the models took* ***95 hours*** *on getting the results for 3-month prediction of the alluvium and bed sediment transport.*

$\Delta x = \Delta y = 1000m$, $\Delta t = 100s$, 1 closed grid at Bay Hap estuary (1 km^2 x 1km^2, $\Delta x = \Delta y = 100m$): *the models took* ***28 hours*** *to get the results for 3-month prediction of the alluvium and bed sediment transport.*

It is faster than many models such as WUP-FIN (computed on Me Kong river), the Vietnamese author's models.

With such the speed of this computation, the models meets the demand of predicting alluvium transport in real time.

❖ **Alluvium transportation**

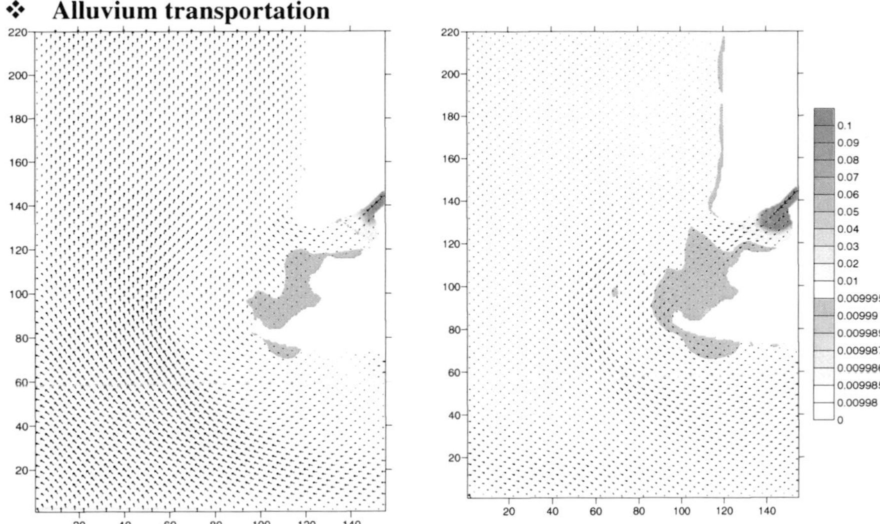

Fig. 6 Alluvium transportation in tidal rise **Fig. 7** Alluvium transportation in tidal fall

➢ **Evaluate**

At Bay Hap estuary, there is a heavy alluvium transportation. In tidal rise, the alluvium runs from the estuary to the ocean. In tidal fall, it runs following with the water flow to the South.

In reality, Bay Hap estuary is a place filled with a lot of alluvium. The models give good results in this area.

❖ **Bed sediment transportation**

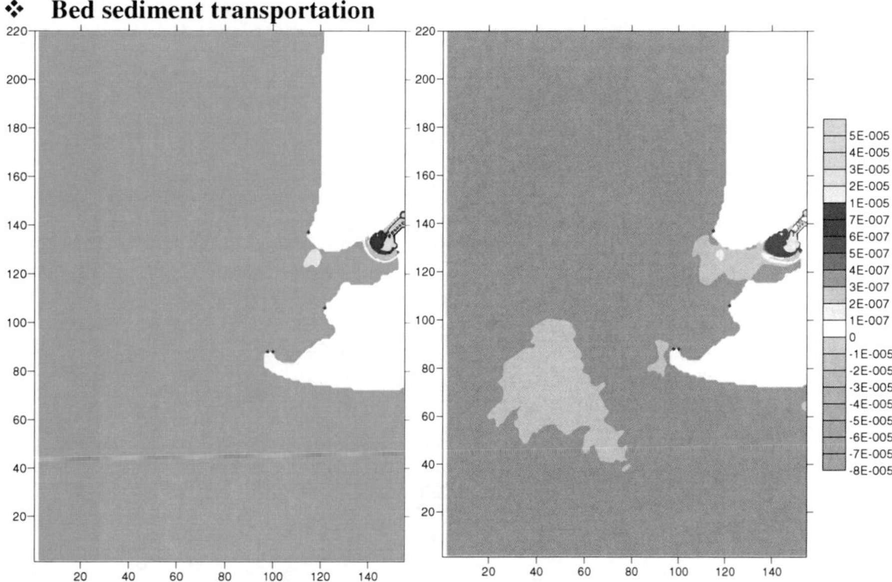

Fig. 8 Bed sediment transportation in tidal rise

Fig. 9 Bed sediment transportation in tidal fall

➢ **Evaluate**

Bay Hap estuary is an accreted area. The results agree with the fact that a large amount of alluvium is added up to this place annually. There is little erosion in the estuary.

3.2 *Simulation Tested Results of the Models*

The alluvium transport model is tested in a squared grid (3000m * 3000m) with some conditions: h=5m; $\Delta t = 10s$;available velocities u=v=0.1m/s; E_x=E_y=4m^2/s; source of alluvium M=5000 kg (x_o=5x, y_o= 5y).

The formula to compute the alluvium concentration from M:

$$C(x,y,t) = \frac{M/H}{4\pi\sqrt{E_x E_y}\,t}\exp\left(-\frac{(x-x_0-ut)^2}{4E_x t}-\frac{(y-y_0-vt)^2}{4E_y t}\right)$$

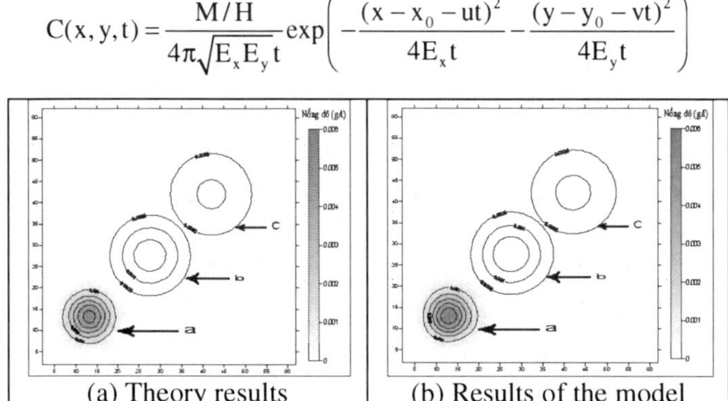

(a) Theory results (b) Results of the model

Fig. 10 Alluvium transport results
(a): after 1hour; (b): after 3 hours; (c): after 5 hours

➢ **Evaluate**

The results get from the models are similar to the ones in theory. This prove that the models can be applied to reality.

3.3 *Test the Model Results with Real Data*

The velocities at some stations are compared with the data from Ca Mau coastal zone.

Table 3 Computed velocities and real ones

Station	Real velocities (m/s)			Computed velocities (m/s)		
	Max	*Min*	*Average*	*Max*	*Min*	*Average*
BayHap	1.271	0.030	0.641	1.395	0.025	0.562
Tai Lan estuary	1.360	0.034	0.492	1.369	0.031	0.573

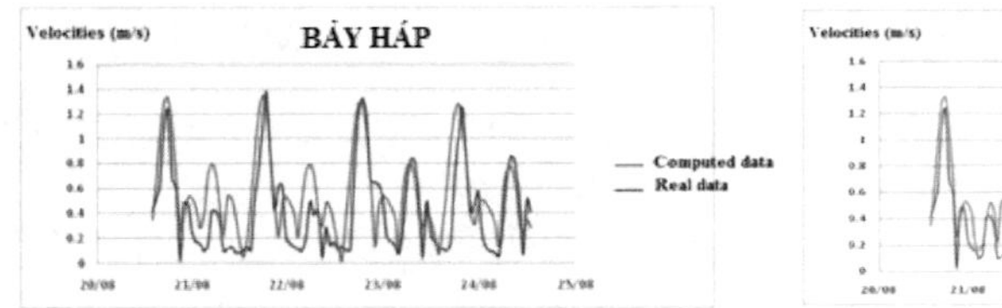

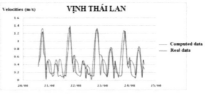

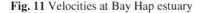

Fig. 11 Velocities at Bay Hap estuary

Fig. 12 Velocities at Thai Lan bay

➢ **Evaluate**

The computed velocities are appropriate with the real data in two aspects: amplitude and phase. This proves that the current model is reliable.

4 Future Developments

The models have been tested in Ca Mau coastal zone. To be more reliable, the model needs to be tested in real accidents with enough data.

5 Conclusions

This paper presents models to compute the accretion and erosion at an estuary. We have built two new algorithms and applied them to the models to get better results than the previous ones: moving boundary to help the models to give the more correct results and nested grid to make them run rapidly. In addition, the results were tested with simulation experiments and the real data in Ca Mau coastal zone. With the reliability and good speed of computation, we can apply them to reality.

References

[1] Downer, C.W., James, W.F., Byrd, A., Eggers, G.W.: Gridded Surface Subsurface Hydrologic Gridded Surface Subsurface Hydrologic Analysis (GSSHA) Model Simulation of Hydrologic Conditions and Restoration Scenarios for the Judicial Ditch 31 Watershed, Minnesota (2002)

[2] Minh, Đ.C., Nhân, N.H.:Thủy triều biển Đông, chương trình nghiên cứu cấp nhà nước KT. 03, đề tài KT.03.03 (1993)

[3] Wolanski, E., Nhan, N.H., Spagnol, S.: Sediment Dynamics During Low Flow Conditions in the Mekong River Estuary, Vietnam (1998)

[4] Hansen, M., de Fries, R.: Detecting long term forest change using continuous fields of tree cover maps from 8 km AVHRR data for the years 1982–1999. Ecosystems in press (2004)

[5] Friedrich, H.J.: Preliminary results from a numerical multilayer model for the circulation in the North Atlantic (2004)

[6] Tsanis, I.: Environmental Hydraulics: Hydrodynamic and Pollutant Transport Models of Lakes an Coastal Waters, vol. 56. Elsevier Press (2006)

[7] Horikawa, K.: Nearshore Dynamics and Coastal Processe. University of Tokyo Press (1988)
[8] van Rijn, L.C.: Principles Of Sediment Transport In Rivers Estuaries And Coastal Seas. Delft Hydraulics (1993)
[9] Hoa, L.T.V., Shigeko, H., Nhan, N.H., Cong, T.T.: Infrastructure effects on floods in the Mekong River Delta in Vietnam (2008)
[10] Bay, N.T., Phung, N.K.: The 2-D model of flow and sediment transportation in a curved open channel. International Colloquium in Mechanics of Solids, Fluids, Structures and Interaction (2002)
[11] văn Hoặc, P., Nhân, N.H.: Nghiên cứu xâm nhập mặn trên sông Đồng Nai phục vụ nhà máy nước 100.000m3/ngày, Tổng cụ Khí tượng thủy văn, phân viện Khí tượng thủy văn tại TPHCM (1993)
[12] văn Hoặc, P.: Báo cáo đề tài: Nghiên cứu tương tác động lực học biển – sông ven biển Cần Giờ phục vụ xây dựng cơ sở hạ tầng cho du lịch TPHCM, Sở Khoa học và công nghệ TPHCM (2004)
[13] Trung tâm khí tương thủy văn phía Nam: Vai trò của thủy triều trong vấn đề ngập lụt tại TPHCM, TPHCM (2000)
[14] Saied, U., Tsanis, I.K.: A coastal area morphodynamics model. Journal of Environmental Modelling & Software 23, 35–49 (2008)
[15] Kowalid, Z., Murty, T.S.: Numerical Modeling of Ocean Dynamics. Advanced Serieson Ocean Engineering 5 (1993)
[16] DHI Software. MIKE 21 Flow Model - Mud transport module. Scientific Background (2007)

Study on the Evaluation of Technological Innovation Efficiency of Equipment Manufacturing Industry in China Based on Improved DEA

Duan Jie, Liu Yong, and Wang Yanhong

Abstract. This paper uses a panel data of statistical panel data over the period 2007-2009, and applies the DEA method to evaluate the efficiency of the equipment manufacturing industry technology innovation in China, by using the improved DEA to evaluate 7 kinds of equipment manufacturing industry technology innovation efficiency, and has achieved the differentiation and sorting of various sub-sector by using the improved DEA model, which solve the problem that when the number of DMU is too small, the DMU can't be ranked. The Empirical result shows that the major categories of equipment manufacturing industry, technological innovation efficiency is generally low. The evaluation results can provide not only a new evaluation method for less decision unit, but also a new idea for the DEA evaluation theory and application.

Keywords: Equipment Manufacturing Industry, Technological Innovation Efficiency, DEA.

1 Introduction

Compared with developed countries, there is still a wide lag for China's equipment manufacturing industry on the development of scale, technical level and economic benefits. It's particularly reflected by the fact that the ability of independent innovation is weak, the core technology is controlled by others, the key

---

Duan Jie · Liu Yong
Northwestern Polytechnical University
e-mail: lwxgd2006@126.com, ly830621@sohu.com

F.L. Gaol (Ed.): Recent Progress in DEIT, Vol. 2, LNEE 157, pp. 11–17.
springerlink.com

technology equipment and core spare parts are relied on import. China's degree of dependence on foreign technology is more than 50%, and is far higher than 5% that is the developed countries'. So it has become urgent to study the current situation of technological innovation efficiency and enhance the level of technological innovation of China's equipment manufacturing industry. This paper evaluates the efficiency of technological innovation of China's equipment manufacturing industry based on DEA model, and has realized the effective differentiation and sorting of various sub-sectors by using the improved DEA model, so it can provide scientific and quantitative references for formulating the follow-up development strategies.

Differently from the existing research[1-7], this paper adds two virtual decision making units to the original mode and constructs the corresponding dual linear programming style with non-Archimedean infinitesimal, so the accuracy of evaluation results is improved, As the number of the equipment manufacturing sub-sectors is relatively small, it has some limitations to use DEA method to evaluate the efficiency of this industry[9-11], so DEA methods or evaluation model need to be improved. Domestic scholars' research on this field is few. The problem that the evaluation unit is less and can not effectively assess and distinguish has been solved, which provides new ideas for the theory and applications for the evaluation.

2 Econometric Models

2.1 The Mathematical Principles of DEA Model

DEA (Data Envelopment Analysis) is put forward by A. Charnes and WW Cooper in 1978, and is an efficiency evaluation method which deals with the problem concerning about multi-input and multi-output. This method compares the relative efficiencies of decision making units and evaluates them by using the mathematical programming. The first model CCR model in the DEA method has been selected.

2.2 Improved DEA Model

When we use the DEA method to evaluate the DMU, it will reduce the ability to differentiate the efficiency of the DMU as the input and output variables increase, so the accuracy of the evaluation result is affected. And when the number of DMU is small, the differentiation and sorting can not been effectively realized. To compensate for the shortcomings of traditional DEA model, this paper selects two years of data to increase the number of the DMU, and adds two virtual decision making units to improve the original model. The improved DEA model has achieved the distinction and sorting of all the DMU.

Two virtual decision making units (DMU_{n+1} DMU_{n+2}) are introduced, DMU_{n+1} represents the optimal decision making unit, the input and output vectors are:

$$X_{n+1} = (x_{1,n+1}, x_{2,n+1}, \dots, x_{m,n+1})^T, Y_{n+1} = (y_{1,n+1}, y_{2,n+1}, \dots, y_{s,n+1})^T$$

The input and output indicators of this DMU values respectively the minimum and maximum of corresponding indicators.

DMU_{n+1} indicates the worst decision unit , the input and output vectors are:

$$X_{n+2} = (x_{1,n+2}, x_{2,n+2}, \dots, x_{m,n+2})^T, Y_{n+2} = (y_{1,n+2}, y_{2,n+2}, \dots, y_{s,n+2})^T$$

The input and output indicators of the worst DMU values respectively the maximum and minimum of corresponding indicators.

The evaluation value of the optimal decision making unit is the objective function. On this basis, establish the following DEA model:

$$\mathrm{D}_{\varepsilon}) \begin{cases} \min[\theta - \varepsilon(\hat{e}^T s^- + e^T s^+) \\ s.t. \sum_{j=1}^{n} \lambda_j x_j + s^- = \theta x_0 \\ \sum_{j=1}^{n} \lambda_j y_j - s^+ = y_0 \\ \lambda_j \geq 0, s^- \geq 0, s^+ \geq 0, j = 1, 2, \dots, n \end{cases} \tag{1}$$

Firstly, we examine the optimal DMU, because it is supposed to be DEA effective, the following equations are available:

$$\theta = 1, s^- = 0, s^+ = 0 \tag{2}$$

Put(3)into(2),we can get two constraints that ensure this DMU is DEA effective:

$$\begin{cases} s.t. \sum_{j=1}^{n+2} \lambda_j x_j = x_{n+1} \\ \sum_{j=1}^{n+2} \lambda_j y_j = y_{n+1} \end{cases} \tag{3}$$

Now we examine the worst DMU, because it is supposed to be non-DEA effective and the worst one, the efficiency value of this DMU: $\theta_{n+2} \to 0$. Put the constraints that ensure it is the optimal DMU into the linear programming of the worst DMU, namely put (4) into linear programming of the worst DMU, it can be gotten:

$$\begin{cases} x_{n+1} + s^- = \theta_{n+2} x_{n+2} \\ y_{n+1} + s^+ = y_{n+2} \end{cases}$$

And the following equation is available:

$$s^- = \theta_{n+2} x_{n+2} - x_{n+1} \tag{4}$$

As $\theta_{n+2} \to 0$ and x_{n+2} is a bounded function, so $\theta x_{n+2} \to 0$, and $x_{n+1} > 0$, the following can be gotten:

$$s^- = \theta_{n+2} x_{n+2} - x_{n+1} \le 0 \tag{5}$$

It can be gotten from (1):

$$s^- \ge 0 \tag{6}$$

From (5) and (6), we know $s^- = 0$. And the constraint of the worst DMU can be gotten:

$$\theta_{n+2} = x_{n+1} / x_{n+2}$$

In summary, the dual linear programming with n +2 DMU is available:

$$(D_\varepsilon)\begin{cases} \min[\theta - \varepsilon(\hat{e}^T s^- + e^T s^+)] \\ s.t. \sum_{j=1}^{n+2} \lambda_j x_j + s^- = \theta x_0 \\ \sum_{j=1}^{n+2} \lambda_j y_j - s^+ = y_0 \\ s.t. \sum_{j=1}^{n+2} \lambda_j x_j = x_{n+1} \quad , \quad \sum_{j=1}^{n+2} \lambda_j y_j = y_{n+1} \\ \theta_{n+2} = x_{n+1} / x_{n+2} \\ \lambda_j \ge 0,\ s^- \ge 0,\ s^+ \ge 0,\ j = 1, 2, \ldots, n+2 \end{cases}$$

3 Empirical Analysis

3.1 *The Selection of Indicators*

Draw on the existing research and followed such the principles as representativeness, comparability and accessibility, we select the following input and output indicators of technological innovation:

3.2 *The Indicators of Technological Innovation*

The input and configuration of innovation resource determines the efficiency and quality of technological innovation, the selection principle is that it can measure the input level.The following input indicators of the innovation resource are selected: R & D number, R & D funding, the number of new developing projects. Output capability of innovative resources is measured by the following indicators: the number of patent applications, new product output.

3.3 Data Sources

The data comes from "China Statistical Yearbook" and "China Statistical Yearbook of Science and Technology", "China Industrial Economy Statistical Yearbook" (2008, 2009, 2010) and the National Bureau of Statistics official website.

3.4 Evaluation Results

To ensure the results of evaluation are accurate, we add two virtual units according to the modified DEA model, and evaluate the efficiency of technological innovation of China's equipment manufacturing industry by using EMS (Efficiency Measurement System) software. The specific results are shown in Table 1.

Table 1 China's equipment manufacturing industry's efficiencies of technological innovation

	Traditional DEA model				Improved DEA model				
	CRS	VRS	SE		CRS	VRS	SE		Ranking
DMU1	0.686	0.723	0.948	irs	0.504	0.758	0.665	irs	6
DMU2	0.622	0.630	0.987	drs	0.525	0.618	0.850	irs	5
DMU3	0.546	0.547	0.997	-	0.441	0.441	1.000	irs	7
DMU4	1.000	1.000	1.000	-	0.810	0.870	0.931	drs	1
DMU5	0.929	0.935	0.994	irs	0.777	0.796	0.976	irs	3
DMU6	1.000	1.000	1.000	-	0.798	0.844	0.945	drs	2
DMU7	0.743	0.823	0.903	drs	0.546	0.682	0.801	irs	4

From the evaluation results, DMU4 and DMU6 are DEA ineffective. Most of the innovation efficiencies of the sub-sectors are low, innovation resources have not been effectively used, the innovation output is not ideal.The most efficient industries of China's equipment manufacturing industry are DMU4 and DMU6. Larger market demand and more industry competition force the industries to do research on the technological development and innovation. And it improves the ability of the technology innovation, so there is the highest efficiency.

The industries with higher innovation efficiency are: Manufacture of Electrical Machinery &Equipment, Manufacture of Measuring Instrument and Machinery for Cultural Activity & Office Work, Manufacture of General Purpose Machinery. The input of innovative resources is sufficient and the innovating mechanism works well, so there is a higher efficiency.The least efficient industries are: Manufacture of Metal Products, Manufacture of Special Purpose Machinery. These two sub-industries is producing or processing equipment for the related industries, the demand is relatively fewer, and is also the end of the industrial chain links. Besides, the technological innovation ability is controlled by the level of technological innovation of their service industries. So the efficiency of these two sub-industries is relatively lower.

When use the traditional CCR model to evaluate, both the efficiencies of DMU4 and DMU6 are 1, which can not effectively distinguish and sort them.

So we add two virtual units (the best unit and the worst unit) and we assume the efficiency of the best unit to be 1 and that of the worst unit to be 0. The original best unit is not the best and the original worst unit is not the worst. The efficiencies of the two original best units are 1 and now they become 0.810 and 0.798, the former is more than the latter, so we can distinguish them. As a result, the innovation of improved DEA model solves the problem that when they both are the best unit, of which is better.

To make a further analysis of the status of technological innovation, we sort the technological innovation ability of China's equipment manufacturing industry based on the improved DEA model, and slack variable is given in evaluation results, it aims to identify the reasons for the low efficiency of innovation (because the input is redundant, or because the output is insufficient).The results are shown in Table 2.

Table 2 The evaluation results of innovation efficiency

	Efficiecy value	Effective-ness	Redundant input s_1^-	Redundant input s_2^-	Redundant input s_3^-	Inadequate output s_1^+	Inadequate output s_2^+
DMU1	0.686	invalid	0	21903	236697	780133	0
DMU2	0.622	invalid	0	98489	91883	1990127	0
DMU3	0.546	invalid	0	9968	0	0	90
DMU4	1.000	Effective	0	0	0	0	0
DMU5	0.929	invalid	0	11405	11856	7679848	0
DMU6	1.000	Effective	0	0	0	0	0
DMU7	0.743	invalid	20545	11068	0	0	199

Manufacture of Transport Equipment and Manufacture of Communication Equipment, Computer and other Electronic Equipment are DEA effective, it shows that the resources for innovation had been effectively used, technological innovation mechanism works well, the innovation output are satisfactory. The other sub-sectors are DEA ineffective, the technological innovation elements have not been fully used, the overall operational efficiency is low, the science and technology achievements transformation need to be strengthened.

DMU1 (Manufacture of Metal Products), DMU2 (Manufacture of General Purpose Machinery) and DMU3 (Manufacture of Electrical Machinery &Equipment) have two input indicators (R&D expenses, the number of developing projects) are redundant, an output indicator (new product output) is insufficient, which shows that the input elements had not been effectively used and the output value of new products need to be improved. DMU3(Manufacture of Special Purpose Machinery and Manufacture of Measuring Instrument) and DMU5 (Machinery for Cultural Activity & Office Work) have two input indicators (R & D number, R & D funding)are redundant, an output indicators(patent number) is insufficient, which shows that these two innovation resources have not been effectively utilized, the output of knowledge production is inadequate.

4 Conclusions

To make sure the evaluation result is accurate; this paper increases the number of the DMU to meet evaluation requirements, and achieves the effective separation and sorting by using improved DEA model. And it solves such the problem that when the number of decision making units is relatively small; some DMU can not been effectively distinguished and sorted. This method can been popularized and applied in such evaluation field. It provides a new way of thinking for the DEA theory and application.

Manufacture of Transport Equipment and Manufacture of Communication Equipment are DEA effective, the technological innovation resources of these industries have been effectively used, the technology innovation output is better. The other industries are DEA ineffective, the input is redundant and the output of innovative resources is inadequate. The evaluation results consistent with the general law: the majority technological innovation resources of China's equipment manufacturing industry have not been effectively used, the scientific and technological achievements transformation is need to be strengthened. The industry with higher technological content and stronger overall competitiveness has higher innovation efficiency, while the innovation efficiency of traditional industry is relatively low.

References

1. Zhao, N., Shao, H., Guo, W.: Empirical Study on Effect of IT Training on Information Transformation. Journal of Intelligence, 53–56 (2010)
2. Zhao, Z.: The New Vision of Effectiveness evaluation Based on DEA-New Model's Establishment and Application in Equipment Manufacturing Industry. Journal of Fujian University of Technology, 404–407 (2010)
3. Wu, L., Chen, W.: Study on the Evaluation of Technological Innovation Capability of the Equipment Manufacturing Industry Based on DEA. Science and Technology Management Research, 45–46 (2009)
4. Gongxin: Evaluation on the Developmental Efficiency of Heilongjiang 's Equipment Manufacturing Industry Based on Composite DEA Model. Journal of Harbin Engineering University, 657–663 (2010)
5. Chen, H.: Comprehensive Evaluation on the Technological Innovation Capability of the Equipment Manufacturing Industry Based on AHP/DEA Model In: Industry Discussion, pp. 117-119 （2009)
6. Zhang, J., Wang, Y.: Measure on the Efficiency of Technological Innovation of Equipment Manufacturing Industry in Shanxi Province Based on DEA. Economist, 220–225 (2010)
7. Chen, W., Man, Y.: Evaluation on Management Efficiency of Equipment Manufacturing Industry in Heilongjiang Province Based on DEA. Science and Technology Management Research, 204–205 (2009)

Design of Distributed Heterogeneous Data Retrieval Engine Using Service Oriented Architectures and Rich Internet Applications

Mutao Huang and Yong Tian

Abstract. As the demand for geospatial data rapidly increases, the lack of efficient way to find and retrieve appropriate data embedded in the mountain of available data on the Internet becomes critical. To meet this challenge, this paper presents a novel design of the data retrieval engine (DRE) targeted at facilitating one stop retrieval of geospatial data from heterogeneous data sources distributed on the Internet. The engine enables users to search, access and visualize the geospatial data of different types by setting query criteria in a map-based web portal. The engine is based on the service oriented architecture. It contains a built-in middleware composed of a harvest server, a central service database and a service search engine. Meta information of Open Geographic Consortium (OGC) web service are harvested by the harvest server and stored in the service database. When searching for particular geospatial data, those cached OGC services providing access to the data can be identified by the service search engine. The core functionalities of the middleware are exposed through a set of web services which can be consumed at client side. Two popular Rich Internet Applications (RIAs) technologies including Microsoft Silverlight and ArcGIS API for Silverlight are fully utilized to develop the web portal. The feasibility and effectiveness of the engine are carefully tested and demonstrated through investigations.

1 Introduction

Geographic information plays a key role in effective planning and decision-making in a variety of application domains. In order to facilitate sharing, discovery and retrieval

---

Mutao Huang · Yong Tian
College of Hydropower & Information Engineering,
Huazhong University of Science & Technology, Wuhan 430074, China
e-mail: `hmt1973@sina.com`, `ytian.world@gmail.com`

F.L. Gaol (Ed.): Recent Progress in DEIT, Vol. 2, LNEE 157, pp. 19–24.
springerlink.com

of geographic information, many organizations devote great efforts towards developing web service standards through which geospatial data are requested and delivered in a standard way no matter what underline infrastructure, platform or programming language are being used. The Open Geospatial Consortium (OGC), an international voluntary consensus standards organization, defines a set of open interface standards and associated encoding standards that can be used to expose and exchange geospatial information. Currently, the most widely used OGC web services include: Web Map Service (WMS), Web Feature Service (WFS) and Web Coverage Service (WCS). They are designed to enable client to retrieve geospatial data encoded in a raster map (e.g., JPG or PNG), features (e.g., GML) and grid (e.g., GeoTiff) respectively. They have been increasingly adopted to serve scientific data [1]. However, the OGC services are hard to be found and utilized because of various heterogeneous problems as well as the complexity of the services' specifications. This is especially true for the users who don't have professional knowledge and skills about geosciences domain.

Taking these issues into account, we propose a novel design of data retrieval engine with the purpose of providing a simple and efficient way through which users are able to easily locate and quickly retrieve geospatial data of interest from heterogeneous data sources. Through combinational use of several popular and powerful web technologies including service oriented architecture (SOA), rich internet applications (RIAs), and OGC web service standards, the engine brings the capability of on-the-fly retrieving geospatial data to users with easy to use web interface and high performance search results. More specially, our approach enables users to obtain geospatial data of different types in a single environment, rather than requiring them to navigate to the data provider's web page, query and download data, exploits OGC web service standards to facilitate discovery and retrieval of geospatial data, and supports seamlessly integrating multiple data layers into single view and provides high performance rendering capability by using RIAs.

2 Architecture Design

In order to provide a scalable framework for the development of web applications and deliver rich, cross-platform interactive experiences for user, the fundamental design of DRE is based on the philosophy of Service-Oriented Architectures (SOA) [4]. Fig. 1 shows the architecture of the system, which is classified into four layer. The bottom layer is composed of online data repositories that store geospatial data and provide access to their data content via web service compliant with OGC protocols. To make use of different protocols, a critical challenge is to resolve heterogeneous issues like syntactic and schematic heterogeneity. To alleviate these heterogeneities, one approach is to develop corresponding APIs (application program interfaces), each API is tailored to an individual protocol.

Upon the bottom layer is the middleware layer which consists of several key infrastructures. The harvest server is used to harvest geospatial web services, preprocess the metadata of the services and cache the processed information into the service database. The service database is in charge of storing and managing the metadata of the harvested services. The service search engine provides functions to search suitable geospatial services based on certain query criteria. The third layer contains several native web services that enable the client to retrieve information stored in the service data database and consume functionalities provided by the middleware. The top layer is the application layer where the geospatial web portal of the DRE resides. The portal is developed by leveraging two popular RIAs technologies including Silverlight and ArcGIS API for Sivlerlight. It offers a set of components that run inside users' web browser. Some of the components are described as follows: (1) the map viewer provides an interactive and map-based interface where users are able to graphically search and retrieve geospatial data. It organizes and displays geospatial data by means of layers. Layers of different type (raster, vector, etc.) could be overlaid upon a base map, with each layer individually configured and styled; (2) the geospatial web service manager enables users to view and manage the list of discovered services and analyze capabilities for a particular service; (3) the data adaptor is able to to retrieve geospatial data via different protocols; (4) the data exporter is used to export retrieved data to users' computer in various formats.

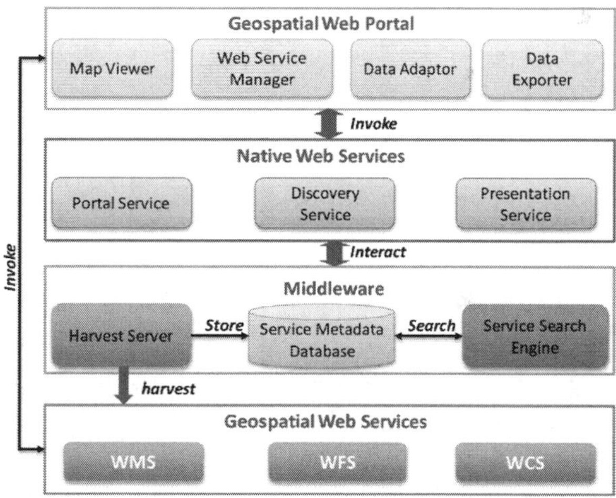

Fig. 1 System Architecture

3 Methodology

(1) Harvesting Metadata of Services

GetCapabilities is a standard Http request defined by OGC specifications, it's implemented by most of OGC web services. The response of this request is a XML document containing information about the service such as service owner, contact information, spatial reference system, geographic extent (i.e., bounding box), and data layers served by the service and so on. By making use of this request, the harvest server is able to automatically extract layer-based metadata and store them in the service database. In this work, SQL Server 2008 is adopted to serve as the service database.

Fig. 2 outlines partial diagram of the data model used to store the service metadata. At the center of the diagram is the Geospatial Service entity. A Geospatial Service entity represents an individual service. It provides the access entry of the service, indicates what service type it is and contains description information about the service such as title, abstract and so on. Specially, it contains a BBox field which utilizes the SQL Server's geographic data type representing the geographic extent on which the data provided by the service overlap. This entity is related to several other entities described as follows: (1) Service Provider. This entity contains information about the provider of the geospatial services, including contact information, industry categories and a list of services it provides; (2) Data Layer. It represents a data layer served by the service. For a geospatial service, the actual data content is exposed by the data layer. In most cases, a geospatial service contains more than one layer, and all the layers are organized in a hierarchy structure. In order to provide more accurate search capability, it's important to cache this rather deep but valuable information. The data layer entity is used to store such information; and (3) Service Operation. It stores standard operations supported by the service, such as GetMap, GetFeature and so on, and indicates the request method (e.g., Http get or post) for each operation.

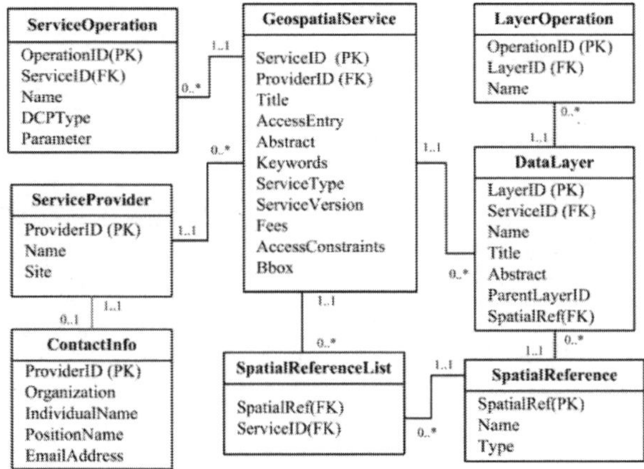

Fig. 2 Partial diagram of the data model used to store service metadata

(2) One Stop Retrieval of Geospatial Data

To acquire geospatial data of interest, the first task is to discover the services that provide access to the data. The procedure of searching for services is outlined as follows: (1) Users specify the query criteria (typically including topic keywords, geographic area and temporal range) in the web portal and submit the search request; (2) Once receiving the request, the search engine is invoked to extract parameters in the request and search through the service database to find out the most appropriate services matching against the query parameters. The service can be identified by its name, title, keywords, spatial reference system, and geographic extent and so on; (3) The discovered services list is returned to the client for presentation. Then users can select an interesting service, view data layers contained in the service and access data for a specified layer.

By taking advantaging of the RIAs technology, the retrieval of geospatial data is accomplished at the client side without the need to communicate with the web server of DRE. This capability makes full use of the client's computation power, thus may greatly accelerate the speed of retrieving information. The procedure of WMS access is taken as an example to demonstrate the approach. WMS specification defines three types of Http request: GetCapabilities, GetMap and GetFeatureInfo. The GetMap is the only way to request a map layer. The GetFeatureInfo provides feature information by identifying a point on a map based on its pixel location. When retrieving a map for a user-selected layer, the system automatically forms a GetMap request and sends the request to the WMS server. The request contains a base path identifying the WMS server and a query string identifying the desired map layer and any optional parameters. The full list of required and optional parameters could be found in the WMS specification. Once receiving the map image, the system creates a new raster layer rendering the image and adds the layer to the map viewer.

(3) Silverlight-based Data Visualization

Microsoft Silverlight provides a cross-browser, cross-platform development environment for building and delivering RIAs for the Web. It's used in this work to implement the DRE' web portal in combination with ArcGIS API for Silverlight. The user interface of the web portal has been designed to be simple and consistent through a single web page that provides a similar look-and-feel as desktop applications. The visualization of geospatial data is carried out by the map viewer. The map viewer has the ability of organizing, displaying and managing geographic information using layers. Spatial primitives like points, lines, polygons and other more complicated geometries are rendered using feature layer. Due to the usage of RIAs, it's easy to attach event handler directly to primitive features, providing highly interactive user interface [5]. In addition, attributes can be associated to individual spatial feature, producing more meaningful map presentation and enabling to apply advanced functionalities such as filtering, costumed rendering style and so on.

4 Case Study

In order to evaluate how well the system can be applied to enhance discovery and retrieval of geospatial data, an investigation was made. The investigation was made to retrieve vector data that represent surface water elements such as rivers and lakes. Fig. 3(a) depicts the graphic user interface of DRE's web portal. An auto-complete search box that automatically suggests topic keywords according to what users input locates the top of the interface. The map viewer occupies the main frame of the interface. Depending on the keyword "water" used in the investigation, the DRE searches for all the OGC services that provide access to this kind of data at first. The resulting services list is displayed in the floating window entitled with "Web Service Manager" in Fig. 3 (a). From the services list, users can select an interesting one and get its capabilities. The data layers contained in the service are displayed in a tree structure located at right part of the Web Service Manager. Fig. 3 (b) displays the retrieved vector data that represent rivers and lakes overlaying on Wuhan city located in the central of China.

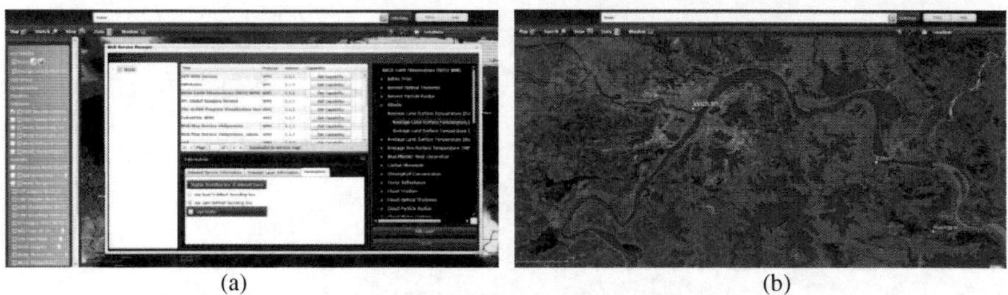

(a) (b)

Fig. 3 (a) Geospatial service discovery results; (b)Visualization of vector data representing lakes and rivers around Wuhan, China

Acknowledgements. This work was supported by a grant from the Special Research Foundation for the Public Welfare Industry of the Ministry of Water Resources (Project No.201001080) and a grant from the National Basic Research Program of China (Project No. 2007CB714107) as well as a grant from the Fundamental Research Funds for the Central Universities (Program No. 2010MS096).

References

[1] Li, Z., Yang, C., et al.: An optimized framework for seamlessly integrating OGC Web services to support geospatial sciences. International Journal of Geographic Information Science (2011), doi:10.1080/13658816.2010.484811

[2] Longueville, B.D.: Community-based geoportals: The next generation? Concepts and methods for the geospatial Web 2.0. Computers, Environment and Urban Systems (2010), doi:10.1016/j.compenvurbsys.2010.04.004

[3] Maguire, D.J., Longley, P.A.: The emergence of geoportals and their role in spatial data infrastructures. Computers, Environment and Urban Systems (2005), doi:10.1016/j.compenvurbsys.2004.05.012

[4] Nachouki, G., Quafafou, M.: MashUp web data sources and services based on semantic queries. Information Systems (2011), doi:10.1016/j.is.2010.08.001

[5] Wood, J., Dykes, J., Slingsby, A., Clarke, K.: Interactive visual exploration of a large spatio-temporal dataset: reflections on a geovisualization mashup. IEEE Transactions on Visualization and Computer Graphics 13, 1176–1183 (2007)

Relationship between Internet Search Data and Stock Return: Empirical Evidence from Chinese Stock Market

Ying Liu, Benfu Lv, Geng Peng, and Chong Zhang

Abstract. Internet search data can be used for the study of market transaction behaviors. We firstly establish a concept framework to reveal the lead-lag relationship between search data and stock market based on micro-perspective of investors' behaviors. Then we develop three types of composite search indices: investor action index, market condition index, and macroeconomic index. The empirical test indicates the cointegration relationship between search indices and the annual return rate of Shanghai composite index. In the long-term trend, each percentage point increase in the three types of search indices separately, the annual return rate will increase 0.22, 0.56, 0.83 percentage points in the next month. Furthermore, Granger causality test shows that the search indices have significant predictive power for the annual return rate of Shanghai composite index.

1 Introduction

As a scarce cognitive resource (Kahneman, 1973), investors' attention has become an important factor in asset pricing and decision making of market transaction. Traditional studies usually use proxy variables to measure the investors' attention, such as trading volumes, the number of news, advertising cost, etc (Barber and Odean, 2008; Hou, Peng, and Xiong, 2008). While the development of Internet provide a direct metric for investors' attention based on Internet search data.

Search engine, as the most general tool to get information from Internet, connects information resources and users' needs. At the same time, it also records their searching behavior. Based on hundreds of millions of search engine users'

Ying Liu · Benfu Lv · Geng Peng · Chong Zhang
Department of Management Science,
Graduate University of Chinese Academy of Sciences, Beijing, China
email: liuy218@126.com, lubf@gucas.ac.cn,
penggeng@gucas.ac.cn, zczoln@hotmail.com

F.L. Gaol (Ed.): Recent Progress in DEIT, Vol. 2, LNEE 157, pp. 25–30.
springerlink.com

records, the Internet search data can reflect the users' concerns and intends, foreshadow their behavior trends and patterns in real lives. In Chinese stock market, the percentage of online transaction in total market transactions has come up to 80% (Ma Guangti, 2009), so it has a certain universality of the stock transaction behavior reflected by Internt search data.

In this paper, we focus on the study of relationship between Internet search data and stock market based on methods of cointegration and granger causality.

2 Literature Review

These studies covered the indicators in three perspectives of microcosmic, mid-scope, and macrocosmic. From the microcosmic aspect, a good instance is epidemic symptomatic detection. Ginsberg, etc (2009) found that the percentage of influenza-like illness (ILI) search query in Google highly correlated with the percentage of ILI physician visits, and then they built a linear model based on the search data for monitoring influenza epidemics, which can estimate the level of influenza activity 1-2 weeks ahead of the traditional surveillance systems. Following this work, Jurgen A. Doornik (2009) extended it to autoregression model with calendar effects, and improved the prediction accuracy. In other empirical analysis, it also got good results of using Internet search data for forecasting movie box office, the popularity of online games and songs, the traffic of website and so on (Sharad, 2009; Heather, 2010).

From the mid-scope aspect, Choi and Varian (2009) did empirical test on the sales of U.S. automobile, housing, travel and other industries, they put keywords search frequency as a new independent variable adding to the traditional time-series models, and found that the prediction accuracy of above industries was significantly improved. Lynn and Erik (2009) also made study on the U.S. housing market and found that the search data had strong predictive power on the sales and prices of U.S. housing. From macrocosmic aspect, current studies mainly focused on the forecast of unemployment rate and private consumption. Askitas and Zimmerman(2009) demonstrated there existed strong correlations between keyword searches and unemployment rate based on monthly German data, and found significant improvements in prediction accuracy by using Google Trends. The similar work had done by Suhoy (2009) and Choi & Varian (2009) to study unemployment rate of Israel and US. As for consumption area, Kholodilin etc (2009) compared the growth rates of the real US private consumption based on both the conventional consumer confidence indicators and the Google indicators, and the results showed that the latter were 20% more accurate than the former. The predictive power of Internet search data was also confirmed by Torsten (2009), Nicolás(2009), and Marta(2009).

3 Data Processing

3.1 Relationship between Internet Search Data and Stock Return

Search data in this paper origin from Google insights, which supplies standardized searches of keywords in special time range. The standardized search data reflect attention of the keyword in Google and we call it search attention-degree. The stock data origin from Wind consulting financial database and the research sample is the closing price in Shanghai Composite Index (code: 000001) from January, 2004 to November, 2009. Data transformation is as follows:

Stock annual return: yt=100*ln(Pt/Pt-12), yt is return in Time t, Pt is closing price in Time t;

Search annual rate of change: xt=100*ln(St/St-12), xt is rate of change in Time t, St is search attention in Time t;

3.2 Keywords Selection and Time Difference Measurement

We choose 131 relevant keywords according to influence factors of stock transaction process. In order to appraise the relationship between search attention-degree variation and stock return, we attach 2 parameters to every keyword: leading order and relativity. If leading order >0, then it represents leading relationship; if leading order =0, then it represents synchronous relationship; if leading order <0, then it reflects hysteretic relationship. Relativity represents the similarity between change rate of search attention curve and stock return curve. The higher the relativity is, the more similar they are.

3.3 Keywords Index Composition

There are two kinds of method for setting weight in Academia: one is method of system assessment, the other is empowering according to the relativity. In this paper, we combine the two methods together and devide antecedents into three categories, composing three types of index: investor behavior index, market quotation index and macro situation index.

4 Emprical Analysis

In this paper, we choose Shanghai Composite Index annual return(y) as dependent variable, investor behavior index(x1), market quotation index(x2) and macro situation index(x3) as independent variable. Firstly, use ADF Test to inspect the stationary character of every index, and the results are showed in table 2: At 0.05 significant level, all sequences receive null hypothesis, indicating sequences were not steady, but in first order difference, test results all reject null hypothesis at 0.01 significant level, indicating they are sequences with one order unit root.

Next use Co-integration Test to test whether there is long-time stable relationship between dependent variable and independent variable: firstly, erect regression equation of independent and dependent variables. Secondly, take residual error of the equation for ADF Test.

When there is Co-integration relationship between dependent variable and independent variable, there is Granger causality between them. Granger causality test could reveal whether one variable could predict another.

In the test, we erect VEC to test Granger causality between variables, and ascertain lag order according to Criterion AIC and FPE suggested by Hsiao(1981).As a result, x1、x2、x3 is the Granger causality of y separately to 5%、1%、10% significant level, their joint survey is notable; y is the Granger cause of x1, x2 at 0.1 significant level, but can't cause x3.

5 Economic Analysis

Co-integration equation describes the long-term equilibrium relationship between three categories of search index and Shanghai Composite Index annual return, and we can estimate the trends of Shanghai Composite Index return according to this relationship.

Intercept: maturity of the stock market. Intercept is -5.94, which means when annual change rate of search index is 0, Shanghai Composite Index annual return is -5.94%. That is Shanghai Composite Index closing price will fall 5.94 percentages relative to last year. It can be concluded that our stock market is nonzero sum, and it takes short-time and speculative transactions as principal.

Coefficient: contribution margin of search index. Coefficients in the equation are all positive, which means there is a positive correlativity between search index and Shanghai Composite Index return. Macron situation index x3(0.83) > market quotation index x2(0.56) > investor behavior index x1(0.22), indicating that the prospects for macron situation most affects stock market, the second is investors' prospects for market quotation, and new private investors' prospects and behaviors less affects stock market.

Granger causality test indicates that the causal relation between search index and Shanghai Composite Index annual return have different characteristics in different keywords. The causal relation between Shanghai Composite Index annual return and investor behavior index, market quotation index is bidirectional, and the significance of search index is better, while macron situation index only unidirectional lead to Shanghai Composite Index annual return, which is consistent with the fact.(p<.1)

6 Closing, Limitations and Future Research

This paper theoretically and empirically analyzes the correlation between Internet search data and stock market, and confirms the predictive ability of keywords index to Shanghai Composite Index. The innovations are as follows:

(1) Theoretical framework and the internal mechanism
It is a new thinking and attempt that making socio-economic forecasting based on the Internet search data. Documents on this field are mostly seen after 2008 and are more focused on the empirical test, ignoring the internal mechanism of this method. A comprehensive intrinsic mechanism explanation is given in our paper: the behaviors of people have performance on both the Internet and the actual markets. The two performances are a reflection of the same thing, so there is some correlation between the two. Screening of the related words and using these keywords with scientific methods we can judge and predict the trend of the stock market. Our theoretical framework on the Internet is also universal with other social and economic behaviors.

(2) Search index as a new data source for stock market analysis and prediction
Compared with traditional studies of stock market, we introduce a processing method for Internet search data, which includes: keywords selection, time difference measurement, and leading search index composition. The search index as a new data source can provide better data basis for empirical analysis and prediction study.

However, there are also some shortage in this paper, although the Internet search data has universality, it isn't the only channel to get information from search engine; it may leave out some valuable information. If the initial keywords sample is not comprehensive, it may affect the reliability and effect. The methods to get initial keywords, combing internet search data and traditional data are the future research.

Acknowledgments. This research is supported by the National Natural Science Foundation of China under Grant 70972104, the National Natural Science Foundation of China under Grant 70772103, Beijing Natural Science Foundation of China under Grant 9083017, and Young Scholar Foundation of Alibaba Corp. under Grant Ali-2010-A-5.

References

[1] Kahneman, D.: Attention and Effort. Prentice-Hall, Englewood Cliffs (1973)
[2] Barber, B.M., Odean, T.: All That Glitters: The Effect of Attention and News on the Buying Behavior of Individual and Institutional Investors. Review of Financial Studies 21(2), 785–818 (2008)
[3] Hou, K., Lin, P., Wei, X.: A tale of two anomalies: The implications of investor attention for price and earnings momentum. Working Paper. Ohio State University and Princeton University (2008)
[4] Ginsberg, Mohebbi, Patel, Brammer, Smolinski, Brilliant: Detecting influenza epidemics using search engine query data. Nature 457, 1012–1014 (2009)
[5] Doornik, J.A.: Improving the Timeliness of Data on Influenza-like Illnesses using Google Search Data. Working Paper (2009)
[6] Tierney, H.L.R., Pan, B.: A Poisson Regression Examination of the Relationship between Website Traffic and Search Engine Queries. Working Paper (2010)
[7] Choi, H., Varian, H.: Predicting the Present with Google Trends, Working Paper, Technical Report. Google Inc. (2009)

[8] Lynn, W., Erik, B.: The Future of Prediction—how google searches foreshadow housing prices and sales. Working Paper (2009)
[9] Suhoy, T.: Query Indices and a 2008 Downturn: Israeli Data. Bank of Israel (2009)
[10] Askitas, N., Zimmermann, K.F.: Google Econometrics and Unemployment Forecasting. Applied Economics Quarterly (2009)
[11] Schmidt, T., Vosen, S.: Forecasting Private Consumption: Survey-based Indicators vs. Google Trends. Technische Universität Dortmund. Department of Economic and Social Sciences (2009)
[12] Marta.: Consumption and Information: An Exploration of Theories of Consumer Behavior using Daily Data. Working paper (2009)
[13] Moore, G.H., Shiskin, J.: Indicators of Business Expansions and Contractions. NBER Occasional Paper No. 103 (1967)
[14] Boehm, E.A.: The Contribution of Economic Indicator Analysis to Understanding and Forecasting Business Cycles. Indian Economic Review 36, 1–36 (2001)

Hybrid Web Service Selection by Combining Semantic and Keyword Approaches

Florie Ismaili

Abstract. The challenge of Web service discovery increases with the remarkable raise of available Web services. The lack of semantic description in current keyword based service search makes it difficult for clients to find a required Web service. The pure semantic search restricts the types of queries that users can perform. In order to solve these problems, in this paper we suggest a new hybrid Web service discovery architecture, which combines the ability to query and reason on metadata of semantic based search, and the flexibility of syntactic based search.

1 Introduction

Service-Oriented Architecture is believed to be an important paradigm for efficient businesses application development. Within SOA, software components that provide a piece of functionality and communicate with each other via message they exchange are called services. Web Services are autonomous and modular applications deployed and invoked over Internet [1, 11].With the increasing number of available Web services, discovery of correct web service according to the user needs has been widely recognized as one of the most challenging problems in the application of Service Oriented Architecture [3].

The objective of the proposed hybrid Web service discovery approach is to improve semantic service retrieval performance by combining the ability to query and reason on metadata of semantic based search and the flexibility of syntactic based search.

The main contributions of the proposed approach are the following:

- Introducing a novel approach for efficiently finding Web services on the Web.
- A paradigm for uniting the diverse standards of XML-based Web technologies like XML, Web services and the Semantic Web.

Florie Ismaili
SEEU, Faculty of Contemporary Sciences and Technologies,
Ilindenska nn Tetovo, Fyrom
e-mail: f.ismaili@seeu.edu.mk

F.L. Gaol (Ed.): Recent Progress in DEIT, Vol. 2, LNEE 157, pp. 31–38.
springerlink.com

- Allowing users to define customized matchmaking strategies and using similarity measures to improve the matchmaking performance.
- Collecting, analyzing and running several experiments on a large dataset consisting of real world Web services.

The rest of the paper is organized as follows:

In section 2 an overview of proposed Hybrid Web service discovery approach is presented. Section 3 discusses the keyword based searching. In section 4 the strict ontology based searching is presented. Section 5 introduces the hybrid searching. Section 6 concludes this paper.

2 Overview of Hybrid Web Service Discovery Approach

Hybrid Web Service discovery approach supports keyword based searching, ontology based searching and hybrid searching by combining both keyword and semantic searching.

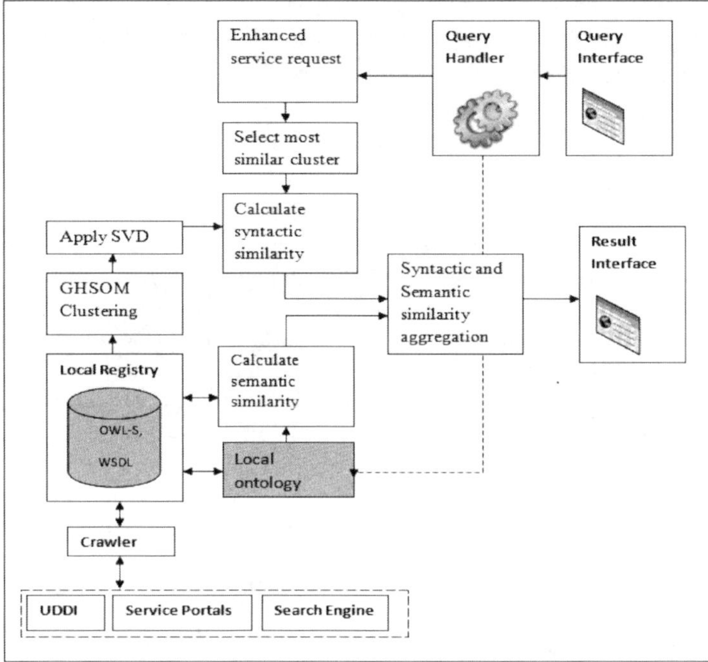

Fig. 1 The Outline of Hybrid Web Service Discovery Approach

The following is an outline of the key steps of the proposed approach:

- A local register server is created which stores WSDL and OWL-S files of web services collected.

- The GHSOM [6] is applied to the service set in order to construct self organizing maps which will be used for SVD [5] matrix construction. This division is done in order to reduce the computational cost of directly applying SVD to the huge number of services.
- Using LSI [5] for scoring and ranking the documents by their relevance.
- Using SemWeb owl-s api [7], for mapping OWL-S documents into Ontology mapping storage in order to find relevant services to the query.
- Calculate syntactic based similarity of Web services.
- Calculate semantic based similarity of Web services.
- Compose syntactic and semantic similarity.
- Sort and return the list of relevant services to the user.

The system supports service discovery by using keyword based, semantic based and hybrid query types. The keyword based search for Web service WSDL files is described in [4], but here we extend the method on hybrid Web service search. First keywords of user query are extracted where all keywords are considered as terms. Each of the extracted terms is expanded using the WordNet [9] to enhance its semantics.

3 Keyword Based Searching

Keyword based Web service discovery procedure begins with taking as input a user query, which is used as search criterion and returns a set of web services matching this query. The process of discovery consists of the following phases: Decomposing the Web Service Corpus and Service Discovery in Latent Semantic Space.

3.1 Decomposing the Web Service Corpus

At the beginning, the Web service collection is divided into smaller groups of related Web Services using unsupervised clustering method GHSOM.

The process starts with pre-processing of service description files. Web service description files are converted into numerical vectors suitable for GHSOM training. The service I/O as well as service description and service name concepts are extracted, where concepts are considered as terms. Here is important to note that some features of the SOMLib[6] are used to create the tf x idf input vectors.

GHSOM is able to cluster the representation vectors, arranged as nodes in hierarchy, where each hierarchy presents a group of the services related according to their semantic similarity. Each hierarchy is used to construct a SVD matrix.

3.2 Service Discovery in Latent Semantic Space

GHSOM is used to divide the Web service description files in different clusters. Each cluster has a centre. Based on Euclidian distance, the similarity between a Query vector and a cluster centre can be calculated. On the next step, the SVD is applied to the cluster whose cluster centre is more similar to the query vector.

Suppose the relevant cluster has n Web Services with m corresponding description files $WS=\{WS_1,\dots,WS_n\}$. So A_{ij} matrix is created with terms as row and service description files as columns. The Latent semantic indexing [5] involves the Singular Value Decomposition technique which replaces the original matrix with a low level rank approximation matrix $A \approx A_k$.

Algorithm 1: Algorithm for syntactic matching
Input: Service request SR_e
Output: Sorted list of services

```
begin
for all clustercenter Kci ∈ Kc do
  AppCluster= MinDis (sim(SRe, Kci ))
end for
return AppCluster
  qv ← EnhanceRequest(query)
  sv ← ReadSingularV alues()
for all term in qv do
  termV ector[i] ← readTermVector(term)
  w[i] ← calculate(Weight)
  qv[i]=termVector[i]*w[i]
end for
for all term in qv do
  dq=q[i]*T[i]
  queryNorm ← calculateNorm(qv)
end for
for all WSi in DD do
//DD is document-by-document similarity table
  wsNorm ← readNorm()
  sum=sum+DD[i]*dq[i]
  sim=sum/(queyNorm+wsNorm)
  Result=Result+(WSi, sim)
end for
return Sort(Result)
```

Formally, for a given service matrix, decomposition of A, can be represented as $A = T_0\, S_0 D_0^t$ where S_0 is rxr diagonal matrix composed of non-zero eigen values of AA^t. The columns of T_0 and D_0 orthogonal eigen matrixes composed of r non-zero values of AA^t. The number of the nonzero values in the diagonal in S_0 represents the rank of matrix A.

The aim here is to reduce the rank of A and obtaining a reduced matrix A_k constructed from the k-largest singular triplets of A. A new k-by-k diagonal matrix S is obtained by deleting the zero rows and columns of S_0. Likewise, the corresponding columns of T_0 and D_0 are removed in order to derive new left singular vector T and a new right singular vector S. The resultant matrix is $A \approx A_k = TSD^t$, where T is the m-by-k matrix whose columns are the first k columns of T_0, D is the n-by-k matrix whose columns are the first k columns of D_0.

Next, the query can be represented as vectors in dimension-reduced semantic space as $Q = A_q^t T S^{-1}$, where A_q is the query's term vector in the original vector space A.

Once such a conversion is achieved, any user query can be compared to existing service documents which produce a ranking vector in the reduced semantic space A_k.

In order to recommend the relevant service documents to a user, the cosine similarity [8] between document vectors and query is used.

4 Strict Ontology Based Searching

To use strict ontology based search, two tasks must be fulfilled. The first task is to define the service's domain ontology in terms of OWL classes, properties, and instances. The second task is to collect or create OWL-S descriptions [2] of the services, relating the description to the domain ontologies.

These descriptions are stored in Service Registry which serves as repository of all OWL-S files. These files are used during matching process where a semantic match between service advertisements and service requests is computed.

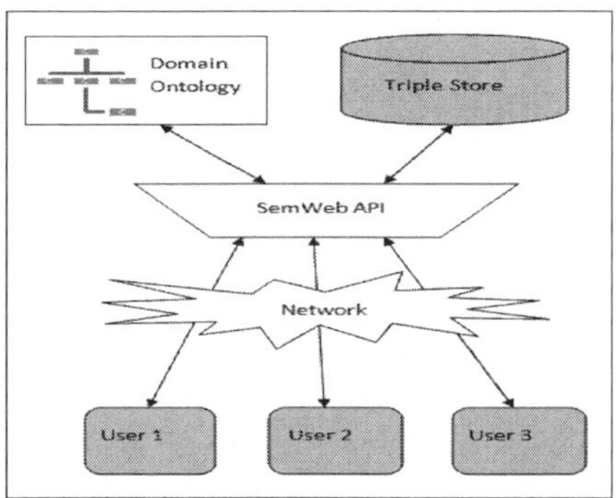

Fig. 2 The Outline of Strict Ontology Based Web Service Discovery

We used SemWeb Library for C# [7] as the developing platform. SemWeb Library for C# is an RDF API for .Net framework developed by Joshua Tauberer. It provides a command-line tool for loading data into a database which is used to load the triples into a MySQL database. At the moment of new Service advertisement registration, OWL-S files are parsed by extracting necessary semantic information and load them into a triple store as triples <S, P, O>.

This API has support for SPARQL interrogations by storing them in special object, Query class objects for local queries and SparqlHttpSource for remote ones.

Once the data is in a triple store, SPARQL allows running all sorts of queries against the data set. This can be performed by configuring the"web.config" which enables to configure the ASP.NET server that will run the SPARQL endpoint.

The figure 2 displays the strict ontology based searching.

As it can be seen, the typical architecture of ontology based search is similar to relational database model. The semantic service request is received from the user, query methods are called to be executed which runs the queries and extracts results which meets the user's requirements.

5 Hybrid Searching

To use strict ontology based search Hybrid Web service discovery combines the previous explained keyword based searching with semantic searching. In this case, instead of the traditional matching on input and output called relaxed algorithm presented by Paolucci [10], the Generalized Cosine Similarity [1] will be used. This algorithm calculates final values of inputs and outputs by calculating matching levels for each input and output.

Formally, let ξ be the set of all advertisements in advertisement repository. For a given Query Q, the matchmaking algorithm returns the set of all advertisements which are compatible, *Match (Q)*.

$$Match(q) = \{A \in \xi \mid compatible(A, Q)\}\,. \tag{1}$$

In this approach Generalized Cosine Similarity Measure is used in order to calculate the similarity between a query and advertisement, which is an extension of relax algorithm:

For given to terms or collection of elements C1and C2, the semantic distance is defined as similarity of concepts in relation *subClassOf.*

In this manner, we should take in consideration the depth of the node and lowest common ancestor (LCA) which is the node of greatest depth that is an ancestor of both C1 and C2, and the semantic similarity can be defined as follow:

$$Sim(C_1, C_2) = \frac{2 * depth(LCA(C_1, C_2))}{depth(C_1) + depth(C_2)} \tag{2}$$

This similarity information will be used during the matching process in rating the input and output matching.

Algorithm 2: Semantic Matching
Input: ServiceRequestInput SRInpj, ServiceRequestOutput SROupj
Output: Sorted list of services

```
begin
for all ServiceRequestInput SRInpj ∈ SRInp do
for all ServiceInput WSInpi ∈ WSInp do
   sim(WSInp, SRInp)
   WSInp[i] = sim(WSInp, SRInp)
end for
SRInp[j] = 1
end for
for all ServiceRequestOutput SROupj ∈ SROup do
for all ServiceOutput WSOupi ∈ WSOup do
   sim(WSOup, SROup)
   WSOup[i] = sim(WSOup, SROup)
end for
SRoup[j] = 1
end for
WS = WSInp +WSOup
SR = SRInp + SROup
for all ServiceAdvertisements ∈ WS – j do
for all input/output parameters in WS – j do
RequestNorm ← calculateNorm(SR)
ServiceNorm ← ServiceNorm(WS)
sum = sum + SR[i] *WS[i]
end for
sim[j] = sum/(RequestNorm + ServiceNorm)
Result = Result + (WSj , sim[j], resultDegree[j])
return Sort(Result)
```

6 Conclusion

In this paper we addressed the problem of improving the process of Web service discovery, by introducing a new approach for efficiently finding Web services on the Web. The basic idea behind the proposed approach is the combination of Web services existing standards, common information retrieval methods, Semantic Web and data mining techniques. The purpose of building such a system is to enable the data to be accessible for both humans and machines.

References

1. Papazoglou, M.P., Traverso, P., Dustdar, S., Leymann, F.: Service-Oriented Computing: State of the Art and Research Challenges. Computer 40(11), 38–45 (2007)
2. W3C, OWL-S: Semantic Markup for Web Services, http://www.w3.org/Submission/OWL-S/
3. Ismaili, F., Sisediev, B.: Web Services Research Challenges, Limitations and Opportunities. WSEAS Transactions on Information Science & Applications 5(10), 1790–1832 (2008) ISSN: 1790-0832
4. Ismaili, F., Sisediev, B., Zenuni, X., Raufi, B.: GHSOM-based Web Service Discovery. In: Proceedings of the European Computing Conference (2010) ISSN: 1790-5117
5. Mi IsLita, http://www.miislita.com/term-vector/term-vector-3.html
6. The SOMLib Digital Library Project, http://www.ifs.tuwien.ac.at/~andi/somlib/
7. SemWeb.NET: Semantic Web/RDF Library for C#/.NET, http://razor.occams.info/code/semweb/
8. Jiawei, H., Kamber, M.: Data Mining: Concepts and Techniques, 2nd edn. Elsevier (2006) ISBN: 13: 978-1-55860-901-3
9. WordNet, A lexical Database for English, http://wordnet.princeton.edu/
10. Paolucci, M., Kawamura, T., Payne, T.R., Sycara, K.: Semantic Matching of Web Services Capabilities. In: Horrocks, I., Hendler, J. (eds.) ISWC 2002. LNCS, vol. 2342, pp. 333–347. Springer, Heidelberg (2002)
11. Information Retrieval, http://www.dcs.gla.ac.uk/Keith/Chapter.7/Ch.7.html

Research of Ontology-Driven Agricultural Search Engine

Wei Yuanyuan, Wang Rujing, Li Chuanxi, and Hu Yimin

Abstract. An ontology-driven retrival method for agriculture search engine is proposed in this paper. An agricultural products ontology was built, with that a semantic expanson algorithm for search keyword was achieved when executing search task. This process supply semantic-level retrieval service for search engine users. The goal is to improve intelligentization of agriculture search engine.

1 Introduction

Ontology is from the field of philosophy, to study the existence nature and composition of objective things. In recent years, with the rapid development of information science, Ontology become a hot topic in the fields of knowledge engineering and information management. The nature of Ontology is modeling for problems or even the whole world, thus establishing a set knowledge system that can be understand by computer and people[1]. Through the description and analysis of relations between concepts, ontology lay a foundation for machines understanding of information and intelligent information process.

Ontology provides a scientific organizations method based on knowledge or concepts. It profoundly reveals the relationship between concepts, is conducive to further knowledge discovery[2]. Support by ontology, the efficiency of knowledge services, such as knowledge search, knowledge accumulation, knowledge sharing, etc, will greatly enhance, real knowledge reuse and knowledge sharing can become a reality.

Laboratory where the author work has been committed to the research of agriculture professional search engine - SouNong (http://www.sounong.net/) study. At present, search based on keywords or content categories can not meet the needs of users. Using ontology technology for the semantic level retrieval is an important research direction of ontology services. It is the focus of our current research work.

Wei Yuanyuan · Wang Rujing
Institute of Intelligent Machines, Chinese Academy of Sciences
e-mail: {yywei,rjwang}@iim.ac.cn

F.L. Gaol (Ed.): Recent Progress in DEIT, Vol. 2, LNEE 157, pp. 39–44.
springerlink.com

An ontology-driven retrival method for agriculture search engine is proposed in this paper. First of all, we construct an agricultural products ontology. The ontology provide the knowledge basis for semantic retrieval of search engine. Semantic expansion of search keyword was achieved when executing a search task. The method reflects the search intelligence, and improves recall ratio and precision ratio.

2 Ontology and Agricultural Ontology

Ontology is a scientific methodology about the ontology-based knowledge representation, knowledge organization and develop of knowledge services. It is a theory of solving problem by knowledge ontology. The set of concept system, relationship system and field axioms constructed by experts in the field, is called ontology, so ontology is an entity [1].

Gruber[4] divided ontology into four categories: top-level ontology, domain ontology, task ontology and the application ontology, under the two indicators of detail level and area dependence.

As an important means of knowledge organization, an ontology generally has Statement, Axiom, Concept, also known as Class, Property, Function, Instance and other elements.

Agricultural Ontology should belong to the scope of domain ontology. Agricultural ontology is the formal representation of relationships between the concepts in agricultural disciplines. It is the semantic basis for future agricultural semantic web. It provides a powerful knowledge base and corpus for the retrieval of semantic web [2]. The goal is to better agricultural knowledge reuse, sharing and processing [3].

3 Construction of Agricultural Products Ontology

3.1 Knowledge Structure of Agricultural Products Ontology

According to the application background of agricultural products ontology, in addition to their own classification system, agricultural products have important linkages with crop pests and diseases, place of origin and other properties. The ontology we build is divided into three classes: agricultural products categories, crop pest and disease, administrative divisions . First of all, these three classes as three sub-ontology, built separately, then by means of ontology integration mechanism, combined into agricultural products ontology. Fig. 1 shows taxonomic hierarchies. The relationships of class and class, instance and instance were established by the definition of object properties and anonymous class. The relationships between the three classes is shown in Fig. 2.

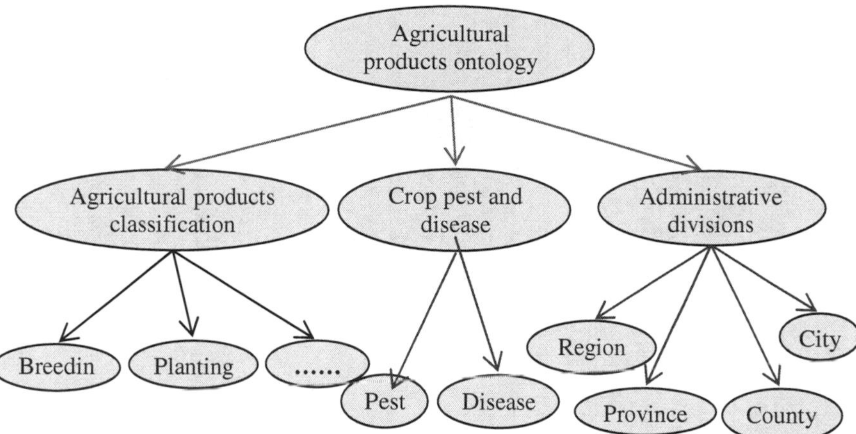

Fig. 1 Ontology taxonomic hierarchies

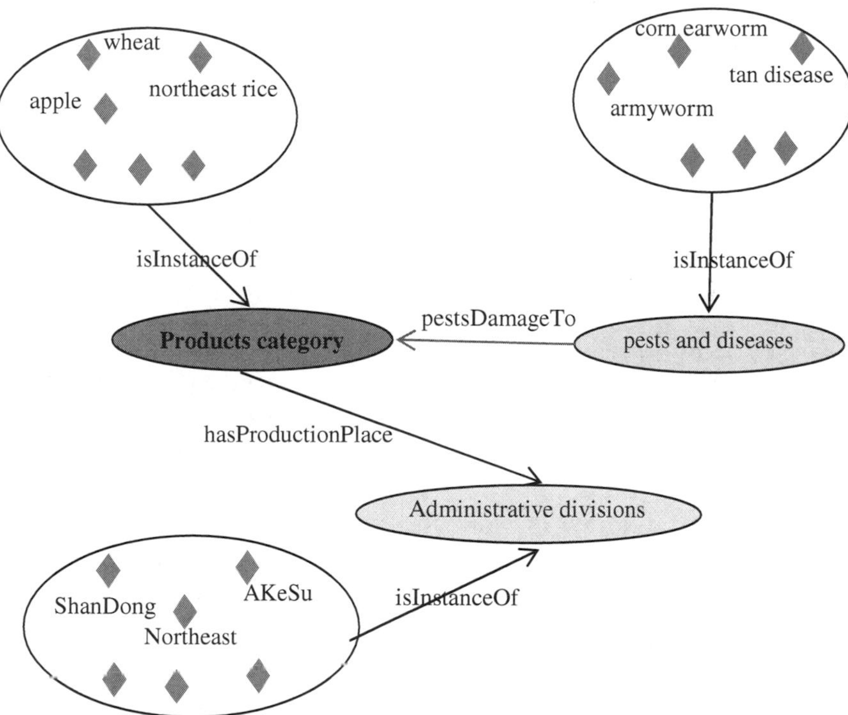

Fig. 2 Relationship of classes

3.2 Basis for the Construction

The construction of agricultural products categories ontology take "China's agricultural professional classification" and "Chinese Library Classification" as the main reference. 2000 species of high frequency category in SouNong database were added as instances, relevant descriptions of instances refer to Baidu Encyclopedia.

Agricultural pests and diseases ontology based on our years of collecting the "basic database of agricultural pests and diseases" to build. Administrative divisions ontology based on the latest information of administrative divisions in the official website of "National Bureau of Statistics" (http://www.stats.gov.cn/).

3.3 Construction Approach

We establish agricultural products ontology using manual and semi-automatic combination approach. Agricultural products categories ontology is manual build using protégé 4. In view of good structure of crop pest and disease module and administrative divisions module, we carry out automatic construction by writing java program.

Protégé 4 is developed by Manchester and Stanford University, it adopt OWL as the ontology language. The tool is a Windows-based ontology construction tools under java development environment.

Data of administrative divisions and agricultural pests and diseases are obtained from the online collection, sorting into structured data. For the two parts of ontology construction, at first we establish the class hierarchy relationship, object properties and data properties in hand, next through programming, automatic generate owl encoding. Auto-complete operations include: addition of an instance, addition of an instance's property values, relationship established between instances, etc.

Finally, agricultural products categories ontology, agricultural pests and diseases ontology, administrative division ontology merge into agricultural products ontology, it includes a total of 674 classes, 39 properties and more than 7000 instances.

4 Ontology-Driven Intelligent Retrieval

Agricultural products ontology is well established, the next main task is how ontology services. Through the open source java development package Jena, we achieve ontology-driven agricultural search engine semantic retrieval.

Through the ontology query language SPARQL query and ontology model user interface, the system complete the user input query keywords semantic expansion. The extended information such as the respective categories, related category,etc, feedback to the user. The way provides users with more comprehensive and convenient search service. Semantic expansion process is shown in Figure 3.

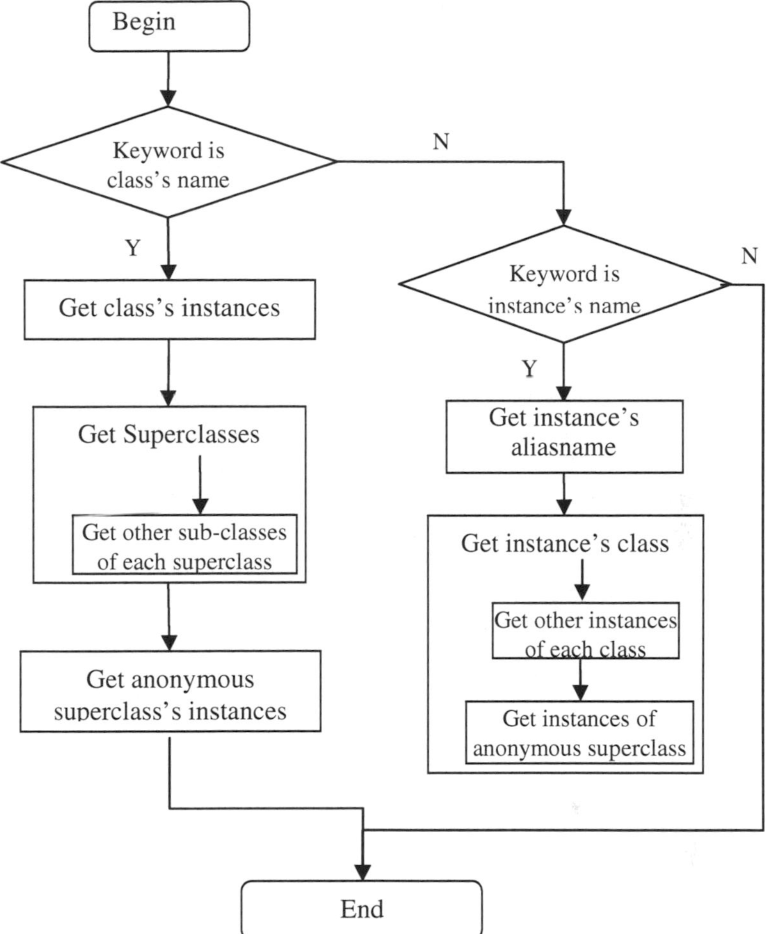

Fig. 3 Flow chart for keywords semantic expansion

After users input query keywords, the search engine performs ontology reasoning, we can get keywords list with expanded semantic. Based on these expanded keywords, search engine guide user locate further query needs. This process reflects the search is intelligent and effectively improves recall ratio and precision ratio.

5 Conclusion

An ontology-driven search method is presented in this paper. Achieving semantic retrieval through ontology-driven agricultural search engine is an important research interest of Agricultural Ontology Services. Based on ontology needs for agriculture search engine, an agricultural products ontology is constructed.

At present, the ontology-driven agriculture intelligent search engine entered a trial run, the next work includes: (1) Further consummate agricultural products ontology. According to search engine running log, modify, add more classes, instances and properties, and continuously enrich ontology content. (2) Explore more ways to apply ontology during the process of ontology application, let agricultural ontology play a greater role in ontology service. (3) Study automatic extension of ontology to enhance ontology completeness.

Acknowledgments. This work was supported by the Knowledge Innovation Program of the Chinese Academy of Sciences.

References

1. Cui, Y.P.: Study of Key Technologies of Agricultural Knowledge Management Based on Ontology. China Agricultural Science and Technology Press (2009)
2. Qian, P., Zhen, Y.L.: Research and Application on Agricultural Ontology. China Agricultural Science and Technology Press (2006)
3. Xie, N.F., Wang, W.S.: Ontology and acquiring of agriculture knowledge. Agriculture Network Information, 12–16 (August 2007)
4. Gruber, T.R.: Toward principles for the design of ontologies used for knowledge sharing. International Journal of Human–Computer Studies 43(5–6), 907–928 (1995)

Efficient Encoding Technique for Strings-Based Dynamic XML Labeling Schemes

Shaorong Feng and Canwei Zhuang

Abstract. Several dynamic XML labeling schemes have been proposed to efficiently process updating in dynamic XML data. In this paper, we focus on one class of these schemes which are using strings of lexicographical order to support dynamic XML. We point out the problems of existing encodings which include memory inefficiencies when initial labeling and achieving labels of sub-optimal size when dynamic labeling. A Full-Tree-based(FT) encoding technique is proposed to overcome these problems. We bring the concept of *self-increase* into strings, and it assures our encoding technique labeling for initial XML with efficient memory usage, which make it possible to process large XML with limited memory. Moreover, the concept of *subtraction* is generalized into lexicographical order for the guarantee of achieving new labels with optimal size when XML frequently updates, which has the advantages of both reducing the storage cost and optimizing query performance. Experimental results confirm that our FT technique provides a new method which is intuitional and efficient for dynamic label schemes.

1 Introduction

XML has become a standard of information exchange and representation on the web. To query XML data effectiveness and efficiency, the labeling schemes are widely used to determine the relationships such as the ancestor-descendant between any two elements. If the XML is static, the current labeling schemes, for examples, containment scheme[1], prefix scheme[2] and prime scheme[3] can process different queries efficiently. However, when XML data become dynamic, a large amount of nodes need re-labeling, which is costly and becomes a bottleneck for XML-Database performance. Designing dynamic labeling schemes to avoid the re-labeling of existing nodes has been recognized as an important research problem.

Shaorong Feng · Canwei Zhuang
School of Information Science and Technology, Xiamen University, Xiamen, China
e-mail: shaorong@xmu.edu.cn, cwzhuang0229@163.com

F.L. Gaol (Ed.): Recent Progress in DEIT, Vol. 2, LNEE 157, pp. 45–50.
springerlink.com

In this paper, we present a Full-Tree based (FT) encoding technique to overcome the problems mentioned above. Two fundamental concepts are generalized into strings based on our FT tree: *next-string* and *strings-subtraction*. We use *next-string* to produce new code directly from current code and thus no table is needed when XML initial labeling. On the other hand, *strings-subtraction* is introduced to achieve new labels of optimal size and thus reuse the deleted labels efficiently when XML frequently updates.

2 FT Tree

2.1 BFT(Binary-FT) Tree and QFT (Quaternary-FT) Tree

BFT encoding is based on a complete binary tree we call BFT tree.

The BFT tree is a full binary tree with the same shape as Tmp-BFT. Binary strings derived from Tmp-BFT are assigned to QFT: Let each binary string in Tmp-BFT be T, and denote the position of last encountered “1” of T as p, then *substr*(T,*1*, p) is allocated to the corresponding node in BFT.

QFT encoding scheme is based by a complete ternary tree we call QFT tree.

The QFT tree is a full ternary tree with the same shape as Tmp-QFT. Ternary strings derived from Tmp-QFT are assigned to QFT: Let each ternary string in Tmp-QFT be T, and denote the position of last encountered “1” or “2”of T as p, then the corresponding position in QFT is assigned the *substr*(T, *1*, p) with every “symbol + 1” (“symbol+1” means if symbol is “0”, then “1”; if “1”, then “2”; and if “2”, then “3”. Likewise, “symbol-1” is of the similar meaning.)

QFT encoding is based on inorder traveling a QFT tree of a certain level. Similar to BFT, “under given level” for QFT is understood as “the (3^l–1) CDQS Codes generated by inorder traveling a QFT tree of given level l”.

Based on QFT tree under a certain level, we define the basic functions for QFT corresponding to those of BFT, which are *I-Index*, *QED-Subtract*, and *nextQED*. The main properties of these concepts are summarizing as follows, and we omit the proofs since there are similar details to that of BFT.

Besides, *nextQED* can be got or gotten by discussing cases if the code is a leaf node or not in QFT tree. The implements of these functions are shown in Algorithm 1.

Algorithm 1. Basic functions for QED
Function 1 ***I_Index***(Q, l)
1. i←1; K←0;
2. **while** i <= size(Q) **do** K←3×K+(Q[i]–1); i++; **endwhile**
3. **while** i<= l **do** K←3× K; i++; **endwhile**
4. **return** K;

Function 2 QED_*subtract*(Q_L, Q_R, l)
1. **return** I_Index(Q_R, l)– I_Index(Q_L, l);

Function 3 ***NextQED***(Q, l)

1. **if** $Q = NIL$ **then return** "$1^{l-1}2$";
2. $k \leftarrow l$–size(Q);
3. **if** $k > 0$ **then return** $Q \oplus$ "$1^{k-1}2$";
4. **else** denote the position of last encountered "1" or "2" as p;
5. **if** *substr*(Q, p, p)="1" **then return** *substr*(Q,1, p-1) $\oplus$ "2";
6. **if** *substr*(Q, p, p)= "2" **then return** *substr*(Q,1, p-1) $\oplus$ "3";

3 BFT Encoding Technique

3.1 Initial Labeling

We introduce how BFT labeling for initial XML.

When encoding a range *N*, the former approach needs to create an encoding table of size *N* first, which is memory inefficient. Our BFT encoding focuses on reducing the encoding table. We design a technique to calculate the *N* codes successively, and then no table is needed.

Let's take a look at how labeling XML using integers firstly. No matter whether scheme is applied, the integers are assigned to certain places in order. That is to say, we first allocate number 1, and then allocate next number 2…and so on. Likewise, when using CDBS to label XML, the CDBS Codes can be assigned in order as well: the first code is assigned first, and then assign the following one… Thus, we just need implement a function of getting next code to label XML.

Suppose encoding a range *N*, we need generated *N* CDBS Codes. From paper [5], the sequence of these codes can be achieved by inorder traveling STB tree of *N* nodes. We denote the index of this sequence as I-Index. Then we illustrate how to get the codes in this sequence one by one.

We first construct an STB tree of *N* nodes, suppose its level is *l*, and then we construct two BFT tree with level *l*-1 and *l*. We call them *STB*, *BFT-1* and *BFT-2* respectively.

Lemma 1. *Suppose the level of an STB tree contained N nodes is l, and the I-Index of the last node in level order is I_Index_of_Last, then l is equal to* $1+log_2N$ *and I_Index_of_Last is equal to* $2\times(N-2^{l-1})+1$.

3.2 Getting a Batch of Middle Codes with Optimal Size

The previous algorithm can't get the middle codes with smallest total size since it is difficult to calculate the right position for the middle code. Based on the concept of *CDBS-subtract* in our BFT, we design a new algorithm (Algorithm 2) to find the codes with smallest total size between two given strings.

Algorithm 2. DynamicEncoding(S_L, S_R, *TN*)
Input: A positive integer *TN*, CDSBS Code S_L, S_R and $S_L < S_R$
Output: *TN* codes satisfied $S_L < S_1 < S_2 < \ldots < S_{TN} < S_R$, and the total size of these codes is smallest
1. define an array *CodeArr*[0, *TN*+1];
2. *CodeArr*[0]←S_L; *CodeArr*[*TN*+1]←S_R;
3. *SubEncoding*(*CodeArr*, 0, *TN*+1);
4. return *CodeArr*[1, *N*];

Procedure *SubEncoding*(*CodeArr*, *L*, *R*)
1. **if** *L*+1>=*R* **then return;**
2. S_M←assignMiddleBinaryStringWithSmallestSize(*CodeArr*[*L*], *CodeArr*[*R*]);
3. denote the first difference position of *CodeArr*[*L*], *CodeArr*[*R*] and S_M as *p*;
4. TMP_L←*substr*(S_L, *p*); //remove the common prefix
5. TMP_R←*substr*(S_R, *p*);
6. TMP_M←*substr*(S_M, *p*);
7. *level*←*max*(size(TMP_L), size(TMP_R), size(TMP_M));
8. d_1←*subtract*(TMP_L, TMP_M, *level*); // $d_1 = S_M - CodeArr[L] = TMP_M - TMP_L$
9. d_2←*subtract* (TMP_M, TMP_R, *level*); // $d_2 = CodeArr[R] - S_M = TMP_R - TMP_M$
10. *M*←*round*($L + [d_1/(d_1+d_2)] \times (R-L)$);
11. *CodeArr*[*M*]←S_M;
12. *SubEncoding*(*CodeArr*, *L*, *M*);
13. *SubEncoding*(*CodeArr*, *M*, *R*);

4 QFT Encoding Technique

QFT optimizes the performance of QED both when XML initial labeling and when XML dynamic updating.

4.1 Initial Labeling

QFT labeling for initial XML is similar to that of BFT.

Lemma 2. *Suppose the level of an QTB tree contained N nodes is l, and the I-Index of the last node in level order is I_Index_of_Last, then l is* $1+\log_3 N$ *and I_Index_of_Last is* $(3\times(N-3^{l-1})+2)/2$.

Algorithm 3. GetChildrenSelfLabel(parent)
Input: An XML node *parent*
Output: Self_Labels for child nodes of *parent*
1. N←GetChildrenNum(parent);
2. *l*←$1+\log_3 N$; //get the level *l*
3. *I_Index_of_Last*←$(3\times(N-3^{l-1})+2)/2$; //get the I-Index of last node
4. *I_Index*←0; // index the *i*th QED code under given *N*
5. *Q*←*NIL*; // *Q* is initialized to null

```
Tmp_Child←Parent.Firstchild( );
while Tmp_Child !=NIL do {
   I_Index ++;                          //get the next I_Index
   Q←NextQED( Q , l );          //get the next code Q
   if I_Index=I_Index_of_Last then { l– –;
   if last symbol of Q is "2" then Q←Q with last symbol change to "3"; }
   assign code Q to Tmp_Child as self_label;
  Tmp_Child ← Tmp_Child.Nextsibling( );    }
```

4.2 Getting a Batch of Middle Codes with Optimal Size

QED can't get the middle codes of optimal size. Based on the concept of *QED-subtract* in our QFT, we can find the codes with smallest total size between two given quaternary strings. The algorithm is with similar details to that of BFT and we ignore it here.

5 Experiments and Results

We experimentally evaluate and compare our BFT and QFT encodings against the previous schemes including CDBS and QED using containment labels. The comparisons of CDBS and QED with the other labeling schemes are beyond the scope of this paper and can be found in [4]. Besides, we don't compare our FT against ST [5] since the ST can also be applied to FT encoding when labeling multiple ranges. We used Dev-C++ for our implementation and all the experiments are carried out on AMD 2.8GHZ with 2G of RAM running on windows XP. The test datasets chooses from real world.

5.1 Performance Study on Frequent Updates

If there are only insertions in updates, both FT and former algorithms guarantee that the inserted codes are of optimal size. We evaluate the performance when there are both deletions and insertions. We test the performance on all the data sets and present the results from Hamlet data set as the others show similar trends.

Hamlet has totally 6,636 nodes, and there are 5 "*act*" sub-trees under root. We delete the first "*act*" sub-tree, and then insert it back. Based on the new file, we delete the second "*act*" and then insert it back too. We repeat this kind five times until all the "*acts*" are updated. After every update, the new Hamlet is the same as the original; however, the labels are different. We evaluate the average label size and max label size of different schemes and show the results in Fig 1.

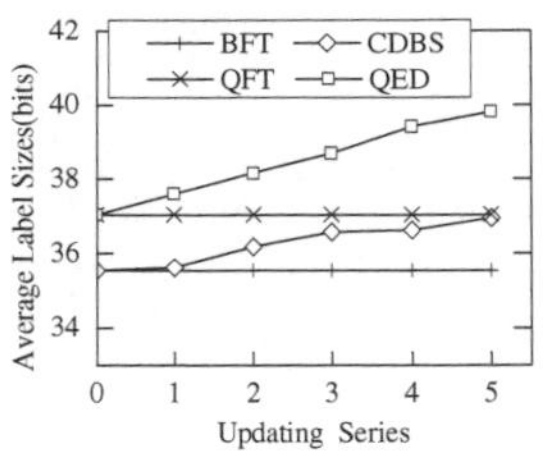

(a) Comparison of average label size

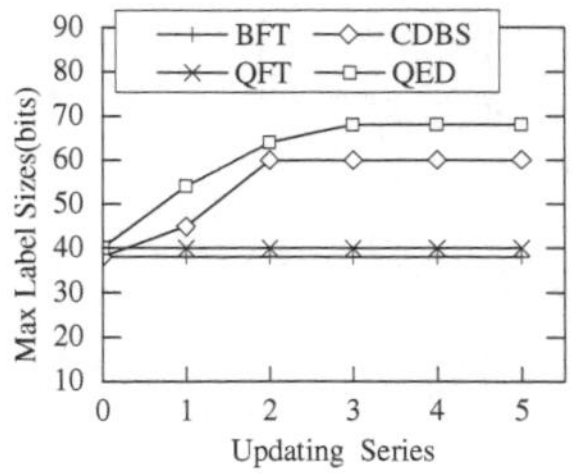

(b) Comparison of max label size

Fig. 1 Performance study on Frequent updates

6 Conclusion

In this paper, we address the problems for dynamic labeling schemes of strings-based. We propose FT technique which can be applied to optimize the performance of both CDBS and QED. Compared with former encodings which are memory-based, our FT encoding technique labels for initial XML with no table needed, which is highly efficient and is able to process very large XML with limited memory. On the other hand, FT provides a technique to achieve new labels of optimal size and thus optimize query performance when XML frequently updates. Another problem of dynamic labeling schemes is that the label size will increase fast if the nodes are always inserted at one fixed place (called skewed insertion [4]), and how to solve this skewed insertion is an interesting future research direction.

References

1. Zhang, C., Naughton, J.F., DeWitt, D.J., Luo, Q., Lohman, G.M.: On Supporting Containment Queries in Relational Database Management Systems. In: SIGMOD (2001)
2. Tatarinov, I., Viglas, S., Beyer, K.S., Shanmugasundaram, J., Shekita, E.J., Zhang, C.: Storing and Querying Ordered XML Using a Relational Database System. In: SIGMOD (2002)
3. Wu, X., Lee, M., Hsu, W.: A prime number labeling scheme for dynamic order XML tree. In: ICDE (2004)
4. Li, C., Ling, T.W., Hu, M.: Efficient Updates in Dynamic XML Data: from Binary String to Quaternary String. In: VLDB J (2008)
5. Xu, L., Ling, T.W., Bao, Z., Wu, H.: Efficient Label Encoding for Range-Based Dynamic XML Labeling Schemes. In: Kitagawa, H., Ishikawa, Y., Li, Q., Watanabe, C. (eds.) DASFAA 2010. LNCS, vol. 5981, pp. 262–276. Springer, Heidelberg (2010)

Cross-Language Peculiar Image Search Using Translaion between Japanese and English

Shun Hattori

Abstract. As next steps of Image Retrieval, it is very important to discriminate between "Typical Images" and "Peculiar Images" in the acceptable images, and moreover, to collect many different kinds of peculiar images exhaustively. As a solution to the 1st next step, my previous work has proposed a novel method to more precisely search the Web for peculiar images of a target object by its peculiar appearance descriptions (e.g., color-names) extracted from the Web and/or its peculiar image features (e.g., color-features) converted from them. This paper proposes a refined method equipped with cross-language (translation between Japanese and English) functions and validates its retrieval precision.

1 Introduction

One example of more niche demands for Image Retrieval, when only a name of a target object is given, is to search the Web for its "Typical Images" [1] which allow us to adequately figure out its typical appearance features and easily associate themselves with the correct object-name, and its "Peculiar Images" [2, 3] which include the target object with not common (or typical) but eccentric (or surprising) appearance features. For instance, most of us would uppermost associate "sunflower" with "yellow one", "cauliflower" with "white one", and "Tokyo tower" with "red/white one", while there also exist "red sunflower" or "black one" etc., "purple cauliflower" or "orange one" etc., and "blue Tokyo tower" or "green one" etc. When we exhaustively want to know all the appearances of a target object, information about its peculiar appearance features is very important as well as its common ones.

As next steps of IR in the Web, it is very important to discriminate between "Typical Images" and "Peculiar Images" in the acceptable images, and moreover, to

Shun Hattori
School of Computer Science, Tokyo University of Technology, 1404–1 Katakura-machi, Hachioji, Tokyo 192–0982, Japan
e-mail: hattori@cs.teu.ac.jp

F.L. Gaol (Ed.): Recent Progress in DEIT, Vol. 2, LNEE 157, pp. 51–56.
springerlink.com

collect many different kinds of peculiar images as exhaustively as possible. In other words, "Exhaustiveness" is one of the most important requirements in the next-generation Web image searches as well as Web document searches. As a solution to the 1st next step, my previous work [3] proposes a novel method to precisely search the Web for peculiar images of a target object whose name is given as a user's original query, by expanding the original query with its peculiar appearance descriptions (e.g., color-names) extracted from the Web by text mining techniques [4, 5] and/or its peculiar image features (e.g., color-features) converted from the Web-extracted peculiar color-names. In order to make the basic method more robust, this paper proposes a refined method equipped with cross-language (translation between Japanese and English) functions and validates its retrieval precision (robustness).

2 Single-Language Method

This section explains my basic method [2, 3] to precisely search the Web for "Peculiar Images" of a target object whose name is given as a user's original query, by expanding the original query with its peculiar appearance descriptions (e.g., color-names) extracted from the Web by text mining techniques and/or its peculiar image features (e.g., color-features) converted from them.

Step 1. Peculiar Color-Name Extraction
When a name of a target object as an original query is given by a user, its peculiar color-names (as one kind of appearance descriptions) are extracted from exploding Web documents about the target object by text mining techniques.

The two kinds of lexico-syntactic patterns which consist of a color-name *cn* and the target object-name *on* are often used as follows:

1. "*cn*-colored *on*", such as "yellow-colored sunflower",
2. "*on* is *cn*", such as "sunflower is yellow".

The weight pcn(*cn*, *on*) of Peculiar Color-Name extraction is assigned to each candidate *cn* for peculiar color-names of a target object-name *on* as follows:

$$\mathrm{pcn}(cn,on) := \begin{cases} 0 & \text{if } \mathrm{df}(\texttt{["}on \text{ is } cn\texttt{"]}) = 0, \\ \dfrac{\mathrm{df}(\texttt{["}cn\text{-colored } on\texttt{"]})}{\mathrm{df}(\texttt{["}on \text{ is } cn\texttt{"]})+1} & \text{otherwise,} \end{cases}$$

where df(["*q*"]) stands for the frequency of Web documents retrieved by submitting the phrase query ["*q*"] to Google Web Search.

Step 2. Color-Feature Conversion from Color-Name
The peculiar HSV color-features cf_p (as one kind of image features) of the target object are converted from its Web-extracted peculiar color-names cn_p by referring the conversion table [6] or [7] in each language.

Step 3. Query Expansion by Color-Name/Feature

Here, we have three kinds of clues to search the Web for peculiar images: not only a target object-name *on* (text-based condition) as an original query given by a user, but also its peculiar color-names cn_p (text-based condition) extracted from Web documents in Step 1, and its peculiar color-features cf_p (content-based condition) converted from its peculiar color-names in Step 2.

The original query (q0 = text: ["*on*"] & content: null) can be expanded by its peculiar color-names cn_p and/or its peculiar color-features cf_p as follows:

q1 = text: ["*on*"] & content: cf_p,
q2 = text: ["cn_p-colored *on*"] & content: null,
q3 = text: ["cn_p-colored *on*"] & content: cf_p.

Step 4. Image Ranking by Expanded Queries

First, the weight $\text{pis}_{q1}(i,on)$ of Peculiar Image Search based on the 1st type of expanded query (q1 = text: ["*on*"] & content: cf_p) is assigned to a Web image i for a target object-name *on* and is defined as

$$\text{pis}_{q1}(i,on) := \max_{\forall(cn_p,cf_p)} \left\{ \text{pcn}(cn_p,on) \cdot \text{cont}(i,cf_p) \right\},$$
$$\text{cont}(i,cf_p) := \sum_{\forall cf} \text{sim}(cf,cf_p) \cdot \text{prop}(cf,i),$$

where a Web image i is retrieved by submitting the text-based query ["*on*"] (e.g., ["sunflower"]) to Google Image Search, $\forall(cn_p,cf_p)$ stands for not completely any pair but each pair of its Web-extracted peculiar color-name cn_p and its converted peculiar color-feature cf_p in Step 2, $\text{sim}(cf,cf_p)$ stands for the similarity between color-features cf and cf_p in the HSV color space [8], and $\text{prop}(cf,i)$ stands for the proportion of a color-feature cf in a Web image i.

Next, the peculiarity of a Web image i for an object-name *on* by the 2nd type of expanded query (q2 = text: ["cn_p-colored *on*"] & content: null) is defined as

$$\text{pis}_{q2}(i,on) := \max_{\forall cn_p} \left\{ \frac{\text{pcn}(cn_p,on)}{\text{rank}(i,on,cn_p)^2} \right\},$$

where $\forall cn_p$ stands for each Web-extracted peculiar color-name cn_p in Step 1, and $\text{rank}(i,on,cn_p)$ stands for the rank of a Web image i in the retrieval results by submitting the text-based query ["cn_p-colored *on*"] to Google Image Search.

Last, the peculiarity of a Web image i for a target object-name *on* by the 3rd type of expanded query (q3 = text: ["cn_p-colored *on*"] & content: cf_p) is defined as

$$\text{pis}_{q3}(i,on) := \max_{\forall(cn_p,cf_p)} \left\{ \frac{\text{pcn}(cn_p,on) \cdot \text{cont}(i,cf_p)}{\text{rank}(i,on,cn_p)} \right\},$$

where $\forall(cn_p,cf_p)$ stands for each pair of its Web-extracted peculiar color-name cn_p and its converted peculiar color-feature cf_p in Step 2.

3 Cross-Language Method

This section proposes a refined method equipped with cross-language (translation between Japanese and English) functions to make the basic method more robust. Fig. 1 and 2 show my Cross-Language Peculiar Image Search (XPIS).

When an English object-name is given by a user, my proposed cross-language method in Fig. 1 runs from English to Japanese language space as follows:

Step 0. translates the user-given English object-name (e.g., "sunflower") into its Japanese object-name (e.g., "himawari") automatically,
Step 1. extracts its Japanese peculiar color-names (e.g., "akairo" and "shiro") from the Web by text mining techniques [4],
Step 2. converts its Japanese peculiar color-names into its peculiar color-features (e.g., :red and :white) by referring the conversion table [7],
Step 3-4. retrieves peculiar images from the Web by its Japanese object-name and its Japanese peculiar color-names and/or its peculiar features.

Meanwhile, my proposed cross-language method in Fig. 2 runs back and forth between English and Japanese language spaces as follows:

Step 0. translates its user-given English object-name (e.g., "sunflower") into its Japanese object-name (e.g., "himawari") automatically,
Step 1. extracts its Japanese peculiar color-names from the Web and translates them into its English peculiar color-names (e.g., "red" and "white"),
Step 2. converts its English peculiar color-names into its peculiar color-features (e.g., :red and :white) by referring the conversion table [6],
Step 3-4. retrieves peculiar images from the Web by its English object-name and its English peculiar color-names and/or its peculiar features.

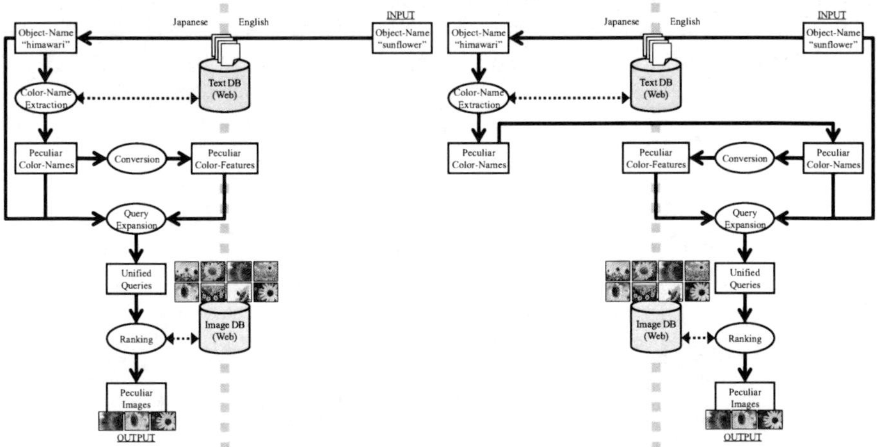

Fig. 1 XPIS to make a one-way trip.

Fig. 2 XPIS to make a round trip.

4 Experiment

This section shows several experimental results for the following eight kinds of target object-names from among four categories to validate my proposed cross-language methods to search the Web for their peculiar images more precisely than my previous single-language method and conventional keyword-based Web image search engines: "sunflower" (typical-color: yellow), "cauliflower" (typical-color: white), "Tokyo tower" (typical-color: red), "Nagoya castle" (typical-color: white), "praying mantis" (typical-color: green), "cockroach" (typical-color: brown), "wii" (typical-color: white), "sapphire" (typical-color: blue).

Fig. 3 and 4 show the top *k* average precision of my proposed cross-language methods, my basic single-language methods, and Google Image Search. They show that my cross-language EJE*q2 method is superior to all the others, and that my cross-language EJE*qX methods to make a round trip from English to Japanese are the best, my cross-language EJ*qX methods to go from English to Japanese (and not to come back) are the second-best (better), and my basic single-language E*qX methods are the worst.

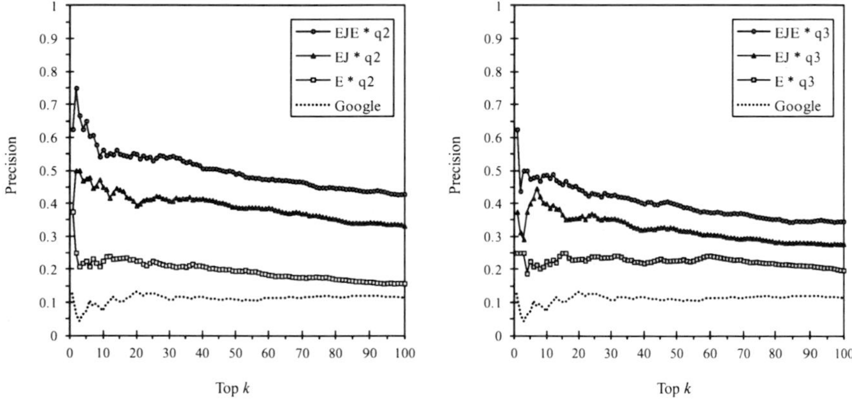

Fig. 3 Top *k* Precision of Google Image Search vs. Peculiar Image Searches (method: X*q2).

Fig. 4 Top *k* Precision of Google Image Search vs. Peculiar Image Searches (method: X*q3).

5 Conclusion

As next steps of IR, it is very important to discriminate between "Typical Images" and "Peculiar Images" in the acceptable images, and to collect many different kinds of peculiar images exhaustively. In other words, "Exhaustiveness" is the most important in the next IR. As a solution to the 1st next step, my previous work proposed a basic method to precisely search the Web for peculiar images of a target object by its peculiar appearance descriptions (e.g., color-names) extracted from the Web and/or its peculiar image features (e.g., color-features) converted from them.

To make the basic method more robust, this paper has proposed a refined method equipped with cross-language (translation between Japanese and English) functions. And several experimental results have validated the retrieval precision (robustness) of my cross-language methods by comparing with such a conventional keyword-based Web image search engine as Google Image Search and my basic single-language method [2, 3]. My proposed cross-language Peculiar Image Search has been about twice as precise as my basic Peculiar Image Search, and about quadrice as precise as Google Image Search, when an English object-names is given as a user's original query for peculiar images.

In the future, I try to utilize the other appearance descriptions (e.g., shape and texture) besides color-names and the other image features besides color-features in my basic single-language and my proposed cross-language Peculiar Image Searches. In addition, I also try to evaluate my proposed cross-language Peculiar Image Searches with translation between English and the other languages besides Japanese, or between Japanese and the other languages besides English.

Acknowledgements. This work was supported in part by JSPS Grant-in-Aid for Young Scientists (B) "A research on Web Sensors to extract spatio-temporal data from the Web" (23700129, Project Leader: Shun Hattori, 2011–2012).

References

1. Hattori, S., Tanaka, K.: Search the Web for typical images based on extracting color-names from the Web and converting them to color-features. Letters of DBSJ (Database Society of Japan) 6(4), 9–12 (2008)
2. Hattori, S., Tanaka, K.: Search the Web for peculiar images by converting Web-extracted peculiar color-Names into color-features. IPSJ (Information Processing Society of Japan) Transactions on Databases 3(1), 49–63 (2010)
3. Hattori, S.: Peculiar image search by Web-extracted appearance descriptions. In: Proceedings of the 2nd International Conference on Soft Computing and Pattern Recognition (SoCPaR 2010), pp. 127–132 (2010)
4. Hattori, S., Tezuka, T., Tanaka, K.: Extracting visual descriptions of geographic features from the Web as the linguistic alternatives to their images in digital documents. IPSJ Transactions on Databases 48(SIG11), 69–82 (2007)
5. Hattori, S., Tezuka, T., Tanaka, K.: Mining the Web for Appearance Description. In: Wagner, R., Revell, N., Pernul, G. (eds.) DEXA 2007. LNCS, vol. 4653, pp. 790–800. Springer, Heidelberg (2007)
6. Wikipedia - List of colors, `http://en.wikipedia.org/wiki/List_of_colors`
7. Japanese Industrial Standards Committee. Names of Non-Luminous Object Colours. JIS Z 8102:2001 (2001)
8. Smith, J.R., Chang, S.-F.: VisualSEEk: A fully automated content-based image query system. In: Proceedings of the 4th ACM International Conference on Multimedia (ACM Multimedia 1996), pp. 87–98 (1996)

VSEC: A Vertical Search Engine for E-commerce

Quan Shi, Zhenquan Shi, and Yanghua Xiao

Abstract. With the explosion of e-commerce sites, querying information residing on these sites with high accuracy becomes a practical issue when designing next generation search engines. The inherent natures of general search engines limit their applications in scenarios where result quality is a critical concern. To overcome this difficulty, we build a Vertical Search Engine for e-commerce (VSEC), which crawls, deep-processes, index information of e-commerce sites of given topics and provide querying service of high quality. VSEC is characterized by two defining features: (1) multi-thread model is employed to speed up the crawling procedure; (2) index building and query answering is implemented on Lucene.Net. Experimental results show that our system has high retrieval efficiency and higher query accuracy and recall rate than general search engine.

1 Introduction

The amount of information residing on WWW has experienced the exponential growth in recent years. By the end of this year, hundreds of millions of web pages have been indexed by search engines. The search results of general search engines usually contain enormous duplicates or unrelated information. Hence, it is hard for general search engines to capture a user's actual search intension, which limits the application of general search engine into scenarios where accuracy is a critical concern. One of theses scenarios is searching for business information on e-commerce sites. The success of a search engine for business information is highly depending on the accuracy of the returned result. For this purpose, we build a Vertical Search Engine for e-commerce （VSEC）, which only index information of e-commerce sites for given topics and provides search result of high accuracy.

Quan Shi · Zhenquan Shi
School of Computer Science and Technology, Nantong University, Nantong, China
e-mail: {sq,szq}@ntu.edu.cn

Yanghua Xiao
School of Computer Science, Fudan University, Shanghai, China
e-mail: shawyh@fudan.edu.cn

F.L. Gaol (Ed.): Recent Progress in DEIT, Vol. 2, LNEE 157, pp. 57–63.
springerlink.com

In this paper, we will elaborate the architecture of VSEC and give a brief introduction about the key techniques used in VSEC. Due to it commercial value, VSEC has been employed for querying business information in some typical enterprise. At the last section, we will use a running example to showcase VSEC in a real application and give some the experimental results about the key performance of the system.

2 The Architecture of VSEC

A general search engine usually consists of the following components: crawlers, indexers, retrieval components, databases and user interfaces. However, in order to return topic related results, topic identifier and keywords dictionary are also necessary components of vertical search engine. Figure1 shows the architecture of VSEC.

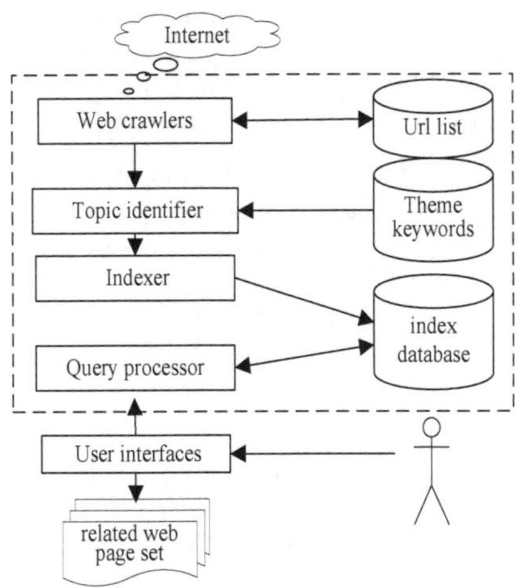

Fig. 1 Architecture of VSEC

(1) Web Crawler
A Web crawler is a computer program that browses the WWW in an automated manner or in an orderly fashion. Web crawlers are mainly used to create a copy of all the visited pages for later processing. In general, it starts with a list of URLs, called the seeds. As the crawler visits these URLs, it identifies all the hyperlinks in the page and adds them to the queue of URLs, called the crawl frontier. Each URL from the frontier will be visited according to a set of policies. We artificially select some e-commercial sites as the seeds that will be fed to the crawler.

(2) Topic identifier
In topic identifier, we calculate the similarity between the target page and the predefined topic using vector space model. All web pages that match the given topics will survive and be saved to the database.

(3) Indexer
Indexer will parse the content of a page survive in the topic identifier, then extract the desired information from the parsed result. All extracted information will be saved to a structured database, such as a relational database. After that, to speed up the querying processing, appropriate index will be built on the structured database.

(4) Query processor
Query processor will accept user's key words and quickly retrieve the documents required by users. The retrieved records will be ranked by the similarity between key words and content of the record.

(5) User interfaces and related database
User interface can provide a visual query input and output environment for the sake of convenient query. Related database are mainly used for storing URL seed list, topic related keywords and related web page set.

3 Key Techniques in VSEC

3.1 Information Acquisition

Information collecting module is responsible for collecting information from internet. The key of designing information collector is how to design a crawler and a web page parser. Crawler is the main tool for search engine to gather data. Since the topic of interest is limited for vertical search engine, the web crawlers only need to visit topic-related web pages.

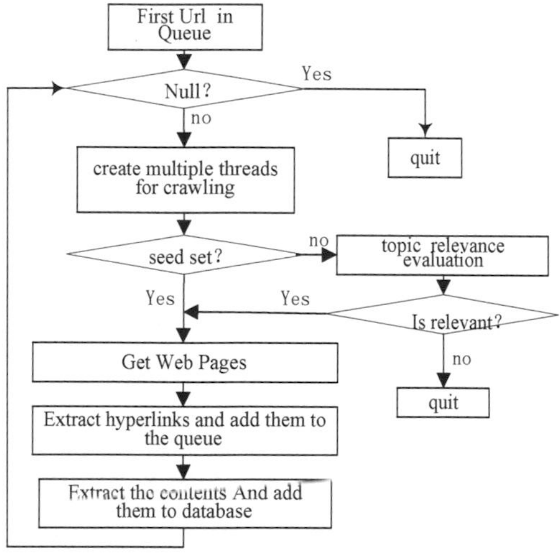

Fig. 2 The flow graph of crawling procedure of VSEC

The detailed process crawling of VSEC is illustrated in Figure2 Effective regular expressions and other configurations are predefined for the crawler. To successfully crawl all web pages of the given topic, we first push URLs of seed set into a queue. Then, start the crawling from the queue. At each step, we get the head of URL queue, if is belonging to the seed set, we retrieve the page and directly extract the information of interest as well as all hyperlinks by predefined regular expression. Otherwise, we need to evaluate the topic of the page. If it is closely related to the given topic, its content and hyperlinks will be processed as those URLs in the seed set. If it is not related to the given topic, we will drop this URL and terminate crawling pages from this website.

Extracted hyperlinks will be inserted into the URL queue. Those URLs belong to a web site in the seed set will be inserted to the head of the queue of URLs. Others will be inserted to the end of the queue. The rationality of this strategy is to crawl web page in seed set as early as possible. To improve performance, we use multiple threads to speed up the processing and made full use of bandwidth. For a targeted URL, it's necessary to retrieve the title, content, hyperlink information in the corresponding web page for further topic relevance analysis and index construction.

3.2 Evaluation of Relevance Score

We adopt vector space model to evaluate the topic relevance of a web page and a given topic.

Given the query vector $Q=(w_{1q},w_{2q,\ldots,}w_{tq})\in R^{t}$ and webpage vector $V_j=(w_{1j},w_{2j,\ldots,}w_{tj})\in R^{t}$, then the similarity between Q and V_j can be measured by:

$$\mathrm{sim}(Q,V_j)=\frac{\sum_{i=1}^{t}(w_{iq}\times w_{ij})}{\sqrt{\sum_{i=1}^{t}(w_{iq})^2\times\sum_{i=1}^{t}(w_{ij})^2}},$$

w_{iq} and w_{ij} are the weights of word i in query vector Q and webpage vector V_j, respectively, and can be computed by traditional TF-IDF algorithm. In other words, the similarity between two vectors can be measured by dot produce of this those two vectors.

Given the relevance score, we now can give the detailed topic relevance calculating algorithm as follows:

(1) Preprocess: retrieve keywords from the web pages coming from the seed web site, and weight each keywords. Then we can construct the feature vector of the given topic, i.e., Q.

(2) Segment the text of the target page into words and extract the featured key words. Calculate the TF-IDF for each key words as the weight of the word.

(3) Segment the title of the target page into words. Combine those words with the result of step 2. Then, we can construct vector V_j.

(4) Calculate the relevance score between Q and V_j.

(5) If the relevance score is greater than the given threshold *T*, the page is regarded as belonging to the topic and will be saved to the database. Otherwise drop this page.

3.3 Index Building and Querying Utilizing Lucene.net

Apache Lucene is a free/open source information retrieval software library, originally created in Java. At the core of Lucene's logical architecture, a document is regarded as the collection of fields of text. Due to this flexibility, Lucene's API is independent of the file format. Lucene.Net is a C# version of the original Lucene project. It has two main functions: index building and querying.

To build the index by Lucene, we first need to define the term and the field, documents. After that, by calling thc add() method provided by class Document, we can add the field to Document. And then, an index structure will be created by calling IndexWrite() method. After index building, documents are organized into a list of segments, each of which represents a complete index section. After these steps, we can write the index onto disks.

The logcial process of query answering of VSEC is as follows. VSEC first parse the input keyword that was submitted users. Based on the parsed result, VSEC perform the key word search on the index and return query results to the user. Using Lucene.net, its detailed process consist of following steps: (1) initialize Lucene search tool IndexSearcher; (2) construct query by using QueryParser to parse the input keywords;(3) construct a DataTable using the result IndexSearcher and return the result to the user as a formalized list.

Fig. 3 An running example of VSEC

VSEC is implemented using C#. All the experiments are conducted on Windows Server 2003, DotNet 2.0 Framework, SQL Server 2005. We select two typical E-business sites as the crawling seed: http://detail.zol.com.cn/ and http://www.pconline.com.cn/. We focus on the information of computer products and digital products. Hence, we use following keywords to describe the topic :“computer”, “mobile phone”, “digital products” ,“desktop”, “server”, “laptop” and so on.

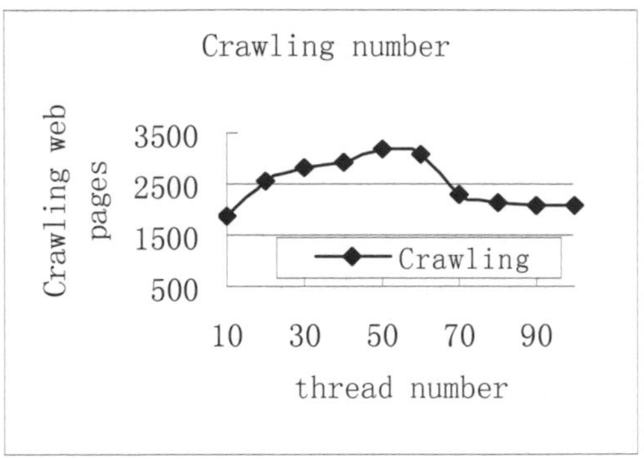

Fig. 4 Crawling throughput as the function of thread number

We first use a running case to showcase VSEC. As illustrated in Figure3, all major information about the laptop computer of type "Y450A-TSI" is correctly and clearly returned to a user if "Y450A-TSI" is input as the search key words.

Note that in VSEC, we use multiple threads to speed up the crawling procedure. Thus, selecting an appropriate thread number plays an important role in the performance of multi-threading crawler. To evaluate the best thread number, we summarize the crawling throughput under different thread numbers. Figure4 shows the result test in the five minute, from which we can see that there exists an optimal number after which the performance of our crawler will decrease. This can be naturally explained since the increment of the thread number leads to a lot of concurrent operations, thus causing extra overheads. Based on this experiment, we finally use 50 as the thread number of our crawler.

4 Conclusion

Vertical search engine orienting on E-business is believed to be the next generation searching business mode. We build such a system called VSEC. There are two typical features of VSEC that distinguish it from other vertical search engines: (1) multiple thread model is used to speed up the crawling procedure; (2) index building and query answer0069ng is implemented on Lucene.Net. Experimental results show our system can collect relatively complete data and can achieve higher query accuracy and recall rate than general search engine.

Acknowledgements. This work was supported by the National Natural Science Foundation of China under grants No.61003001, No.61171132; Specialized Research Fund for the Doctoral Program of Higher Education No.20100071120032; the Natural Science Foundation of Jiangsu Province No.BK2010280.

References

[1] Liu, Y.Q.: Research and design of vertical search engine. Computer Applications and Software 127(7), 130–132 (2010) (in Chinese)
[2] Zhao, Y., Teng, G.F., Zhang, Y.X., He, D.M.: The agriculture information vertical search engine design based on Internet. Journal of Agricul Tural University of He-Bei 32(6), 125–128 (2009) (in Chinese)
[3] Pan, B., Xu, L.L.: Study of chinese blog search engine. Computer Engineering and Design 31(8), 1718–1721 (2010) (in Chinese)
[4] Hung, X.J., Xia, Y.J., Wu, L.D.: A text filtering system based on vector space model. Journal of Software 14(3), 435–442 (2003) (in Chinese)
[5] Li, G., Song, W., Qiu, Z.: AJAX + Lucene built the search engine, pp. 218–286. People Posts and Telecommunications Press, Beijing (2006) (in Chinese)
[6] Liu, G.L., Liu, J.F.: Research and implement of vertical search engine. Journal of Intelligence 28(10), 144–147 (2008) (in Chinese)

Knowledge Based System for Intelligent Search Engine Optimization

Héctor Oscar Nigro, Leonardo Balduzzi, Ignacio Andrés Cuesta, and Sandra Elizabeth González Císaro

Abstract. This paper presents a Knowledge-Based System that using heterogeneous inductive learning techniques and domain knowledge representation, has the major aim of supporting the activity of SEO (Search Engine Optimization). The system arises from the need to answer the following questions. Is it possible to position a web site without being an expert in SEO? Is it possible for a SEO tool to indicate what factors should be modified to position a web site? It attempts to answer both questions from a Domain Knowledge Base and an Inductive Knowledge Base by which the system suggests the most appropriate optimization tasks for positioning a pair [keyword, web site] on the first page of search engines and infers the positioning results to be obtained.

1 Introduction

The method used by search engines to sort the search results is based on algorithms that are not public knowledge. Guidelines are known for optimizing a web site that involves changing certain factors tending to improve its positioning. Being unknown algorithms, SEO activity is empirical and is based on the experience of experts. Despite the experience, it is not possible to know exactly which the most effective actions to improve positioning are. So a "trial-and-error" empirical model is applied, which cost of implementation is usually very high. The success of the activity is further compromised due to increased competition and constant changes in sorting algorithms of search engines.

Héctor Oscar Nigro · Leonardo Balduzzi · Ignacio Andrés Cuesta ·
Sandra Elizabeth González Císaro
Department of Computer Sciences and Systems, Exacts Sciences Faculty,
Universidad Nacional del Centro de la Provincia de Buenos Aires,
Paraje Arroyo Seco s/n, Campus Universitario, B7001BBO Tandil, Bs. As. Argentina
e-mail: oscarnigro@speedy.com.ar, lbalduzzi@gmail.com,
machocuesta@gmail.com, sagonci@exa.unicen.edu.ar

F.L. Gaol (Ed.): Recent Progress in DEIT, Vol. 2, LNEE 157, pp. 65–72.
springerlink.com

The market has several SEO tools that provide values of the factors affecting the positioning of a web site. The right analysis, interpretation and modification of these values may allow the web site to be located in the top of the search engines. To do this work, it is essential to have a deep knowledge of the definition of each evaluated factor, know its domain, scope, relevance, metrics and how to modify it to obtain better positioning.

SEO is one of the leading and most influential activities in the field of online marketing, which is why there is a wealth of information and tools designed to train and support the community of experts in this activity. The challenge of SEO expert is to position a web site in search engines whose sorting algorithms he ignores. Therefore, he faces problems such as high uncertainty about the success of the results, the high cost of scaling his work and slow and costly learning due to the constant changes of search engine algorithms. SEO is rapidly evolving. Search engines are constantly changing their algorithms, and new media and Technologies are being introduced to the Web on a regular basis [3].

Goals of a SEO application centered on a knowledge-based system are: a) Intelligently support the process of positioning in search engines, making the domain knowledge to be dependent on a system, instead of on persons; b) Provide a list of actions to implement in order to position a keyword for a particular web site. c) Identify the difficulty of positioning a particular keyword. d) Infer the position to get by the keyword in the search engine after applying the proposed optimizations. e) Estimate the organic traffic that would generate the keyword when the web site is positioned. f) Dynamically adapt to changes in the search engine algorithms. g) Rank by relevance the factors affecting positioning.

2 Background

Most available information focuses on identifying and describing the factors affecting positioning. A factor is a particular feature related to a web site, web page or keyword that must be evaluated through one or more quantitative metrics. For example, the factor "keyword use in <title> tag" can be evaluated through the "density" metric (total number of occurrences of the keyword in the title over total number of words in the title) and the "existence" metric (which can take the value 1 or 0 according to whether the keyword is found or not in the title of the web page).

In this sense, there are several studies and publications that categorize the factors and describe their metrics, even some of them rank the factors by their relevance on the positioning. Every two years, SEOmoz surveys top SEO experts in the field worldwide on their opinions of the algorithmic elements that comprise search engine rankings [4]. As a result of the survey, an updated ranking of the factors affecting positioning is published.

There is also bibliography describing quality guidelines for optimizing a web site designed to guide webmasters to create and maintain a web site so it is more likely to be well positioned in the search engine for certain keywords. In [6] it is stated that "following the best practices outlined below will make it easier for search engines to both crawl and index your content".

Although different sources have several items in common, there is not still a unified classification of the factors affecting positioning. Analyzing the similarities, factors can be grouped according to the following features: 1) *Web-site's factors*: are those that are related to the structure, organization and updating of a web site and its relationship to other web sites. For example: "Inbounds links" (number of links pointing to a web site from other web sites), "Page Rank" (popularity indicator of a web site). 2) *Web-Page's factors*: are those related to the structure, organization and updating of a given web page and its relationship to other web pages of the web site. For example: "use of subtitle tags <H1>, <H2> ... <Hx>". 3) *Keyword's factors*: are those that are related to the nature of the keyword. For example: "Popularity" (number of times the keyword is searched in the search engine in a given period).

3 Clustering the Rules

Clustering is a data mining technique used to place data elements into related groups without advance knowledge of the group definitions. Popular clustering techniques include k-means clustering and expectation maximization (EM) clustering 1]. The reason to use clustering is that each tuple or case study [keyword, web site] only participates in the creation of rules of the cluster to which it belongs, without affecting the rest. This considerably reduces the "noise" that exists when tuples are not clustered and thus obtain more efficient rules.

To perform clustering, the k-means algorithm available in the WEKA package is used [5]. The objective is that the algorithm can identify clusters of data subsets to work with them individually. Clusters are generated by using a subset of factors comprising the 14 factors that provide greater information gain according to the algorithm InfoGainAttributeEval. The result is 5 clusters that group similar tuples.

Once clusters are generated, it is very valuable to analyze the characteristics of each of them to discover their meaning. Cluster meaning should describe the context of each positioning case (tuple) under study. The analysis consists of calculating the average arithmetic of the values of each factor of the tuples of each cluster and compared them to the average arithmetic of the values of each factor of all tuples used for clustering. The factors distinguishing the cluster can be grouped into factors related to the degree of competitors' optimization and factors related to the keyword competition (InQuotesSearch, InAnchorSearch, AllInTitleSearch, CompetitionGoogle).

To determine the membership of a tuple to a cluster, the Euclidean distance between the tuple and the centroid of each cluster is obtained. A tuple will belong to that cluster with which it has lower Euclidean distance.

4 Knowledge-Based System Applied to SEO

A knowledge-based system is a software system capable of supporting the explicit representation of knowledge in a specific domain and of exploiting it through appropriate reasoning mechanisms in order to provide a high-level problem-solving performance [7]. Developed SEO application (Fig. 1), centered on a knowledge-based system, and comprises:

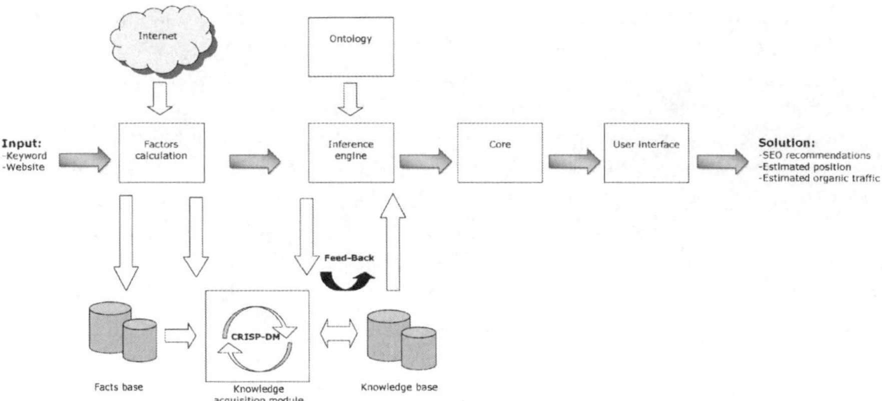

Fig. 1 System architecture

1. **Facts base** is continuously updated by real positioning cases. For each case, the system calculates a set of factors affecting the positioning and related to a keyword, a web site, and competitors' web sites. Each case is labeled positioned in the event that the web site is among the top 10 search results of that keyword, and is labeled as not positioned otherwise. A knowledge base that contains the domain knowledge represented by inductive rules. A rule is a conditional structure that logically relates the information contained in the antecedent part with other information contained in the consequent part. The rules of the system are the result of extracting knowledge of facts base using a heterogeneous set of Data Mining models (J48, JRip and Part).

2. **Inductive knowledge base** contains the knowledge acquired by rules obtained from data mining algorithms. A rule is a conditional structure that logically relates the information contained in the antecedent part (conditions on positioning factors) with other information contained in the consequent part (keyword position in the search engine results). The rules of the system are the result of extracting knowledge of facts base using a heterogeneous set of data mining models. There are certain attributes that influence the positioning of a web site. The study of the different factors of an individual web site is not enough to infer the different positioning strategies of a search engine. The fundamental problem is to determine how the various relevant factors are related to each other and the relevance they have in the final positioning. The rule generator inductive models generalize the set of training examples in the form of rules that can be evaluated to classify new instances Will be used different models of inductive learning, including: J48, JRip and Part [2, 8]. Each rule obtained through Data Mining models, consists of logical conditions refer to the analyzed positioning factors. Each rule is associated with a cluster and has indicators of reliability, support, estimated position and cost. Indicators are continuously updated, providing feedback to the system with new knowledge.

3. **Domain knowledge base** represented in part by an ontology, whose function is to indicate under what conditions should be considered a rule rather than another. Beside to model and conceptualize the knowledge, the ontology will have as main objectives to filter rules and add knowledge in them. Although the rules obtained from inductive models allow inferring the possible strategies of search engines to sort search results, there is SEO domain-specific knowledge that cannot be obtained automatically. Obtaining this information is not trivial; a methodology for achieving this is based on interviews and surveys of SEO experts. From these interviews can be determined, conceptualizations, taxonomies of factors, ranges of possible values, relations between them, etc. The best way to represent this information is through an ontology. The positioning factors can be classified as a taxonomy of positioning factors (Fig. 2).

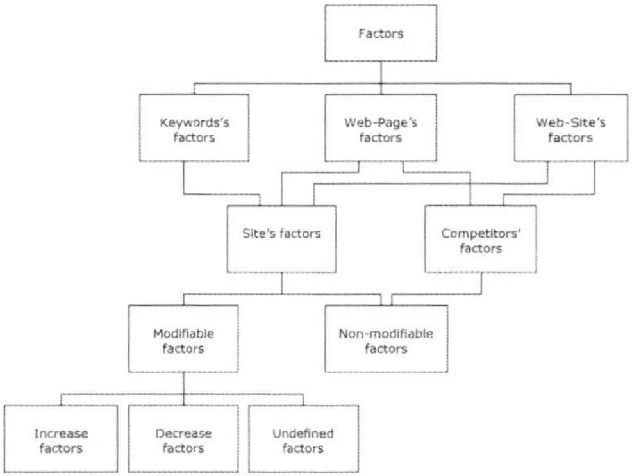

Fig. 2 Taxonomy of positioning factors

They are: 1) *Web-site's factors*: are those related to the structure, organization and updating. 2) *Keyword's factors*: are those related to the nature of the keyword. 3) *Site's factors* (factors of the analyzed site): are those whose values are given by the pair [keyword, website]. 4) *Competitors' factors*: factors whose values are given by the analysis of the first 5 competitors' web sites. 5) *Modifiable factors*: are those whose value can be changed by optimization actions. 6) *Non-modifiable factors*: are those whose value cannot be changed by optimization actions. 7) *Increase factors*: are those whose value should be increased. 8) *Decrease factors*: are those whose value should be decreased. 9) *Undefined factors*: are those whose value is undefined whether to increase or decreased.

The following rules of the ontology, which respond to business rules, are incorporated into the knowledge based system through knowledge of the domain expert (SEO in this case).Is a set of rules that can be expanded or eliminated according to the evaluation done through interviews with experts. Restrictions of the factors are conditions on the values of the factors that should be met not causing the rule to be discarded. Some of them are: 1) CalculatePRCriterion must be <= a CalculatePR. 2) If CalculatePRCriterion value needs to be higher than CalculatePR current value, the value of CalculatePR must be increased. 3) For TotalDMozEntries factor, actions that increase its current value over 10 units should not be suggested.

The utilization of ontology is essential to add SEO domain expertise knowledge in rules obtained from the knowledge base. The ontology is intended to exclude those rules that are non-applicable for different reasons. Listed are some of the causes: 1) the impossibility of performing an action to change the value of a "non-modifiable" factor. 2) The impossibility of performing an action of an absurd change. It means the requirement of changing the value of a factor in the opposite direction to its optimal value. 3) The impossibility of performing an action that not comply with "business rules".

The following example shows how the ontology discards rules from the knowledge base. Example: let (CompGoogle=572000, PR=3; TSInYahoo= 24, ..., Pos=1) be the input tuple. The rules of the knowledge base that match with the input tuple are:

Rule 1: *CompetitonGoogle < 1356000; CalculatePR < 2; ExactDensityInTitle >= 0.66 -> Position = 0; Reliability: 0.81%; Support: 320.* This is discarded as it requires modifying the Page Rank of the site below 2 when currently value is 3. CalculatePR is an Increase factor; so its value only should be changed for a value over the current one, otherwise would be an absurd.

Rule 2: *CompetitonGoogle < 2532000; CalculatePRCriterion >= 2; TotalSiteInboundsYahoo >= 4230 -> Position = 0; Reliability: 0.79%; Supporte: 510.* This is discarded as it requires increasing the number of incoming links to the site (TotalSiteInboundsYahoo) over 4230, which would entail a non-admissible effort in hours according with the business rules.

Rule 3: *AllInTitleSearch < 159000; ExactDensityInTitle >= 0.33; ExactDensityInBodyText >= 0.14 -> Position = 0; Reliability: 0.78%; Support: 492.* This is accepted by the ontology since there are no restrictions on the actions to implement. This requires increasing the exact density in the title of the page (ExactDensityInTitle) over 0.33 and increasing the exact density in the text of the page (ExactDensityInBodyText) over 0.14.

4. **Inference engine** models human reasoning process from the information in the knowledge base. The engine is in charge of analyzing an input case and determine from all the rules of the knowledge base, which are the candidates to form the solution to be displayed as output. This check will be conducted from the matching between the values of the factors in the entry case and the rules of the knowledge base, while respecting the constraints expressed by the ontology.
5. **Knowledge acquisition module** allows adding, deleting and modifying elements of knowledge (rules) in the system. Being a dynamic context, its presence is essential because the system works properly only if it keeps its knowledge updated. The module allows this maintenance automatically, registering in the knowledge base the changes occurred.
6. **Core module** gets the data processed by the inference engine and is responsible for identifying the degree of difficulty of the solutions, inferring the position to have the keyword in the search engine after applying the proposed solution, estimating the organic traffic to be generated by the keyword in the analyzed web site and getting the ranking factors sorted by relevance.

7. **User interface** displays in natural language the optimization tasks that should be applied to position a keyword of the analyzed web site.

5 Experimental Results

Are denominated success cases at those cases (keyword, website) that during positioning campaign (six weeks in total) the website reaches a top ten position of the search engine for the keyword and at least 66% of the optimization tasks recommended by the system were completed.

Whereas, failure cases are those for which the 100% of the optimization tasks were completed and never can rise to the top ten ranking of the search engine. We observe that exists different methods (I, II, III, IV) in each period of time in which were implemented. Each method has changed the statistical inference algorithms used and its parameters; the inference engine of the knowledge base was updated, class of clusters were added and only more efficient rules were maintained, eliminating those whose support and reliability did not exceeded the thresholds of 200 and 70%, respectively. The total historical result considering all the methods has been 74% of success, from 703 analyzed cases and 518 succeeding cases.

Table 1 Experimental results

Method	Period	Language	Total cases	Success cases	% of success cases
I	2008	Total	168	93	55%
II	Q1 / Q2 2009	Total	55	32	58%
III	Q3 / Q4 2009 and Q1 2010	Total	191	154	81%
IV	Q2 / Q3 2010 and Q4 2010	Total	289	239	83%
		Spanish	257	215	84%
		English	25	21	84%

In order to evaluate the rule generation and its application, the experiments were based on 42320 training vectors to generate rules for the knowledge base. Then the rules were tested and evaluated with 27179 case studies.

To create inductive models, one file per cluster was generated and three inductive models were generated for each cluster with the corresponding vector. Over one month 4634 rules were evaluated by 27179 tuples of real cases, which modify initial values of reliability and support of each rule. After evaluation arose a significant number of rules with a support higher than 200, this is due to the large number of instances that have evaluated them and modified their reliability and support values and rules below an acceptable level of reliability were discarded.

There is an improvement of 10% in the rule generation of acceptable quality (above 70% reliability) of methods that use clustered rules over those methods that use non-clustered rules. While the rules are evaluated with new cases, they increase their support and, in 50% of the cases, their reliability is reduced below the threshold in the first two to three months, thereby are discarded. Among others reasons, it is supposed by changes in the algorithms of search engines. The average stay of a rule is six months, if that test is made with new cases every month.

6 Conclusions

This chapter has proposed using a knowledge-based system for positioning web sites in search engines. The system has been designed and implemented using heterogeneous inductive learning techniques and an ontology for knowledge representation, with the aim of supporting SEO activity.

An ontology for SEO domain has been specified, defining different concepts, its taxonomy and relationships between them. Although the rules obtained from inductive models allow inferring the possible strategies of search engines to sort search results, the knowledge of SEO domain experts that cannot be extracted automatically must be represented. The best way to represent this information is through an ontology. It's expected to continue adding knowledge, based on interviews with SEO experts, to enhance relationship between factors and concepts.

The results obtained allow concluding that the use of a defined ontology decisively improves the quality of the system. Failure to use the ontology for the system would bring the risk of absurd solutions to an SEO expert, either because of going against the logic of domain knowledge, or being impossible to implement because of its high cost. The results obtained in the cases analyzed were positive. The ability of system evolution and learning allows optimizing the productivity and effectiveness of the SEO process.

References

[1] Chapple, M.: Clustering (2001), `http://databases.about.com/od/datamining/g/clustering.htm` (cited February 11, 2011)

[2] Cohen, W.W.: Fast Effective Rule Induction. Paper Presented at the Twelfth International Conference on Machine Learning, Tahoe City, CA (1995)

[3] Enge, E., Spencer, S., Stricchiola, J., Fishkin, R.: The Art of SEO. O'Reilly Media (2009)

[4] Fishkin, R.: Search Engine Ranking Factors (2009), `http://www.seomoz.org/article/search-ranking-factors` (cited February 13, 2011)

[5] Frank, E., Witten, I.: Generating Accurate Rule Sets Without Global Optimization. Paper Presented at the Fifteenth International Conference on Machine Learning. Morgan Kaufmann Publishers, San Francisco (1998)

[6] Google's Search Engine Optimization Starter Guide. Google, page 1 (2008)

[7] Guida, G., Tasso, C.: Design and Development of Knowledge-Based Systems. From Life Cycle to Methodology. John Wiley and Sons Ltd., Chichester (1994)

[8] Quinlan, R.: C4.5: Programs for Machine Learning. Morgan Kaufmann, San Mateo (1993)

A Novel Research Methodology for Gastronomic Blogs

Hsiu-Yuan Wang, Tung-Jung Chan, Yu-Sheng Chang, and Ching-Mu Chen

Abstract. Blogs are effective on spreading eWOM to provide hospitality and tourism promoters and different from traditional ones, but has big potential to create enormous impact on hospitality or tourism promotion. Although the growth rate of blogs is impressive, less research effort has been devoted to investigating the influence of gastronomy blogs. Assessing gastronomy blogs from readers' perspective can be a great value to both researchers and practitioners and it can be useful to researchers in developing and testing theories relating to gastronomy blogs, enable researchers to justify gastronomy blogs as an effective way in the context of hospitality marketing, and help to understand the drivers behind gastronomy blogs to drive readers' behavioral intention. By developing such a study, restaurant executives is able to better justify their web-based promotional activities if they devote a significant portion of their organizational budgets to these activities, as well as to understand how to use gastronomy blogs as part of their business strategies for greater culinary destination promotion. Therefore, the main purpose of this study is to explore the novel research methodology deriving from gastronomy blogs influenced blog readers' behavioral intention to taste local gastronomy.

Keywords: eWOM, gastronomy blogs.

1 Introduction

From the inspiring taste desire point of view, both of the two proposed constructs that are experiencing appeal and generating empathy had a significant, positive

Hsiu-Yuan Wang
Department of Hospitality Management, Chung Hua University, Hsinchu, Taiwan
e-mail: hywang@chu.edu.tw

Tung-Jung Chan
Department of EE, Chung Chou University of Science and Technology, Taiwan
e-mail: jung@dragon.ccut.edu.tw

Yu-Sheng Chang · Ching-Mu Chen
Department of CSIE, Chung Chou University of Science and Technology, Taiwan
e-mail: {ysc,kchen}@dragon.ccut.edu.tw

F.L. Gaol (Ed.): Recent Progress in DEIT, Vol. 2, LNEE 157, pp. 73–78.
springerlink.com

effect on behavioral intention to taste. First, in line with prior work on marketing [1], *t*he results revealed *that* experiencing appeal was considered to be a vital factor that can inspire blog readers' taste desire, for it presented a significant influence on potential readers' intention to taste local food and beverages. This means that the majority of our respondents care about whether or not the presentation, the photos and videos of gastronomy blogs can make them have the feeling of excitement, curiosity, attraction and even being persuaded so as to prompt their taste intention. Visual impressions of local food such as fresh, clean and delicious food well displayed and prepared attract customers and play a critical motivational role leading to their local food choice [2]. Thus, local restaurant firms could cooperate with blog solution providers to provide easier blog interface for loyal customers to post food pictures and positive comments on firms' websites, which help potential customers to experience sensory appeal, send a clear message to them and convey a total emotional experience at both overt and subliminal levels. Previous research supported the concept that blogs help to generate empathy so as to arouse individual behavioral intention to have a direct experience [3]. So, it is believed that gastronomy blogs with pictures and videos that can induce readers to generate emotional or intellectual identification with blog writers are more likely to encourage high intention to visit author-described gastronomic location.

On the facilitating interpersonal interaction side, social influence created a salient effect on behavioral intention to taste in the situation of gastronomy blogs. This is in line with prior research on technology diffusion which indicated that social influence plays an important role *in* shaping an individual's behavioral intention [4-6]. Interpersonal communications have long been recognized as influential in the hospitality and tourism industry. With the improvement of Internet technologies, increasing numbers of customers are using the Internet to seek information and to conduct transactions online. By gastronomy blogs, writers who had positive customer experience can spread eWOM [7-8] that appear in an unprecedented large scale, potentially forming new dynamics in the market.

2 Research Methodology

In this study, we developed instruments based on prior studies. Responses to the items in the group of inspiring taste desire that are experiencing appeal and generating empathy, the category of forming taste awareness that are providing image, delivering knowledge and presenting guides, the kind of facilitating interpersonal interaction that are social influence, cyber-community influence, and behavioral intention to taste were measured on a 7-point Likert scale from 1 represented as strongly disagree to 7 represented as strongly agree. The experiencing appeal construct was measured by four items [1-2], [9]. The generating empathy construct was composed of four items [3], [10]. The providing image construct comprised five items [11-12]. The delivering knowledge construct consisted of four items [2], [11]. The presenting guides construct contained four items [13-14]. The social influence construct was measured by four items [15-16] and the cyber-community influence construct was gauged by four items [17-18]. Lastly, the behavioral intention to taste construct was composed of three items adapted from [6]. To make sure that important aspects are not overlooked, we performed experience surveys on the measures with three professionals in hospitality management field and four graduate students who had regular experience in reading gastronomy blogs. They were asked to comment on list items that corresponded to the constructs, including scales wording, instrument length, questionnaire format, and ambiguous question. After careful examination, the items were slightly revised so the wording is more precise to constitute a complete scale for this study. Consistent with prior research on social and human behavior, the questionnaire also contained demographic questions for further understanding the background of our research participants. Furthermore, the original question items were in English; however, we invited a bilingual expert to translate them into Chinese to ensure the validity of the questionnaire.

Because of the lack of a reliable sampling frame, it is difficult to conduct a random sampling for all the readers of gastronomy blogs in Taiwan. Thus, in this study we adopted a non-random sampling technique which is convenience sampling to collect the sample data in June 2009. To make the results stable, we mainly gathered sample data from five international or local organizations in Taiwan: Aerospace Industrial Development Corporation (AIDC), MassMutual Mercuries Life, National Chiao Tung University, China Medical University Hospital and IBM Taiwan. Respondents were first asked whether they had ever read gastronomy blogs; if they replied in the affirmative, they were asked to participate in the survey. The respondents were instructed to answer the questions based on their prior experience of reading certain gastronomy blogs. This served to relate the survey respondents to certain gastronomy blogs. For each question, respondents were asked to circle the response that best described their degree of agreement. On this basis, a convenience sampling was implemented and a sample of 329 usable responses was obtained from a variety of respondents with different computer, Internet or blog experiences. A total of 42.9% of the respondents is male. The respondents had an average of 11.52 years of computer experience (S.D. = 5.7), and 8.98 years of Internet experience (S.D. = 4.06). Additionally, 68% of respondents had experience in blogs over 6 months, and 81.1% of them had a degree at the college level or above.

3 Data Analysis and Results

A confirmatory factor analysis via AMOS 7.0 was conducted to test the measurement model. Six common model-fit measures were employed to assess the model's overall appropriateness of fit: the ratio of χ^2 to degrees-of-freedom (df.), goodness-of-fit index (GFI), adjusted goodness-of-fit index (AGFI), normalized fit index (NFI), comparative fit index (CFI), and root mean square residual (RMSR), and to attain a better model fitness, four items were eliminated due to low or cross factor loadings. As shown in Table 1, all the model-fit indices exceeded their respective common acceptance levels suggested by previous research, thus demonstrating that the measurement model revealed a fairly good fit with the data collected. We could therefore go forward to evaluate the psychometric properties of the measurement model in terms of reliability, convergent validity, and discriminate validity.

Table 1 Fit indices for measurement and structural models

Fit index	Recommended value	Measurement model	Structural model
χ^2/d.f.	<3.00	1.562	1.562
GFI	>0.80	0.909	0.909
AGFI	>0.80	0.879	0.879
NFI	>0.90	0.932	0.932
CFI	>0.90	0.974	0.974
RMSR	<0.10	0.041	0.041

GFI=goodness-of-fit index; AGFI=adjusted goodness-of-fit index; NFI=normalized fit index; CFI=comparative fit index; RMSR=root mean square residual.

Reliability and convergent validity of the factors were calculated by composite reliability, and by the average variance extracted as shown in Table 2. The composite reliabilities can be calculated as follows: (square of the summation of the factor loadings)/ {(square of the summation of the factor loadings) + (summation of error variables)}. The interpretation of the resultant coefficient is similar to that of Cronbach's alpha. Composite reliability for all the factors in the measurement model was above 0.76, thus demonstrating all were greater than the bench mark of 0.60 suggested by [19]. The average extracted variance were all above the recommended 0.50 level [20], which implied that more than one-half of the variances observed in the items were accounted for by their hypothesized factors. Convergent validity can also be evaluated by observing the factor loadings and squared

Table 2 Reliability, average variance, and discriminant validity

Factor	CR	1	2	3	4	5	6	7	8
1.Experiencing appeal	0.919	0.739							
2.Generating empathy	0.866	0.630	0.684						
3.Providing image	0.877	0.448	0.457	0.588					
4.Delivering knowledge	0.766	0.195	0.258	0.331	0.622				
5.Presenting guides	0.865	0.392	0.423	0.503	0.340	0.617			
6.Social influence	0.895	0.411	0.393	0.346	0.144	0.483	0.681		
7.Cybercommunity	0.897	0.354	0.421	0.514	0.335	0.498	0.454	0.685	
8.Behavioral intention to	0.920	0.510	0.537	0.494	0.194	0.487	0.471	0.498	0.851

CR = Composite Reliability

Diagonal elements are the average variance extracted. Off-diagonal elements are the shared variance.

multiple correlations from the confirmatory factor analysis. Based on suggestion in [20], factor loadings greater than 0.50 were deemed as very significant. All of the factor loadings of the items in the research model were greater than 0.70. Also, squared multiple correlations between the individual items and their a priori factors were high. Thus, all factors in the measurement model had adequate reliability and convergent validity.

To test discriminate validity, this study compared the shared variance between factors with the average variance extracted of the individual factors. This analysis exhibited that the shared variances between factors were lower than the average variance extracted of the individual factors, thus confirming discriminate validity as shown in Table 2. In brief, the measurement model demonstrated adequate reliability, convergent validity, and discriminate validity.

4 Conclusions

In terms of the effect on forming taste awareness, we found that two variables that are providing image and presenting guides had a significant positive influence on behavioral intention to taste. Regarding providing image, the results indicated that if gastronomy blogs can help readers to establish a clear and complete impression that is in correspondence with a gastronomic location, the extent of promoting their intention to visit the place is higher. In the past, gastronomic tourism was likely to mean dining at a renowned restaurant such that the food quality, service quality, eating environment and comfort level will be key points for customers' making decisions. This study confirms previous researchers' argument that sharing true views on destinations help website visitors to develop an image and then increase people's visit intention. Thus, when marketing local and regional cuisine, travel companies can encourage gastronomers or cuisine lovers to offer advice and recommend prestigious restaurants in their own or company-sponsored blogs, which can attract more readers to taste the local cuisine promoted.

References

1. Rust, R.T., Oliver, R.L.: Should we delight the customer? Journal of the Academy of Marketing Science 28(1), 86–94 (2000)
2. Kim, Y.G., Eves, A., Scarles, C.: Building a model of local food consumption on trips and holidays: A grounded theory approach. International Journal of Hospitality Management 28(3), 423–431 (2009)
3. Lin, Y., Huang, J.: Internet blogs as a tourism marketing medium: A case study. Journal of Business Research 59, 1201–1205 (2006)
4. Moore, G.C., Benbasat, I.: Development of an instrument to measure the perceptions of adopting an information technology innovation. Information Systems Research 2(3), 192–222 (1991)
5. Venkatesh, V., Davis, F.D.: A theoretical extension of the technology acceptance model: Four longitudinal field studies. Management Science 45(2), 186–204 (2000)

6. Venkatesh, V., Morris, M.G., Davis, G.B., Davis, F.D.: User acceptance of information technology: toward a unified view. MIS Quarterly 27(3), 425–478 (2003)
7. Kaikati, A.M., Kaikati, J.G.: Stealth marketing: How to reach consumers surreptitiously. California Management Review (4), 6–22 (2004)
8. Thorson, K.S., Rodgers, S.: Relationships between blogs as eWOM and interactivity, perceived Interactivity, and parasocial Interaction. Journal of Interactive Advertising 6(2), 39–50 (2006)
9. Huang, L., Chou, Y., Lin, C.: The Influence of reading motives on the responses after Reading Blogs. Cyberpsychology & Behavior (3), 351–355 (2008)
10. Kim, S.S., Agrusa, J., Lee, H., Chon, K.: Effects of Korean television dramas on the flow of Japanese tourists. Tourism Management 28, 1340–1353 (2007b)
11. Cohen, E., Avieli, N.: Food in tourism: Attraction and impediment. Annals of Tourism Research 31(4), 755–778 (2004)
12. Susskind, A.M., Chan, E.K.: How restaurant features affect check averages. Cornell Hotel and Restaurant Administration Quarterly 41(6), 56–63 (2000)
13. Julienne, A.Y., Joelle, B.Y., Brennan, E.Y., Loren, G.Y.: Adolescent fast food and restaurant ordering behavior with and without calorie and fat content menu information. Journal of Adolescent Health 37(5), 397–402 (2005)
14. Murphy, J., Forrest, E., Wotring, C.E.: Restaurant marketing on the worldwide web. Connell Hotel and Restaurant Administration Quarterly 37(1), 61–71 (1996)
15. Ashforth, B.E., Mael, F.A.: Social identity theory and the organization. Academy of Management Review 14(1), 20–39 (1989)
16. Sparks, B.: Planning a wine tourism vacation? Factors that help to predict tourist behavioural intentions. Tourism Management 28, 1180–1192 (2007)
17. Hsu, C., Lin, J.C.: Acceptance of blog usage: The roles of technology acceptance, social influence and knowledge sharing motivation. Information & Management 45, 65–74 (2008)
18. Litvin, S.W., Goldsmith, R.E., Pan, B.: Electronic word-of-mouth in hospitality and tourism management. Tourism Management 29(3), 458–468 (2008)
19. Bagozzi, R.P., Yi, Y.: On the evaluation of structural equation models. Journal of Academy of Marketing Science 16(1), 74–94 (1998)
20. Hair, J.T., Anderson, R.E., Tatham, R.L., Black, W.C.: Multivariate data analysis with readings, 3rd edn. Macmillan, New York (1992)

Semantic Network Active Directory Service System

Sergio Beltran, Adrienne Lam, Jorge Estrada, Aleksander Milshteyn,
John Paul Adigwu, Alexander Alegre, Charles Liu, and Helen Boussalis

Abstract. The project focuses on the implementation of a Semantic Network utilizing an Active Directory Service System in providing real-time access of relevant information within closed communities, ranging from educational standpoints to secure server networks where queried information must be protected. A distributed server model is implemented to accommodate an expanding Semantic Network. However, the added scalability poses a challenge in the synchronization of data across multiple servers. To provide a uniform Semantic Network and efficiently utilize the computational power of this distributed server platform, an innovative software architectural model is presented. Such endeavors fulfill on-demand Internet-based client requests for aerospace information exchange [1]. This exchange is based on synchronization of an Extensible Markup Language (XML) [2] directory and file structure, which tracks content and metadata across the distributed server network.

1 Introduction

Technical information can be organized to form, disseminate, and extend knowledge bases. A Semantic Network has been proposed by the NASA-sponsored Structures Pointing and Controls Engineering University Research Center (SPACE URC)[1] to organize and render the contents of technical information through prudential and professional guidance to facilitate literature search functions.

The proposed Semantic Network represents concepts of knowledge as nodes in a tree-structured hierarchy. Here, class-inclusion relations determine node

Sergio Beltran · Adrienne Lam · Jorge Estrada · Aleksander Milshteyn ·
John Paul Adigwu · Alexander Alegre · Charles Liu · Helen Boussalis
California State University of Los Angeles
e-mail: {sbeltran00,adrienneslam}@gmail.com

F.L. Gaol (Ed.): Recent Progress in DEIT, Vol. 2, LNEE 157, pp. 79–85.
springerlink.com

interconnection. Previous research revealed that current students are deemed "technically savvy." However, there is no evidence in the serious literature that these people are experts in searching, nor have their search skills improved with time. One area of concern is the way young people evaluate, or rather, fail to evaluate information from electronic sources. Students who are lacking the information literacy may be easily overwhelmed by the Internet search, and fail to find useful information. In addition, by simply storing bookmarks of separate web pages based on the existing World Wide Web, the information, even stored in hypertext format, is organized deductively according to the author's perspective. The readers have to store the entire link in which, perhaps, most of the information is not needed. Meanwhile, readers may not efficiently use the bookmarks generated by another reader; the other readers have to re-digest all the information linked by the bookmarks.

A concept of Semantic Web [3] is proposed so information is organized based on a machine-recognizable matter. Semantics of the information is implied when such information is generated and stored. Thus, the search engines will be dramatically more powerful to precisely, concisely, and efficiently provide the users with the right information. Moreover, the users who acquire information may also trigger the Semantic Web to categorize and organize such information for enhancing the search capability in the future. Thus, the users may develop the Semantic Web in a collaborative matter. However, by nature, it is impossible to focus on broad topics on the Semantic Web. Serious users who need technical information may lack motivation to use and contribute to the generation of information since the contents are too generic. Here, a collaborative information generation and dissemination model is used to implement a Semantic Network for "the birds of a feather"; i.e., communities with similar research and educational interests (Fig. 1). The Semantic Web loses its effectiveness if the content becomes too broad in scope. In a search engine there are no node connections and thus no context, but with too many node connections to too many ideas, it would become just as ineffective.

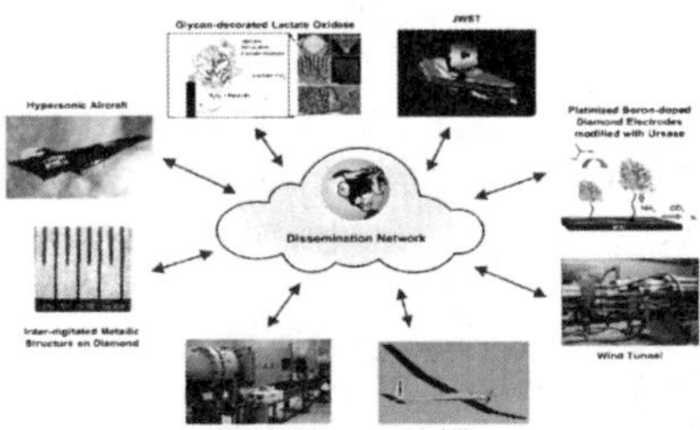

Fig. 1 Semantic dissemination network for communities with NASA-centered interests

To support the growth of content in the Semantic Network and provide wide accessibility to users, a distributed server model was deemed best for the task. As more information nodes are added and new connections are made, additional servers can be incorporated into the server network. However, using multiple servers with different sets of users would yield varying content on each server. Essentially, each server would contain its individual version of the Semantic Network. This lack of synchronization among servers undermines the goal of the Semantic Network and reduces its effectiveness.

To address this issue and achieve uniformity across the Network, a method or model must be developed to track changes and update the entire Network in real-time. The architectural model that overcomes these challenges is the Active Directory Service System. To fully understand this process, first we look at underline key technologies, and then proceed with an introduction of Semantic Network information scraping, followed by Active Directory Service System and finally case studies.

2 Decentralized Information Server

The DIS is a high performance parallel and distributed processing server, which supports efficient, on-demand information dissemination. The DIS platform is aimed at the Science, Technology, Engineering, and Math (STEM) information repositories, and accesses for research and educational purposes. Users are able to generate, integrate, publish, and access information of different formats related to aerospace engineering, digital engineering, control engineering, material science, life support systems, and nanotechnology. A task of keyword match is performed based on tuple-space technology. However, the power of the service platform lies in the "metadata" generated through the efforts of individuals on information analysis, integration, and synthesis. The metadata serves to facilitate the understanding, characteristics, and management usage of data, and hence provide the "semantics" of the information.

Active Directory Service System will leverage the Decentralized Information Server (DIS) architecture using tuple space programming paradigm as its framework. Both technologies play key roles to the utilization of Active Directory Service System.

The educational platform will be developed based on a DIS technology. The architecture of a DIS is shown in Fig. 2. Administrative Servers will serve as connection nodes between cluster networks. Clients will be able to connect to local Servers in order to access the Semantic Network. However, Server protocols will dictate data transmission to remote computer clusters. As such, in order to access remote information, clients will rely on the node connections of servers.

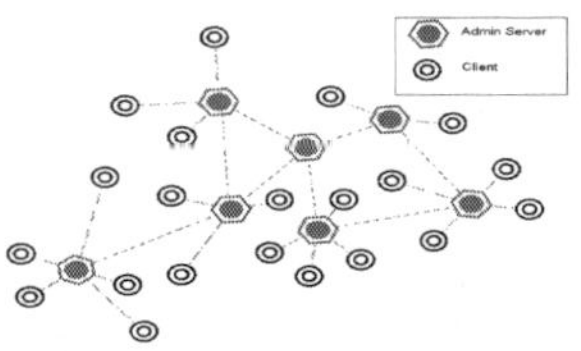

Fig. 2 Decentralized Information Server technology cluster diagram

3 Semantic Network Information Scraping

The "semantics" of the information is a major component that separates the Semantic Network from a simple repository or a storage system. Since it is the user who scrapes content and provides metadata, the success of the server platform will rely on the ease and efficiency of generating such metadata. Referring back to Fig. 2, it should be noted that as users scrape and upload new content, they are only updating their local server. The process of scraping, generation of metadata, and the role of the Active Directory will be described more in detail in the next section. Currently, information is stored in human- but not machine-recognizable format. This can be seen in the search engines which are mostly limited to keyword matching. It makes the automation of semantic-based information organization very difficult. The SPACE URC endeavors to develop the mechanism of generating, and publishing metadata using a scraper toolkit to provide the ontology for automatic deductive reasoning in the Semantic Network. The generated metadata provides a context for each node of information. A common Internet search engine will perform keyword matching, while node connections in the Semantic Network provide additional context to a search.

An example of information scraping task based on a James Webb Space Telescope (JWST) is shown in Fig. 3[4]. During the scraping operation, a user will receive organized data of primary components, subcomponents and their relationship to JWST. Each node in the tree can be a subset of another node, or branch out to additional nodes as in Fig. 3. The connections provide the context that distinguishes this structure from traditional search engines, mentioned earlier.

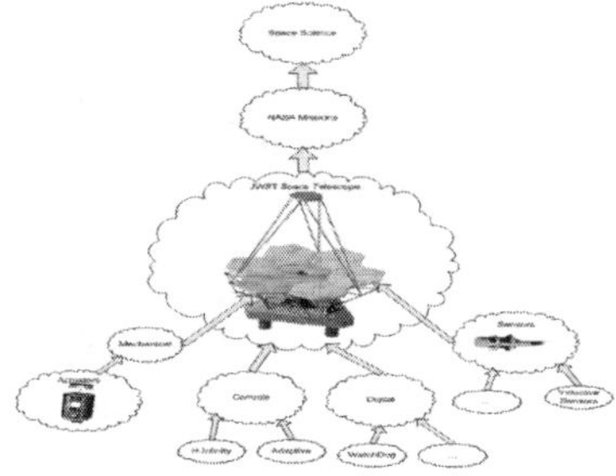

Fig. 3 Relationship of the JWST components is represented using XML architecture tree

4 Scraping Tool

In order to facilitate the XML Architecture Tree and manage Semantic Network information, tools and protocols must be initiated. This section will introduce new software that will handle those contents. The metadata, or data about other data, refers to the descriptive words and phrases that define the content itself. In terms of the Semantic Network, this information includes but is not limited to, basic file information such as name, size, type, date created/modified and other vital document properties. Thus, the success of the Semantic Network server platform will rely on the efficiency of generating such metadata.

A SPACE URC Scraping tool is currently being developed which will allow specific mediums (i.e. books, lectures) to be objectized into smaller meaningful, and more manageable components. The tool will be capable of handling various file types and will facilitate the generation and publication of objectized tags as metadata for the proposed Semantic Network.

5 Active Directory

The addition of a directory service offers a solution, providing current information on all tags residing within the network. This information will include the tag server location, tag hierarchical tree structure, and the name of the files that make up the tag. Using the XML schema, a directory service will be implemented within each network cluster server. The design will store and maintain current information on the location of all networked metadata tags. This will allow each of the servers to easily obtain information from any interconnected computer cluster.

5.1 Tag Request Generation and Application

When a client or server makes a request to another server, a tag request utilizing the XML schema must be generated. The "type" attribute of the <tuple> tag indicates the type of request. The requested semantic tag follows this information. Encapsulated within are the name, IP address, and port of the requesting party. This allows processed metadata to be routed back to the requester.

A history tag is added to the tag request when sending information back to the requester. Here, the history of the tuple's route through the network is passed. This allows data to be traced back to a specific source. As such, clients are able to verify the sources of obtained data. As a tuple enters a server the history tag is updated.

5.2 Tag Request Case Scenarios

Since data on the Semantic Network is distributed across multiple servers, a client should be able to make a request to a local server for data that is residing in a remote location; To simplify a client-side interaction, a user needs to request for an object from one of the distributed servers. After connection protocols are established, a server controller processor accepts the tuple request and places it into a local tuple space region. An idle worker then retrieves the tuple from the tuple space region and searches for the tag in the Active Directory Service System. If the information is found within the local server, the worker processor will access the local database to extract the necessary files and transmit them back to the client as displayed on Fig. 4. However, if the request is found to reside within a remote server on the network, the worker will forward the request to the appropriate remote server. The remote server's controller processor will intake and deposit the request into the server's tuple space region. Idle worker processor within the remote server will then retrieve, search, and transmit the necessary information back to the local server.

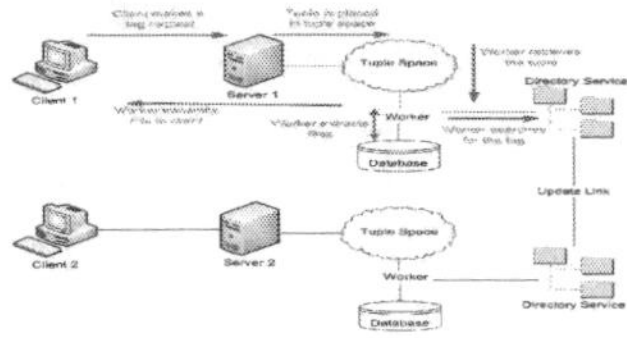

Fig. 4 Tuple request in Active Directory Service System

5.3 *File Server Update Analysis*

The graph shown in Fig. 5 represents data collected from inter-server file processing. The average time measurements were taken to determine how much time it takes for one server to scan its local directory and transmit collected information to another server. Current evaluation was focused on a one to one server connection with multiple clients receiving updated information. The information received on server contains listing of new or modified files and their respective locations on another server.

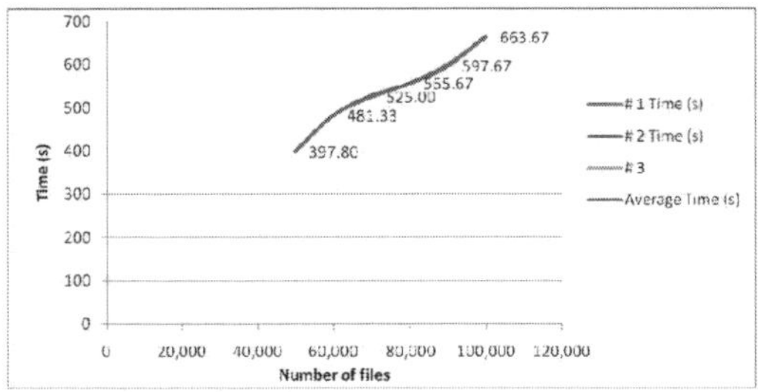

Fig. 5 Active Directory Inter-server File Processing

6 Conclusion

The model of the proposed Semantic Network Active Directory Service System platform will allow for the expedited access to relevant information within a closed community. Targeted communities may range from educational standpoints, where ideas and concepts may be interchanged freely, to large secure network systems, where queried information must be protected. In both cases, the Semantic Network Active Directory Service System will allow for the real-time exchange of current data and metadata throughout the distributed server system. Future efforts will revolve around the fault tolerance and performance analysis of networked systems.

References

1. Alegre, A., Estrada, J., Coalson, B., Milshteyn, A., Boussalis, H., Liu, C.: Development and Implementation of an Information Server for Web-based Education in Astronomy. In: Proceedings of the International Conference on Engineering Education, Instructional Technology, Assessment, and E-learning (December 2007)
2. Tolksdorf, R., Liebsch, F., Minh Nguyen, D.: XMLSpaces.NET: An Extensible Tuplespace as XML Middleware
3. Steyvers, M., Tenenbaum, J.B.: The Large-Scale Structure of Semantic Networks: Statistical Analyses and a Model of Semantic Growth. Cognitive Science Journal 29, 41–78 (2005)
4. Hawick, K., James, H., Pritchard, L.: Tuple-Space Based Middleware for Distributed Computing. Technical Report DHPC-128 (2002)

Lost in Translation: Data Integration Tools Meet the Semantic Web (Experiences from the Ondex Project)

Andrea Splendiani, Chris J. Rawlings, Shao-Chih Kuo, Robert Stevens, and Phillip Lord

Abstract. More information is now being published in machine processable form on the web and, as *de-facto* distributed knowledge bases are materializing, partly encouraged by the vision of the Semantic Web, the focus is shifting from the publication of this information to its consumption. Platforms for data integration, visualization and analysis that are based on a graph representation of information appear first candidates to be consumers of web-based information that is readily expressible as graphs. The question is whether the adoption of these platforms to information available on the Semantic Web requires some adaptation of their data structures and semantics. Ondex is a network-based data integration, analysis and visualization platform which has been developed in a Life Sciences context. A number of features, including semantic annotation via ontologies and an attention to provenance and evidence, make this an ideal candidate to consume Semantic Web information, as well as a prototype for the application of network analysis tools in this context. By analyzing the Ondex data structure and its usage, we have found a set of discrepancies and errors arising from the semantic mismatch between a procedural approach to network analysis and the implications of a web-based representation of

Andrea Splendiani · Chris J. Rawlings · Shao-Chih Kuo
Biomathematics and Bioinformatics Dept., Rothamsted Research, Harpenden, United Kingdom
e-mail: {andrea.splendiani,chris.rawlings}@rothamsted.ac.uk, shaochihkuo@gmail.com

Robert Stevens
School of Computing Science, University of Manchester, Manchester, United Kingdom
e-mail: robert.stevens@manchester.ac.uk

Phillip Lord
School of Compting Science, Newcastle University, Newcastle upon Tyne, United Kingdom
e-mail: Phillip.Lord@newcastle.ac.uk

F.L. Gaol (Ed.): Recent Progress in DEIT, Vol. 2, LNEE 157, pp. 87–97.
springerlink.com

information. We report in the paper on the simple methodology that we have adopted to conduct such analysis, and on issues that we have found which may be relevant for a range of similar platforms.

1 Introduction

In this paper we describe a simple methodology used to examine a graph based data resource so that it can be transformed to a representation suitable for the Semantic Web. Such simple methodologies are needed if Semantic Web technologies are to be used as widely as possible. The web has been a revolutionary technology to exchange and integrate information represented in natural language that has enabled the development of new means of communication and interaction. Now the web is evolving into a platform that also supports the integration and exchange of machine processable information. This platform has the potential to enable radical new approaches in the way we make sense of information. It has been the object of active research, from the Semantic Web [1] to its more recent development as Linked Data [2].

An increasing number of resources are now available on the Semantic Web, either exporting their information in standard languages such as the Resource Description Framework (RDF [3]), or directly providing information servers that respond to standard query protocols (SPARQL [4]). In addition, a number of key players are committing to either publish their information on the Semantic Web, or to support some related forms of structured knowledge publication and consumption via the web, including national governments (UK [5], US[6]) and leading enterprises (Facebook [7], Google [8], Yahoo [9]).

The availability of these information resources is complemented by the increasing number of tools and web systems that natively support the creation of information that is ported to a Semantic Web framework. We cite as examples, tools such CouchDB (couchdb.apache.org) and Neo4j (neo4jorg), web resources such as Freebase (freebase.com). Refer to NoSQL [10] for further details.

Now that 'a' *de-facto* Semantic Web is a reality, it is time to consider how it can be exploited and which software tools are needed to reap benefits from it. As the Semantic Web (as well as the traditional web), is founded on a graph-based representation of information, tools and methods for the analysis, manipulation and visualization of graphs are first candidates for this purpose, and we are witnessing the first examples in this direction, such as Gremlin (gremlin.tinkerprop.com) or RelFinder [11]. Most of these tools, however, have been developed following assumptions that do not necessarily apply to a web based representation of information and both their information engineering approach, and their usage need to be adapted for the Semantic Web context. This is particularly true for several tools which have been developed in the domain of Life Sciences for the analysis of biological networks [12], some of which (e.g.:Ondex[13], Cytoscape[14]) are inherently domain independent. It should be noted that the Life Sciences present a number of information related issues that makes the Semantic Web an ideal solution, and computational biologists and bioinformaticians have been among the most enthusiastic adopters of these technologies [15]. The size of the user community and potential impact of the Semantic Web in the Life Sciences makes it a most attractive domain to deploy Semantic Web graph-based analysis and visualization tools (cfr. RDFScape [16]).

When considering software which could be made available for users of the Semantic Web, we wish to ask how we can evaluate if the usage of such tools is inconsistent with the principles of representation in the Semantic Web and what are the main aspects of the information engineering in these applications that may need to be adapted to ameliorate any conflicts?

To answer these questions, we have focused on the Ondex data integration platform. Ondex is a data integration and analysis platform that has been developed, starting in 2005, for research in systems biology and the Life Sciences in general, but which is inherently domain independent. Its information engineering design is based on a graph data structure and on the use of ontologies to characterize the graph entities. Among other graph based tools developed in the Life Sciences, Ondex is unique for the precise semantic characterization of information and for its focus on graph-based data integration. Furthermore, the information design of Ondex resembles that of RDF data in the Semantic Web, and a number of issues that users of Ondex deal with are essentially the same issues posed by information in a Semantic Web context (e.g., linking with provenance or evidence attached to the entities).

We have developed a simple methodology to analyze the correspondence between the intended semantics of these graph-based tools, and the semantics resulting from their usage on web-data. This method was designed with a view to adapting Ondex to work within the Semantic Web and to learn lessons that would inform users and developers of other network analysis systems that are similar to Ondex in their intended usage. From the application of this methodology to Ondex, and learning by the experience of its usage in the last 5 years, we have highlighted a set of issues that are likely to be found in other network analysis systems when these are ported to the Semantic Web. To our knowledge this is the first time that a systematic assessment of the semantic mismatch between the data model and usage of pre-existing graph-based analysis tools and those of the Semantic Web is attempted.

1.1 The Ondex Data Structure, Life Sciences, and the Semantic Web

Life Sciences data is characterized by its complexity, its high interrelatedness, its heterogeneity, and by a multitude of naming and identity issues [17, 18, 19]. Graph based models are a natural fit, as they are in many disciplines, to deal with these problems. From metabolic pathways to ecosystems to anatomies, graphs are a convenient means to capture these relationships. Data in many forms can be represented as a graph and the schema-less approach adopted in RDF and other such representations offers a means of integrating data through connections that can be as strong or weak as the applications require.

Ondex has taken this approach by providing a graph based model, a collection of parsers that transform various bioinformatics resources into its graph representation and plugin modules to perform further data reduction and analysis. The graph model of Ondex uses nodes to represent concepts and edges to represent relationships between them. Both concepts and entities can be characterized via a type (that can be organized in an simple ontology), via an arbitrary set of attributes, and via a set of predefined attributes which support the representation of identifiers, information provenance, evidence and context.

Thus Ondex's data model has an intuitive correspondence to that used by RDF, and its parsers are *de facto* equivalent to mappers from the original resources to RDF.

There are, however, aspects of the Ondex representation, and its usage, that make its use on the Semantic Web not as simple as it might be. For instance its development has led to inconsistencies in the interpretation of how the Ondex parsers transform the data into the graph, one notable example being different interpretation of *provenance* information.

Our goals, therefore, are two-fold: First to develop a normative Ondex data model and map its transformation to RDF; and second to be able to describe a normative model against which the builders of Ondex parser can model their transformations. Both objectives are relevant to adaptation of other network based analysis tools to the Semantic Web.

2 Semantic Analysis

We have developed a simple methodology to examine the semantics of a graph based data integration and analysis tool, and to guide the transformation of its representation and usage as to be suitable for the Semantic Web. In this methodology, we first list all elements that make up the data structure of the system, and for each of these elements, we execute a set of steps, which result in a document, for each element, which describes in natural language: the intended semantics of the element, its actual semantics in current practice, a definition, recommendations for best practices and recommendations for future developments. The documents are then circulated among stakeholders for comments and iterative refinements of the proposal. We illustrate the steps that compose this methodology using its application on Ondex as an example, and in particular focusing on the *CV* data element.

2.1 *Methodology*

2.1.1 Listing Data Structure Elements

We first list the elements that make up the data structure of the system. In the case of Ondex, these are *Concept*, *Relation*, *ConceptClass*, *RelationType*, *Generalized Data Set*, *CV* (Controlled Vocabulary), *Accession* (identifier), *EvidenceType*, *Context* (an extensive explanation of the Ondex data types can be found in the detail of our analysis, available at [20]). In the remainder of this example we will focus on *CV*.

2.1.2 Definition of the Intended Semantics

We first elaborate a concise, informal definition of the intended semantics of a data structure (e.g.: *CV*). This definition is based on the answers to two questions:"what is a *CV* ?" and "when do you use a *CV* ?". Answers to these question are derived from the official documentation and from interviews with interested parties (developers and core users). We enrich this definition with a few examples of values that are assigned to this data structure. The application of this step results, in our example, in :

- *Definition*: Describes the bioinformatics origin of Concept or accessions of Concepts (It is intended to represent provenance information);
- *Examples*: UNIGENE, GO, unknown, AFFYMETRIX, BROAD, NWB (these are all examples of which are assigned to *CV*).

2.1.3 Observation of the Actual Usage

We then analyze the actual usage of the data structure element, with the help of a domain expert, to find patterns in the attribution of values to the data structure, values that are inconsistent with its intended usage, and degenerate usages. We compile a set of observations for each of the patterns, inconsistencies or degenerate usages that are found.

In the case of Ondex, and for other platform that present a plugin architecture, the inspection of the code of plugins provides an easy way to perform such analysis. Some of the observations found for *CV* are:

- *CV* is often in association with an identifier (*Accession* in the Ondex data structure) to characterize its scope. This happens both in parsers that extract information from ontologies such as the Gene Ontology (e.g.: *CV*=GO) or some databases such as the the Unigene DNA sequence database (e.g: *CV*=Unigene) and plugins that perform mapping operations.
- sometimes when *CV* is associated to a *Concept*, it is assigned values that refer to the database from which the information was extracted, rather than to the domain of identifiers for the concepts in the database. This is for instance the case for the parser for the ATRegNet database of plant transcription factors. (e.g.: *CV*="ATReg-Net").

- *CV* is sometimes assigned values that indicate the format from which some information was extracted, such as in Network Workbench (NWB) format (e.g. *CV*="NWB").
- *CV* is assigned an arbitrary identifier in plugins that need to distinguish between different graphs.

2.1.4 Analysis of the Actual Usage and Normative Definition

Following the observations in the previous step, we elaborate a second concise definition for the semantics of the data element, on which we base the development of recommendations in the following steps. This definition also traces the relations between the data structure analyzed and elements of RDF that it more closely represents. In the case of *CV*:

CV is currently used with several distinct meanings:

- when used in association with an identifier, *CV* has the meaning of a namespace, and characterizes the scope of the identifier.
- when used with a *Concept*, *CV* has the meaning of provenance.

2.1.5 Recommendation for Best Practice

We elaborate a set of best practices, that are intended to restrict the possible usage of the data structure to keep it coherent between users and with a Semantic Web representation. Best practices are designed to not require any change in the code base, and take into account the observations previously derived. In our example:

- Usage of *CV* in association to an *Accession*:
  When used as a namespace, *CV* should be assigned values that correspond unambiguously to the resources that provide a definition for the identifier. A pair (*CV*, *accession*) should be semantically equivalent to a URI. In particular the following usage should be avoided: *CV* that are not specific enough, for instance that correspond to a family of ontologies (e.g. OBO) rather then a single ontology, to which identifiers are specific (e.g.; GO), *CV* that correspond to a technology used to generate data (e.g.: Affymetrix) or to the institute providing the data (e.g.: Broad).
- Usage of *CV* in association with a *Concept*:
  When *CV* is used to represent information about provenance, it is intended to indicate the last source that asserted this information. In the case of information originating from a database, *CV* is intended as the most specific authority that is responsible for the validity of the data (this is often the last data source from which this concept is derived).
- Any other usage of CV is discouraged.

2.1.6 Recommendation for Future Development

We then present recommendations for future evolutions of the data structure, that would help in enforcing the best practices and would enable further integration with RDF:

- *CV* should be split into two distinct elements, corresponding to the meaning of “Namespace” and “Provenance”.
- Values for the “Namespace” element should be associated with one or more effective namespaces that may be used to generate common URIs for the concept.

2.1.7 Request for Comments

Finally, all the specifications produced for the data structure element are circulated to interested parties for feedback, which can lead to new observations and further refinement.

3 Results

The analysis that we have conducted on the Ondex data structure definition and usage highlights a series of issues that are not limited to this platform, as they relate to typical assumptions behind the usage of simple network based analysis platforms, and their incongruence with a Semantic Web based representation of information. We list here the most relevant issues we have found, with a brief discussion of the risk they pose to make a consistent usage of network based analysis tools on Semantic Web knowledge bases.

3.1 Scope of Information

In Ondex, a *Concept* (the equivalent of a resource in RDF) has an identifier that is an integer generated when a graph is imported into the system. A similar behavior can be found in Cytoscape. Both Ondex and Cytoscape support the annotation of a Concept (or node in the Cytoscape terminology) with identifiers, that can then be used to derive identities between Concepts in different graphs (or in different versions of the same graph). This is typical of a procedural, document based, data integration strategy where the ‘document’ provides an implicit scope for the validity of the information that it represents.

In a web based context, it is important to explicitly define the scope of validity of identifiers of resources and of the information relative to these resources. This is because in a distributed web environment, it is not possible to import all the information before being able to ‘name’ and ‘access’ the entities included.

In the Semantic Web framework, URIs act as identifiers with a global scope that allow direct access to the relative information. It is also important to explicitly define the context of validity of information, as the implicit context provided by a document has a limited validity in a web framework, where information can easily be filtered and recombined.

3.2 *Information Basis*

When using a graph based data integration and analysis platform, it is a tempting practice to use the graph for all information, without making a distinction between the different basis for particular types of information. For instance it is common practice in an Ondex plugin to represent, in the same graph, information that is based on 'knowledge external to the system', information that is based on the results of an analysis of the graph (e.g.: measures of betweenness and centrality of nodes) and sometimes information that is based on the specific instance of the platform (e.g.: graph coordinates for a given layout). This happens despite Ondex providing support for typing concepts and relations via a simple ontology definition. Other platforms are, in general, even more vulnerable to this 'congestion' of the graph. This usage of the graph data structure is acceptable in a procedural framework, where there is a starting point that holds only 'knowledge external to the system', that is replicated in the system and never altered in its original representation, and where information later added to the graph have the implicit scope of the task that is being carried out. In a web based framework, however, it is necessary to distinguish information that persists beyond the specific task carried out, information that is dependent on a specific subset of information (i.e.: it is invalidated when this subset is altered) and information that is not shared, but specific to a given instance of execution of a tool.

3.3 *Cardinalities of Relations*

Most network-based analysis tools, including Ondex, apply to the network a data modeling approach that is typical of object oriented (or framework based) systems and that is not consistent with a web based representation of information. This is particularly evident in the case of relations. A tool like Ondex (or Cytoscape) will expect that, if for a given concept the same property is asserted twice, with two different values, the second value for this property will override the first. This is in contrast with a web based representation of information where there is no limit on the number of values that a property can be assigned for a given resource.

3.4 *Objectification of Entities*

Another inconsistency that arises when an 'object oriented' paradigm is applied to Semantic Web resources derive from the fact that, when entities are represented via objects, there is an additional entity (the object) that has its own identifiers (the pointer). This can have subtle consequences, in particular for the implementation of graph manipulation plugins. For instance, a plugin can refer to the 'first' or in general to the 'n-th' property asserted on a concept, simply by retaining its pointer, and it can base its computations on this ordering. There is not an equivalent of the 'first' or 'n-th' property asserted, in the Semantic Web framework.

3.5 Datatypes

While some platform such as Cytoscape limit datatypes to a limited set of basic types (strings, integers, booleans), platforms like Ondex allow datatypes of arbitrary complexity (in practice, they allow serializations of Java objects). This can limit interoperability of systems for two reasons. First, other systems may not be able to reconstruct an arbitrary Java object. Second, and more importantly, data types are semantically opaque: complex datatypes provide information without an explicit characterization of its meaning.

3.6 Over-Specification

Finally, we have observed that much imprecision stems from an over-specification of the data-structure. In order to cope with characteristics as 'provenance' and 'evidence' of information, often the data structure require information, which cannot be guaranteed to be meaningful for the heterogenous nature of data on the web. For instance Ondex requires information on provenance and evidence for all *concepts* and *relations*, where provenance is intended to characterize the source of data (see discussion on *CV* in the previous section) and evidence its validity. Clear definitions for provenance and evidence apply to only a subset of the information that can be represented in Ondex. For instance, what is the evidence of an ontology term ? Or what is the provenance of a value that is the result of a numerical analysis ? Furthermore, users may simply not know the original data sources in the detail that is necessary to assign correct evidence and provenance information. The result is a set of uninformative entries, ranging from the generic "imported from microarray-database" to the ambiguous "unknown".

4 Conclusions

We have developed a simple, practical methodology to assess how the documented semantics of a data integration tool differs from its actual usage and, more specifically, where the semantic definition of the data structures of these tools is underspecified to cope with distributed information on the web. While this simple method has been devised to support the integration of Semantic Web functionalities in Ondex, it describes a general approach that can be of help to the adaptation of a variety of similar network based analysis tools to operate on the Semantic Web.

Ondex exhibits problems of systems that have grown in an *ad hoc* manner that have under-specified semantics and roles for their data models. The result are graphs that themselves have barriers to integration. Our simple, practical approach to normalising the project's understanding of its own data-model will have obvious benefits within Ondex. As a preliminary result, it has enabled us to define a mapping between a subset of the Ondex data-model and that of the Semantic Web. This has been the basis for the development of an Ondex prototype which can consume and produce information in RDF. Within the Ondex experience is a simple message that just creating a graph does not mean integration; a common integration pattern must be used. Otherwise, we have integration of format that is still difficult to use.

Learning from the Ondex experience, we have identified problems that are common to similar tools. We hope that this experience will help to improve the information design of the next generation of data integration, analysis and visualization platforms that will help in fulfilling the promises of the Semantic Web.

References

1. Berners-Lee, T., et al.: The Semantic Web. Scientific American (May 2001)
2. Bizer, C., et al.: Linked Data - The Story So Far, to be published in the International Journal on Semantic Web and Information Systems, Special Issue on Linked Data
3. Manola, F., Miller, E.: RDF Primer, World Wide Web Consortium (W3C) recommendation (February 2004), `http://www.w3.org/TR/2004/REC-rdf-primer-20040210/`
4. Prud'hommeaux, E., Seaborne, A.: SPARQL Query Language for RDF, World Wide Web Consortium (W3C) recommendation (January 2008), `http://www.w3.org/TR/rdf-sparql-query/`
5. HM Government, data.gov.uk, `http://data.gov.uk/sparql`
6. United States Government, data.gov, `http://www.data.gov/semantic/index`
7. Open Graph protocol, Facebook, `http://developers.facebook.com/docs/opengraph`
8. Adida, B.: Google Announces Support For RDFa, RDFa Blog, `http://rdfa.info/2009/05/12/google-announces-support-for-rdfa/`
9. Mika, P.: RDF and the Monkey, Yahoo Developer Network Blog, `http://developer.yahoo.com/blogs/`
10. Edlich, S.: No-SQL movement Blog, `http://nosql-database.org`
11. Heim, P., Lohmann, S., Stegemann, T.: Interactive Relationship Discovery via the Semantic Web. In: Aroyo, L., Antoniou, G., Hyvönen, E., ten Teije, A., Stuckenschmidt, H., Cabral, L., Tudorache, T. (eds.) ESWC 2010. LNCS, vol. 6088, pp. 303–317. Springer, Heidelberg (2010)
12. Pavlopoulos, G.A., et al.: A survey for visualization tools for biological networks analysis. BioData Mining 1, 12 (2008), doi:10.1186/1756-0381-1-12
13. Koheler, K., et al.: Graph-based analysis and visualization of experimental results with. Bioinformatics 22, 1383–1390 (2006), doi:10.1093/bioinformatics
14. Shannon, P., et al.: Cytoscape: a software environment for integrated models of biomolecular interaction networks. Genome Research 13, 2498–2504 (2003), doi:10.1101/gr.1239303
15. Neumann, E.: A Life Science Semantic Web: Are We There Yet? Sci. STKE 2005, pe22 (2005), doi:10.1126/stke.2832005pe22
16. Splendiani, A.: RDFScape: Semantic Web meets Systems Biology. BMC Bioinformatics 9(suppl. 4), S6 (2008), doi:10.1186/1471-2105-9-S4-S6

17. Goble, C., Stevens, R.: The State of the Nation in Data Integration. Journal of biomedical Informatics 41, 687–693 (2008), doi:10.1016/j.jbi.2008.01.008
18. Karp, P.: A Strategy for Database Interoperation. Journal of Computational Biology 2, 573–586 (1995)
19. Davidson, S.B., et al.: Challenges in Integrating Biological Data Sources. Journal of Computational Biology 2, 557–572 (1995)
20. Splendiani, A., et al.: Ondex Semantics Specifications, `http://ondex.svn.sourceforge.net/viewvc/ondex/trunk/doc/semantics/`

Multi-classification Document Manager

A Rich Ontology-Based Approach for Semantic Desktop

Paulo Maio, Nuno Silva, Ricardo Brandão, Jorge Vasconcelos,
and Fábio Loureiro

Abstract. We propose a lightweight technological system for managing user's documents according to multiple classification dimensions. The core of the proposal is the application of ALC-expressive ontologies for capturing the multi-property-based classification of documents. The ontology is then responsible for representing (i) the properties that serve for the document classification (e.g. authors, subjects, types), and (ii) the classes of documents specified based on the properties' values of the documents. Once the ontology is populated with data captured from the documents via parsers/analyzers, an inference engine logically classify documents according to the classes.

1 Introduction

A large amount of our information, both in the professional and private domains, is stored in the form of files on our personal computers. These are manipulated by the so called file managers (also known as file browsers and navigators) or specific applications. From the user perspective, current file systems are based on two principles. First, documents are classified according to a single hierarchy: the subdirectory structure. Second, each document is given a single fixed name that is the way users indicate the document to access.

Furthermore, the physical dimension of the documents (e.g. location, serialization) is abstracted by applications (e.g. email client), but preventing or complicating the relation between documents. For example, an email message is related to a specific presentation document in the file system.

Instead, users require an integrated logical view of the documents regardless of their physical location, serialization or manipulating application, which current

Paulo Maio · Nuno Silva · Ricardo Brandão · Jorge Vasconcelos · Fábio Loureiro
GECAD and Department of Informatics,
School of Engineering – Polytechnic of Porto Porto, Portugal
e-mail: {pam,nps,jrmjb,1060479,1070987}@isep.ipp.pt

F.L. Gaol (Ed.): Recent Progress in DEIT, Vol. 2, LNEE 157, pp. 99–106.
springerlink.com

file-system model and file managers are inadequate to fully satisfy. The so called Semantic Desktop applications [1, 2] emerged in this context, providing an integrated view of several sources and types of document regardless of their physical dimension, but instead considering their content and logical dimensions, thus treating every document as first order citizen. Yet, semantic desktop applications still lack the following features:

- Multi-classification of documents, both formally and informally according to the documents' meta-data, and not only by its physical location in the hierarchy (e.g. the music documents would be listed in the music folder independently of where the file is physically located, but instead dependent on their content);
- Browse documents by multiple paths, i.e. not the (single) physical subdirectory hierarchy but instead several hierarchies representing different classification dimension which turns to be a graph (e.g., a music document would be accessible directly not only by its type but also by its subject);
- Search for documents according to their content, classification, location, etc., and maintain the search constantly updated as a new class of documents (e.g. all documents whose subject is "DEIT" and created after July 2010).

Summarizing, we envisage a system application that allows the management of document such the users do not need to deal with the physical dimension of the document, nor being limited to the single classification provided by the file system.

We propose a lightweight technological system for managing the user's documents according to multiple classification dimensions. The core of the proposal is the application of description logics [3] ALC-expressive ontologies [4] for capturing the multi-property-based classification of documents. The ontology is then responsible for representing (i) the properties that serve for the document classification (e.g. authors, subjects, types), and (ii) the classes of documents specified based on the properties' values of the documents. Once the ontology is populated with data captured from the documents via parsers/analyzers, an inference engine logically classify documents according to the classes' conditions.

The following section further characterizes the problem and the context. The section 3 introduces and describes the proposed system architecture. The section 4 details the process of applying ontologies and inference engines in the system for capturing rules and classify documents. The section 5 presents the experiments and evaluation of the proposed system. Finally the section 6 discusses the evaluation results and in the light of that suggests some directions for future work.

2 Context

Current mainstream file systems are externally exposed as a single hierarchical structure of directories (folders) and files. Besides the physical location, files and folders are annotated with meta-data related to the physical dimension (e.g. creation date, author, access rights). Recent file managers (e.g. Windows Explorer) are able to read application specific meta-data from files and allow its manipulation (Fig. 1). Yet, file managers are unable to:

- Multi-classify files (or documents) and access files accordingly;
- Search for files and save the query as an always update folder;
- Manage files and specific applications document (email, contacts, appointments, bookmarks) as first-order citizens, thus preventing their association with files.

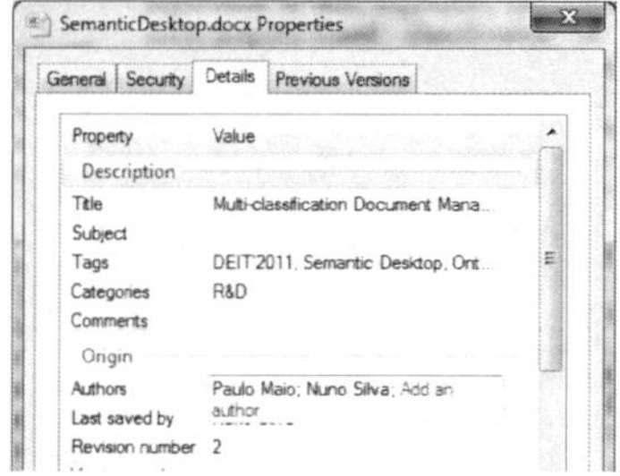

Fig. 1 Snapshot of the properties dialog box of Windows Explorer

Most of the file systems encompass the concept of logical link between folders and files. In most of the file systems, a link (shortcut in Windows OS) is a physical file whose content is a reference to another physical file. Consequently, there is no logical classification of the referred physical file, but instead a physical specification of the classification. From the user perspective this process is neither intuitive nor easy (to create but especially to maintain). This considerably differs from the hyperlink concept popularized in the Web page, where the link is in fact a part of a document. This small difference causes an enormous difference in terms of usability and implementation (i.e. it is not necessary to have a physical file as a hyperlink to another file/document).

When a web page is created online (i.e. when it is requested), the content (e.g. links to other documents) not only depends on the available document but also on the user request. Based on this observation, the envisaged document manager has to provide similar features in respect to the search and classification of documents, i.e. (i) allow searching the repository based on several parameters (the user request), and (ii) classify the documents (available documents/data) according to the query (the user request).

3 Architecture

The proposed system architecture is depicted in Fig. 2. The core components are described next:

- the source repositories, which contain the documents to manage;
- a set of analyzers, that are responsible for reading the sources, distinguish individual documents and extract respective meta-data. These analyzers depend on

the type of the source repositories. E.g. an analyzer for NTFS is different from an analyzer for Ext2 or from an email client content analyzer;

- the internal repository, is where the document meta-data and classification rules are stored for internal use;

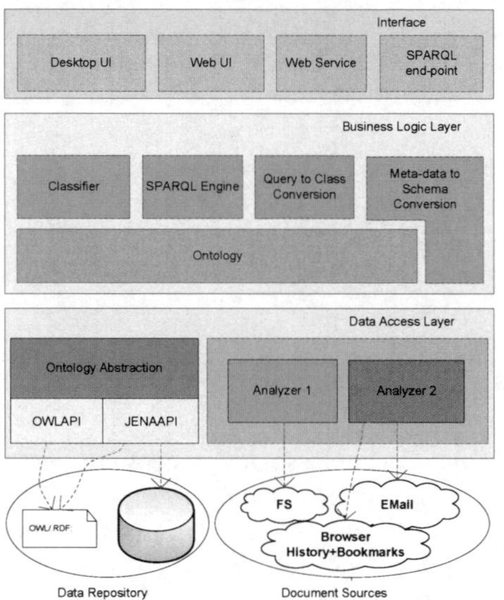

Fig. 2 System architecture

- the ontologies (i) model schema respecting the meta-data generated by the analyzers and (ii) capture the classification rules and queries. Hence, ontologies depend on the analyzers capabilities to extract meta-data from source repositories and form the classification and queries initiatives. The role of ontologies is further described in section 4;
- the classification engine (classifier) is a generic, off the shelf inference engine, responsible for the actual document classification. The schema and classification rules from the ontologies are applied by the classifier for classifying the documents according to their meta-data;
- the system interface provides the mechanisms for different entities to use the system. It should provide ways to (i) query, (ii) change classification rules and (iii) update documents meta-data. Not only the human user is considered in the process (desktop/web application and SPARQL) but also other systems (through web services and SPARQL).

4 Classification

This section describes the details of the document multi-classification feature. This feature includes three processes: (i) define the classification rules, (ii) convert the user defined queries into classification rules, (iii) the actual document classification. As introduced in previous section, these processes are founded in two core components: ontologies and classifier. In this particular case, the classification process will be responsible for:

- Retrieving all documents for a given specific class(ification);
- Realization of documents, i.e. determine the classes documents belongs to;
- Determining whether documents statically, user-based classified do not violate descriptions and axioms described by the classification rules.

The classification process will allow two other non-functional requirements:

- Check the satisfiability of a class(ification), i.e. determines whether a document can exist that would be classified as such. Checking satisfiability permits determine whether the rule generated (either by the user or by the "Query to Rule") is valid;
- Determines the subsumption of class(ification), i.e., determines whether class(ification) A is more general than class(ification) C. This allows defining hierarchical relation between defined classes.

Ontologies will be used by the classifier in runtime for classify the documents. Ontologies are characterized according to several dimensions [5] including the dimension of the community using it and the size and dynamics of the domain. In this context though, the formalism and expressivity level, which implies the reasoning complexity, are the main dimensions to consider. While ontology technology is around for many years, development boosted with the advent of the semantic web [6] in the last decade, giving rise to technology appropriate in this context.

The semantic web ontology language (OWL) [7] proposes several dialects depending on the required expressivity and reasoning complexity. OWL DL is a dialect that adopts Description Logics principles, namely, high expressivity while maintaining decidability. Decidability is a fundamental requirement because it is necessary to classify documents in finite time: its reasoning complexity is NExpTime-complete. It is our conviction that ALC-expressive ontologies are sufficient for capturing the required rules of the envisaged scenarios. Being a less expressive than SHOIN, the system can also benefit from its smaller reasoning complexity (i.e. ExpTime-complete), which is more adequate for online user-oriented systems.

In order to capture the document classification rules it is necessary to define modeling principles that fit the requirements. Ontology design patterns are commonly used for many different modeling problem types [8]. For the problem in hands, we have decided to adopt the ontology design patterns called "definable class" and "primitive class".

Definition 1 (Primitive class). $A \sqsubseteq C$ (A is a subset of C). I.e. A is an atomic class that is a subclass of the complex class C. This defines the necessary conditions (defined in C) of class A. Consequently, the instances of a primitive class must be explicitly defined as such.

Definition 2 (Definable class). $A \equiv C$, (A is equivalent to C), where A is an atomic class (concept in DL terminology) and C is a complex class. The $\equiv$ symbol determines that any instance that satisfies the right-side defined conditions is an instance of class A.

A complex class is defined in term of the Attribute Language with Complements (ALC) whose constructs are: ⊥, A, ¬C, C∧D, C∨D, ∃R.C, ∀R.C, where:

- ⊥ is the empty set or an inconsistency;
- A is a class name (atomic class);
- C and D are complex classes;
- R.C, R is a role whose range is of type C.

Next is presented an example of an ALC-expressive ontology for the document multi-classification domain:

1. MusicDoc ≡ Doc ∧ ∃hasType.(MP3∨WMA)
2. MyDoc ≡ Doc ∧ ∃hasAuthor.(ME)
3. Doc ⊆ ∃hasType.Type ∧ ∃hasAuthor.Person
4. Person ≡ ME ∨ ELVIS
5. ME ∧ ELVIS≡⊥
6. Type ≡ MP3 ∨ WMA ∨ HTM
7. MP3∧WMA≡⊥;MP3∧HTM≡⊥;WMA∧HTM≡⊥

The first class (classification rule) defines MusicDoc class(ification) as a definable class that encompasses all the Docs whose extension is either MP3 or WMA. The second class defines MyDocs as all Docs whose author is ME. Third class defines Doc as anything that as type and an author. The fourth and fifth lines define a set partition for Person and lines six and seven define a set partition for Type.

Consider now the following individuals and property value definition according to the previous ontology:

1. d1∈Doc, hasAuthor(d1,ME), hasType(d1,MP3)
2. d2∈Doc, hasAuthor(d2,ELVIS), hasType(d2,MP3)

Through the classification process, the classifier would infer the following:

1. d1∈MusicDoc, d1∈MyDoc
2. d2∈MusicDoc
3. MusicDoc≡{d1,d2}
4. MyDoc≡{d1}

5 Experiments

The experiments envisage evaluating the proposal and determine (i) how intuitive the multi-classification of documents is and (ii) how satisfactory the query to classification conversion is. For that we fully developed the data access and the business layers as described, and developed the desktop UI (Fig. 3a) and the web UI (Fig. 3b) of the interface layer. The desktop UI application combines the hierarchical physical structure of the source repositories, with the user defined classifications. The web UI application instead, abandoned the hierarchical view and presented the user defined classes of documents only. We prepared the experiments in three laptop computers with documents from the file system and email messages.

The experiments ran for three groups of 3 persons with 3 different levels of proficiency (Basic, Medium and High-level proficiency using file managers). None of the users was previously familiar with the documents or the application. Additionally, we defined three tasks. In task 1 the user is required to search for the documents for a specific project (subject). The expected result is a set of files and email messages. In task 2 the user is required to define a new folder BlueBerry whose documents' colour is Blue and smell like berries. It was expected that the user creates a primitive class whose necessary conditions correspond the specified characteristics. In task 3 the user is required to search for blue documents. It was expected the user creates a definable class BlueDocs subsumed by the BlueBerry class.

Fig. 3 Application screenshots: A) Desktop UI and B) Web UI

Table 1 depicts a summary of the experiments. Columns A reflect the success accomplishing the task (either y or n). Colunms C reflect the user perception of the complexity of the task. Columns S reflect the user facility understanding the task results. Colunms C and S are valued in the range 1-4. The evaluation results of the Web UI are depicted in shaded cells whereas Desktop UI results are depicted in clear cells.

Table 1 Summary of experiments

U	P	Task 1						Task 2						Task 3					
		A		C		S		A		C		S		A		C		S	
1	B	Y	Y	2	2	3	3	Y	Y	2	2	3	3	Y	Y	2	2	1	3
2	B	N	Y	3	2	1	2	Y	Y	4	2	3	3	Y	Y	4	4	1	3
3	B	Y	Y	3	1	2	3	Y	Y	3	2	3	4	Y	Y	3	2	1	4
4	M	Y	Y	1	1	4	4	Y	Y	1	1	4	4	Y	Y	1	1	2	4
5	M	Y	Y	1	1	4	4	Y	Y	2	1	4	4	Y	Y	2	1	2	4
6	M	Y	Y	2	2	4	4	Y	Y	2	2	4	4	Y	Y	2	1	2	4
7	H	Y	Y	1	1	4	4	Y	Y	1	1	4	4	Y	Y	1	1	3	4
8	H	Y	Y	1	1	4	4	Y	Y	2	1	4	4	Y	Y	1	1	3	4
9	H	Y	Y	1	1	4	4	Y	Y	1	1	4	4	Y	Y	1	1	4	2

6 Conclusions

Considering the evaluation results, we conclude that:

- The users are able to understand the concept of multi-classification;
- The users accomplish simple classification of documents based on their characteristics;
- The more the users are proficient with file managers and understand the hierarchical concept, the more they are able to understand and deal with multi-classification;
- The users do not like or do not understand subsumption relations between defined classes. This is clear from the comparison of satisfaction columns on task 3. Because on the Web UI app the subsumption relation was not represented, users understood and were satisfied with results. Instead, in Desktop UI application, they did not understand the result. The exception is user 9 that understood the subsumption relation and did not liked the fact that it was not represented in the Web UI app.

Accordingly, it is necessary to research more on the need for representing subsumption relation between user defined classes as it might be a factor contributing for misunderstandings. Also, because multi-classification conceptually gives rise to a graph, it would be interesting to analyze the effect of such representation on the user.

Acknowledgments. This work is partially supported by the Portuguese MCT-FCT project COALESCE (PTDC/EIA/74417/2006).

References

1. Papailiou, N., Christidis, C., Apostolou, D., Mentzas, G., Gudjonsdottir, R.: Personal and Group Knowledge Management with the Social Semantic Desktop. In: Proc. Collaboration and the Knowledge Economy: Issues, Applications and Case Studies (2008)
2. Decker, S., Park, J., Quan, D., Sauerman, L.: The Semantic Desktop - Next Generation Information Management & Collaboration Infrastructure. In: Proceedings of Semantic Desktop Workshop at the ISWC, Galway, Ireland (2005)
3. Baader, F.: The description logic handbook: theory, implementation, and applications. Cambridge University Press (2003)
4. Description Logic Complexity Navigator, `http://www.cs.man.ac.uk/~ezolin/dl/`
5. Hepp, M., de Leenheer, P., de, M.: Ontology management: semantic web, semantic web services, and business applications. Springer, New York Inc. (2007)
6. Lee, T.B., Hendler, J., Lassila, O.: The Semantic Web. Scientific American 284, 34–43 (2001)
7. Dean, M., Schreiber, G.: OWL Web Ontology Language Reference, `http://www.w3.org/TR/owl-ref/`
8. Ontology Design Patterns. org (ODP) - Odp, `http://ontologydesignpatterns.org/wiki/Main_Page`

Investigating the Potential of Rough Sets Theory in Automatic Thesaurus Construction

Gloria Virginia and Hung Son Nguyen

Abstract. This paper presents the result of initial study about implementation of rough sets theory in generating a thesaurus automatically from a corpus. The main objective of this study is to investigate the relation between keywords (defined by human experts as highly related with particular topic) and the sets generated based on rough sets theory. Analysis was conducted into comparison results of all available sets. We concluded that implementing rough sets theory is a rational way to automatically construct a thesaurus, as it can enrich a concept and proved to be able to cover the keywords given by the human experts.

1 Introduction

Thesaurus is a type of lightweight ontology which provides some additional semantics in their terms relations, e.g. synonym relationships, and does not provide an explicit hierarchy [6]. In Information Retrieval system (IR), it is a significant tool used as a controlled vocabulary in indexing process and as a means for query expansion, such as in [7] and [3].

Constructing an ontology automatically has been studied for years. In [1], Crouch and Yang reported that the thesauri generated automatically based on document collection clustering substantially improved the retrieval effectiveness in their four test collections, although the implementation of term discrimination value theory (used to differentiate the classes produced between the useful-thesaurus-classes and the non-useful-thesaurus-classes) was unsuccessful. A study of Patry and Langlais in [10] presented an approach to automatically generate term extractor from a training

Gloria Virginia · Hung Son Nguyen
University of Warsaw, Faculty of Mathematics, Informatics and Mechanics,
Banacha 2, 02-097 Warsaw, Poland
e-mail: gloriavirginia@gmail.com, son@mimuw.edu.pl

F.L. Gaol (Ed.): Recent Progress in DEIT, Vol. 2, LNEE 157, pp. 107–112.
springerlink.com

corpus as well as proposed a way of combining some statistical metrics in order to extract the terms more efficient than when they were used in isolation.

The major issue of constructing a thesaurus automatically is identifying the semantically related terms. Term co-occurrence is one way that has been studied since 1960 [8]. Considering the semantic relatedness between words, rough set theory received our attention as it is a mathematical approach to vagueness [12]. Moreover, it has been successfully implemented in numerous areas of real-life applications [5].

The following section give a brief explanation of tolerance rough set model. Before delineating the phases of study, the experiment data is expounded. We discuss our findings at Sect. 5, then make the conclusion and propose some future works.

2 Tolerance Rough Set Model

Introduced by Pawlak [11] in 1982, rough set theory expresses vagueness of concept by means of a *boundary region* of a set. Suppose we have a concept, then the idea is to approximate the concept by two descriptive sets called *lower* and *upper approximations*. Intuitively, the lower approximation consists of all elements that *surely* belong to the set, the upper approximation consists of all elements that *possibly* belong to the set, whereas the boundary region consists of all elements that *cannot be classified uniquely* to the set or its complement, by employing available knowledge [12].

Tolerance rough set model (TRSM) is an extension of Rough Sets Theory introduced by Kawasaki, Nguyen, and Ho in [4] as a tool to model document in text mining. Basically, this method came from the *generalized approximation space* using *tolerance relation* described by Skowron and Stepaniuk in [13].

In order to enrich the document representation in terms of semantics relatedness, TRSM creates tolerance classes of terms and approximations of subsets of documents. The tolerance classes of terms in T was based on the co-occurrence of index terms in all documents from D, where $D = \{d_1, d_2, ..., d_N\}$ is a set of text documents and $T = \{t_1, t_2, ..., t_M\}$ is a set of index terms from D. Then, a weight vector is used to represent each document $d_i = \{w_{i,1}, w_{i,2}, ..., w_{i,M}\}$, where $w_{i,j}$ denotes the weight of term t_j in document d_i.

Defining the TRSM means defining the tolerance space $\mathbb{R} = \{U, I, \nu, P\}$ suitably for the information retrieval problem. Due to page limitation, we recommend [2] or [9] for detailed explanation of definition. After all, the lower approximation $L_{\mathbb{R}}(X)$, upper approximation $U_{\mathbb{R}}(X)$, and boundary region $BN_R(X)$ of any $X \subseteq T$ in tolerance space $\mathbb{R} = (T, I, \nu, P)$ are as follow

$$L_{\mathbb{R}}(X) = \{t_i \in T \mid \nu(I_\theta(t_i), X) = 1\} \tag{1}$$

$$U_{\mathbb{R}}(X) = \{t_i \in T \mid \nu(I_\theta(t_i), X) > 0\} \tag{2}$$

$$BN_R(X) = U_R(X) - L_R(X) \tag{3}$$

3 Experiment Data

We used ICL-corpus which consists of the first 1,000 emails of *Indonesian Choral Lovers* (ICL) Yahoo! Groups of Indonesian choral community. Each document of ICL-corpus was assigned topic(s) by choral experts (described in [15]) and this process yielded 127 topics.

During annotation process, in addition to decide topics, keywords were determined for each document in order to express the high related words with the given topic. We treated these keywords as the body text of each document in the second corpus, named *WORDS-corpus*. Thus, the WORDS-corpus also consists of 1,000 documents and its document id is in accordance with document id of ICL-corpus.

Assuming that each topic given by the experts are a concept, we consider the agglomeration of keywords for each topic as the terms variants of the concept that semantically related. Therefore, the WORDS-corpus became the ground truth of this study.

4 Main Phases of the Study

There were three phases conducted in this study, those were *extraction*, *rough sets*, and *evaluation*. Fig. 1 depicts the whole process including the resulted sets of each phase. The rectangle represents the phase while the circle represent the result.

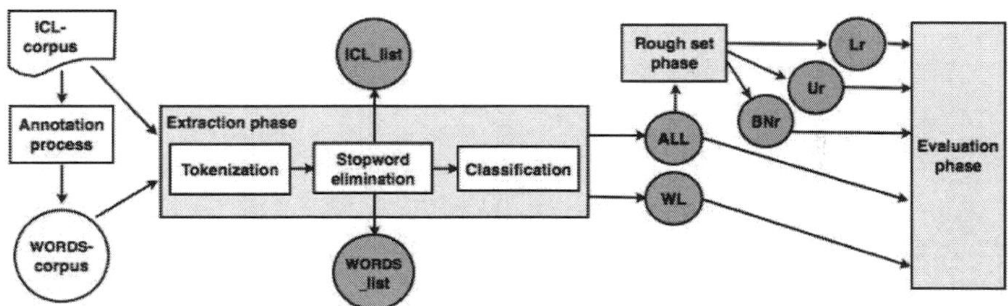

Fig. 1 Main phases of the study

From *extraction phase* it was identified that there were 12,363 unique words in ICL-corpus (called *ICL_list*) and 4,281 unique words in WORDS-corpus (called *WORDS_list*). We then classified our corpora based on the 127 topics and considered all terms appears in each class as the terms of its representative vectors. A frequency matrix of topic-term was created based on these classes. The resulted classes of ICL-corpus (called *ALL*) were then further processed in rough set phase whereas resulted classes of WORDS-corpus (called *WL*) was used later in evaluation phase.

In *rough set phase* the upper set *Ur*, lower set *Lr*, and boundary set *BNr* of each class from ICL-corpus were generated using (1), (2), and (3) respectively; assume that *RS* refers to those three sets. In order to ensure this job could be done quickly, the occurrence binary matrix (OC), co-occurrence matrix (COC), and tolerance matrix (TOL) were created based on algorithm described in [9]. For evaluation purpose, RS were generated with co-occurrence threshold θ between 1 to 30.

Finally, comparison between all available sets were conducted in *evaluation phase* to get the amount of terms appear in both compared sets. From these comparisons we got values for each topic, then the average computed for each θ value.

5 Analysis

Comparison between ICL_list and WORDS_list shows that ICL_list consists of almost all of WORDS_list terms (99.6%). Analysis of 17 words of WORDS_list which not appear in ICL_list shows that it is caused by typographical error (7 words, all appear in ICL_corpus), informal words (5 words), derived words (4 words), and foreign language (1 word) that emerge in both corpus. For at least 8 of them should be resolved by the stemming process which has not been employed in this study.

The difference between ICL_list and WORDS_list is 8,082 terms, that is 65.37% of ICL_list. Further data related with this are comparison results between ALL and WL, which calculate the same term for each topic. We noticed that the average percentage of same word between ALL and WL in each topic is only 14.56%.

Table 1 Average of same word between sets

Set	Ur (%)	Lr (%)	BNr (%)
ALL	100.00	5.00	95.00
WL	97.74	4.89	92.85

Table 1 shows the results of sets comparison. The value in each cell is the average of number of same word between two sets. The value of ALL-Ur in Table 1 (100%) could be used as an indicator that the RS sets were generated in right manner. Ur is the upper approximation of ALL set hence should consist of all elements belongs to the ALL set. With regard to the number of same terms between ALL and Ur, there are only 42.35% of ALL terms appear in Ur. From this, we can say that the implementation of rough sets theory definitely enrich a concept.

The other values of the first row of Table 1 show that only few terms of ICL_corpus (about 5%) actually could be classified as belong to specific topic while most of them (about 95%) cannot be classified uniquely into a specific topic. These values are somehow similar with comparison results of WL and RS, presented at the second row of Table 1, although the Ur does not cover all terms of WL (only 97.74% of WL). This is possibly the case happen in classification, that limited number of terms could be considered precisely belong to a particular class while numerous of them are in uncertain condition.

Regarding that WL consists of keywords define as highly related with particular concept, the high value of WL-Ur (97.74%) and the small value of WL-Lr (4.89%), attest that topic assignment is beyond the written terms on a text, i.e. the number of terms being considered during decision of a topic is more than the number of terms written on a text, hence automatic topic assignment task cannot be simply relied on the particular written text. Related with the previous finding that average percentage of same word between ALL and WL in each topic is only 14.56%, then reduction of index term seems to be compulsory.

The experiment to alter the co-occurrence threshold θ value in range 1 to 30 shows a great improvement in decreasing the number of Ur, as it is clearly depicted in Fig. 2. From this figure, we can see that the dramatic change starts to be stable at θ value around 19. Analyzing the other graphs (i.e. the graph of all comparison made) yielded similar result, thus it is suggested to set the θ value ≥ 19. This finding is supported by the high average of same word between WL and Ur that is still larger than 90%, up to θ value 40.

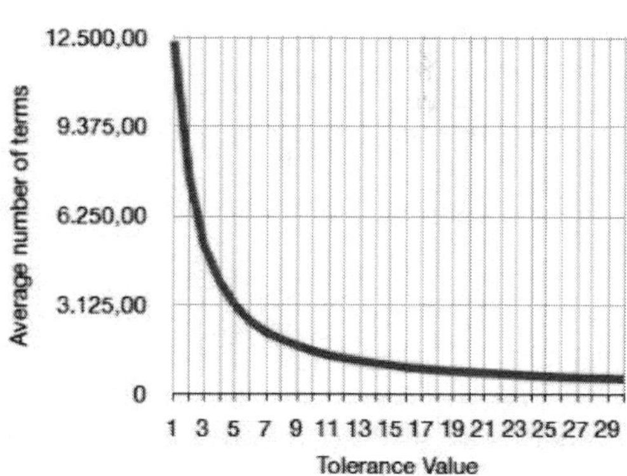

Fig. 2 Total terms of Ur. Total number of terms in upper sets decrease dramatically at the beginning and then become stable at tolerance value around 19.

6 Conclusion

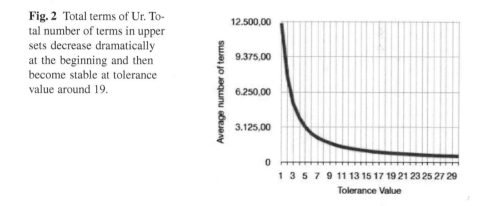

By employing the TRSM, some sets consist of terms were generated from ICL-corpus and WORDS-corpus. Comparison between resulted sets were conducted in order to get the total number of terms that occur on each topic in both sets being compared. We analyzed the average value of each comparison with θ value in range 1 to 30.

We concluded that implementing rough sets theory is a rational way in order to automatically construct a thesaurus, as it can enrich a concept and proved to be able to cover the keywords given by the human experts. However, further study that employs Indonesian stemming and feature selection in rough set theory are requisite.

Acknowledgements. Specific Grant Agreement Number-2008-4950/001-001-MUN-EWC from European Union Erasmus Mundus External Cooperation Window EMMA, research grant of Duta Wacana Christian University, Indonesia, and grants from Ministry of Science and Higher Education of the Republic of Poland (N N516 368334 and N N516 077837).

References

1. Crouch, C., Yang, B.: Experiments in automatic statistical thesaurus construction. In: Proc. The 15th Annual International ACM SIGIR Conference on Research and Development in Information Retrieval, pp. 77–88. ACM Publisher, New York (1992)
2. Ho, T.B., Nguyen, N.B.: Nonhierarchical document clustering based on a tolerance rough set model. International Journal of Intelligent System 17, 199–212 (2002)
3. Imran, H., Sharan, A.: Thesaurus and query expansion. International Journal of Computer Science & Information Technology (IJCSIT) 1, 89–97 (2009)
4. Kawasaki, S., Nguyen, N.B., Ho, T.-B.: Hierarchical Document Clustering Based on Tolerance Rough Set Model. In: Zighed, D.A., Komorowski, J., Żytkow, J.M. (eds.) PKDD 2000. LNCS (LNAI), vol. 1910, pp. 458–463. Springer, Heidelberg (2000)
5. Komorowski, J., Pawlak, Z., Polkowski, L., Skowron, A.: Rough sets: a tutorial. In: Rough Fuzzy Hybridization: A New Trend in Decision-Making, pp. 3–98. Springer, Singapore (1998)
6. Lassila, O., McGuinness, D.: The role of frame-based representation on the semantic web. Technical Report KSL-01-02, Knowledge System Laboratory, Standford University
7. Lee, H., Lin, S., Huang, C.: Interactive query expansion based on fuzzy association thesaurus for web information retrieval. In: Proc. of the 10th IEEE International Conference on Fuzzy Systems, vol. 3, pp. 724–727 (2001)
8. Maron, M.E., Kuhns, J.K.: On relevance, probabilistic indexing and information retrieval. Journal of the ACM 7, 216–244 (1960), doi:10.1145/321033.321035
9. Nguyen, H.S., Ho, T.B.: Rough document clustering and the Internet. In: Pedrycz, W., Skowron, A., Kreinovich, V. (eds.) Handbook of Granular Computing, ch. 47, pp. 987–1003. John Wiley & Sons Ltd. (2008), doi:10.1002/9780470724163
10. Patry, A., Langlais, P.: Corpus-based terminology extraction. In: 7th International Conference on Terminology and Knowledge Engineering (TKE 2005), pp. 313–321 (2005)
11. Pawlak, Z.: Rough sets. International Journal of Computer and Information Science 11, 341–356 (1982)
12. Pawlak, Z.: Some Issues on Rough Sets. In: Peters, J.F., Skowron, A., Grzymała-Busse, J.W., Kostek, B.z., Świniarski, R.W., Szczuka, M.S. (eds.) Transactions on Rough Sets I. LNCS, vol. 3100, pp. 1–58. Springer, Heidelberg (2004)
13. Skowron, A., Stepaniuk, J.: Tolerance approximation spaces. Fundam. Inf. 27, 245–253 (1996)
14. Vega, V.B.: Information retrieval for the indonesian language. Master thesis. National University of Singapore (2001) (unpublished)
15. Virginia, G., Nguyen, H.S.: Automatic ontology constructor for Indonesian language. In: Proc. 2010 IEEE/WIC/ACM International Conference on Web Intelligence and Intelligent Agent Technology (WI-IAT 2010), pp. 440–443. IEEE Press (2010), doi:10.1109/WI-IAT.2010.122

Sentiment Analysis in Colombian Online Newspaper Comments

Diego Pérez-Granados, Carlos Lozano-Garzón, Alexandra López-Urueña, and Claudia Jiménez-Guarín

Abstract. The Web 2.0 evolution has renovated the way in which the producers and consumers of information interact. Online newspapers have allowed readers to express their views of the news being read, by publishing a comment. Recognizing that the aggregation of comments can produce valuable information about the topic being commented is the main contribution of the present work. It presents the development and the results of NOA, a web application intended to detect opinion tendencies, using sentiment analysis and NoSQL technology. It produces an aggregated, qualitative and quantitative tendency. NOA also diagnoses the quality of the comments from a Web source. It is tested on three online newspapers, gathering the information required to make a valuable opinion. Conclusions are withdrawn from the quality of the comments and the commenting culture in general.

1 Introduction

Sentiment analysis, or opinion mining, is the consolidation of information available on the web, aimed to structure aggregated opinions of several speakers towards a specific subject [1]. It encompasses a broad area of information mining and natural language processing to appraise a judgment as positive or negative, and determine emotional state of the writer and subjectivity of the opinion.

A comment is a short value judgment about a specific topic, which can be a piece of news, image or any common document. Although, it is acknowledged that the information being commented contains value for the reader, the comments that follow the piece of information can generate value by themselves, especially when taken into account in an integrated way. From the aggregation of diverse opinions, qualitative and quantitative tendencies can be detected, and a general

Diego Pérez-Granados · Carlos Lozano-Garzón · Alexandra López-Urueña · Claudia Jiménez-Guarín
Dep. of Syst. and Comput. Eng., Univ. de los Andes, Carrera 1 Este # 19A-40, 111711 Bogotá, Colombia
e-mail: {df.perez29,ca.lozano968,la.lopez77, cjimenez}@uniandes.edu.co

F.L. Gaol (Ed.): Recent Progress in DEIT, Vol. 2, LNEE 157, pp. 113–119.
springerlink.com

opinion of the public is unveiled. A newspaper comment is an opinion about a body of news or a response to another comment within the discussion. The comment characterizes the speaker, as it expresses the point of view from which favorable and unfavorable sentiments can be inferred.

This paper describes NOA (News Opinion Analyzer), a web based application that uses crawling and syntactic analysis, aimed at aggregating opinions expressed in newspaper comments to provide means to access information with no arbitration. The main contribution of the project is the way by which sources are added, comments are gathered and processed and the conclusions that are withdrawn from the analysis of the semantic responses.

The interest of NOA comes from the time consuming process of mining single opinions and generating an integrated response [2]. It is worth noting that the proposed architecture allows for high scalability as multiple sources can be added dynamically with simple parameterization. The choice of these sources responds to the need of diagnosing a particular geographic region in a language different than English. The specific criteria are explained in detail in Section 3.

The rest of the paper is organized as follows. Section 2 sketches a general description of the application. Section 3 describes the information sources and their quality. In Section 4 a detailed description of the architecture and solution is given. Section 5 shows the implementation as Section 6 presents the conclusions, recommendations and future work.

2 General Description

The objective of the present work is to evaluate in an aggregated manner "what the web is thinking" about a specific topic in an environment of collaboratively created content. Special effort was directed towards detecting "garbage comments" such as spam. Therefore, a method for detecting useless comments was enforced so that measurements could be taken about the quality of the sources.

To access these sources, the user of NOA inputs a text term to trigger the web search. The response he gets aims to provide information about the search term expanded with related terms and favorability within the available sources. NOA is a web search engine which provides two specific services: detecting tendencies about the general opinion of the public and diagnosing the quality of the comments of an online newspaper. The main consideration to choose the sources is their homogenous and accessible contents.

NOA harvests and classifies the information, which is later stored in a NoSQL database [3]. In order to provide an appropriate general opinion, the database was populated and calibrated to offer a relevant response. Then, requests are processed, gathering the necessary comments and elaborating a response.

In order to generate value from online newspaper comments a web based application was developed to access the aggregated opinion. The proposed mashup aims to consolidate an aggregate opinion about a given subject. The main steps are: Warm up (or Training), Pre-request (composed by Harvest, Classification and Storage) and Request (divided in Retrieval and Presentation).

3 Online Newspaper Sources

Abundant effort has been directed towards sentiment analysis in English [4]. The choice of sources is restrained to a particular country, in order to test natural language processors in other regional contexts; also, classification requires homogeneous and comparable sources that characterize a specific context. Colombia fits well for the project because is highly polarized politically, a developing Internet culture of about 17 million internet users at a growth rate of 46.4% as of December 2008 [5]. The interest is then to evaluate the Colombian opinion about their reality expressed in newspaper comments, knowing that analogous studies have not been conducted. Since the polarity classification aims to detect language patterns, three Colombian sources were chosen because they are published in Spanish, widely known, web available and internationally recognized. Finally, the comments must be embedded in the HTML structure of the web page. Curiously, the widest read and most visited and commented national online newspaper, El Tiempo, was not chosen because it does not match the latter condition.

El Espectador (http://elespectador.com) is a tabloid newspaper with daily national circulation and online publication. It is the oldest and second most read newspaper in the country, with around 750.000 unique visitors daily [6]. This traffic generates a great amount of comments per day, and it is not necessary to sign in to post a comment. It is remarkable that most of the posts are destructive critiques and/or offensive statements.

Silla Vacía (1) focuses mainly on Colombian politics and is only published online, and signing is necessary for commenting. This news website has a very specific audience, interested in the Colombian power relationships and key participants aiming to do "good journalism". The exclusiveness of the target audience is enforced by the academic tone and objective journalism, which leads to healthy discussions and self-moderation, with scarce apparition of derogatory comments.

El Colombiano (http://www.elcolombiano.com) is a leading daily regional newspaper. Although the journal possesses great affluence from its target region, few readers comment on the online news. The comments are published by registered users and consist of brief remarks that generally sketch the opinions of this region.

4 Design and Architecture

The software architecture is based on a three layer model, using Web 2.0 tools. The constraints are imposed by the developing context: use of technologies with academic license, scalability for harvesting and storing and natural language processing to evaluate content. Figure 1 show the component diagram. The correspondence between the process and the architectural components is:

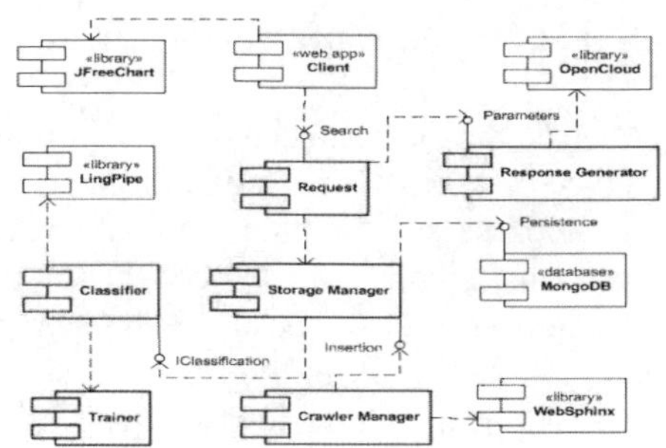

Fig. 1 Component Diagram. The Application architectural components and their relations are presented. It includes libraries and external components. The main components are red boxed.

Harvest: A crawler was implemented to scan for new comments (which were extracted using regular expressions over HTML tags) by providing a source root address and without leaving the domain. Since the syntaxes for a comment in each webpage are different, the crawler was parameterized for each source, specifying its particular HTML syntax. The Crawler Manager module is responsible for keeping every instance of the crawler and uses a timer to run the crawlers daily.

Classification: Once the comments are gathered, they are clustered before its insertion to the database. A Classifier was implemented using the LingPipe [7] sentiment analysis capabilities. It does polarity classification by constructing a dynamic language model classifier using an n-gram model.

The Classifier was previously trained with 500 comments, unambiguous in polarity, and sorted into four categories: positive, negative, neutral and spam. Most of them include endemic and colloquial language, and systemic orthographic mistakes. After sorting each training comment into one of the categories, a probability to a sequence of words to belong to a particular classification is assigned. Ironic or sarcastic comments were excluded for training. Training allows more sources to be added dynamically; the Classifier should be refined to reflect the possible chains of words expected in the comments that are going to be analyzed. This training can be done every time a crawler runs.

Storage: The dynamic nature of sources needs a highly scalable and flexible schema database. Additionally, as only insertions and no update operations were required, eventual consistency was tolerated. A NoSQL database was chosen, because it profits high-performance and replication tools. The stored object is composed by: its author, publication date, body, title of the article, source, URL and classification. Figure 2 shows a Gane-Sarson data flow diagram (DFD) [8], which exhibits how the comments are harvested and inserted in the database.

Retrieval: User search terms are matched by the Retriever using the body of stored news titles or comments. This retrieval by content is supported by a dynamic indexing of the documents, as search terms are added as they are used. The first time, that a search term is used, the retrieved documents are indexed by the search term. Then, a collection of common search terms is stored. To keep the database scalable and efficient, only the most used search terms are indexed.

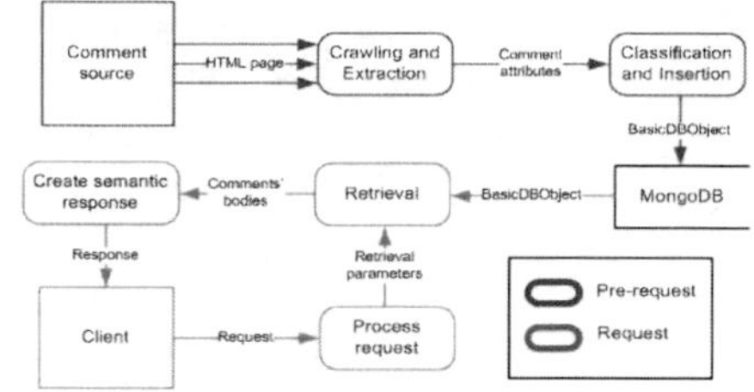

Fig. 2 Data flow diagram. It presents how information flows from its source to its final consumer, receiving a semantic response

Presentation: The response allows the user to create an opinion of what the commentators are saying about an introduced search term. It is twofold: The first part is quantitative; it revolves around the concept of approval rating. The goal is to present the user with a fraction of the retrieved documents that correspond to the four categories established above. The answer determines to what extent the comments that were retrieved have a positive tone. The second part of the response is directed at establishing topics that commentators usually relate to the topic that is being searched: a relevance tag cloud [9] is created from the documents that were retrieved. In this case the relevance is given by the number of apparitions of every word and the tag cloud is constructed from determining the frequency of each word that appears on the retrieved documents and presenting the top words in a weighted way.

5 Implementation and Results

NOA is a fully implemented search engine which allows the user to input the topic which wants to be analyzed retrieving by content newspaper comments in Spanish. Before any request, a process of harvesting is done. Recollection of the comments for further analysis was supported by WebSphinx (Website-Specific Processors for HTML INformation eXtraction) library for java. The crawler had a good performance, gathering about 10.000 comments per hour, when scanning the three selected sources at a time. The Classifier executed an experiment with 100 comments showed 82% efficacy.

A secondary objective was to conduct an analysis to choose a suitable database. MongoDB was chosen. It fits the technical requirements, provides java support and open source license, and extensive documentation.

The Storage Manager finds the comments containing the search term in the body or the title of the corresponding news. Two pattern recognition metrics were obtained from another 100 comments, with 56.1% of precision, and 59.6% of recall, because the terms were matched exactly when searched. Synonyms, analogous words and nicknames could refine these metrics.

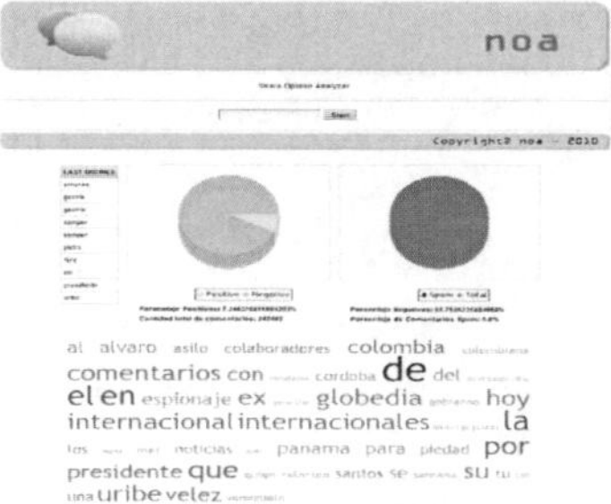

Fig. 3 Web application with results. It presents the search engine with a response for a search term.

The problem of identifying what the comment is talking about is tackled by establishing content-based retrieval rules, which can be varied to obtain different responses. The specific rules used during the implementation do not guarantee that the retrieved comments are directed at making an opinion about the search term, but instead the elaborated response corresponds to the documents which refer directly or indirectly to the search term. This introduces a bias as retrieved documents can sometimes contain search words that were used as example or opposition. A large enough sample was taken from the documents, ameliorating the impact on the significance of the results, to generate a response.

Figure 3 shows the web user interface. The JFreeChart [10], and OpenCloud [11] libraries were used. NOA was stabilized by running the Crawler during a month, from which the last set of testing was done from the data gathered during two days from November 28thof 2010 with around 242.000 comments. The conclusions and analysis of results was done with this stabilization.

6 Conclusions, Recommendations and Future Work

The conclusions withdrawn from the analysis of results are divided into two main fields. First, regarding the implementation, NOA permits simple and focused access to a vast number of comments, which would be otherwise unreachable to the user. To achieve this, NOA uses dynamic addition of sources and dynamic training of a polarity classifier with configurable classifications integrating multiple sources. The elaboration of a further semantic response is to be taken as step towards Web 3.0, in which source selection is personalized and interactive.

On the other hand, some conclusions were withdrawn from the quality of the sources itself. It was concluded from the analysis of experimentation, that the negative comments systematically were much greater than the positive ones. Further analysis demonstrated that obtaining positive comments beyond 10% was exceptional. Detailed analysis evidenced that most of the comments are destructive criticism which include coarse language and personal insults. In this sense, the qualitative results are biased since the sources do not provide a random sample since the comments have negative connotations regularly. Under those circumstances, NOA could to be tested over a corpus that can provide a statistically significant response, even though these types of analysis bring light to evidence problems in the usage of online discussion threads.

As a recommendation, the Colombian online newspapers are encouraged to generate healthy commenting environment by establishing house rules, monitoring the comment repositories, and generating tools for reporting offensive comments.

References

1. Casa Editorial El Tiempo: EL TIEMPO fortalece su liderazgo, según el EGM - Archivo - Archivo Digital de Noticias de Colombia y el Mundo desde 1.990 - eltiempo.com. In: eltiempo.com, HYPERLINK `http://www.eltiempo.com/archivo/documento/MAM-4091310` (accessed August 10, 2010)
2. JFreeChart: JFreeChart Homepage. In: JFreeChart. HYPERLINK, `http://www.jfree.org/jfreechart/`
3. Alias-i: LingPipe Homepage. In: LingPipe. HYPERLINK, `http://alias-i.com/lingpipe/`
4. OpenCloud: OpenCloud - Tag Cloud Java Library. In: OpenCloud. HYPERLINK, `http://opencloud.mcavallo.org/`
5. Secretaría de Prensa - Presidencia de la República Colombia: SP NOTICIAS - Presidencia de la República de Colombia. In: Presidencia de la República de Colombia, HYPERLINK, `http://web.presidencia.gov.co/sp/2009/marzo/13/10132009.html` (accessed March 13, 2009)
6. Chris, G., Trish, S.: Structured Systems Analysis: Tools and Techniques. Prentice Hall, Englewood Cliffs (1979)
7. Denecke, K.: Using SentiWordNet for multilingual sentiment analysis. In: IEEE 24th International Conference on Data Engineering Workshop, Cancún, México, vol. I, pp. 507–512 (2008)
8. Hung-Yu, K., Zi-Yu, L.: A Categorized Sentiment Analysis of Chinese Reviews by Mining Dependency in Product Features and Opinions from Blogs. In: 2010 IEEE/WIC/ACM International Conference on Web Intelligence and Intelligent Agent Technology, Toronto, ON, vol. I, pp. 456–459 (August 2010)
9. Leavitt, N.: Will NoSQL Databases Live Up to Their Promise? Computer 43(2), 12–14 (2010)
10. Seifert, C., Kump, B., Kienreich, W., Granitzer, G., Granitzer, M.: On the Beauty and Usability of Tag Clouds. In: 12th International Conference Information Visualisation, London, UK, vol. I, pp. 17–25 (July 2008)
11. Wiebe, J.: Tracking point of view in narrative. Journal Computational Linguistics 20(2), 233–287 (1994)

Privacy Tradeoffs in eCommerce: Agent-Based Delegation for Payoff Maximization

Abdulsalam Yassine, Ali Asghar Nazari Shirehjini, Shervin Shirmohammadi, and Thomas T. Tran

Abstract. Collecting and analysis of personal information are among the most far-reaching development in e-commerce. Such endeavor, however, has resulted in the unprecedented attrition of individual privacy. This paper describes an agent-based system that allows consumers to benefit from the dissemination of their personal information. The implementation and analysis of such system are provided in this paper.

1 Introduction

Although the Internet has brought unquestionable benefits to people around the globe and helped make our society connected like never before, it has also made it possible for unknown characters to surreptitiously watch users in every click they make when they visit their favorite websites [2]. Online tracking technologies such as HTTP cookies, Flash cookies, Web bugs, bots, etc. allow service providers to monitor users' behavior as well as scrutinize their lifestyle and personal habits and preferences. Not only that, but also the data the users reveal to complete a transaction (such as name, address, email, telephone numbers, etc.) is often sold to third parties for marketing and advertisement purposes [1].

According to Cavoukian [2], violation of consumers' personal information is "an external cost or negative externality, often created by online businesses, therefore the cost of resolving the problem should be borne by online businesses." While many approaches have been proposed, such as those of [3][4] and [5], to assure consumers that their personal information in online transactions is protected, none offers the necessary guarantees of fair information practices.

Abdulsalam Yassine · Ali Asghar Nazari Shirehjini · Shervin Shirmohammadi · Thomas T. Tran
Distributed and Collaborative Virtual Environment Research Laboratory
School of Information Technology and Engineering (SITE), University of Ottawa, Canada
e-mail: {ayassine,anazari,shervin}@discover.uottawa.ca, ttran@site.uottawa.ca

F.L. Gaol (Ed.): Recent Progress in DEIT, Vol. 2, LNEE 157, pp. 121–127.
springerlink.com

In our previous study [6], we presented analysis of agents that allow consumers to benefit from the dissemination of their personal data in eCommerce. In this paper, we are re-presenting those agents in a combined multi-agent system with extensions that are related to analysis, implementation, and experimental results.

2 Utility of Sharing Personal Information

Consider a consumer with identity I, and several attributes $\Lambda = \{\Lambda_1, \Lambda_2, ..., \Lambda_n\}$ containing private information. Consider also an online service provider SP with target intention to acquire a set of personal information $\Phi \subseteq \Lambda$ and to sell a service K. Every customer has to reveal a certain level of information Λ in order to receive the service K; this level is the same for all customers. However, consumers incur privacy costs when their information is violated. Studies have shown that consumers differ in their concerns for their privacy [3]. We use the parameter $\beta \in [0, 1]$ to capture the privacy-based consumer differentiation. In addition, the privacy cost to an individual differs from one type of information to another. We let the privacy cost to vary with Λ and β according to the functional form $\Psi(\Lambda, \beta)$, where $\Psi(\Lambda, \beta)$ characterizes the magnitude of privacy sensitivity resulted from the composition of different private data attributes Λ and the cost β (privacy cost or risk cost) of revealing them. Let consumers' utility (CU) equal the value θ of the service K at a price P less the privacy cost of sharing personal information; i.e., the consumer utility of receiving a service K equals:

$$CU = \theta - \Psi(\Lambda, \beta) - P \quad (1)$$

It is clear from the above equation that the privacy cost is a negative externality borne by the individual whose privacy is invaded. Intuitively, we want to associate high benefit/compensation with $\Psi(\Lambda, \beta)$ that allow high identification of I (the identity of the consumer) given Λ and high risk β. Next, we present the proposed system.

3 The Agent-Based System

3.1 High Level View

Fig.1 depicts the high level architecture of the agent-based system. The agents in the system takes on the responsibility of helping consumers valuate their personal data objects that are perceived to be valuable, in order to capitalize on them for the goal of maximizing their market value once the consumer decides to reveal them.

To explain how the system works, consider the following scenario: Alice is willing to share her personal data preferences with certain online service providers for a discount value or a reward fee. But Alice has certain requirements before she consents to complete the transaction:

1) She wants complete information about the service provider's privacy practices and its trustworthiness
2) She wants to determine the level of risk involved in the transaction based on information from (1)

She wants the system to valuate her data combination risk based on her perceived risk of each private data object

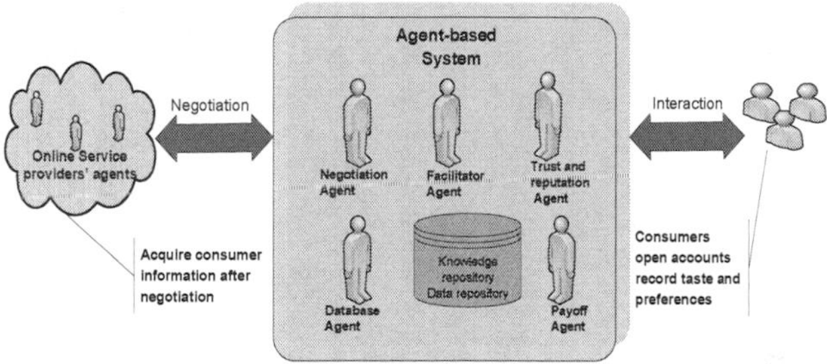

Fig. 1 High level view of the multi-agent system architecture

3) She wants her reward or discount value to be valuated based on the involved risk from (3)
4) She wants the party that negotiates on her behalf to be strategic during the negotiation process so she can get the maximum benefit
5) She wants to have the privilege of accepting or denying the final offer

To satisfy Alice's requirements the system employs several agents where each one performs a specific task. The details of these agents and their functionality is described below.

3.2 Trust and Reputation Agent

As mentioned earlier in the example scenario, in order for Alice to assign privacy risk weights to her private data objects, she wants to know the competency of the online business with respect to privacy and private data handling. Therefore, it is essential to assess the service provider's trustworthiness based on privacy credential attributes which the online provider has obtained (e.g., trustmark seals, privacy certificates, contents of the privacy statement, encryption mechanisms, etc.). In this paper, the rating of each attribute is based on studies on consumer research analysis such as those of [9] industry studies such as the study in [10]. These studies analyze the impact of different trusting attributes on consumers' online purchasing behavior and their perceived trust. For example, in [9] and [10] the "subscription to privacy auditing service" and " use of encryption mechanism" have high impact on the perceived trust according to consumer research analysis.

The attributes and their classifications are stored in a database dedicated for the privacy credential ratings. The trust and reputation agent determine the privacy credential score and the privacy report for each service provider (implementation of such report is shown in section 4).

3.3 Payoff Agent

The payoff agent determines the payoff value by constructing a risk premium that is valuated in response to the amount of potential damages that might occur (cost of risk) with respect to the risk exposure $\Psi(\Lambda, \beta)$. In this manner, the compensation paid to the consumer is justified, at least in part, from the damages that might occur. This approach is widely used in the financial and auto insurance industry [15]. The standard model is presented in [15] and shown below.

$$E(U_q) = \Psi(\Lambda, \beta) \bullet E(\Re) \tag{2}$$

where $E(U_q)$ is the expected return given to the consumer against revealing their private data record q, and $E(\Re)$ is the risk premium which is a positive value.

Equation (2) is simply saying that consumers' demand for revealing their personal information is measured based on the perceived risk and the market risk premium. The market risk premium is a monetary value estimated by looking at historical premiums over long time periods. It is assumed to follow a random process that fluctuates over time following a geometric Brownian motion with general solution as follows (the detailed derivation of the solution is presented in [6]):

$$E[\Re_t] = \Re_o e^{\mu t} \tag{3}$$

where μ is the drift value and $\Re_0$ is a known value at time 0.

3.4 Negotiation Agent

Having an agent negotiating with an online service provider on behalf of a group of consumers would be in a better position to bargain over the revelation of their personal information. In our setup of the negotiation agent, we adopt the work of Bo et al. [11] in defining a negotiation strategy. The negotiation agent and the service provider's agent are assumed to have a different time preference, i.e. their time deadline. In this paper, we adopt the "sit-and-wait" strategy proposed by [11]. The "sit-and-wait" strategy is used in the case when negotiation does not devaluate over time (which is the case of personal information). It has been proven by [11] that this strategy is the dominant strategy for an agent using time-dependent negotiation process. The details of the negotiation protocol and the offer construction are provided in [6].

4 Prototype Implementation

The implemented of the system is inspired by the scenario presented in section 3. The agents that will work on behalf of Alice are implemented using JADE: Java Agent Development Environment. Through the system, Alice can open an account and record her personal information and preferences essentially by using a detailed subscription form. The information about the service providers are presented to Alice in a form a rating score of privacy practice on a scale of 1 to 5 stars, such that 1 star means low privacy protection and 5 means high privacy protection.

4.1 Negotiation Experiment

This experiment examines the effect of the risk premium variation on the overall benefit of both the service provider and the consumers. The variation in the market risk premium is simulated by incrementing the drift value by 0.1, thus increasing the risk premium gradually. For this experiment, we chose two travel agencies that offer similar services and similar prices. The service providers are Escapes.com and Itravel2000.com; for convenience we name them SP_1 and SP_2 respectively. Results from the experiment show that service providers with high privacy credential rating achieve higher utility than service providers with lesser privacy credential rating. In Fig. 2.B, when the market risk premium is high (i.e., when the drift value is 2), service provider SP_2 acquired a benefit value equal to $18370 compared to $0 for service providers SP_1 in Fig.2.A. The reason behind this significant difference is that 334 consumers completed the transaction in the case of SP_2 (shown in Fig. 2.B). The 334 consumers were expecting a payoff less than what offered by the service provider. This is based on the valuation of their private data given their perceived privacy risk of service provider SP_2.

At drift values = (1, 1.1, and 1.2) which represent low risk premium, both service providers were able to accumulate a benefit depending on their type, i.e., their privacy credential rating. However, as the risk premium increases to a moderate level (1.4, 1.5, and 1.6), service provider SP_1, for instance, accumulated $0 benefit (shown in Figure 2.A) because all consumers at that level were expecting higher payoff as the privacy risk increases. Such payoff is unaffordable by service provider SP_1. However, SP_2 was able to keep a higher benefit at a high market risk premium. The experiment shows that consumers who are reluctant to complete transactions because they perceive high privacy risk impact the service providers' benefit as they ignore their services.

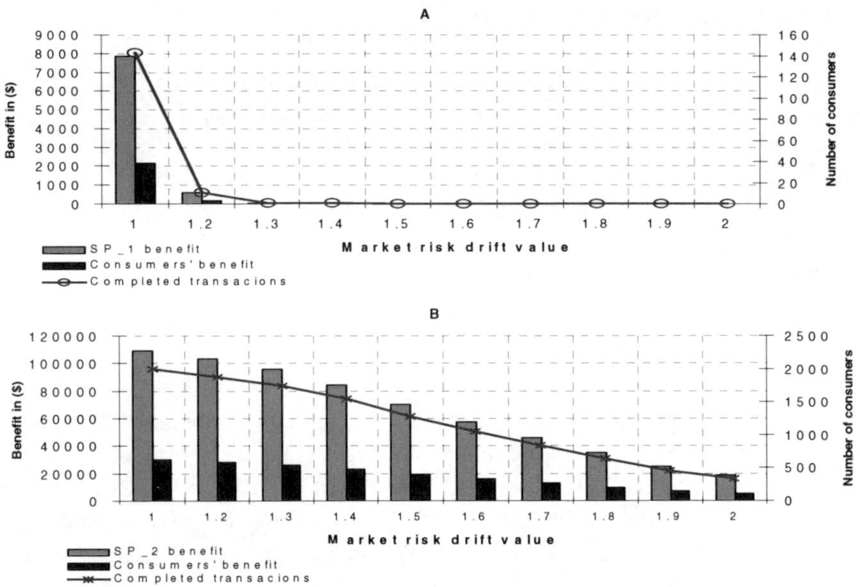

Fig. 2 Benefit of service providers and consumers. (A) Performance of service provider 1. (B) Performance of service provider 2

5 Conclusion and Future Work

This paper presented key ideas of an agent-based model that will assist consumers to benefit from their personal information revelation. Our future plan is to engage real scenarios with actual users to analyze individuals' awareness of privacy trade-offs and their true valuation of privacy.

References

1. Taylor, C.R.: Consumer privacy and the market for customer information. RAND Journal of Economics 35(4), 631–650 (2004)
2. Cavoukian, A.: Privacy as a negative externality: The solution Privacy by Design. In: WEIS Workshop on the Economics of Information Security in London, U.K. (2009)
3. He, Y., Jutla, D.N.: Contextual e-negotiation for handling of private data in e-commerce semantic. In: Web Proceeding of the 39th Hawaii International Conference on System Sciences, HICSS 2006, vol. 3, pp. 62a–62a (2006)
4. Blažic, A.J., Dolinar, K., Porekar, J.: Enabling privacy in pervasive computing using fusion of privacy negotiation, identity management and trust management techniques. In: Proceedings of the First International Conference on the Digital Society, ICDS (2007)

5. Preibusch, S.: Implementing Privacy Negotiations in E-Commerce. In: Zhou, X., Li, J., Shen, H.T., Kitsuregawa, M., Zhang, Y. (eds.) APWeb 2006. LNCS, vol. 3841, pp. 604–615. Springer, Heidelberg (2006)
6. Yassine, A., Shirmohammadi, S., Tran, T.T., Nazari, A.A.S.: Intelligent Agents for future personal information markets. In: IEEE/ACM/WIC/IAT, August 31- September 3 (2010)
7. Julia, G., Egelman, S., Cranor, L., Acquisti, A.: Power Strips, Prophylactics, and Privacy. Oh My! Symposium on Usable Privacy and Security, SOUPS, PA, USA (2006)
8. Kimery, K.M., McCord, M.: Third-party assurances: Mapping the road to trust in e-retailing. Journal of information Technology Theory and Application 4(2), 63–82 (2002)
9. Sha, W.: Types of structural assurance and their relationships with trusting intentions in business-to-consumer e-commerce. In: Electron Markets, vol. 19, pp. 43–54. Springer (2009)
10. Hussin, A.R., Macaulay, L., Keeling, K.: The importance Ranking of Trust Attributes in e-Commerce website. In: 11th Pacific-Asia Conference on Information Systems (2007)
11. An, B., Sim, K.S., Tang, L.G., Miao, C.Y., Shen, Z.Q., Cheng, D.J.: Negotiation agents' decision making using Markov Chains. SCI, vol. 89, pp. 3–23. Springer, Heidelberg (2008)

An Improved Kernel Trojan Horse Architecture Model

Mingwei Zhao and Rongan Jiang

Abstract. As a new kind of Trojan horse which combines with the kernel Rootkit technologies, kernel Trojan horse has received a great mount of people's attention and been used a lot. However, the sensitive property of kernel Trojan which follows traditional architecture model is fully exposed to the security software, and needs kernel concealment module to complete all the hidden works, thus the concealment module is too large, easily detected by security software. Based on the analysis of Trojan collaborative concealment model, this paper improves the traditional architecture model and introduces a lightweight concealment module of pure kernel Trojan horse architecture model. Furthermore, an example which adopts the improved model is present in this paper. The experimental results verify the feasibility and efficient of the improved model.

1 Introduction

The traditional kernel-level Trojan horse which is achieved by combining both kernel level and user level components conceals itself through modifying the system kernel structure using kernel Rootkit. However, with the emergence new detection technologies, for example, memory searching and cross-view, the flaws of kernel-level Trojan horse using traditional architecture model have been revealed: There are lots of code which are running on the user-level totally exposed to the underlying security software which can easily detect the abnormalities in system by hooking the underlying system function; the user-level code will leave more sensitive properties in the system which need kernel-level Rootkit to hide by modifying a mass of operating system structure. The more it modifies, the more easily for security software to detect.

In this paper, we first analyze the conspiring concealment of Trojan horse, then point out the defect of traditional Trojan horse architecture model. After that we

Mingwei Zhao
Department of Computer Science and Technology,
Faculty of Electronic Information and Electrical Engineering,
Dalian University of Technology
e-mail: mwzhao@126.com

F.L. Gaol (Ed.): Recent Progress in DEIT, Vol. 2, LNEE 157, pp. 129–134.
springerlink.com

improve the traditional model and introduce a lightweight concealment module of pure kernel Trojan horse architecture model. Finally an example is given to show the effectiveness of the improved model.

2 Kernel-Level Trojan Horse

2.1 Kernel-Level Trojan Horse Functional Structure

After implanted into the target host kernel-level Trojan horse is required to achieve a long-term effective control of target host. Since Trojan horse must be loaded upon system reboot, an auto-start module should be involved to achieve this requirement. Another important feature of Trojan horse is concealment. Trojan horse is required to work smoothly without the awareness of user which means it has to hide the information about itself to avoid the detection of antivirus software, so a concealment module must be consist of. Controllability is another characteristic of Trojan horses, to achieve controllability, we need to establish a stable remote connection to accept remote commands and send back the results, so communication module is also necessary. The last important module is the function module which is used to achieve the specific tasks and collect important information on target host.

2.2 Conspiring Concealment Technology

After implanted to the target host, kernel Trojan horse is bound to use varieties of techniques to hide its whereabouts, to avoid being discovered and to prolong survival as much as possible. Conspiring concealment is defined as a number of Trojan horses or children Trojan horses cooperate with each other to accomplish the concealment and long-term latency of whole Trojan horse. In the operating system, the object performance is based on its features and attributes. For example, a running program is usually composed of program file directories, program files, processes, communication link status and other properties and functional characteristics. To hide these performance characteristics, Trojans need to hide the corresponding attributes. As Trojans often contain more than one property, they usually adopt the collaboration of multiple programs or program components to achieve the concealment of attribute features.

3 Improved Module and Its Implement

Based on the study of kernel Trojans functional structure and conspiring concealment techniques, the hidden principle of kernel-level Trojans is analyzed as below. Each module of Trojans cooperates together to complete two functions, collecting information on target host and hiding the Trojan. The module which collects information is defined as task module and the module that hides the whole Trojan is defined as concealment module. Specific, concealment module's responsibilities include hiding task module and hiding concealment module itself. Concealment module can be divided into multiple sub-modules and each sub-module hide one sensitive property of task module. What's more, sub-modules cooperate with each other to complete the hidden of the concealment module, so that the whole kernel Trojan can totally be invisible.

The number of task module attributes determines the number of concealment sub-modules. The more the attributes are, the more concealment sub-modules are needed to hide corresponding attributes. To make it worse, the large number of sub-modules makes the concealment module even larger. On the other hand, with the increasing number of concealment sub-modules, more cooperation between them is needed to hide each other. A vicious circle is constituted. With such a large concealment module, it is inevitable that concealment module might make some noise even it is in the kernel. Security software can detect it and inform user that malice software exists.

From another aspect, the task module properties, more precisely, should be called the properties which are exposed to security software. After hidden by concealment module, these sensitive properties won't be perceived by security software. Moving forward a single step, the reason that task modules properties exposed to security software is because of security software running on the lower level of system, so that every move of task module can not escape the monitoring of security software. If we transfer those properties to the same level or even lower level compared to the security software, in another form, security software can not detect it while neither needs us to cut down the functional properties of the task module. Using the method mentioned above, the number of concealment sub-modules is significantly reduced while providing the same function.

Accordingly, this paper aims at the point that the task module is totally exposed to security software under traditional architecture model and focus on the problems that too large the concealment is and too much kernel data are changed. After that, it introduces a new model which has a lightweight concealment module and totally implemented in pure kernel-level.

3.1 Improved Conspiring Concealment Model

Let *S* be the traditional kernel Trojan, *P1* be the set of user-level components, *P2* be the set of kernel-level components, *R1* be set of concealment relationship between user-level components and kernel-level components, *R2* be the set of concealment relationship of kernel-level components. In the traditional architecture model, kernel Trojan can be expressed in this way $S = \{P1, P2, R1, R2\}$. However, in the improved model, only kernel-level components are involved. Let *T* be the Trojan which uses the improved architecture model, it can be defined as $T = \{P2, R2\}$. The conspiring relationship of lightweight concealment module Trojan model is described as following:

Let L^* be the kernel code sequence, *P2* be the set of Trojan components, $P2 \in L^*$. Both concealment module and task module are involved in *P2*. Let *F* be the set of task module components, *H* be the set of concealment module components, $F = \{F_i \mid i=1, 2, 3, \cdots,m\}$, $H = \{H_i \mid i=1, 2, 3,\cdots,n\}$.

$$r,\hat{r},r',\hat{r}'\in R_{\circ}$$
$$\wedge E(r)P2\approx E(\hat{r})P2$$
$$\wedge r\sim\hat{r},H_i\in L^*,\exists t$$
$$\wedge[\![E(r)P2]\!]r\xrightarrow{t}r'$$
$$\wedge[\![E(\hat{r})P2]\!]r\xrightarrow{t}\hat{r}'$$
$$\begin{cases}\wedge r'\sim\hat{r}'(t\in H)\\ \wedge r'\nsim\hat{r}'(t\notin H)\end{cases}$$

From the formal description we can see that in the improved conspiring model, task module and concealment module are both completely achieved in kernel-level. The sensitive property collection and conspiring relationship are both reduced, which simplify the conspiring concealment relationship. While realizing its own function, task module fully uses the kernel Rootkit technique to hide itself. By spreading out the implementation of concealment to reduce the noise of concealment module, the concealment of Trojan horse is greatly improved.

3.2 Improved Architecture

1. Automatic Start Module: This module is to ensure that the target system can automatically load the Trojan after reboot like the traditional one does. Which make it different is that it is implemented in kernel-level. Therefore thesensitive properties of this module will not be exposed to the security software.
2. Feature Module: This module is responsible for finishing tasks assigned by the remote control terminal. In the traditional framework, this module is running in the user mode, in the form of process or other object which can be easily detect by the security software. However, in the improved architecture model, this module is implemented in kernel-level, existing in kernel in the form of driver. Since it is running on the same layer as security software, only few properties needs to be hide by the concealment module to evade detection of security software.
3. Communication Module: This module is supposed to communicate with remote control host surreptitiously. The traditional communication module is realized in user mode and bypass firewalls or other security software through a variety of network techniques. Nevertheless, most of these methods rely on imperfect configuration of the firewall. Although you can pass through the firewall, communications activities are still exposed to supervision of a firewall.

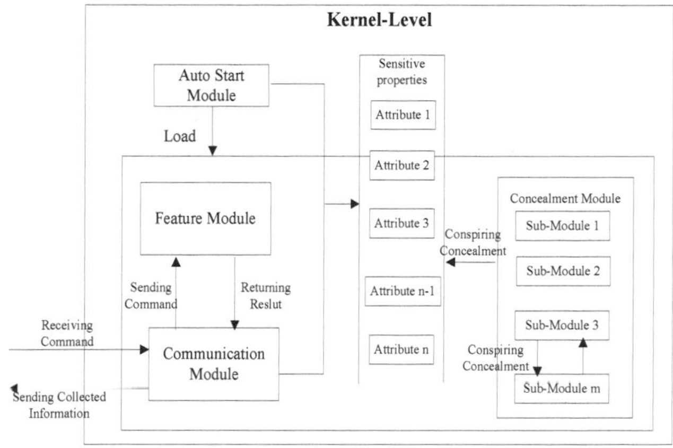

Fig. 1 Improvend Kernel-levle Trojan model Structure

4 Experimental Result

To prove the superiority of our model framework, BOES and Kang's Rootkit Trojan which is both based on the Traditional Trojan horse Model are used here to compare to. Two advanced anti-rootkit software and ZoneAlarm are used here to test the concealment of the three Trojans. Results are showed in table 1.

The symbol '×' means failing to bypass security software while the symbol '√' presents succeeding.

Table 1. Compare of three Trojans

	Cross-view anti-Rootkit	Memory search anti-Rootkit	ZoneAlarm Firewall
BOES	×	×	×
Kang's Rootkit	×	×	×
Pure kernel Trojan	√	√	√

Experimental results show the feasibility of the model architecture and superiority over the traditional concealment architecture model.

5 Conclusion

In terms of the study on Trojan conspiring technique, we find that the concealment of Trojan should not only be the job of concealment module but the job of every module. This paper improve traditional architecture model based on the point. In the improved framework, kernel Rootkit is fully used and every module accomplishes some conceal work, hence the concealment module has been greatly reduced which strengthen the concealment of whole Trojan. However, there are various windows operating system versions and in each version the kernel code is not quite the same, so compatibility and stability is the further research direction that will be carried out.

References

1. Lacombe, E., Raynal, F., Nicomette, V.: Rootkit modeling and experiments under Linux. Journal in Computer Virology 4, 137–157 (2008)
2. Wang, J.: doi: 10.1109/ICIME.2010.5478178
3. Gong, G., Li, Z.-J., Hu, C.-J., Zou, Y.-K., Li, Z.-P.: Research on Stealth Technology of Windows Kernel level Rootkits. Computer Science 37, 59–62 (2010)
4. Liu, D., Gan, Z.: Research on Concealment Technology of kernel-based Trojan Horse Under Windows. Microprocessors 3, 41–44 (2009)
5. Kang, Z.-P., Xiang, H., Hu, H.-B.: Research and practice on concealing technology of Windows' Rootkit. Computer Engineering and Design 28, 3334–3337 (2007)
6. Zuo, L.-M., Jiang, Z.-F., Tang, P.-Z.: Concealing Technology of Windows Rootkit and Integrated Detection Method. Computer Engineering 35, 118–120 (2009)
7. Zhang, X.-Y., Qing, S.-H., Ma, H.-T., Zhang, N., Sun, S.-H., Jiang, J.-C.: Research on the concealing technology of Trojan horses. Research on the concealing technology of Trojan horses. Journal of China Institute of Communications 25, 153–159 (2004)
8. Hoglund, G., Butler, J.: Rootkits: Subverting the Windows Kernel. Addison-Wesley Professional, Boston (2005)

Automatically Resolving Virtual Function Calls in Binary Executables

Tao Wei, Runpu Wu, Tielei Wang, Xinjian Zhao, Wei Zou*, and Weihong Zheng

Abstract. Call graph plays an important role in interprocedural program analysis methods. However, due to the common exist of function pointers and virtual functions in large programs, call graphs used in current program analysis systems are usually incomplete and imprecise, especially in analysis systems for binary executables. In this paper, we present a scalable and effective approach to automatically resolve virtual-function calls in executables. For the benchmark used in previous studies, our approach resolved almost 100% of reachable virtual function call-sites, whereas CodeSurfer/x86 resolved about 82%.

1 Introduction

A call graph (CG) represents calling relationships between program's procedures and is a fundamental data structure for interprocedural program analysis. A large number of source code level call graph construction algorithms have been proposed. However, with the increasing need of directly analyzing executables, how to construct call graphs on binary code level becomes more and more important.

When source code is not available to end users, the state-of-the-art techniques can not handle this problem effectively yet. In this paper, we present a novel scalable and effective approach to automatically detect and resolve virtual function calls(VFC) in executables.

Tao Wei · Tielei Wang · Xinjian Zhao · Wei Zou
Peking University
e-mail: {weitao,wangtielei,zhaoxinjian,zouwei}@icst.pku.edu.cn

Runpu Wu · Weihong Zheng
China Information Technology Security Evaluation Center
e-mail: {wurp,zhengwh}@itsec.gov.cn

* Corresponding author. This research was supported in part by National Natural Science Foundation of China (Grant No. 61003216).

F.L. Gaol (Ed.): Recent Progress in DEIT, Vol. 2, LNEE 157, pp. 135–140.
springerlink.com

In summary, this paper makes the following contributions:

- A scalable and effective approach to automatically detect and resolve virtual function calls from executables.
- In contrast to traditional symbolic execution [5], our symbolic execution algorithm is more scalable. Experiment results show our approach can effectively track the propagation of objects.
- We have implemented a prototype system of our approach, named AutoCG. An empirical study of applying AutoCG to many programs shows that AutoCG is able to resolve 100% of reachable virtual-function call-sites in the benchmark used in previous studies.

2 Problem Statement

An object-oriented program generally consists of various kinds of functions, such as virtual functions, imported functions, etc. The function calling information becomes implicit when a binary executable is compiled from the source program and is hard to resolve.

When source code is available, type inference techniques can resolve a part of virtual function calls. However, type information is usually not available in executables. Therefore, to resolve VFCs, the address of the object, the virtual table and the target virtual function must be computed at the virtual function call site.

To resolve the virtual-function calls, information about the addresses of objects and the correct `vfptr` field must be available at a virtual function call site. For a static-analysis algorithm to determine such information, it has to track the propagation of object information(e.g., objects′ layout, the `vfptr` fields). However, the number of paths in real-world software is too large to explore.

Programs typically make extensive use of dynamically linked libraries, which are not available in the executables. This is a big challenge for dataflow analysis using symbolic execution technique. For example, these imported functions may influence the balance of stack by purging the parameters in the stack in binary program execution, and may also have side-effects.

3 Our Approach

At a high level, our approach consists of four main steps. The first step is to convert an executable to an equivalent intermediate representation (IR) form. The second step is to identify the constructors that are able to create objects. The third step will leverage symbolic execution technique to track the propagation of objects and their memory layout (such as `vfptr` fields). The forth step is to detect and resolve the indirect call caused by virtual function call.

3.1 Conversion to IR

AutoCG disassembles the executable, obtains the disassembled information of the executable, such as code section, data section, imported functions, and so on. AutoCG can create a control flow graph for this program. Meanwhile, AutoCG will perform loop structure identification [8] and branch structure identification [9].

AutoCG converts the assembly code into an equivalent intermediate representation(IR). We designed an SSA-like IR, named PANDA (Program ANalysis Dedicated to ASM). We convert x86 assembly into our PANDA representation based on our previous work [8, 9]. Due to the space limitations, we omit the detail of description and implementation on PANDA language here. More information about PANDA could be found in [7].

3.2 Identification of the Constructors

In this step, AutoCG identifies class constructors (construction functions). The techniques are based on our observation that constructors have some common patterns. Usually, a constructor has two feature:

1) The function has a pointer-like parameter, say *arg1*: i.e., the function will dereference *arg1* (or with some displacement). The pointer-like parameter usually corresponds to `this` pointer in source code.

2) More importantly, the variable *arg1* (or with some displacement) will be assigned an immediate value, which points to a beginning of a function pointer array which locates in the data section. Although different compilers may have different implementations, a class with virtual functions will be associated with a function pointer array. The element in the array points to different virtual functions.

We can identify class constructors based on the two features. In this step, our approach also detects and collects the information about all virtual tables and the virtual functions in the executables.

3.3 Symbolic Execution

Traditional symbolic execution [4], which maps a variable to a symbolic value, is usually a path-sensitive analysis. Instead, fixed-point-based dataflow analysis (such as Constant Propagation), which maps a variable to an element in a lattice, needs to merge over all paths (MOP) information to compute a fix-point. Our basic idea is that we can take advantage of the fact that we use symbolic execution method to track the object memory layout, and use a merge function to reduce program states.

There is no notion of variables in binary programs. In order to simulate program execution symbolically, our approach needs to build and maintain a symbolic memory which is responsible for mapping symbolic memory addresses M to symbolic values V. However, different from tradition symbolic execution methods that map each variable to a single symbolic value, our symbolic execution maps a variable to a set of symbolic values.

Give two symbolic value sets, $vs1$ and $vs2$, the following operators are defined for value set operations.

- $vs1 \sqcup vs2$: return the union(join) of value-sets $vs1$ and $vs2$. For instance, assume $vs1 = \{x, y\}$ and $vs2 = \{1, 2\}$, $vs1 \sqcup vs2 = \{x, y, 1, 2\}$;
- $vs1 + c$: return $\{x + c, for\ \forall x \in vs1\}$; For instance, assume $vs1 = \{x, y\}$, $vs1 + 1 = \{x + 1, y + 1\}$
- $vs1 + vs2$: return $\{x + y, for\ \forall x \in vs1,\ \forall y \in vs2\}$; for instance, assume $vs1 = \{x, y\}$ and $vs2 = \{z, 1\}$, $vs1 + vs2 = \{x + 1, x + z, y + 1, y + z\}$. Similarly, we can define the operator $- \star /$ on value sets;
- $[vs1]$: returns $\{[x], for\ \forall x \in vs1\}$. For instance, assume $vs1 = \{x, y\}$ and the memory map M has entities $x \mapsto 1$ and $y \mapsto z$, $[vs1]$ will return $\{1, z\}$.

Compared with existing path-based symbolic execution, our algorithm is more scalable because we execute symbolically every statement only one time when a function `f` is examined. Our algorithm will ignore branch statements, and directly examine their two branches. But for the node with two or more direct predecessors, our approach will collect and merge symbolic states from all the predecessors. When meeting an indirect call, our symbolic execution algorithm may compute the possible target function addresses of the indirect call. Note that loops are executed only once in our method.

4 Evaluation

We implemented a prototype system of AutoCG. We performed experiments on 14 C++ programs, presented in Table 1. These programs were originally used in [6] and [3] to evaluate their algorithms for virtual function resolution in binaries. We also present the number of functions in the third column. In addition, we measured the number of import functions and indirect call sites, showed in the fourth and fifth columns respectively.

Table 1 Benchmark List

program	# x86 insts	Procs	import funcs	indirect
NP	2694	217	59	6
Primes	2788	175	61	2
family	2982	180	60	3
vcirc	2856	179	58	4
office	3763	208	82	4
trees	4113	212	60	3
deriv1	4937	245	83	18
chess	4392	238	61	1
objects	4428	247	61	23
simul	4506	252	60	3
greed	4485	247	60	17
shapes	5346	265	79	12
ocean	5650	280	61	5
deriv2	6646	253	83	56

Table 2 Experiment Results

Program	Codesurfer/x86 VS AutoCG						
	$\perp$	1	2	$\geq$3	$\top$	% Reachable call-sites resolved	Time(s)
NP	0 : 0	0 : 0	6 : 6	0 : 0	0 : 0	100 : 100	1 : <1
Primes	1 : 1	1 : 1	0 : 1	0 : 0	1 : 1	**50 : 100**	<1 : <1
family	0 : 0	3 : 3	0 : 0	0 : 0	0 : 0	100 : 100	1 : 1
vcirc	0 : 0	4 : 4	0 : 0	0 : 0	0 : 0	100 : 100	<1 : <1
office	0 : 0	4 : 4	0 : 0	0 : 0	0 : 0	100 : 100	<1 : <1
trees	1 : 1	0 : 0	0 : 1	0 : 1	2 : 0	**0 : 100**	9 : 9
deriv1	8 : 8	8 : 7	2 : 3	0 : 0	0 : 0	100 : 100	4 : 7
chess	0 : 0	0 : 1	0 : 0	0 : 0	1 : 0	**0 : 100**	16 : 5
objects	18 : 19	0 : 0	4 : 4	0 : 0	1 : 0	**17 : 100**	2 : 1
simul	2 : 0	0 : 0	0 : 3	0 : 0	1 : 0	**0 : 100**	6 : 1
greed	6 : 6	10 : 11	0 : 0	0 : 0	1 : 0	**59 : 100**	10 : 4
shapes	4 : 5	4 :4	3 : 3	0 : 0	1 : 0	**58 : 100**	10 : 12
ocean	3 : 0	0 : 1	0 : 0	0 : 4	2 : 0	**0 : 100**	17 : 32
deriv2	33 : 30	22 : 22	0 : 1	0 : 3	1 : 0	**39 : 100**	2 : 18

CodeSurfer/x86 [1, 3] uses the Value-Set Analysis(VSA) approach to recover variable-like entities from binary programs, and then translates x86 binary code into an IR which can be analyzed by the CodeSurfer System. The experiment results in [3] showed CodeSurfer/x86's ability to resolve virtual function calls in binary programs listed in Table 1. We also chose these programs and performed our experiments on a machine with the Intel(R) Core(TM) 2 CPU (2.40GHz) and 2 GB of RAM running Linux.

Our results are presented in Table 2. The "Program" column indicates the program name. "$\perp$" is the number of unreachable indirect call-sites caused by virtual function invocation (i.e., dead code). "$\top$" is the number of reachable indirect calls, but targets are unknown. "%Reachable call-sites resolved" column gives the virtual function resolution rate at the reachable indirect call-sites. The columns "1", "2" and "$\geq$3" columns indicate the distribution of the number of target functions at the virtual function call-sites. For example, "2" shows the number of indirect call-sites that have two target functions. Note that while the numbers before ":" are CodeSurfer/x86's results [3], the numbers after ":" are AutoCG's results.

Our approach has resolved 100% of reachable indirect calls induced by virtual functions, whereas CodeSurfer/x86 has resolved 82% of such calls. We find that CodeSurfer/x86 did not resolve any indirect call whose targets is more than 2. However, AutoCG can resolve all of those indirect calls.

5 Discussion and Related Work

Many algorithms that resolve virtual-function calls in C++ programs have been proposed, such as [6]. However, these algorithms need source codes.

As source code is not always available to end users, the state-of-the-art techniques have to dynamically run the program [2], or use interval-based abstract interpretation execution [1] to determine the possible targets of indirect calls. The former depends heavily on program inputs, and the latter could not resolve VFCs well because object type information is usually missed during fixed point iteration.

The main weakness of our symbolic execution approach is that the number of symbolic values that a symbolic address points to may grow exponentially. However, we found that most of symbolic values have little effect on object memory layouts. Thus, we only kept limited symbolic values that an address maps to. Our approach can reduce the unnecessary values that are not related to objects and the addresses of virtual tables, virtual function addresses,etc.

6 Conclusion

We present a novel approach to automatically locate and resolve virtual-function calls in executables. Our approach first converts the executable into our own intermediate representation, and then identifies the constructors, after that tracks the propagation of the type information of objects using symbolic execution method, finally detects and resolves the indirect call caused by virtual function calls. We implemented our technique in the AutoCG tool and evaluated our approach on many programs. The results show our approach can effectively detect and resolve reachable virtual-function call-sites.

References

1. Codesurfer/x86, http://www.grammatech.com/research/products/CodeSurferx86.html
2. Uqbt:a resourceable and retargetable binary translator, http://www.itee.uq.edu.au/~cristina/uqbt.html
3. Balakrishnan, G., Reps, T.: Recency-abstraction for heap-allocated storage. In: Proc. Static Analysis Symposium, pp. 221–239 (2006)
4. Godefroid, P., Klarlund, N., Sen, K.: Dart: directed automated random testing. In: Proc. the 2005 ACM SIGPLAN Conference on Programming Language Design and Implementation, pp. 213–223 (2005)
5. King, J.C.: Symbolic execution and program testing. Communications of the ACM 19(7) (1976)
6. Pande, H.D., Ryder, B.G.: Data-flow-based virtual function resolution. In: Proc. Static Analysis Symposium, pp. 238–254 (1996)
7. Wang, T., Wei, T., Lin, Z., Zou, W.: Intscope: Automatically detecting integer overflow vulnerability in x86 binary using symbolic execution. In: Proc. 16th Annual Network & Distributed System Security Symposium (2009)
8. Wei, T., Mao, J., Zou, W., Chen, Y.: A new algorithm for identifying loops in decompilation. In: Proc. Static Analysis Symposium, pp. 170–183 (2007)
9. Wei, T., Mao, J., Zou, W., Chen, Y.: Structuring 2-way branches in binary executables. In: Proc. 31st Annual International Computer Software and Applications Conference, pp. 115–118 (2007)

Research of Network Intrusion-Detection System Based on Data Mining

Shijun Yi and Fangyuan Deng

Abstract. In this paper, the algorithm for data mining of intrusion detection system has been improved and optimized so as to achieve intelligent detection of network data. Winsock2 SPI is used during the design to intercept data in the network, and the method of "session filtering" is adopted to filter network packets. The system consists of modules of control rules and intelligent detection, etc. According to actual detection, the system is capable of displaying network connection status on a real-time basis, effectively controlling application programs and intelligently detecting network data.

1 Introduction

With the continuous development of science and technology, the network has become a country's most crucial political, economic and military resources as well as a symbol of national power. Moreover, the development of network has also continuously changed people's work style and lifestyle and made information acquisition, transmission, processing and utilization more efficient and rapid [1]. The network has become an important part in human life. However, its negative effects also occur with the continuous expansion of network application. The network has made it possible for hackers, industrial spies and malicious intruders to infringe upon and manipulate some key information so that network security issues may occur. Hackers joined to invade some of the world's largest and hottest websites, such as Yahoo and AOL.com and made their servers unable to provide service properly; hackers also attacked some Chinese human rights websites and changed their homepages into those with anti-government speeches; besides, computer viruses also develop rapidly [2]. Network security now faces serious challenges, and it is a core issue of informatization to solve network security issues.

Shijun Yi · Fangyuan Deng
Network Center, Chengdu Electromechanical College, Chengdu, China
e-mail: ysj1967@yahoo.com.cn, vax11@163.com

F.L. Gaol (Ed.): Recent Progress in DEIT, Vol. 2, LNEE 157, pp. 141–148.
springerlink.com

Intrusion detection is essentially a process of electronic data processing. Collected security audit data are analyzed and processed with the pre-determined method, and then whether the system is intruded or not is determined according to the analysis results. Collected data are usually system logs or network event logs. The pre-determined method is the analytic technique adopted by the analytic engine, usually a matching technique or statistic analytic technique in a certain mode or a combination of the two [3]. Intrusion detection usually includes three steps, i.e. information collection, data analysis and response (passive response and active response). Accordingly, an IDS usually consists of functional modules of information collection, data analysis and response, etc., which provide the functions of the above 3 steps.

The contents of information collection and module collection include statuses and behaviors of system, network and user activities. Data sources adopted for intrusion detection usually include several aspects: system logs (including audit records of operating system, and system logs and application logs of operating system), abnormal changes of directories and files, abnormal behaviors during program execution and network packets, etc. The data analysis module is the core of IDS, which synchronizes, organizes, sorts and classifies raw data, conducts detailed analysis of various types, extracts the hidden system activity characteristic or mode and uses that for IDS to judge whether any intrusion occurs[4]. The IDS, upon discovery of any intrusion, shall make timely response according to the pre-determined strategy, including disconnecting the network, recording the event and sending alarms.

2 Data Mining Methods for Intrusion Detection

Main data mining methods for intrusion detection include the data classification analysis method, the correlation analysis method, the cluster analysis method and the sequential pattern analysis method.

2.1 Correlation Analysis Method

Correlation rules are the inter-dependent relationship among data. The task of correlation analysis is to discover from the database the strong rules whose credibility and supportiveness are both greater than the given values. The correlation analysis method is used to discover the mutual relationship among coordinated intrusion events and derive some unknown coordinated attack modes so as to form new coordinated intrusion rules. Common correlation analysis algorithms include: DHP, Apriori, and AISIISETM.

The mining of correlation rules algorithms includes two steps:

a) Discover all frequent itemsets. Use the minimum supportiveness given by the user to discover all frequent itemsets, i.e. all itemsets whose supportiveness is not less than min_sup; when some frequent itemsets may have inclusion relations, the itemsets that are not included by other frequent itemsets are called large frequent itemsets.

b) Generate correlation rules from the frequent itemsets. Use the minimum confidence degree (min_con) given by the user to discover the correlation rules whose confidence degree is not less than min_con from each maximum frequent itemset.

2.2 Data Classification Analysis Method

The classification analysis method is the most frequently applied method of data mining. Classification is to find a concept of category for description. It represents the overall information of data of that category, i.e. the connotation description of that category, usually represented in the mode of rules or decision tree. The connotation description of a category includes characteristic description and distinctive description [5]. Characteristic description is to describe the common characteristics of objects of a category, while distinctive description is to describe distinctions among categories. The task of classification analysis in the system is to analyze the security events in the security event database, accurately describe the categories of all events and derive the classification rules of events. Common classification algorithms include: CART, ID3, and C45.

The procedures of data classification are:

- Obtain the training data set; the data records in the data set have the same data items as those of data records in the target database;
- In the training data set, each data record has the known type identification for correlation;
- Analyze the training data set, extract the characteristic attribute of its data record and derive the accurate description model for each type;
- Use the derived type description model to classify the data records in the target database or derive optimized classification models (classification rules).

2.3 Cluster Analysis Method

Data clustering is to classify the physical or abstract objects into several groups. Within a group, objects are highly similar, while objects are less similar among groups [6]. The input set of cluster analysis is a set of unlabelled records without any classification. Clustering is to reasonably divide record sets and describe different grades with an implicit or explicit method. Clustering analysis is capable of analyzing a set of unclassified security events, and, according to the pre-determined classification rules of classification analysis, categorizing the large number of scattered security events into the security event sets that describe intrusion behaviors. Common clustering algorithms include: CLARA, PAM, and BIRCH [7].

2.4 Sequential Pattern Analysis Method

The sequential pattern analysis is similar to the correlation rules analysis. Its purpose is also the mining of the relationship among data. The difference lies in that correlation analysis is used to mine the correlation among different data items of data records, while sequential analysis is to discover the correlation among different data records [8]. The purpose of sequential analysis is to mine sequential patterns from the transaction database and the maximum sequence meeting the minimum supportiveness requirement specified by the user. That sequence pattern must be the highest sequence.

The mining of sequential patterns is usually done in five steps:

a) Sort phase: Use the transaction body as the primary key and the transaction time as the secondary key, sort the original database and convert into a database of customer sequences;
b) Litemset phase: Discover all large itemsets (L) and map large itemsets into a set of adjacent integers. Each large itemset corresponds to an integer;
c) Transformation phase: Replace each transaction of customer sequences in the database with the litemset (mapped integer) included in the transaction;
d) Sequence phase: Use the litemset to mine the sequential pattern;
e) Maximal phase: Discover the maximal sequence sets of all the sequential patterns.

In the field of intrusion detection, many intrusion activities of attackers are in an order of time and of a time sequence. For example, when a hacker attacks, he/she usually first scans the system ports and then discovers the bugs, followed by intrusion. Therefore, the sequential pattern analysis can be used to mine the relationship among intrusion activities.

3 Functional Design of System Modules

The Winsock2 SPI technology is adopted by the system to intercept network data and display the filtered data on a real-time basis; control rules are capable of filtering hazardous network behaviors; the algorithm of data mining is to extract data from the database, mine them and detect the mined results; the module of statistic analysis is to gather information of the database and feed back statistic results to the user; the module of log query is capable of providing the user with the function of searching for historic records and saving any record in which he/she is interested. See Fig. 1 for the general structure of the system.

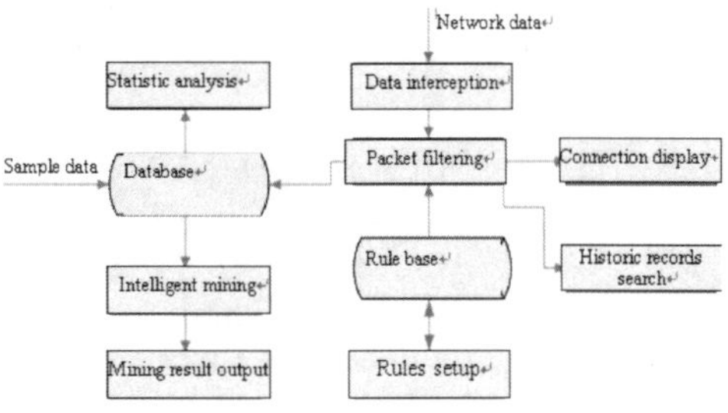

Fig. 1 General Structure of the System

The functional design of system modules is as follows:

a) Module of data interception: intercept the network packets with the functions in Winsock2 SPI.

b) Module of packet filtering: filter the intercepted network packets. It is the core of this design. SPI provides the interface functions for setting up work modes and control rules. According to the work modes and control rules, filtering rules are used to filter passing packets.
c) Module of connection display: provide the monitoring interface of network session (session contents include the session establishment time, session source and destination addresses, session input and output flows, direction and session processes). In case of any blocked package, tips are sent based on the set parameters, and the user may independently set up the work mode (release, rejection or enquiry).
d) Module of historic records search: record network sessions/packets into log files. Packet information can be searched for according to the time range. Saving to another address is available.
e) Rule base: manually add the control rules, such as the manual addition or deletion of IP address ranges and monitoring port ranges; manually add, modify and delete control rules.
f) Database: store attributes and information of sample data and real-time network data, and provide those information for the statistic analysis module and the intelligent mining module.
g) Module of statistic analysis: conduct statistic analysis of data within a given period in the database, and feed back the information in diagrams such as bar charts and pie charts to the user, e.g. the top 10 most-accessed IP addresses within a given period.
h) Module of intelligent mining: extract sample data from the database, apply the improved algorithm to conduct mining treatment of data, extract user behavior characteristics and rules, and combine and update the obtained rules to establish rules. Use the rules and the module of data mining for mining of real-time data, and use the obtained rules for matching to detect whether any anomaly exists.

4 System Implementation

4.1 Design of the Module of Data Interception and Packet Filtering

The module of data interception and packet filtering is the basis of this design. Its accuracy, reliability and efficiency of data interception directly influence the system performance. Winsock2 SPI is a low-layer-system-oriented programming interface provided in the Winsock system components. Upper-layer application programs do not need any concern about the details of Winsock implementation; transparent services are provided for upper-layer application programs. When the user intercepts the data packets, the compiled SPI program is installed into the system. And then, all Winsock requests will be first transmitted to user programs. Without interfering with the proper work of the network system, the program monitors the network communications and conducts security setup according to various filtering rules.

The work procedures of Winsock2 SPI are as follows:

a) Upon the startup of the application program, first load the provider of layered services in the current directory, and initialize the communication channel between user interface and DLL;
b) Upon success of (1), open 527safe.cfg – the control rules setup file under the current directory, and read the work mode, control rules and filtering conditions into the RAM.
c) The user interface, through the channel function, sends interface window handles, control rules and filtering conditions and other elements to trigger the service provider to work properly.
d) The service provider records the passing packets, sends messages to the user interface through channel functions, then notifies the user interface to fetch packets from the service provider, and finally displays information in the network session list of user interface and updates the historic records.
e) The user updates the control rules and filtering conditions; the user interface notifies the rules change of the service provider, works again according to the rules and updates the control rules file.
f) Before exiting the application program, complete the recording of current work mode and the updating of logs.

4.2 Design of the Module of Rule Base

The module of rule base is to set up the filtering rules. Winsock2 SPI is a DLL program, which works between API and DRIVER and provides services to the upper-layer application program. According to its characteristics, the global work procedure for the rule base is prepared. Trough this part, IP addresses, ports, application programs and fixed rules can be set up. See Fig. 2 for the design procedure of control rules. The priority levels of control rules matching in this design are: IP addresses are matched first, followed by the matching of ports and finally the matching of application programs.

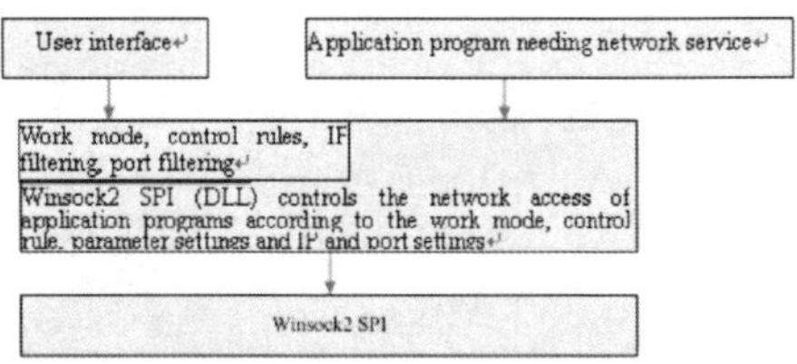

Fig. 2 Design Procedure of Control

4.3 Design of the Module of Intelligent Mining

The module of intelligent mining is to serve the function of offline detection. See Fig. 3 for the procedure of the module of intelligent mining. Its function can be mainly divided into data preprocessing, data mining and intelligent detection.

a) Data preprocessing

Data stored in the database are in a specific data format. They can not be directly processed with the algorithm of data mining, so they must be converted and processed. Data stored in the database contain important attributes of network data, such as the starting and ending time of connection, service type, protocol type, byte numbers from source port to destination port, and byte numbers from destination port to source port.

b) Data mining

After the data format that can be processed with the algorithm of data mining is obtained through preprocessing, conduct data mining; the results of mining are rules expressed in integers. Then, rules like the following are obtained: (service=http, flag=syn), (service =http, flag=syn) → (service =http, flag=syn)[0.86,0.02,2]. The rules express that: when flag is syn, after two https are connected, the possibility of the occurrence of a second same link within 2s is 86%, and the occurrence probability of that mode is 2%. The above mined sequential mode is the frequent segment mode of syn flood. Save sample data into the database; mine the sample data; store the mined rules.

c) Intelligent detection

The function of intelligent detection is to match rules, i.e. matching the output results produced by the module of data mining with the current rules. If the matching is successful, some anomaly is deemed to occur; if no matching is successful, the network is deemed to function properly.

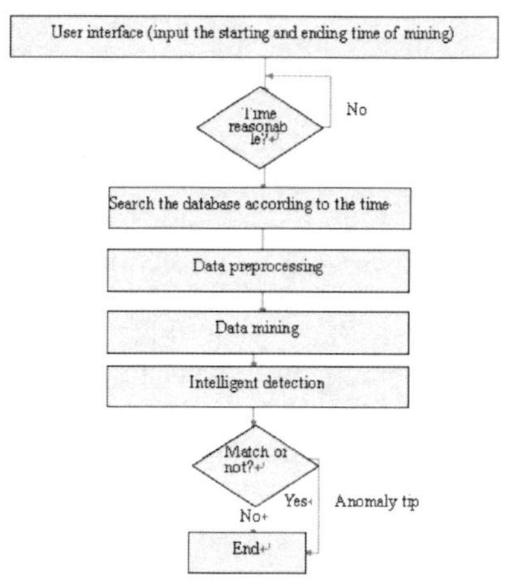

Fig. 3 Procedure of the Module of Intelligent Mining

- Form of rules: In the module of data mining, non-sample data are extracted from the database and preprocessed into the format of integers; mined rules in the format of integers will not be converted into the format of characters and will be directly matched with the standard rules of integer format, thus reducing the length of rules, decreasing the byte numbers to be matched and then somewhat improving the efficiency.
- Matching algorithm: The rules matching of this system differs from the general rules matching. A general rules matching is to match some characteristics among all the data in the network packets, requiring a great amount of computation. The mode in this system is unlike the normal mode. After the extracted information is sorted and preprocessed, its rules have been significantly simplified. Therefore, the proportion of the mode matching part in the system computation load is smaller than the common mode-matching-based intrusion detection system. Rules obtained from data mining in the system are somewhat similar, i.e. many rules character strings are the same. According to the above analysis, the multiple-mode matching AC-BM algorithm is adopted. The BM algorithm is a single mode matching, while the AC algorithm does not adopt the heuristic strategy for hopping, thus leading to poor running efficiency. The AC-BM algorithm, which combines the BM algorithm

and the AC algorithm, may utilize the advantages of the two, and its efficiency is also better than the BM algorithm and the AC algorithm.

5 System Operation Results

The main function of intelligent detection is to conduct intelligent mining of the recorded data so as to discover whether any abnormal information exists. In actual detection, input the desired mining starting and ending time. If the input time is unreasonable, the system will give you tips for re-input; the system will search the database according to the time interval, analyze each set of values for mining, and finally match the mined rules with the original rules to judge whether they match or not. According to the matching results, it will judge whether any anomaly exists in the system. See Fig. 4 for the operation results.

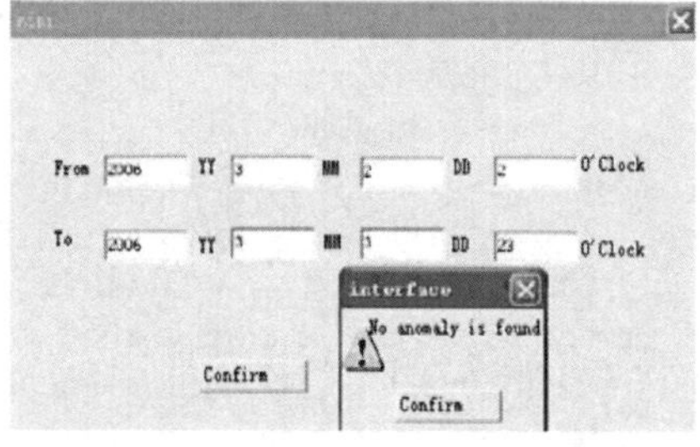

Fig. 4 Interface of Intelligent Detection

The real-time monitoring system of computer network security, as a security protection technology, provides real-time monitoring of the system, intelligently detects the recorded information with the data mining algorithm, and timely discovers any abnormal network behavior. With the increasingly higher requirement of network communication technology security, the intrusion detection system can provide security service from a perspective of three-dimensional, deep and multi-level defense of network security, so it is certain that people will attach more and more importance to it.

References

[1] Nie, X., Zheng, D., Gui, J.: Security Research of Certificate Authentication. Computer Engineering 30(10) (2007)

[2] Xu, H., Zhang, W., Wu, Z.: Study on Intrusion Detection System. Microcomputer Development 13(1) (January 2008)

[3] Zhu, Y.: Windows Firewall and Network Packet Interception Technology. Publishing House of Electronics Industry, Beijing (2006)

[4] Ma, J., Wang, T.: Method of Data Privacy and Integrality in .NET Framework. Aeronautical Computing Technique 25(4) (2006)

[5] Geng, H., Chen, Q.: Research and Pure JAVA Implementation of Web-based Autonomous Mobile Telerobotic System. Journal of Tongji University (8) (2007)

[6] Subramanian, M.: Network Management Principles and Practices. Addison-Wesley (November 2008)

[7] Rose, M., McCloghrie, K.: Strncture and Identification of Management Information for TCP/IP-based Internets (SMI) RFC 1155 (May 2008)

[8] Wijnen, B., Presuhn, R., McCloghrie, K.: View-based Access Control Model (VACM) for the Simple Network Management Protocol (SNMP), RFC 2575 (April 2007)

Classification Algorithm for Filtering E-mail Spams

Lixin Fu and Geetha Gali

Abstract. In this paper we propose an incremental spam e-mail filtering using a modified Naïve Bayesian classification that is simple, adaptable and efficient. Our email spam filter is a hybrid filter that combines the advantages of the various filtering techniques. We also illustrate the effectiveness of our filtering scheme by simulations.

1 Introduction

An unsolicited commercial e-mail from someone without a pre-existing business relationship is called a spam. Most of the spams are commercial emails. Emails are usually classified as hams and spams. Ham emails are legitimate emails we want to receive whereas spam emails are unwanted emails. Most spam filters applied empirical rules to distinguish between ham and spam emails. Typical email anti-spam filters are categorized into origin based filters, traffic analysis based filters, and content-based filters.

The goal of this paper is to propose a model for an incremental spam filtering and test the model using three training schemes: the model trained on which has considered as Whitelist, the model trained on consideration of Bulk email and special characters on receiving an email which is forwarded to more than 30 email ids, and the model trained on calculating the probability of already seen words and unseen words. Our work uses a modified Naïve Bayesian classifier based on a single word representation since it has good performance, simplicity and auto-adaptability. The rest of the paper is organized as follows. The significance of anti-spamming and related work is explained in Section 2. Three layer spam filter is proposed in Section 3. Section 4 illustrates some simulations. Finally, Section 5 concludes and suggests future work.

2 Related Work

There has been phenomenal growth in the number of emails spreading among the internet users recently and so did spam emails. Spam emails consume network

Lixin Fu · Geetha Gali
Computer Science, UNCG, 167 Petty Building, Greensboro, NC 27401, USA
e-mail: lfu@uncg.edu

F.L. Gaol (Ed.): Recent Progress in DEIT, Vol. 2, LNEE 157, pp. 149–154.
springerlink.com

bandwidth, occupy space on the hard disks and also reduce the productivity of recipients at work. Spam filtering can cause significant financial impact on the economy. Ferris Research estimates that spam has cost businesses $35 billion in 2007. And that's only in the US. Globally, the cost would have been approximately $100 billion, according to Ferris research [2 – page 3].

There are many spam filters available but most of the spam filters use Bayesian technique and Content Based technique. Bayesian Filters are helpful after training period, Bayesian filter have to be trained from known ham and spam emails. During this time filter will extract words from spam and ham emails and store them in a database. When analyzing a new message, the message is split into words and each word is given a value [1]. When a central filtering is needed at a central point on the network such as the proxy server or internet router, content based filtering is more useful.

Definitions:
False Positive (FP): A spam filter identifies a legitimate message as spam, then that is the worst kind of error for a spam filter to make.
False Negatives (FN): A spam filter fails to identify a spam as spam message; this is a lesser problem than False Positives. The ideal spam filter has to produce zero false positives and zero false negatives.

3 Three Layer Spam Filter

We propose a personalized hybrid spam filter. A new incoming message is classified by a classifier built from the previous messages. The features of the tested message are added to the training set, so that the features of the training set are always up-to-date.

Three-Layer Spam Filter Technique

If filter receives a message then filter process the mail authentication by checking the sender email address in whitelist database. If email address is found in whitelist then filter will send message to inbox directly by updating the stats file. If not, then filter will check for every significant word in message with keyword probability file and then calculate the p(s) and p(h) ratio. If p(s) > p(h) then filter will decide the message as spam as explained in [4]. If not then filter will proceed to check for special characters in every significant word. Using pattern matching if any significant word matched with the keyword then it is assumed as spam keyword. By this the hybrid filter will decide whether the received message has to go to spam folder or inbox folder.

Terminology:
p (spam): probability of spam; p (ham): probability of ham; M is a message

Algorithm:
Input: Message (e-mail) M
Output: Decision on a message (M) if it is a spam or ham

Three-Layer-Spam-Filter (M):
Initialize: p(spam) = p(ham) = 0.5

```
/* Automatic 3 layer filter */
 if ( senders e-mail in white list )
        updateStats( M, 1 );    send e-mail to Inbox
else
        calcProbRatio( M, p(spam), p(ham) )
        if (  p(spam)  >  p(ham) )  /*  spam e-mail */
              updateStats( M, 2 );     send e-mail to spam
        else  /* ham e-mail */
              updateStats( M, 1);      send e-mail to Inbox
/* User authentication of an e-mail in Inbox */
if ( user authenticates e-mail as spam )
     if ( senders e-mail in white list )
          remove from whitc list
     else  send e-mail to spam
     end
     updateStats( M, 3 ) /* decrement ham counts and increment spam counts */
end
```

Input: Message M
Output: p(spam) , p(ham)
calcProbRatio(M, p(ham), p(spam)):

```
    Initialize p(spam) = p(ham) = 1
                T1 =  # of spam messages
                T2 = # of ham messages
    for (  each significant word in M )  /* ignore insignificant words in M like
as, in, for, to, is, this, etc  */
        if (  any special chars) /* like @, $, |, &, %, *, #  */
            ignore special characters and match the word with the keyword in
keyword list from database and identify the word in database
  /*  eg vi@gra matches Viagra in keyword list */
        else
            if (word is not in the keyword list in the database )
                continue;
            else
                  X1 = keyword_count [word][spam]
                  Y1 = keyword_count[word][ham]
                  p(word1) =  (X1/T1) / (  (X1/T1) + (Y1/T2) )
                  p(word2) =  (Y1/T2) / (  (X1/T1) + (Y1/T2) )
                  p(spam) = p(spam) * p(word1)
                  p(ham)   = p(ham)* p(word2)
```

Input: Message (M), flag
Output: p(spam) , p(ham)
updateStats(M, flag) /* flag = 1 if e-mail is ham
flag = 2 if e-mail is spam
flag = 3 if e-mail in inbox has been authenticated as spam by user */

```
for ( each word in M ) /* only significant word in M */
   if (  word is in keyword list database )
        if ( flag is equal to 3 )
            keyword_count [word][ham] --;  keyword_count [word][spam] ++;
       else keyword_count [word][flag]++;
   else   add_keyword( word, flag )

if (  flag equal to 3 ) /* correct the count */
   ham_count --;     spam_count --;
else if ( flag equal to 1 )     ham_count ++;     total_count ++;
else    spam_count ++;     total_count ++;
```

4 Simulations

Consider a trained database, with a keyword list and message count tables.

Table 1 Keyword list table

Keyword	**Spam_count**	**Ham_count**
viagra	15	1
mortgage	25	4
cheap	10	1
free	20	12
pills	20	6
adult	15	6
dating	20	4
re-finance	15	2
investor	25	5

Table 2 Message totals table

# Messages	# spam	# ham
300	250	50

Table 3 Whitelist

Praveen@gmail.com
mary@uncg.edu
sweety@yaho.com
customerservice@bestbuy.com
John.smith@microsoft.com
scott@live.com

Incoming Message with which looks like the one below:

---------------------------- Original Message --------------------
Subject: RE: Your affordable Mortgage
From: "scott.phillips@best-mortgage.com" scott.phillips@best-mortgage.com
Date: Mon, April 19, 2010 16:47
To: "geethagali@gmail.com"

I hope you enjoined the last mortgage lowban you got from our company. We strongly recommend to re-finance at a che@p 3.4 % rate and decrease your monthly payment. Please visit this link to apply online.

Thank you.
Sandra
Personal Broker Group

Words of interest: re-finance, che@p, mortgage

Step 1: Check if the sender e-mail address, scott.phillips@best-mortgage.com is in white list.

Step 2: We now use out trained tables to calculate the probability of spam and the probability of ham.

Scan each word which are of interest to us (ignore words like I, You, the, are, our, is, hyperlinks, images, non-English alphabets etc) . If we can't find the word in the keyword list table, ignore the word and continue. In our case, the words that are of interest are mortgage, re-finance and cheap. Now calculate p(spam) for the message as follows,

p(spam)=p(re-finance)*p(cheap)*p(mortgage)in spam mails

= ((15/250) / ((15/250) + (2/50))) * ((10/250)/((10/250) + (1/50))) * ((25/250) / ((25/250) + (4/50)) = 0.22

p(ham) =p(re-finance)*p(cheap)*p(mortgage) in Ham mails

= ((2/50) / ((15/250) + (2/50))) * ((1/50)/((10/250) + (1/50))) * ((4/50) / ((25/250) +(4/50)) = 0.058

Note that we ignored special character '@' and matched the word "cheap" from the keyword list table and used its frequency to calculate the p(cheap)

Step 3: since p(spam) > p(ham) , we conclude that this email is a spam

Step 4: we train our filter to update the tables, keyword list and message totals table. We increment the # spam e-mails count in message totals table by one and the # of total emails count by one. We now scan the spam e-mail for any spam keywords and update the keyword list table with the new keywords we find and increment their count by one.

5 Conclusion and Future Work

In summary, in this paper we offer a three-layer comprehensive anti-spamming email filtering strategy that combines whitelist, special character processing, and naïve Bayesian filtering to provide a more versatile and better quality filtering. We plan to investigate other anti-spamming technologies such as SMTP server securing, email authentication and cryptographic signatures to further strengthening email filtering functions.

References

1. Sahami, M., Dumais, S., Heckerman, D., Horvitz, E.: A Bayesian approach to filtering junk e-mail. In: Proc. of AAAI Workshop on Learning for Text Categorization, AAAI Technical Report WS-98-05 (1998)
2. Ferris Research, `http://www.ferrisresearch.com` (last accessed March 31, 2010)
3. Weston, J.: Support Vector Machine. In: Tutorial, 4 Independence Way, Princeton, USA
4. Zhang, L., Zhu, J., Yao, T.: An Evaluation of Statistical Spam Filtering Techniques. ACM Transactions on Asian Language Information Processing 3(4), 243–269 (2004)
5. Issac, B., Jap, W.J., Sutanto, J.H.: Improved Bayesian Anti-Spam filter-Implementation and Analysis on Independent Spam Corpuses, Swinburne University of Technology. IEEE, Kuching (2009)
6. Junejo, K.N., Karim, A.: Automatic Personalized Spam Filtering Through Significant Word Modeling. In: IEEE International Conference on Tools with Artificial Intelligence (2007)

DPEES: DDoS Protection Effectiveness Evaluation System

Haipeng Qu, Lina Chang, Lei Ma, Yanfei Xu, and Guangwei Yang

Abstract. Implemented Distributed Denial of Service(DDoS) protection effectiveness evaluation system (DPEES) in Linux system is proposed in this paper to meet the needs of DDoS attack test and related defense experiments. DPEES can provide a variety of DDoS attack test with different type, intensity and characteristics, and evaluate the results of the attack. This system generates DDoS flows by multiple gigabit network cards, a few hosts with multiple network cards can send a large DDoS attacking flow, simulating a network environment which contains large scale puppet machine in different regions, this will make full use of local resources. Experimental results show that the proposed system can provide convenient environment for the DDoS attack test, defense and the evaluation of the defense effect.

Keywords: DDoS, multiple network adapter, multiprocess, evaluation.

1 Introduction

Distributed Denial of Service（DDoS）is a common attacking type. In DDoS, a large scale of zombie networks in different regions send numerous service requests to the target server, or make the target network flooded with the IP packets. This attack can exhaust the server's resources and block the bandwidth. As a result, the target networks or servers are paralytic and can not provide normal services to legitimate users.

So far, there are a great many products for DDoS defense on the market. How to evaluate the anti-DDoS products' capabilities has become the key issue for users. At present, testing and evaluations for anti-DDoS products are lack of unified and valid methods, because the model of DDoS attack is hard to implement, the hardware requirements is high, and network topology is also complex. Now there are many

Haipeng Qu · Lina Chang
College of Information Science and Engineering,
Ocean University of China, Qingdao
e-mail: quhaipeng@ouc.edu.cn, chang4979@yahoo.cn

F.L. Gaol (Ed.): Recent Progress in DEIT, Vol. 2, LNEE 157, pp. 155–161.
springerlink.com

stress testing tools that are widely used, but their performances are varied, just sending simple TCP or UDP packets. The fake packets are not flexible to look like as true as the normal ones. The stress testing tools are not able to measure the result of the test results. Some tools can be only applied in the traditional networkk but can't well satisfy the under gigabit network environment. At home some companies have to buy some special but very expensive generators from other foreign country to simulate the DDoS flows. Some models have been analyzed[1] and raised such as DDoS-DATA[2] and some improvement in Deter/EMIST[3]. In this paper DDoS protection effectiveness evaluation system was proposed under the Linux system. This system provides a variety of DDoS attack test with different type, intensity and characteristics, evaluates the results of the attack, and also supports graphical interface, user management, audit logs, and other extension modules.

The rest of this paper is organized as follows. The design and describe of DPEES in section 2 and section 3. The analysis of the performance of the flow generator is described in section 4. An evaluation experiment is shown in section 5.

2 The Design of DDoS Protection Effectiveness Evaluation System

2.1 The Goal of DPEES

DPEES is designed to provide convenience when testing some anti-DDoS products. To reduce the cost of the hardware, this system use multiple gigabit network cards to send a large scale DDoS packets, simulating a real DDoS environment. Considering the system should be secure, ontrolled, extensible and simple, the design has to follow the tips as follows:

DPEES can flexibly control the whole attack process, such as setting the type of DDoS, the flow rate, the numbers of the gigabit network card used to send packets, and the degree to fake IP address.

a) DPEES can sign the DDoS packets but not affect the testing result. This will contribute to analyze the capability of the anti-DDoS products, and make an accurate evaluation.
b) DPEES can provide the service for user management, operational control and audit function to prevent the unauthorized users' operation.
c) DPEES can offer a graphical interface for setting and showing testing results, which is easy to understand for users.

2.2 The Structure of DPEES

DPEES is mainly composed by main controller, flow generator and the analyzer three parts. Main controller is designed to support both Linux and windows platform; flow generator and analyzer are completed in Linux environment. The three parts of DPEES is shown in Figure 1:

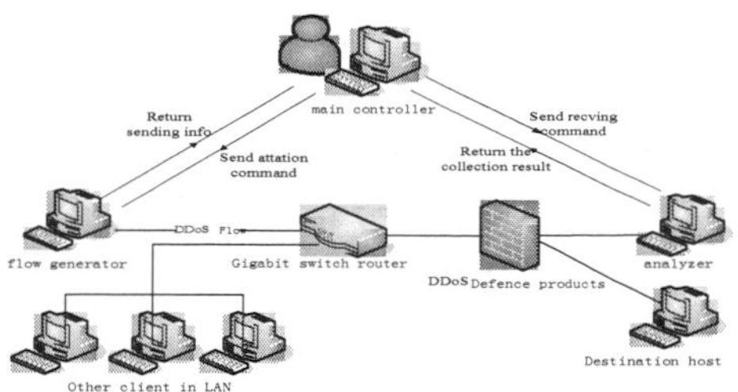

Fig. 1 The structure of DPEES

The main controller sets the behavior of the flow generator and analyzer. The main controller can customize DDoS packets and the DDoS flows, the start attacking time when to start attacking, and some other properties like these. Then the main controller sends these setting to the flow generator, and also notifies the analyzer some related settings about the attacking time and so on, so the analyzer can analyze the anti-DDoS product's performance in time. When the attacking ends, the flow generator sends the message about the whole sending process to the main controller while the analyzer sends the collected data to the main controller. And the main controller compares data from the two and then concludes the evaluation results.

3 The Design and Implement of Submodule

3.1 Main Controller

The main controller manages the whole attacking process, and evaluates the attacking result according to the data send by other two parts. The main controller adopts the B/S structure, supports Linux and windows system. The main controller is divided into four sub modules:

User management: DDoS testing is easy to bring adverse effect to the network. In order to prevent unauthorized users from starting DDoS attacking, the main controller's user management bases on different roles, and has a strict user authentication.

Packet settings: main controller can set the flow analyzer what packets to send, such as attacking packet type(syn flood, icmp flood, udp flood, ect), attacking flows, attacking packet size and flags, the degree to fake IP address written in the packet, the numbers of network card used to attack, the port to attack and so on. Through these setting, user can flexibly start a DDoS attack, and simulate a real DDoS attacking environment.

Evaluating module: After a DDoS attack, flow generator and analyzer both return their own data about this attack. Using the data, evaluating module can conclude a judgment and send the result to the graphical interface.

Audit log: This submodule is designed to record the users' behavior that enters the DDoS protection effectiveness evaluation system.

3.2 Flow Generator

Flow generator is used for generating DDoS attacking packets. There are many tools to simulate DDoS attacks, such as pktgen under Linux system, open source software like tcpreplay, stream3o and TFN2K. Although these tools can satisfy some needs of simulating DDoS attacks, the packets they send are simply faked, with a few functions and bad expansibility .Any more, they require quite a number of computers to send the DDoS packets, the hardware requirement is very high.

The key point when designing DDoS protection effectiveness evaluation system is the flow generator. The flow generator is designed under the Linux environment in C language, and it can provide the functions as follows:

a) In order to reduce the hardware requirements, flow generator can use multiple network cards in a host computer to create a large flow. Also the generator can detect the multiple network cards automatically, including their IP address, MAC address and their active states. All this information will be returned to the main controller, and users can choose which network cards to use.
b) Flow generator can simulate most of the known DDoS attacks, like network layer attacks (ARP, ICMP, IGMA flood)transport layer attacks (syn flood, TCP ACK flood), application layer（HTTP,DNS flood）and so on.

3.3 The Analyzer

The analyzer mainly collects information behind the anti-DDoS product. Sometimes if the packets are marked, the analyzer can collect the information about the marked packets received on the analyzer side. After the attack stops, the analyzer returns the information to the main controller. If the analyzer is installed on the victim host, then the analyzer can analyze the states of the host, such as the response speed of the TCP and HTTP connection. Considering that the DDoS attacks large flows, the analyzer uses the zero-copy technology to captures the packets in gigabit environment.

In order to evaluate the anti-DDoS product effectively, the analyzer need to collect data as follows:

- Not-filter numbers: the number of the packets that are not filtered by the anti-DDoS products.
- Mis-filter numbers: the number of normal packets discarded by the anti-DDoS products.
- Throughput rate: through the flow generator sending DDoS packets with different flows to test anti-DDoS products' throughput limit.

4 The Performance of the Flow Generator

The key point of the performance of DDoS-protection effectiveness evaluation system is the performance of the flow generator. This paper describes the performance of the flow generator in detail by conducting some experiments. In the experiment, four hosts are employed, one is quad-core host with two PCI-E gigabit network cards, one is used for the analyzer, one is for the victim host and the last one is the main controller. The main controller and the victim host is under the windows environment, the other two are under Ubuntu 9.10 environment. The victim host use apache as the http server.

Syn flood, UDP flood, icmp flood and smurf flood are four normally used DDoS attacks. In experiments,different types are adopted according to the condition. Some tests with low requirement about data mainly adopt the syn flood attack.

The flow generator uses multithread to send large scale of DDoS attack packets. In order to make the best use of signal host, this experiment started syn flood to observe the relationship of the numbers of the threads and the utilization rate of the CPUs.

Table 1 The relationship between threads and CPUs‘ utilization rate

Thread numbers	cpu1	cpu2	cpu3	Cpu4	Spedd (Mb/s)
1	8%	8%	8%	96%	1. 3
2	93%	93	10	10	2. 6
3	93%	93%	50%	50%	3. 6
4	90%	90%	90%	90%	4. 6
5	62%	62%	62%	62%	2. 1
6	60%	60%	60%	60%	2. 1
7	60%	60%	60%	60%	2. 1
10	60%	60%	60%	60%	2. 1

Judging from the Table 1, the efficiency of CPUs do not increase together with the numbers of the threads. When there are four threads, the CPUs' utilization is highest, at the rate of 90%; when the numbers of threads is less than four, the jobs are focused on one or two cups, and the utilization of the other cups are below 50%, even no more than 10%; when the number is more than four, even ten, the average utilization of each CPU is 60%, has no big difference, but each one is not fully used. Four is just the number of the CPUs. Conclusion from these experiments is that, the best number of the threat is the number of the CPUs. So the flow generator can automatically exam the information about the host, and set the number of the thread according to the number of the cups. In the following experiments, the number of threads is four in default to make the result best.

5 Experiments and Results Analysis

This experiment tests the resistance to syn flood attack of Linux system, Windows XP system and the HFC algorithm.

There are four hosts inthis experiment. One is used for starting the syn flood attack, other two are installed Linux system and Windows XP system; the last one is install Ubuntu 9.10 and implemented with the HFC algorithm. A gigabit router is used to connect the attacker and the victim machines.

Because the syn packet is fixed size, in this experiment, so only part of the content in the packet need to be faked, such as the source IP address, SEQ bit and TTL bit, and so on .The flooding speed is about 100000 packets /s .

Table 2 The different states of different system when suffering SYN flood attacking

	The ave utilization	Response time	Effect of response(%)
Syncookies	85	60ms-3s	0, 20. 2v-
Window xp	100	No response	No response
HFC	70	45ms-4s	6. 5v+, 11. 4v-

In the above Table 2 , v+ stands for the not-filter rate, and v- stands for the mis-filter rate. There may be a big different result if the HCF's implement varied a little. So, here the experiment just makes a simple test. It shows that Linux system has a better resistance than windows system, because syncookies do not need some buffers to cache information, only if the CPU is still working, the requests from the client could be response. The HCF algorithm is better than syncookies again, but this algorithm has a high mis-filter rate, still need to be improved.

6 Conclusion

A DDoS protection effectiveness evaluation system DPEES is proposed and implemented in this paper. This system can simulate kinds of DDoS attacks and evaluate the performance of anti-DDoS products. The experiments in this paper prove that DPEES can decrease hardware costs and offer convenience for DDoS simulation. But this system is still need to be improved, make fuller use of the CPUs and network cards, and enhance the rate of sending packets. Also there are still some details need to be completed about how to evaluate a product, the analyzer of the DDoS protection, an effective evaluation system needs a better improvement.

References

[1] Hardaker, W., Kindred, D., Ostrenga, R., et al.: Justification and Requirements for a National DDoS Defense Technology Evaluation Facility. Network Associates Laboratories, Tech. Rep.: 02-052 (2002)

[2] Blackert, W.J., Gregg, D.M., Castner, A.K., et al.: Analyzing Interaction Between Distributed Denial of Service Attacks and Mitigation Technologies. In: Proceedings of the DARPA Information Survivability Conference and Exposition. IEEE Press, Washington D.C. (2003)

[3] Benzel, T., Braden, R., Joseph, A., et al.: Experience with Deter: A Testbed for Security Research. In: Proceedings of the 2nd International Conference on Testbeds and Research Infrastructures for the Development of Networks and Communities, Barcelona, Spain: [s. n.] (2006)

[4] Jonathan, L.A.: Resisting SYN: Food DoS Attacks with a SYNCache. In: Proceedings of the BSD Conference, San Francisco, CA, USA: [s. n.] (2002)

[5] Jin, C., Wang, H., Shin, K.G.: Hop-count filtering: an effective defense against spoofed DDoS traffic. In: Proceedings of 10th ACM Conference on Computer and Communications Security (2003)

Host Based Detection Approach Using Time Based Module for Fast Attack Detection Behavior

Faizal Mohd Abdollah, Mohd Zaki Mas'ud, Shahrin Sahib, Asrul Hadi Yaacob, Robiah Yusof, and Siti Rahayu Selamat

Abstract. Intrusion Detection System (IDS) is an important component in a network security infrastructure. IDS need to be accurate and reliable in order to detect the intrusive behaviour of a packet that travelling through the network. With the current technological advancement attack on network infrastructure has evolve to a new level and to make IDS sensitive enough to detect the new attack, the detection framework need to be frequently updated. Both the fast attack and slow attack mechanism has become the subset of phases inside the anatomy of attack. Each of the attack mechanism has their own criteria and fast attack is the important type of attack that need to be considered as any late detection of the fast attack can cause a major bad impact to the organization. Therefore, there is a need to identify a suitable technique to detect the fast attack and based on this, this paper introduce a static threshold using statistical and observation technique for detecting the fast attack intrusion that is within one second time interval. The Threshold selected was based on the real network traffic dataset and verified using classification table on real network traffic.

1 Introduction

In the last decade, there has been a revolution in the wired and the wireless networking. The revolution also changes the attack mechanism to exploit the network infrastructure. The exploitation of the network has becoming quite alarming

Faizal Mohd Abdollah · Mohd Zaki Mas'ud · Shahrin Sahib · Asrul Hadi Yaacob ·
Robiah Yusof · Siti Rahayu Selamat
Faculty of Information and Communication Technology, Univeristi Teknikal Malaysia,
Hang Tuah Jaya, 76100 Durian Tunggal, Melaka
email: faizalabdollah@utem.edu.my, zaki.masud@utem.edu.my,
shahrinsahib@utem.edu.my, asrulhadi.yaacob@mmu.edu.my,
robiah@utem.edu.my, sitirahayu@utem.edu.my

F.L. Gaol (Ed.): Recent Progress in DEIT, Vol. 2, LNEE 157, pp. 163–171.
springerlink.com

especially with the help of the freely available attack tools on the Internet [1]. With the availability of the attack tools, novice attacker can launched a sophisticated attack with just a little bit knowledge and as a result, the growth of the incident reported due to the security breach by NISER [2] has roughly parallel with the evolution of the Internet. Thus, it is necessary to protect vulnerable machines from being compromised. One method to secure the machine is by implementing security mechanism such as IDS, where it can reduce the possibilities of security breach from happening inside the organization [3]. Consequently the confidentiality, integrity and availability of the organization properties can be protected.

Understanding the anatomy of an attack is important before developing an IDS. An attack can be disserted into 5 phases which are reconnaissance, scanning, gaining access, maintaining access and covering tracks [4]. The first two are initial stages for the attacker getting information from the potential vulnerable machines. These phases can be categorized into two, which are fast attack and slow attack. Fast attack can be defined as an attack that uses a large amount of packet or connection within a few seconds [5, 6, 7]. Meanwhile the slow attack can be defined as an attack that takes a few minutes or a few hours to complete [8]. Detecting the fast attack is very useful to prevent any early attack on the network and may help to reduce the possibilities of further actions such as gaining access, maintaining access and covering tracks. Zhang and Leckie [9] also stated that there is a strong need to detect the intrusion activity such as scanning which can be classified as fast attack as quickly as possible inside the network. Unfortunately, fast attack detection technique required a suitable threshold mechanism to increase the detection accuracy. Selecting suitable technique still become a major issues need to be tackle by the IDS developer since it is widely used by the current IDS development [10, 11].

To overcome this challenge, this paper focuses on selecting a threshold mechanism based on one second time interval in detecting the fast attack. By introducing this threshold, it helps reducing the false alarm generated by the IDS and at the same time increase the accuracy of detection. The rest of this paper is organized as follows. Section 2 discusses the related work in detecting the fast attack intrusion. Section 3 discusses the methodology used to select the suitable threshold based on one second time interval. Next, Section 4 presents the result and analysis. Finally, Section 5 presents the conclusion and possible future extension of the work.

2 Related Works

Intrusion detection can be divided into three types which are host based intrusion detection system, network based intrusion detection system and hybrid based intrusion detection system. Although the intrusion detection can be divided into three types, the main goal for each of them is the same which is intrusion detection. Intruder detection system is a system to detect attacks, or to classify them as unwanted authorized login, regardless of their success [12]. The detection method used by intrusion detection system can be classified as anomaly based detection and signature based detection. Signature-based IDS is also known as misuse detection approach IDS [13]. Signature based system is a system which contains a number of attack description or signatures that are matched against a stream of audit data looking for evidence of modeled attack [6]. The audit data can be gathered from network traffic or an application log. Meanwhile, the anomaly based system identifies the intrusion by identifying traffic or application which is presumed to be different from normal activity on the network or host [14]. Both of these approaches have their own disadvantages. False alarms generated by both systems are a major concern and is identified as a key issue and the cause of delay of further implementation of reactive intrusion detection system. Although both system have their own drawbacks, anomaly based detection has capabilities to recognize new attack inside the network without a need to update new rules [15]. This capability requires appropriate value of threshold to distinguish between the normal and abnormal behavior of the fast attack activity. Thus, introducing a fast attack threshold based on the observation technique to select an appropriate threshold is required to reduce the false alarm generated by the anomaly based detection for the fast attack detection.

There are two techniques that can be used in selecting the appropriate threshold to distinguish between the normal network traffic and abnormal network traffic. The techniques are static threshold value and dynamic threshold value. This research will focus on static threshold value because selecting static threshold is very useful to prevent the intrusion activity before the attacker begins to launch the attack [16]. In identifying the static threshold, there are multiple techniques used by previous researcher. Hussain et al, [17] used 60 connections per second from source IP address as one of the criteria to identify the intrusion. The selection of the threshold which is 60 connections per second was purely based on the observation. The detailed process of the observation is not clearly stated in this research. Furthermore, we argue that the selection of 60 connections per second can be further delay for the fast attack detection technique. Therefore, this research insists that selecting the suitable threshold within one second time interval may help to detect the intrusion activity as soon as possible. Kanlayasiri et al, [18] also used static threshold mechanism in identifying the portscan activity. The researcher used rule-based approach combined with static threshold to identify the attacker who launched the portscan attack. The threshold was set to 20 connections and does not use time interval to calculate the threshold. Therefore, the threshold value used the slow attack approach to identify the attack. Furthermore, the selection of 20 connections per second as a threshold is based on the observation and the process of the observation is not stated clearly. The researcher also suggests that the threshold selected can be adjusted manually. KDDCUP99 [19] has introduce time based feature to detect the fast attack intrusion activity. Unfortunately, the feature construct in KDDCUP99 used 2 second time interval where this research used one second time interval to detect the fast attack intrusion activity. The difference of one second between KDDCUP and proposed threshold give a good contribution in fast attack detection system.

Gates and Damon also used static threshold mechanism in identifying the attacker [20]. The researcher used mean and standard deviation from a normal data of a host to distinguish between the normal and abnormal data. The mean and standard deviation was computed using observation data for one week. The system will raise an alarm if the packet exceeds two standard deviation. This technique had a weakness due to the higher threshold value selected in this research. Therefore this research will introduce a threshold value based on one second time interval to detect the fast attack intrusion activity. It will use observation technique and classification table approach to validate the result. Furthermore the threshold selected will be tested using a various attack tool for further validation process. The next section will discuss about the methodology used for the observation technique and classification approach.

3 Methodology

A new framework for fast attack detection is necessary prior to detection of the fast attack intrusion can be done. For that reason, the framework together with a suitable fast attack detection module which consists of time based module has been developed [6, 7]. The time based detection module was based on one second time interval which is the main objective of the research. This research is based on one set of data which is Darpa99[21] and it was used in this project as a reference to identify the normal behavior of the network traffic especially on selecting the normal threshold of the network traffic.

Fig. 1 illustrated the process flow of the threshold selection using the statistical and observation technique on real network traffic dataset. The threshold is computed based on the normal and abnormal network traffic captured from one of the agencies and then verified using classification table. Based on the statistical and observation approach, we manage to conclude that maximum value that the normal behavior of host connection is 3 connections per second. Therefore we select 4 connections per second as abnormal behavior in detecting fast attack focusing on host based malicious activity. After the suitable threshold is choose, the result is verified using the classification table to prove that the threshold selected is suitable in detecting the fast attack intrusion activity. The next section will discuss about the result and analysis of the threshold selection.

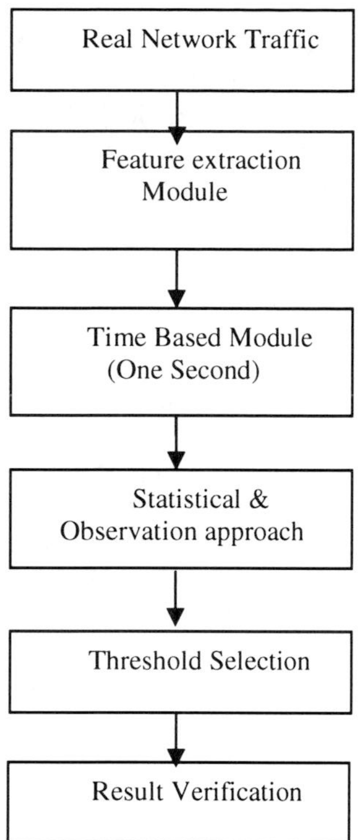

Fig. 1 Threshold selection technique

3.1 Logistic Regression Model

Binary logistic regression is a form of regression which is used when the dependent variable is a dichotomy and the independents are any type [22]. The logistic regression can be used in order to achieve one of the two main objectives which are explanation or prediction. The explanation will reveal the significance of the variable in predicting the outcome variable. Meanwhile, in the prediction perspective, it will predict the normal or abnormal behavior of the event. In this research, we will adapt both the objectives which are explanation or prediction. Using the explanation objective, the influence of the feature in detecting the fast attack can be revealed. Meanwhile, the prediction objective will focus on identifying the appropriate threshold for detecting the fast attack.

Before selecting the suitable threshold from the logistic regression model, identifying the fitting of the model in necessary. The purpose of determining the fit of the logistic model is to assess how effectively the model describes the outcome variable [23]. If the model fits, then a conclusion can be made that the model is suitable and is good in predicting the outcome variable. Therefore the accuracy of the detection also becomes higher. There are two tests can be used to assess the fit of the model. The first test is called likelihood ratio test [22]. The likelihood ratio test can be used on two purposes which are to test the fit of the model and to test the contribution of the individual predictor. The likelihood ratio test is also called model chi-square test. The model chi-square test will test the difference between -2LL (-2 Log Likelihood) for the full model and -2LL for the initial chi-square in the null model. The null model is also called the initial model which involves only the constant.

Besides using the likelihood ratio test, the percentage of correct prediction is used to assess the model. The percentage of the correct prediction can be interpreted by using a classification table. The classification table is the most appropriate test if the test objective is based on the classification [23]. Therefore the classification table is chosen as one of the test used to assess the model. Using the classification table, the percentage of the detection attack rate and detection normal rate can be calculated. Furthermore, error rate can be calculated in the classification table also. The error rate of the classification table can be divided into two categories which are false positive and false negative. False positive means that the number of errors in which a normal event is considered as an attack event. Meanwhile, false negative means the number of errors in which the attack event is predicted to be normal, but is in fact an attack. Below is an example of the classification table. The calculation of the detection rate and error rate is also shown below.

Table 1 Example of the Classification Table

Classified		Predicted	
		Normal	Attack
Observed	Normal	a	b
	Attack	c	d

$$\textit{Detection Attack rate} = d/(c+d) \tag{1}$$

$$\textit{False Positive} = b/(b+d) \tag{2}$$

$$\textit{Detection Normal rate} = a/(a+b) \tag{3}$$

$$\textit{False Negative} = c/(a+c) \tag{4}$$

$$\textit{Overall Detection rate} = (a+d)/(a+b+c+d) \tag{5}$$

The above calculation will be used in the next subsection for assessing the model based on the classification table. The detailed explanation on assessing the model is presented below.

4 Results and Analysis

The approach in this research is verified using real time network traffic captured from one of an agency in Malaysia. From the network traffics there are 105 connections has been declare as normal connection while 108 connection is an abnormal connection. The result analysis is based on the classification table generated by the logistic regression model. The model also suggested the 3 connection per second as a suitable threshold for detecting fast attack intrusion activity focusing on host based attack. For the Classification table, there are two model involve which are null model and full model. The Null Model is a model which has only a constant value without any mechanism to distinguish between the attack and normal connection. Meanwhile the Full Model is model generated after the predictor is involved in the detection. The predictors include focus on the number of connection based on one second time interval. Table 1 and 2 show the result of the Null Model and Full Model.

Table 2, shows that the null model managed to predict 50.7% correctly in classifying the overall percentage but the false positive was also very high at 49.3%. Thus, the model was useless in detecting the normal network traffic and will affect the network security in the organization in terms of massive log generated from the model. The massive log gave extra burden to security administrator in verifying the intrusion activity.

Moreover, the model did not have capabilities to detect the normal event because using constant, the model assumed most of the data were attack. After the predictor was included inside the model, the detection accuracy was high and reduced the false alarm as depicted in table 3. The model had capabilities to predict 98.1% correctly in classifying the attack and only 1.9% false negative generated from the full model. Meanwhile for the normal data, the full model was able to predict 99% correctly and false positive is 0.9%. The model had better prediction and had capabilities to distinguish the difference between the attack and normal traffic. Furthermore, the overall percentage of the classification table for the null model was 50.7 %. The result of the overall percentage increased to 98.6 % after the full logistic regression model was applied to the data. As a conclusion, the increase in of the correct percentage for the classification between the attack and normal indicate that the model is suitable, fits and good in predicting the normal and abnormal behavior. Therefore, it validate the result that the model generated from the logistic regression suggested that 3 connection per second can be used to detect the intrusion behavior especially for fast attack detection system.

Table 2 Null Model

Classified		Predicted	
		Normal	**Attack**
Observed	**Normal**	0	105
	Attack	0	108

Detection Attack rate = 0%, False Negative = 6.4%, Detection Normal rate = 100%, False Positive = 49.3%, Overall Detection rate = 50.7%

Table 3 Full Model

Classified		Predicted	
		Normal	**Attack**
Observed	**Normal**	104	1
	Attack	2	106

Detection Attack rate = 98.1%, False Negative = 1.9%, Detection Normal rate = 99%, False Positive = 0.9%, Overall Detection rate = 98.6%

5 Conclusions and Future Work

In this research we manage to select the suitable threshold for detection host based attack for fast attack intrusion activity. Although the selection of the threshold was based on the observation technique but the classification table approach have proven the threshold can be used to detect the intrusion behavior. Additionally the detail process of the observation technique also has been reveal which most of the previous research do not insist to do so.

For future work, the researcher would like to identify the new technique to select the suitable threshold for fast attack detection system. The future technique will be implemented on real network traffic from various sites to validate the result. In addition, the timeliness parameter also will be assessed to identify the accurate threshold for fast attack detection.

Acknowledgments. This research has been supported by Universiti Teknikal Malaysia (UTeM) Melaka and Malaysia Government under FRGS Fund.

References

1. McHugh, J., Christie, A., Allen, J.: Defending Yourself: he Role of Intrusion Detection System. In: Proceeding of IEEE, Software (2000)
2. Niser (2008), `http://www.niser.gov.my`
3. Microsoft, Ruth, A., Hudson, K.: Security + Certification: CompTIA Exam SYO-101. Microsoft Press, USA (2003)
4. Module for CEH (2009)
5. Lazarevic, A., Ertoz, L., Kumar, V., Ozgur, A., Srivastava, J.: A Comparative Study of Anomaly Detection Schemes in Network Intrusion Detection. In: SIAM International Conference on Data Mining (2003)

6. Faizal, M.A., Asrul, H.Y., Shahrin, S.: An Earlier Detection Framework for Network Intrusion Detection System. In: Proceeding of the Second International Conference on Advances in Information Technology, Bangkok, November 1-2 (2007)
7. Abdollah, M.F., Yaacob, A.H., Sahib, S.: Improved Fast Attack Detection Model for Network Intrusion Detection. In: Proceeding of International Conference on Engineering and ICT, UTeM (2007)
8. Lee, W.: A Data Mining Framework for Constructing Feature and Model for Intrusion Detection System. PhD thesis University of Columbia (1999)
9. Zhang, D., Leckie, C.: An Evaluation Technique for Network Intrusion Detection Systems. In: Proceeding of the First International Conference on Scalable Information Systems, Hong-Kong (June 2006)
10. Bro (2009), `http://www.bro-ids.org`
11. Snort (2009), `http://www.snort.org`
12. Allen, J., Christie, A., Fithen, W., Mc Hugh, J., Pickel, J., Stoner, E.: State of the Practice on Intrusion Detection Technologies. Technical Report on Networked Systems Survivability Program. University of Carnegie Mellon, Pittsburgh, USA (2000)
13. Levitt, K.: Intrusion Detection: Current Capabilities and Future Directions. In: Proceeding of the 18th Annual Computer Security Applications Conference. IEEE (2002)
14. Wang, Y., Huang, G.X., Peng, D.G.: Model of Network Intrusion Detection System Based on BP Algorithm. In: Proceeding of IEEE Conference on Industrial Electronics and Applications. IEEE (2006)
15. Tandon, G., Chan, P.K.: Weighting versus Pruning in Rule Validation for Detecting Network and Host Anomalies. In: Proceeding of KDD 2007 Conference. ACM, USA (2007)
16. Idika, N., Mathur, P.A.: A Survey of Malware Detection Technique. In: Proceeding of Software Engineering Research Center Conference, SERC-TR286 (2007)
17. Hussain, A., Heidermann, J., Papadopoulos, C.: A Framework for Classifying Denial of Service Attacks. In: Proceeding of 2003 ACM SIGCOMM, Germany (2003)
18. Kanlayasiri, U., Sanguanpong, S., Jaratmanachot, W.: A Rule Based Approach for Port Scanning. In: Proceeding of Electrical Engineering Conference, Thailand (2000)
19. KDDCUP99 dataset (2009), `http://kdd.ics.uci.edu/databases/kddcup99/kddcup99.html`
20. Gates, C., Damon, B. (Cpt.): Host Anomalies from Network Data. In: Proceeding from the Sixth Annual IEEE SMC (2005)
21. Darpa99 (2009), `http://www.ll.mit.edu/`
22. Field, A.: Discovering Statistic Using SPSS, 2nd edn. Sage Publication, London (2005), Schuyler W.Huck
23. Hosmer, D.W., Stanley, L.: Applied Logistic Regression, 2nd edn. John Wiley and Son Inc., USA (2000)

Virtual Machine Based Autonomous Web Server

Mohd Zaki Mas'ud, Faizal Mohd Abdollah, Asrul Hadi Yaacob, Nazrul Muhaimin Ahmad, and Erman Hamid

Abstract. Enterprises are turning to Internet technology to circulate information, interact with potential customers and establish an e-commerce business presence. These activities are depending highly on Web server and maintaining good server security has been a requirement for avoiding any malicious attacks especially web defacements and malware. Web server administrators should be alert and attentive to the status of the server at all time. They need to be persistent in monitoring the server in order to detect any attempted attacks. This is an advantage for a web server that is maintained by a big company that has a big budget to hire a knowledgeable web server administrator, for a new established small company it will only burden their expenses. To overcome this problem, this paper proposes a low cost system called Autonomous Web Server Administrator (AWSA) that is fully developed using open source software. AWSA combines several computing concepts such as Virtual Machine, Intrusion Detection System and Checksum. AWSA offers a Virtual Machine based Web server that has the ability to automatically detect intrusions and reconstruct corrupted data or the file system without any human intervention.

1 Introduction

Today, the current explosion in Internet trading may have eventually contributed to cast doubt on the trustworthiness of the services. Trust is at the heart of each and every transaction of online businesses. To establish the trust, the company not

Mohd Zaki Mas'ud · Faizal Mohd Abdollah · Asrul Hadi Yaacob ·
Nazrul Muhaimin Ahmad · Erman Hamid
Faculty of Information and Communication Technology, Univeristi Teknikal Malaysia, Hang Tuah Jaya, 76100 Durian Tunggal, Melaka
e-mail: zaki.masud@utem.edu.my, faizalabdollah@utem.edu.my, asrulhadi.yaacob@mmu.edu.my, nazrul.muhaimin@mmu.edu.my, erman@utem.edu.my

F.L. Gaol (Ed.): Recent Progress in DEIT, Vol. 2, LNEE 157, pp. 173–182.
springerlink.com

only needs to persistently declare the infrastructure is secure but also needs to have the tools to let the customers establish trust by themselves.

As online business sites become attractive targets for the customers, Phishing scams, identity fraud and website spoofing have become some of the most lucrative business models of today's cyber-criminals. According to Malaysian Computer Emergency Response Team (MyCERT) the total number of DDOS and intrusion incidents report are increasing from 60 reports in September 2000 to 1661 reports in September 2010 [1]. Nowadays, anyone can attack Internet sites using readily available intrusion tools and exploit scripts that capitalise on widely known vulnerabilities.

Despite stiff challenges from attackers, the influenced of World Wide Web to broaden market potential across geographical boundaries has led major companies or even Small Medium Enterprises (SMEs) to seriously expand their operations and marketing strategies over the Internet. However, a major concern of these companies is to put aside a huge investment on the security system in order to build trustworthiness among customers and to protect its image.

For that reason this paper presents a novel framework of Autonoumous Web Server Administrator (AWSA), a Virtual Machine (VM) based WWW Server, to enhance the trustworthiness of online businesses. It proposes the combination of an Intrusion Detection System (IDS) and file system integrity to detect and prevent attacks against services running on a VM and maintaining the services after the attacks. In summary, AWSA offers the following advantages: (1) the ultimate solution for SMEs to get involved in online business. All AWSA components utilize open source applications which are relatively low cost, scriptable and modifiable, (2) to maintain the services running on a VM from outside (host system), thus keeping the IDS safe, out of reach from intruders, (3) to detect any malicious data modification that the intruders have made through file system integrity checking, and (4) maximizing the state of the VM which can be saved, cloned, encrypted, moved or restored. It allows AWSA to re-construct the corrupted disks or files on the VM by transferring duplicated data from the host system.

The rest of the paper is structured as follows. Section 2 describes the Virtual Machine (VM) concept. Section 3 presents the architecture of a self-protected AWSA. Section 4 explains the details of our implementation of AWSA and discusses the prototype performance. Section 5 concludes the paper with a brief summary.

2 Virtual Machine

2.1 Virtual Machine Environment

A Virtual machine (VM) is defined as an efficient and isolated duplicate of a real machine [2]. This environment is created by using a Virtual Machine Monitor (VMM) which provides a second layer on a machine for another operating system to run on. VMM reproduces everything from the CPU instructions to the I/O devices in software of the operating system it runs on. Virtualization in a VM involves mapping of virtual resources, for example, the registers and memory, to real hardware resources. It also uses the host machines instructions to carry out the actions specified by the VMM. This is done by emulating the host Instruction Set Architecture (ISA).

The underlying operating system is called the host operating system [3]. VMM runs on top of the host operating system. Hence it can hold a second or more operating system on it. The operating system which runs on VMM is called the guest operating system. Fig. 1 illustrates the virtualization concept of VM. The host operating system and the guest operating system can be either the same or different type of operating system. For instance, Windows can run as the host operating system and Linux as the guest operating system.

VM and virtualization research have been going on for nearly thirty years; IBM VM/370 was among the first VMM applications introduced. VMWare[4], Virtualbox[5], and Xen[6] are some examples of more recent VMMs that have been developed. Another close environment which also provides virtualization is called an emulator. The difference is that the emulator only let one guest operating system to run on the host system. An example of an emulator is Qemu [7], which is used in AWSA development.

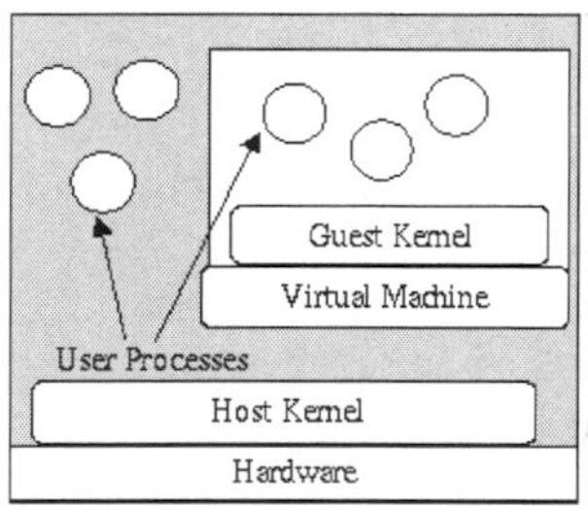

Fig. 1 Virtual machine environment

2.2 *Advantages of Virtual Machine*

VM technology is widely used to run operating systems or software developed specifically for one platform on another platform. For AWSA, this can be manipulated to increase a server security, others are exploited as below:

Compatibility: VMM can run any operating system on top of it without any modification being.

Isolation: VMM provides a complete layer of virtual hardware to the guest operating system, the guest operating system communicates directly with the hardware without going through the host operating system, thus providing a strong isolation between the host and guest operating system. This indeed will protect the host operating system and any host application from being manipulated by an intrusion on guest operating system.

Security: VMM is written in a few lines [7], which make it a simple program. Due to its small code size, VMM provides a better trusted computing base than the guest operating system itself. This narrow interface restricts intruder actions from invading the host operating system.

Inspection: VMM can have access to any operations in the virtual machine; CPU instructions, all memory actions, I/O device and I/O controllers operations. This feature will let the host operating system to view any operations taking place in the guest operating system via VMM and provide a mechanism for monitoring intruder activities.

2.3 Related Work

The main objective of this paper is to propose a system that can sustain intruder invasion on a server and fix any corrupted data in the server without any intervention from an administrator. Similar work has been done in ReVirt[8] and Terra[9]. These two researches manipulate VM as a monitoring mechanism for any intrusion that might occur on the system this is done by modifying some of the code in VM. Hence make it different to the work done in this paper. AWSA only take the advantage of Isolation features in VM without making any changes to the code.

AWSA have a similar concept to Livewire [10], where IDS is used to monitor VM from it host operating system. Using this method IDS is place outside the VM which will protect the IDS itself by making it out of reach from intruders attack. The difference that AWSA has over these two is the next actions that take place after the intrusion occur. In their work, IDS is use to only monitor and reporting the incident in a log file or database and the next action will have to depend on the administrator's responsibility for the server, in contrast attack analysis and necessary recovery action take is done automatically in AWSA.

VM provides isolation between the host operating system and guest operating system. Due to its small and narrow code of VMM and also the power of VMM to monitor the process of guest operating system, VM gives an advantage for AWSA and it component (IDS and checksum) to be invisible from the intruder attack. On the other hand, the monitoring and report generated by IDS will alert AWSA to take further action on the intrusion made to the server where cheksum will act as the integrity checker if any of the data is corrupted.

In general AWSA have the ability of isolating potential target with it defense system, monitoring of intruder as well as reporting their activities, data integrity checker and finally reconstruction of compromised data. By considering all this as the features of AWSA, we can compare the difference and similarities of AWSA with the other related work. The comparison of AWSA with ReVirt, Terra and Livewire is summarized in table 1.

In conclusion, AWSA is developed by improving the works that have been done by researchers and developers of the system mentioned in this chapter. In general AWSA is only putting the bit and pieces provided by each element mentioned above into a complete puzzle that reinforces the security of a server.

Table 1 The Difference and Similarities of AWSA with Other Related Works

System / Features	AWSA	ReVirt	Terra	Livewire
Monitoring and reporting	✓	✓	✓	✓
Isolation	✓	×	×	✓
Data integrity Checker	✓	×	×	×
Data Reconstruction	✓	×	×	×

3 AWSA Architecture

3.1 AWSA Framework

As online business becomes an attractive target to cyber-criminals, self-protection against attacks is becoming an indispensable attribute of web servers. To achieve self-protection, a web server has to monitor itself, analyze the information obtained, plan a defense against detected attacks, and execute the plan. Therefore, in this work, we adapt these characteristics to design the AWSA. It is a combination of 6 main components: (a)Isolation, (b)Network Monitoring, (c)Logger (d)Analyzer (e)Integrity Checker and (f)Data reconstruction, as shown in Fig 2.

The purpose of isolation in AWSA is to keep the rest of the protection mechanism away from the web server, making them invisible to the attacker's eyes. Thus, it protects the other functions from being manipulated by the attackers. To facilitate this component, a VM is used to encapsulate the web server away from the host operating system. Meanwhile, network monitoring is used to view the incoming data packets that go into the web server. It also acts as a triggering mechanism to warn the other components to start taking action if it detects suspicious data packet going through the network traffic. Intrusion Detection System (IDS) [10] [11] is chosen to equip AWSA as network monitoring tool.

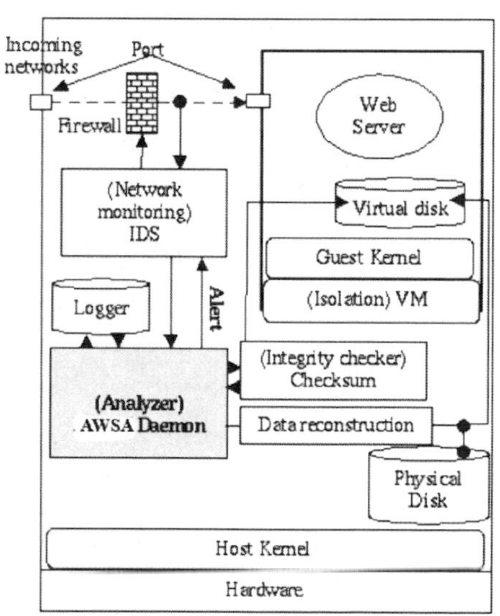

Fig. 2 The system architecture

Logger is the next process executed in AWSA to respond to the alert issued by the previous function. Logger is a database that keeps important information such as date and time of the attack, IP address of the attacker, alert priority generated by the activities, and classification of attack that are captured from the suspicious data packets. The gathered information is used by the analyzer to determine if the data packet is used for intrusion or not, if it is found to be a data packet from an attacker then it will be blocked.

The analyzer also has the ability to compare the original value of the checksum with the generated value from data the integrity function. The integrity checker is able to perform checksum once an intrusion is detected by the analyzer. AWSA's next function is to perform a data reconstruction if the comparison of data integrity by the analyzer found that the data stored is compromised.

3.2 AWSA Daemon

Our main focus in this work is to develop the analyzer or simply called AWSA daemon. It is considered as the heart of AWSA and controls the flow of the system. The processes done by AWSA daemon is depicted in Fig 3.

Upon receiving a triggering signal from IDS, AWSA daemon updates the alert into the database. The alert generated by IDS is based on the risk level caused by an attack. Each risk has its own priority. Table 2 shows the risk.

Once the database updated, AWSA daemon analyzes the data to look for a potential attack, this is done by counting the attack that is originated from similar attackers and similar risk level. The result obtained is then compared with a predetermined threshold value. The threshold is used as an indicator to catch any suspicious attacker with reasonable doubt and at the same time, minimizing the false alarms, which can burden any security system, especially the log file.

Table 2 Prioritization of the IDS alert

Priority	Attack Risk Level
Priority 1	High
Priority 2	Medium
Priority 3	Low

Once the comparison value is over the threshold, the suspicious attacker alert is erased from the attacker's log and the attacker's IP address is temporarily blocked from accessing the services offered by the web server. This will prevent a continuous attack from the same potential intruder. It also prevents the attacker log file from getting bigger as well as to reduce the log processing time.

The next process is the execution of the data integrity checker by checksum. Checksumming in the server helps to detect whether important files have been tampered with. Any changes made by any malicious programs installed illegally or unauthorized modification to the file without the knowledge of file system is checked by checksum soon after the detection of suspicious attack by the AWSA daemon. The checksum process is executed only on the files system of the guest operating system as well as the important data files used by the Web server and not on all the files.

The checksum result is then compared with the original checksum value stored in the AWSA daemon log. If differences are detected, it shows that the files have been tampered with and the next process is activated. When there is a difference between the compared values, AWSA starts implementing its reconstruction mechanism. This is the advantage of utilizing AWSA; it provides automatic recovery of the corrupted files. Therefore, the file system on the VM can operate without any interruption to the web services and intervention from the web server administrator.

Finally, after certain duration of time, the blocked IP address is removed from the list, and the user can access the web services again. However, if they attempt to attack the web server again, AWSA is ready to repeat all these processes.

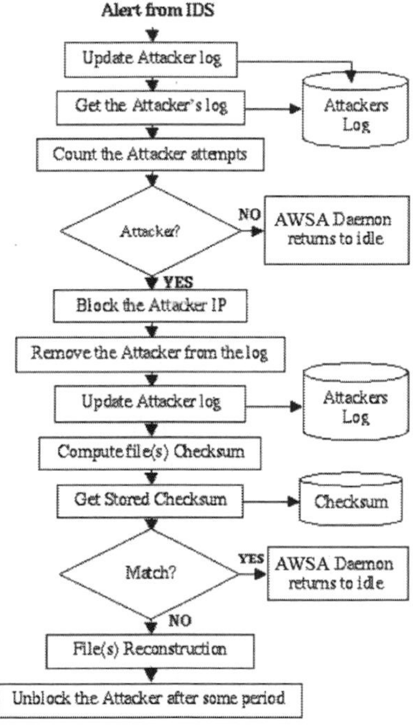

Fig. 3 Flowchart for AWSA daemon

4 Implementation and Result

A prototype was implemented on a Linux platform, using QEMU for virtualization. Linux Fedora core 3 and Slackware Linux were deployed as host and guest operating system respectively. The entire server was being monitored by SNORT

2.0.2 [12], a public-domain IDS tool. Swatch [13], the active log file monitoring tool was used to alert the AWSA. The verification of the data integrity was done by utilizing MD5 checksum. The hardware used in the experiments was a standard PC system (AMD Athlon XP 1600 CPU, 256 MBytes RAM).

The performance of AWSA is measured by two aspect; the durability in facing continuous web attacks and the response times taken to (a) detect and block attacker's IP address, (b) verify the integrity of data, and (c) reconstruct the compromised data. The durability tests were done by exposing the server to several web attack tools, namely Adore [14][15], Brutus AET2[16], Nmap, Smurf2k and Angry IP scanner. Each of the attacks is successfully detected by AWSA and the attackers IP address was blocked.

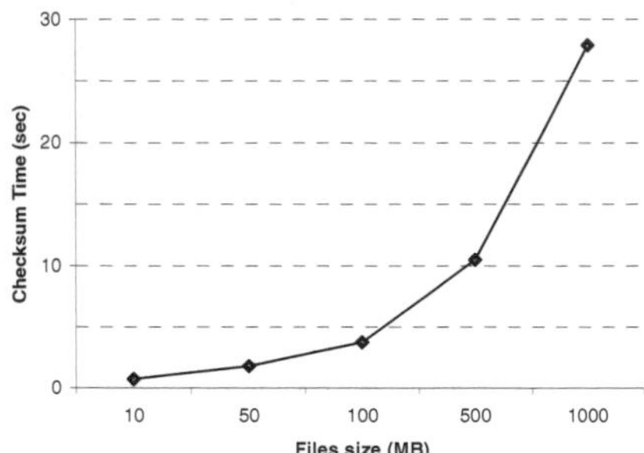

Fig. 4 Integrity checking time

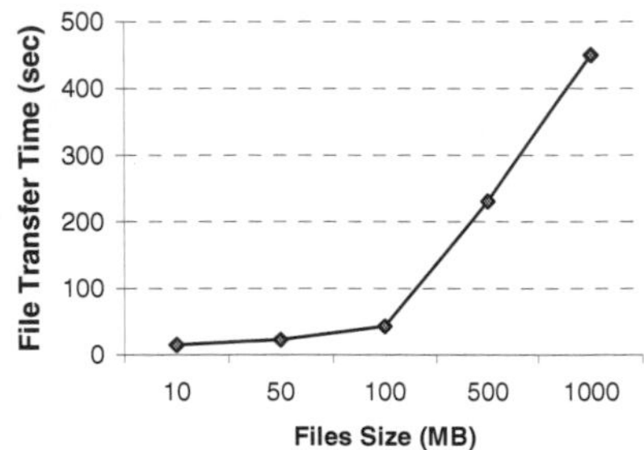

Fig. 5 File recovery time

On average, it took 86 sec for AWSA to impose the access restriction. This Average Blocking Time is taken starting from the time AWSA daemon detects the numbers of attacker's attempts exceed the predetermined threshold until the AWSA daemon alienates the attacker. Fig. 4 illustrates that AWSA daemon took less than 30 seconds to perform the integrity test on 1 Gigabyte of data and the file system of the server. The obtained result shows the time taken by AWSA to perform a checksum on several files with various sizes. It is a good indicator for estimating the period of performing a checksum on the particular files sizes. The final evaluation of AWSA is the file recovery time which indicates the time taken by AWSA to reconstruct the compromised data to its original state. Fig. 5, clearly depicts that files with larger sizes would consume more time in transferring the contents to the guest operating system.

5 Conclusions

This paper proposes the idea of Autonomous Web Server Administrator (AWSA), a VM-based Web Server that provides an autonomous attack detection and data recovery system. AWSA is a script-controlled daemon and mainly consists of Intrusion Detection System (IDS) and file integrity checking. The IDS is used to detect any malicious attacks from the intruders. IDS is kept isolated from the Web Server, therefore, it is inaccessible to VM processes and cannot be subverted by intruders; offers high attack resistance to IDS itself; and provides great visibility to the monitored system. File system integrity checking is deployed by AWSA to detect any attacker's signatures left in any files if they successfully break-in into the system. Once the corrupted data is identified, the AWSA automatically recovers any damaged files by transferring the original contents stored in the host system.

In the near future this work will be extended in finding a better IDS architecture available to be incorporated with AWSA and to extend a research on providing the integrity checker for the virtual machine itself. This research work is also towards finding a process of providing a forensic mechanism in order to investigate how an attack is occurs so that countermeasures can be taken for future attack.

Acknowledgments. This research has been supported by University Technical Malaysia Melaka under the short grant project.

References

1. CyberSecurity Malaysia, MyCERT Incident Statistics (September 2010) (On-line), `http://www.mycert.org.my/en/services/statistic/mycert/2010/main/detail/725/index.html`
2. Popek, G., Goldberg, R.: Formal Requirements for Virtualizable Third Generation Architectures. Communications of the ACM 17(7), 412–421 (1974)
3. Dunlap, G.W., King, S.T., Cinar, S., Basrai, M.A., Chen, P.M.: ReVirt: Enabling Intrusion Analysis through Virtual-Machine Logging and Replay. In: Proceeding of 2002 Symposium on Operating System Design and Implementation (2002)
4. VMware Inc., VMware Workstation (April 05, 2009) (Online), `http://www.vmware.com/`
5. Sun Microsystem, Virtualbox (April 05, 2009) (Online), `http://www.virtualbox.org/`
6. Barham, P., Dragovic, B., Fraser, K., Hand, S., Harris, T., Ho, A., Neugebauer, R., Pratt, I., Warfield, A.: Xen and the art of virtualization. In: Proceedings of the Nineteenth ACM Symposium on Operating Systems Principles, pp. 164–177 (2003)
7. Bellard, F.: QEMU CPU Emulator (January 1, 2009) (On-line), `http://fabrice.bellard.free.fr/qemu/`
8. Garfinkel, T., Paff, B., Chow, J., Rosemblum, M., Boneh, D.: TERRA: A virtual Machine-Based Platform Trusted Computing. In: SOSP 2003, pp. 193–205 (2003)
9. Dunlap, G.W., King, S.T., Cinar, S., Basrai, M.A., Chen, P.M.: ReVirt: Enabling Intrusion Analysis Through Virtual-Machine Logging and Replay. In: Proceeding of 2002 Symposium on Operating System Design and Implementation (2002)
10. Garfinkel, T., Rosenblum, M.A.: Virtual Machine Introspection Based Architecture for Intrusion Detection. In: Proc. Network and Distributed System Security Symposium (2003)
11. Laureano, M., Maziero, C., Jamhour, E.: Intrusion Detection in Virtual Machine Environments. In: Proc. EUROMICRO Conference, vol. 30 (2000)

12. Sourcefire Inc. Snort – The Open Source Network Intrusion Detection System (October 1, 2009) (On-line), http://www.snort.org
13. Sourceforge, Swatch (October 1, 2009) (On-line), http://swatch.sourceforge.net
14. Wichman, R.: Linux Kernel Rootkits (November 21, 2009) (On-line), http://la-samhna.de/library/rootkits/index.html
15. Packet Storm (November 21, 2009) (On-line), http://packetstormsecurity.org.pk/UNIX/penetration/rootkits/index.html (retrieved)
16. Hoobie Inc. Brutus - The remote password cracker (October 10, 2009) (On-line), http://www.hoobie.net/brutus/

Generation of IPv6 Addresses Based on One-to-Many Reversible Mapping Using AES

Nashrul Hakiem, Akhmad Unggul Priantoro, Mohammad Umar Siddiqi, and Talib Hashim Hasan

Abstract. The proliferation of enterprise wireless network raises the security concern in any organization despite the unarguable benefits it brings about. At the same time, the initiative to migrate from IPv4 (Internet Protocol version four) to IPv6 (Internet Protocol version six) is gaining momentum across the globe to resolve the IP address depletion problem as well as reaping the benefit of it. This research proposes a new scheme to manage IPv6 addresses in an enterprise wireless local area network (WLAN) which may be incorporated into DHCPv6 (Dynamic Host Configuration Protocol for IPv6) software. In this scheme each user is assigned a dynamic IPv6 address that is generated cryptographically. Each time a user tries to access the network, different IPv6 address will be given which is generated using CFB (Cipher Feedback) mode of AES (Advanced Encryption Standard) algorithm, whereby there is a one-to-many reversible mapping between user and IPv6 addresses. In this way, it is possible for the network administrator to identify users in real time from their IPv6 address although a user's IP address regularly changed. Dynamically changing IP address will impede an external network adversary's effort to track user's online behavior, thus enhancing privacy.

Keywords: Address Management, Advanced Encryption Standard, Cipher Feedback, DHCPv6, IPv6, WLAN.

1 Introduction

IP address management is one of the important aspects of network management to improve the security and enforce the network policy set up by organizations of

Nashrul Hakiem · Akhmad Unggul Priantoro · Mohammad Umar Siddiqi · Talib Hashim Hasan
Faculty of Engineering, International Islamic University Malaysia, Kuala Lumpur, Malaysia
e-mail: {g0729261@student.,unggul@,umarsiddiqi@, talib@}iium.edu.my

F.L. Gaol (Ed.): Recent Progress in DEIT, Vol. 2, LNEE 157, pp. 183–189.
springerlink.com

today. In WLAN environment a feasible solution is to use DHCP server to automate the IP address assignment whenever a network client wants to connect to the network. DHCP server assigns IP address to requesting client from the pool of IP address configured by the network administrator. Even though a DHCP server may keep records of which IP address is assigned to which client and when, it is a tedious task for a network administrator to identify user from the log. Whenever traffic anomaly or policy breach or internally-generated security threat, it is difficult to pinpoint the culprit. Expensive devices and software as well as the expert opinion must be called upon for investigation.

This research proposes a new scheme to manage IPv6 addresses in an enterprise wireless local area network (WLAN). In the proposed scheme, each user is assigned an IPv6 address which is generated cryptographically. In our proposal the objective of cryptographically generating address is to assert user's address ownership by the network administrator (non-repudiation of address). Each time a user tries to access the network, different IPv6 address will be given to the user which is generated cryptographically. This will also enhance the security and privacy of end users since they can not be tracked by their IP addresses by external network adversaries, i.e., network adversary external to the local area network, This becomes an important issue since in IPv6, the Internet is an end-to-end transparent network, unlike in IPv4 whereby users are mostly hidden behind a firewall [1].

The remainder of this paper is organized as follows. Section 2 describes briefly the background related to this research. Section 3 explains the proposed IP address generation mechanism. Section 4 gives the simulation results. Section 5 concludes the contribution of this paper.

2 Background

In IPv6, 128 bits are used to specify the address of a node which format is shown in Figure 1.

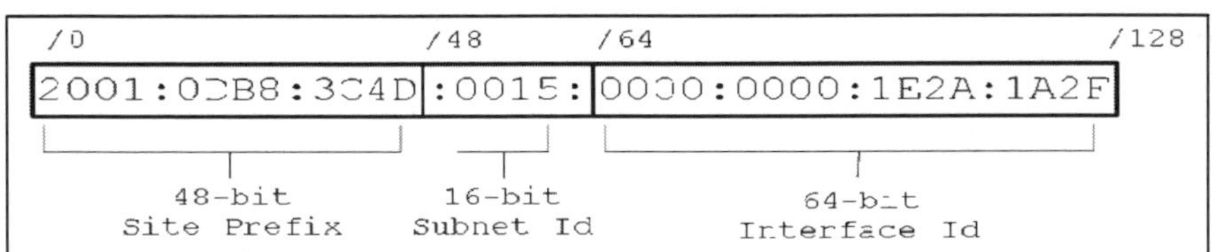

Fig. 1 IPv6 address format

The first 48 bits are allocated for network address which is globally unique, thus globally routable. The following 16 bits are allocated for subnet prefix within a network to allow network administrator defining the desired internal network hierarchy. The remaining 64 bits are allocated for Interface Id for nodes. This proposal focuses on a mechanism to generate this Interface Id in order to simplify the IP address management in enterprise WLAN.

Stateless autoconfiguration is intended for ease of deployment of unmanaged network or network whose nodes have limited capability such as sensor networks, RFID's as well as ad hoc networks [2]. The stateless mechanism allows a host to generate its own addresses using a combination of locally available information and information advertised by routers [3].

In stateless auto configuration IP address can be derived from MAC address [2] or other mechanism such as Cryptographically Generated Address (CGA) [4] and its extension, Multi-key CGA (MCGA)[5]. As extended usage of CGA, it has been proposed to secure other protocols such as the Mobile IPv6 (MIPv6) protocol [6-7]. Further enhancement was done to support multiple hash function [8].

There is stateful autoconfiguration which is widely used in managed network in which hosts obtain interface addresses and other configuration information from a server. DHCPv6 simplifies network administration and requires minimum or no manual configuration of DHCPv6 clients [9]. Thus a DHCPv6 client may be given different IP address each time it requests to the DHCP server. Although DHCPv6 server maintains the binding information and can be logged for later retrieval, it (the binding information) is only known to the DHCP server. Any network node requiring the information must send queries to the DHCP server. Recently there is an increasing interest to merge the goodness of DHCPv6 and CGA [10].

IPv6 address configuration mechanisms previously described do not facilitate IPv6 address owner detection in real time which may be of interest to enterprise wireless network administrator for various reasons. In the next section, we describe a novel IPv6 node Interface Id generation mechanism that supports IPv6 address owner identification in real time. The mechanism takes advantage of the huge size of IPv6 address.

3 Proposed Mechanism

3.1 Framework Development

There is a framework [11] to generate a dynamic IPv6 address which can be mapped to each individual user in real time using a many-to-one mapping from the IP address space to the user space. In that framework [11], user Id bits are distributed within Interface Id in fixed positions instead of using cryptographic method. It is shown to have a higher degree of correlation or lower Hamming distance property. This might be exploited to infer IPv6 address owner by network adversaries which may raise further security and privacy concerns. Therefore, it is desirable to use cryptographic method for increasing the Hamming distance values of the generated IPv6 addresses.

3.2 Cipher Feedback

To generate one-to-many reversible IPv6 address mapping one may use Simplified AES (S-AES)[12]. It produces good avalanche effect (about 50%) as expected [13]. However Simplified-AES is for academic purposes. It is an educational rather than a secure encryption algorithm [14]. It is breakable (vulnerable) with linear calculations [14]. Therefore, we should use AES which is resistant against all known attacks instead of using S-AES[15]. There are some modes of operations for encryption, one of them is Cipher Feedback (CFB). It allows variable size of input and output.

3.3 Proposed Mechanism

Interface Id format is depicted in Figure 2. It contains of 6-bit checksum, static Universal/ Local (1 bit) and Group/ Individual (1 bit), 48-bit encrypted user Id, and 8-bit of key Id. Each of key Ids has 128 bits length which can be retrieved via look up table.

checkSum	ug	encryptedUserId	keyId
/0	/6	/8	/56 /64

Fig. 2 Proposed Interface Id format

The 48-bit encrypted user Id is derived from 18-bit user Id as depicted in Figure 3.a. This user Id is concatenated with 30-bit random to form 48-bit plain. It is encrypted using CFB mode of operation for AES encryption. Initial Value (IV) and key are using 128-bit as standard AES size.

User Id decryption algorithm can be seen in Figure 3.b. Encrypted 48-bit user Id is decrypted using CFB-AES to produce 48-bit plain text. The first 30 bits of the 48-bit decrypted original plain text are then eliminated to obtain 18-bit user Id. Initial Value (IV) and key are used as IV and key for CFB-AES encryption.

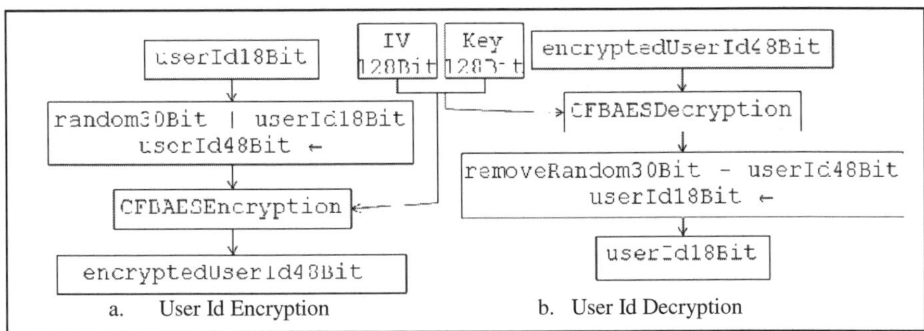

Fig. 3 User Id Encryption & Decryption

Generation of Interface Id format having checksum, U&G bit, encrypted user Id and key Id is shown in Figure 4.a. In the first step, it gets 18-bit user Id data and then generates 8-bit random number to obtain 128-bit key Id from a lookup table. Next step is generating encrypted user Id using flow chart as shown in Figure 3.a and finally we generate the checksum to construct the Interface Id completely as shown in Figure 2.

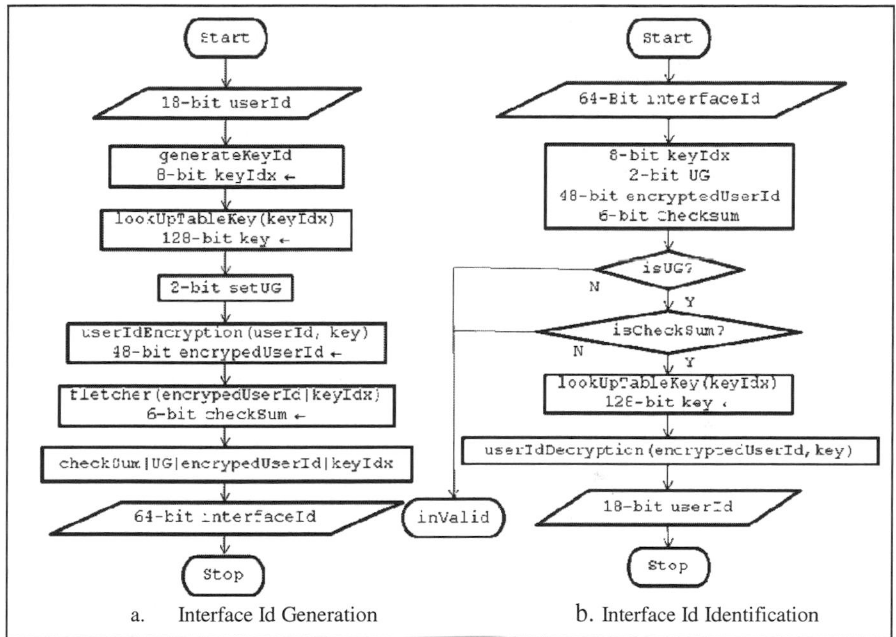

a. Interface Id Generation b. Interface Id Identification

Fig. 4 Interface Id Generation and Identification

The steps to verify Interface Id's ownership are shown in Figure 4.b. Firstly, split the Interface Id data and then check 7th and 8th bit as Universal and Individual bit respectively. If the value is "00", continue to the next step, otherwise stop identifying. Next step is comparing checksum value with checksum generated. If the value is matched, continue to the next step, otherwise stop identifying. Eventually, it gets user Id using flow chart as shown in Figure 3.b.

4 Simulation Results

A simulation based example of Interface Id which has been generated using the mechanism proposed here is shown in Table 1. The simulation software is developed and compiled using Java™ Standard Edition version 1.6.0. The simulation results are saved to text file which can be opened and analyzed using office Spreadsheet. In the example, 512 different Interface Ids are generated belonging to the same user Id, which is an octal number 123456 (18 bits).

Table 1 Generated Interface Id

64-bit Interface Id 6 Digits Octal User Data = 123456 18-bit User Data = 001010011100101110 7th & 8th bit (u & g) is set to 00.	
No	**64-bit Interface Id**
1	1100010001010000111001101110100100010010101101100001000010110100
2	1000010000000101100011001111111101101010110001001100110101010010
3	1011000001010001000010011110100011100011111101111100000110110011
4	1111000010101100100110111010010110011011000100101011100010101010
5	0101010001111100110000011101010101100110101000110011101110100101
...	. . . .
508	1101010001001000111000110010000101011101100010110001010110010100
509	1110000010111110001001110110111011000111001101100100011010000111
510	0111000000011101111001111111100011111101010000110100101001000000 0
511	1100110000100011101110100010001011011010001001010010000011001100
512	0101110000001010000000000011101100110000101111111111101111001111

5 Conclusion

IPv6 address management in enterprise wireless network is of interest to network administrator. It helps network administrator to efficiently manage the network and improve the controllability and visibility of the network that in the long run will increase the ROI of ICT investment and compliance to regulations via policy enforcement. A mechanism to generate IPv6 addresses for enterprise wireless network is proposed that facilitates real-time user identification from their IPv6 addresses. The generated IPv6 address is embedded with user Id so that users can be identified by using Cipher Feedback (CFB) mode of AES. Checksum values are embedded to reduce the probability of randomly generated address coinciding with valid IPv6 addresses generated by the proposed mechanism.

The proposed IPv6 address generation mechanism can be used to simplify the recognition of users by their IPv6 addresses. This is an important part of network policy enforcement. Today more and more ICT services and applications are migrating to web based services [16] in which controlling internet (including intranet) traffic using ports is not feasible anymore. Being able to recognize users by their IPv6 addresses can automate the decision whether to grant or deny access to certain internal and external servers. This will enhance the existing security measures already implemented in the network.

References

1. IETF, Privacy Extensions for Stateless Address Autoconfiguration in IPv6, in RFC 4941. Network Working Group (2007)
2. Amoss, J.J., Minoli, D.: Handbook of IPv4 to IPv6 Transition, Methodologies for Institutional and Corporte Networks. Auerbach Publications, Danvers (2008)
3. IETF, IPv6 Stateless Address Autoconfiguration, in RFC 4862. Network Working Group (2007)
4. IETF, Cryptographically Generated Addresses (CGA), in RFC 3972. Network Working Group (2005)

5. Kempf, J., Wood, J., Ramzan, Z., Gentry, C.: IP Address Authorization for Secure Address Proxying Using Multi-key CGAs and Ring Signatures. In: Yoshiura, H., Sakurai, K., Rannenberg, K., Murayama, Y., Kawamura, S.-i. (eds.) IWSEC 2006. LNCS, vol. 4266, pp. 196–211. Springer, Heidelberg (2006)
6. O'Shea, G., Roe, M.: Child-proof authentication for MIPv6 (CAM). SIGCOMM Comput. Commun. Rev. 31(2), 4–8 (2001)
7. IETF, Applying Cryptographically Generated Addresses and Credit-Based Authorization to Mobile IPv6, in Internet-Draft draft-arkko-mipshop-cga-cba-03.txt. Network Working Group (2006)
8. IETF, Support for Multiple Hash Algorithms in Cryptographically Generated Addresses (CGAs), in RFC 4982. Network Working Group (2007)
9. IETF, Dynamic Host Configuration Protocol for IPv6 (DHCPv6), in RFC 3315. Network Working Group (2003)
10. IETF, Interactions between CGA and DHCPv6. Network Working Group (2007)
11. Hakiem, N., Priantoro, A.U., Siddiqi, M.U., Hasan, T.H.: IPv6 multi generated address for enterprise wireless Local Area Network. In: IGCES 2008, Johor Malaysia (2008)
12. Musa, M., Schaefer, E.F., Wedig, S.: A Simplified Rijndael Algorithm And Its Linear And Differential Cryptanalyses. Santa Clara University, Santa Clara (2002)
13. Hakiem, N., Priantoro, A.U., Siddiqi, M.U., Hasan, T.H.: Generation of cryptographic one-to-many mapping IPv6 address using S-AES. In: 2010 International Conference on Information and Communication Technology for the Muslim World (ICT4M), Jakarta Indonesia (2010)
14. Davod, M.S., Khaleghei, B.H.: On the vulnerability of Simplified AES Algorithm Against Linear Cryptanalysis. IJCSNS International Journal of Computer Science and Network Security 7(7) (2007)
15. Stallings, W.: Cryptography and Network Security, Principles and Practices, 4th edn. Pearson Prentice Hall (2006)
16. Richard, M.: 10 Future Web Trends (2007)

A Study on Privacy Preserving Data Leakage Prevention System

Jinhyung Kim and Hyung Jong Kim

Abstract. DLP (Data Leakage Prevention) system is an essential for protection of valuable information asset of companies and organizations. To prevent the data leakage, security administrator should monitor the employee's network access and it causes violation of privacy. This research deals with the privacy violation issues in DLP system and shows how we can manage the trade-off relation between DLP and privacy protection. Especially, we are proposing a design of privacy preserving DLP system.

1 Introduction

Recently, there have been incidents of business information leakage around the world including South Korea's companies. In addition, those incidents notably were committed by internal employees or researchers. Due to these circumstances, the deployment of DLP (Data Leakage Prevention) system become an inevitable alternative that organization's owners have to consider [1][2].

The deployment of DLP system implies that the employees' network access is monitored and logged by the administration group of organization. Even though the aim of the monitoring is recognition of the abnormal behaviors, the normal behaviors of usual employees will also be revealed to the administration group. Among the normal behaviors, there can be private information that the employee does not want to disclose.

The motivation of this work is the privacy violation of DLP system and we are aiming at eliciting the reasonable management scheme that is preventing the data leakage and preserving the employees' privacy. In addition, as the technical approaches, we are proposing a couple of methods that can be applied to preserve the privacy of employees during the DLP process.

This paper shows usual technical steps of DLP and available technical approaches that can preserve the employees' privacy. The technical approaches

Jinhyung Kim · Hyung Jong Kim
Seoul Women's University, 621 Hwarangno, Nowongu, Seoul, 139-774, Korea
e-mail: {jinny,hkim}@swu.ac.kr

F.L. Gaol (Ed.): Recent Progress in DEIT, Vol. 2, LNEE 157, pp. 191–196.
springerlink.com

include log anonymizing using hash function and DLP process automation. Although those technical approaches can help to protect the privacy, it is inevitable to reveal some part of private information. Due to this limitation of the technical approach, we need to consider the trade-off relationship between DLP level and privacy protection level. This research remarks the trade-off relationship and shows how the relationship can be represented as a system of privacy preserving DLP system.

The remaining part of this paper is organized as follows. The section 2 of this paper shows some background knowledge of this work such as DLP's process and log anonymizing. In section 3, the technical approaches for preserving privacy and trade-off relationship representation are presented. The section 4 shows how this concept and technical approaches can be embodied as a system of DLP preserving privacy.

2 Background Knowledge

The DLP provides the prompt countermeasures that block up the leakage of the important information by monitoring the data transmission through the company networks. In addition, it sees if the transmission is meeting to the data protection regulations of companies. The DLP has several essential steps for stopping the data leakage. Above all, the company which wants to deploy the DLP should have the definition of critical information. In addition, critical data registration should be conducted. In this step, DLP searches the critical information from organization's information asset and registers the information. The registered data should be managed very carefully and whenever any actions of disclosing the registered important information are detected, DLP blocks the transmission or writes log about the transmission.

The private information of employees of an organization should be protected based on the laws and regulations which are legislated by government or related global organization. All kinds of personal information should be collected or shared only if it is agreed by the information owners[4][5].

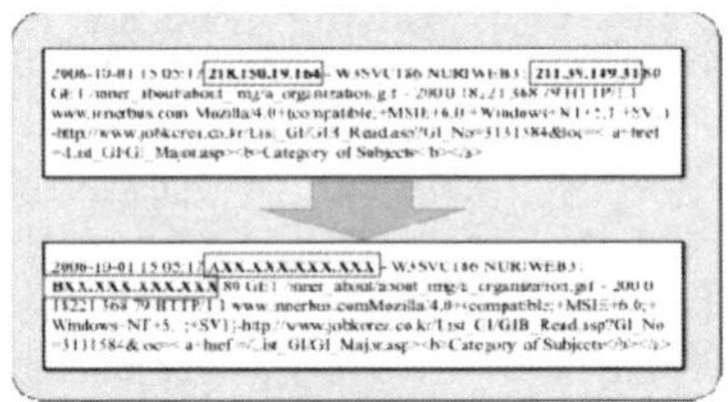

Fig. 1 Example of log anonymizing.

Log anonymizing is a technology for deleting or replacing with special character of some part of log information without the loss of meaning [3][6]. The Fig. 1 shows an example of this technology. It shows that the IP address of web access log is anonymized by replacing the address with character 'AXX.XXX.XXX.XXX'. Even though there are replacements of some part of log, the meaning of the log should not be changed for identification of abnormal behavior of certain users. In addition when the abnormal behavior is recognized, the anonymized data should be able to be recovered to original data. For this purpose, the encryption and hash function can be technical alternatives.

3 Technical Approaches for Proposed System

3.1 *Trade-off Relationship*

There are trade-off relationship between DLP system's purpose and privacy protection. In an organization, if we want to prevent the data leakage strictly, probably, we need to tolerate violation of privacy. If there is a union in the company, it can be an issue that should be discussed with owner or people in higher position. The discussion between two parts is going to be about the portion of information that is supposed to be monitored. The Table 1 shows an example of e-mail content which has DLP and privacy aspect.

Table 1 Trade-off relationship in E-mail Header.

DLP Aspect	Monitoring Target	Privacy Aspect
Mail Receivers can be a competitor of the company	To: Receiver's Email	Mail receiver can be private information in a certain situation.
Cced and Bcced e-mail can be a competitor of the company	Cced and Bcced e-mail Address	
If there is a data leakage, the subject can be summary of information which is leaked.	Subject	It may contain the summary of e-mail.
It can be the identification of important information discloser.	From : Sender's e-mail	If the mail body contains private information, this information can be used to identify the owner of the information.
It contains the data leakage occurrence time.	Date: E-mail sending time	Users may reveal their lifestyles through the time stamps.
It reveals how the disclosed information is formatted.	Content-Type: Text/Plain, Multipart	It does not have any relation with privacy

Since the trade-off relationship exists in every unit of monitored information, if we have the table which shows the two aspects, the union people and owner group can have discussion about monitoring targets. In this work, we are aiming at developing the scoring method for each data unit which can be monitored. The scoring method shows the DLP level and privacy level for each data unit. After the monitoring target is selected, the organization's DLP level and privacy protection level can be calculated. The Table 2 is an example DLP level and Privacy Violation Level. The level has four (4) grades. The level 4 is highest value and it means the most critical information for DLP and the level 4 also means the most trivial in the privacy protection aspect.

Table 2 Example of Level of DLP and Privacy Protection Level.

DLP Level	Email Data	Privacy Protection Level
3	To:	2
3	Cc:	2
3	Bcc:	2
4	Subject:	1
2	From:	3
3	Date:	2
1	charset="euc-kr":	4

3.2 DLP Process Automation

One of the technical approaches for privacy preserving DLP is process automation. The aim of automating the DLP process is minimizing the human's interactions with the system. Even though the DLP is monitoring the private information, if the monitoring process is done in an unmanned manner, it can hardly disclose the private information until the data leakage does not occur. To accomplish the automated data monitoring, there should be a knowledge base which contains the intelligence about handling data leakage.

There are several methods which can be used to represent the intelligence of data leakage detection such as the regular expression, finger printing, and hash function. The regular expression is for representation of the critical information in flexible way. The finger print is also a representation of sensitive information which can be included in disclosed data. The hash function can be used to generate the message digest of a file. The message digest can be used to detect the disclosure of sensitive files by comparing with the saved one.

3.3 Anonymizing Personal Identification Information

The other technical alternative for privacy preserving DLP is anonymizing the personal identification information in log data. The aim of this technical approach is to change the log's user identification information with unidentifiable information. The Fig. 2 shows the concept of the anonymizing the log data.

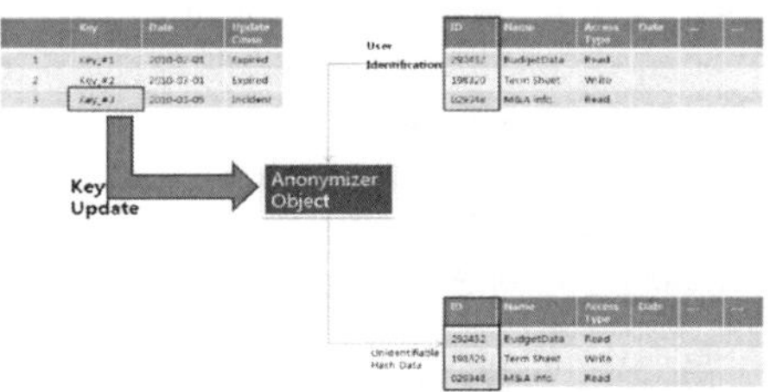

Fig. 2 Log Anonymizing Method for Privacy Protection

Each entity of log data has the user identification information. The identification information is very critical for privacy because people can recognize a certain user's behavior using the information. The anonymizer object of Fig. 2 is working for making the identification information unidentifiable. For this purpose, the anonymizer needs a hash function's key which is used to generate the message digest.

The goal of anonymizing is preventing the administrator from recognizing a real user of an anonymized identifier. If we use a fixed key for the time being, people can be familiar with a certain anonymized identifier. In our suggestion, the key value can be changed based on the strategy. Since the anonymizing method can use different key for the same user identifier, it becomes hard to recognize the real user of anonymized identifiers. For this scheme, we need the key update log which is retrieved whenever it is needed. Table 3 shows the 4 fields of key management history. Sometimes an administrator needs a key which is not currently used. In this case, the administrator should look up the key history table to find the proper key. The key creation and destroy time are used for finding the proper key.

Table 3 Fields and explanations of key history Table

	Field name	Explanation
1	Key Value	The key value which is used for hash value generation
2	Creation Time	The key creation time
3	Destroy Time	The key destroy time
4	Changing Reason	The reason of changing keys (ex. 1. Incident, 2. Expired)

The key changing reason also contains the useful information. If there is a certain incident which reveals the key, the key should be changed as an action item of the incident handling.

3.4 System Integration

Fig. 3 shows the system architecture. There are two physically separated systems. One is DLP system and the other is agent software in client system. The knowledge of fingerprints, regular expression and exact file matching is stored in DB&KB. In addition each knowledge representation method has its own user interface for editing the knowledge such as critical information's keywords, patterns and hash values.

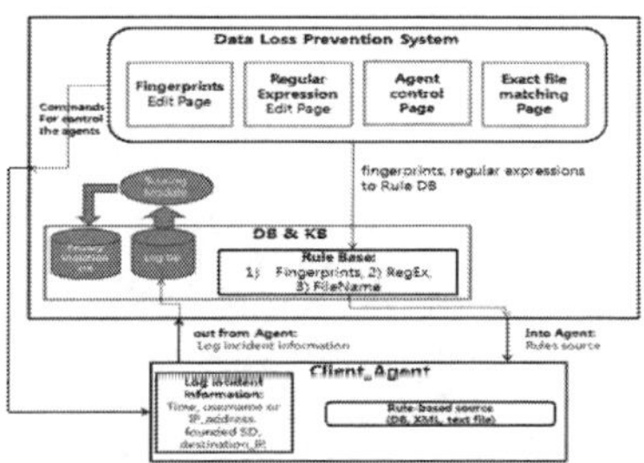

Fig. 3 Architecture of DLP System

Client_Agent modules are installed in client systems to collect each user's packets and it has knowledge patterns which are retrieved from the DLP system. The detection log data is located in client system and DLP system at the same time. In addition, the log data is anonymized using the method mentioned in previous section.

To handle the privacy issues, there is a scoring module for calculation of level of privacy and security. Whenever the rules are fired, the security and privacy scores are accumulated and the scores are used for how many times the private information is retrieved by DLP system.

4 Conclusions

This paper introduced a trade-off relationship between DLP process and privacy protection. In addition usual technical steps in DLP are presented and also the available technical approaches that can preserve the employees' privacy are introduced. The main technical approaches that we are suggesting are finding trade-off relationship for each monitoring target data, automation of DLP process and log anonymizing. At last, suggested overall design of privacy preserving DLP system shows how the technical approaches can be deployed in a server and clients system.

Acknowledgments. This work was supported by National Research Foundation of Korea Grant funded by the Korean Government (2009-0068361).

References

1. Hiroshi, S., Kazuo, Y., Ryuichi, O., Itaru, H.: An Information Leakage Risk Evaluation Method Based on Security Configuration Validation. IEICE Technical Report 105(398), 15–22 (2005)
2. Chou, S., Liu, A., Wu, C.: Preventing information leakage within workflows that execute among competing organizations. Journal of Systems and Software 75(1-2), 109–123 (2005)
3. Reiter, M.K., Rubin, A.D.: Anonymous Web transactions with Crowds. Communication of ACM 42(2), 32–48 (1999)
4. Agrawal, R., Srikant, R.: Privacy-preserving data mining. ACM SIGMOD Record 29(2), 439–450 (2000)
5. Hommel, W.: Policy-Based Integration of User and Provider-Sided Identity Management. In: Müller, G. (ed.) ETRICS 2006. LNCS, vol. 3995, pp. 160–174. Springer, Heidelberg (2006)
6. Agrawal, D., Kesdogan, D., Pham, V., Rautenbach, D.: Fundamental Limits on the Anonymity Provided by the MIX Technique. In: Proceedings of 2006 IEEE Symposium on Security and Privacy (May 2006)
7. Jensen, C., Sarkar, C., Jensen, C., Potts, C.: Tracking Website Data-Collection and Privacy Practices with the iWatch Web Crawler. In: Proceedings of SOUPS (Symposium on Usable Privacy and Security) (July 2007)
8. Krishnamurthy, B., Malandrino, D., Wills, C.E.: Measuring Privacy Loss and the Impact of Privacy Protection in Web Browsing. In: Proceedings of SOUPS (Symposium on Usable Privacy and Security) (July 2007)
9. Clark, J., van Oorschot, P.C., Adams, C.: Usability of Anonymous Web Browsing: An Examination of Tor Interfaces and Deployability. In: Proceedings of SOUPS (Symposium on Usable Privacy and Security) (July 2007)
10. Karger, P.A.: Privacy and Security Threat Analysis of the Federal Employee Personal Identity Verification (PIV) Program. In: Proceedings of SOUPS (Symposium on Usable Privacy and Security (July 2006)

The Trojan Horse Detection Technology Based on Support Vector Machine

Jie Qin and Hui-juan Yan

Abstract. This paper presents a Trojan-detection system model based on Support Vector Machine. First, while monitoring the system, this strategy establish system call sequences in accordance with its system calls function in the system, and convert into supporting vector machine-readable tags, and place in the data warehouse for support vector machine extracted as the feature vectors. And to determine the abnormal behavior of testing procedures to determine whether it is Trojan by classifying the detected program behaviors based on the support vector machine classifier. Experimental results show that, comparing with the existing technology of Trojan horse detection, this method has better performance in detection time and detection of known and unknown Trojan horse attacks. Besides, it has higher accuracy, and takes up very little system resource.

1 Introduction

In recent years, researchers have gradually introduced new intelligent processing technology to the Trojan detection. However, existing methods detect Trojans according to the Trojan horse dynamic execution and feature information, not only for the detection of Trojan horses "after", but also difficult to achieve the detection of the unknown Trojan.

Support vector machine (SVM) is based on a statistical learning theory, its biggest advantage is based on the principle of structural risk minimization, to maximize the generalization ability of learning machine [1]. In addition, because SVM is a convex optimization problem, the local optimal solution is globally optimal solution, and other learning algorithm is unable to do so. Support vector machine is applied to Trojan detection, so as to achieve in the absence of sufficient prior knowledge, still has the goal of better detection results, so that the Trojan detection system has better

Jie Qin · Hui-juan Yan
College of Information Science and Engineering,
Henan University of Technology, Zhengzhou, China
e-mail: qinjie@163.com

F.L. Gaol (Ed.): Recent Progress in DEIT, Vol. 2, LNEE 157, pp. 197–202.
springerlink.com

detection performance. The rest of this paper is organized as follows: first, the basic principle of SVM is introduced, then Trojan detection methods based on SVM-classification are proposed. Finally, experimental results are given.

2 Support Vector Machine

The core idea of SVM: by a pre-selected nonlinear mapping (kernel function), the input vector mapped to a high dimensional feature space, in this space the optimal separating hyperplane is constructed.

To detect Trojan horses in computer system, we must distinguish between legitimate programs and Trojan horse programs, so the nonlinear SVM detection model is the most important procedures to classify these procedures behavior, it is the process of classification system call sequence determine the process of normal or abnormal. Support vector function is a good solution to distinguish between legitimate programs and Trojan, it solves the optimal classification problem by separating hyperplane.

In this paper, the standard class-2 SVM is used to detect Trojan. First, assume that there is such a marked training samples. $\{(x_1, y_1), (x_2, y_2), \cdots, (x_n, y_n)\}$, where, i=1,2, …,n, $x_i \in R^*$, and $y_i \in \{-1,1\}$ is a class marker[2].

We can create standard linear SVM model function:

$$\begin{cases} \min \frac{1}{2}\left(w^T \cdot w\right) + C\sum_{i=1}^{k} \xi_i \\ s.t.\ y_i\left(\left(w^T \cdot x_i\right) + b\right) \geq 1 - \xi_i, i = 1, 2, \cdots k \\ \xi_i \geq 0, i = 1, 2, \cdots k \end{cases} \tag{1}$$

Here, C is the regularization parameter, and C>0, ξ_i is the slack variable, $W \in R^n$ is the hyperplane normal vector, b is the classification threshold, it can be calculated from any support vector. After the introduction of Lagranger multiplier α_i (=1,2,…,k), we can get the dual form:

$$\begin{cases} \max \sum_{i=1}^{k} a_i - \frac{1}{2}\sum_{i=1}^{k}\sum_{j=1}^{k} a_i a_j y_i y_j \left(x_i x_j\right) \\ s.t.\ \sum_{i=1}^{k} a_i y_i = 0 \end{cases} \tag{2}$$

After solutions to these problems, the corresponding classification function can be obtained:

$$f(x) = sign\left[\left(\sum_{i=1}^{k} a_i y_i \left(x, x_i\right)\right) + b\right] \tag{3}$$

For Trojan detection, it is to monitor the data. The data were analyzed to determine the normal procedure or Trojan. It can be seen as a Trojan detection classification problem [3], first, from a large number of monitoring data extract some statistical data, and determine its classification. When the monitoring data satisfy certain conditions that it is the Trojan, otherwise, it is normal data. In this way, Trojan detection put into the classification problem in pattern recognition, and some pattern recognition theory can be used to solve the Trojan detection problem. The monitor data collected by Trojan detection system can be treated as data points of n-dimensional space. Each dimensional representatives of a feature attribute of the data. Using the kernel function change data from the z-dimensional space by inner product mapping to a higher dimensional space, implicitly, to find a hyperplane to classify the data. Therefore, the basic idea is the application of support vector machine training data through the kernel function mapping from low-dimensional space to high-dimensional space, so that the data in high dimensional space can be linearly separable, and to find an optimal classifier, and then using the classifier to detect the monitored data, so as to achieve the purpose of Trojan detection.

3 Trojan Detection System Based on SVM

The Trojan detection system based on SVM consists of six main modules: Trojan behavioral data set library, data acquisition module, data pre-processor module, SVM training module, Trojan detection module, system response module. As shown in Fig.1.

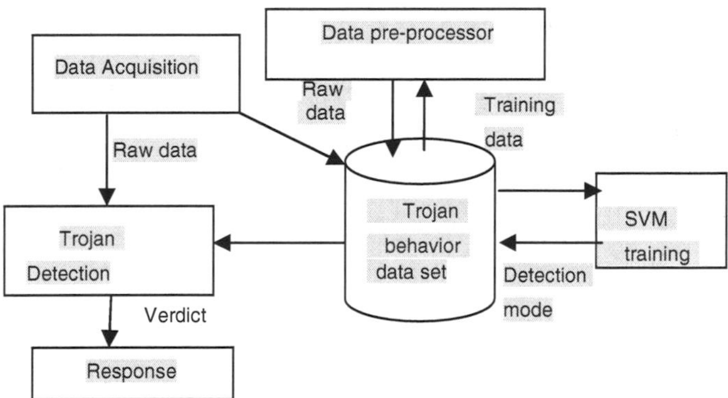

Fig. 1 Trojan detection system based on SVM

The function of data acquisition module is collect data in the process of realtime monitoring. Data pre-processor module is responsible for preprocessing data from the Trojan behavior data set, to extract the features of the data to meet the needs of the various modules. Since SVM classifier only on the number of the same dimension vector classification, but the length of data in the system is different, and may not be the digital type, the quantized data must be changed into digital vector form so that support vector machine can identify. The Trojan behavior data set module is a variety of data storage, such as the original data from data acquisition, the data after preprocessing will used by other modules, and testing mode after SVM training, and some of the detection rules. The function of SVM training module is extracted the sample behavior from Trojan behavior data set. The behavioral characteristics of Trojan horse is trained to learn, and reveals a new Trojan detection mode; Trojan detection module use the training data from SVM training module, and compare with real-time monitoring data. System response module use the results obtained from the Trojan detection module to determine a response, such as delete, block and so on.

The whole process mainly consists of two stages: training stage and testing stage. First, the training data (known normal data and Trojans) and test data are transformed into vector type through the data pre-processor, which can be identified by SVM classifier. In the training phase, SVM training is used to train data, and to get the corresponding parameters and the support vectors. In the Trojan detection phase, the preprocessed test data was classification by formula (6), and the results were submitted to Trojan detection phase, to make final judgments.

4 Experiments and Analysis

1000 legal procedures and 1000 Trojans procedures ware used in this experiment. Among them, the legal procedures were downloaded from Internet, Trojan information were downloaded from the website of the Trojan library, security vendor Symantec:http://www.symantec.com/business/security_response/index.jsp, including the Gray Pigeons, Trojan Downloader, Game Thief and a series of Trojans. For simulation, the first data set collected from a randomly selected part of the data as training data, and then taking three samples from data set D1, D2, D3, as the test data. Training data 400, of which 300 legal procedures and 100 due Trojans. Three test data sets were 200, including 126 legal procedures, 74 Trojans, respectively. The system of these legal procedures and Trojans call sequence were detected, and support vector machine classification model was used on these data, after repeated classification, classifier obtained the optimal solution ,and then the tested data was determined whether Trojan program.

4.1 Data Preprocessing and Selection of Short Sequences

The purpose of data preprocessing is get the short sequences of system calls. Experimental data is handled by the process of system calls. System call for each process name in the system call list (mapping file) has its corresponding index value, such as the '6 'means' close'. Here, the main window with a length of q in the process execution sequence to be implemented on a slide sequence of short sequences of system calls. System call short sequence reflects the process execution order of relationship between system calls. If you select short sequence length is too short, you can lose some system call sequence information, and if the length is too long to lose the system status of the implementation of local information, can not correctly reflect the normal and abnormal sequence of calls in case of local conditions. Professor W. Lee [4] studies the problem of data length choice from the perspective of information theory, and deduced that the most appropriate system calls, the short sequence length of 6 to 7. In our study, choice 7 as the short system call sequence length.

Processing every attribute in the experimental data, and change these attributes to numeric. First, a sequence should be extracted for each system call, and make it reasonable description for the SVM training module to identify. For each system call a number was given, the number was calculate by sum of the weighted values of past sample values of each system call; the weighted coefficient was definite by the means of square prediction error minimum principle.

4.2 Experimental Results

The standards of evaluation of the program tests are detection rate and false negative rate. Here are the formula of evaluation standards.

Detection rate = (The number of detected Trojans/ The total number of Trojans) ×100%

False negative rate = (The total number of Trojans -The number of detected Trojans/ The total number of Trojans) ×100%

False alarm rate = (the number of Normal data is considered to be Trojan/ the number of Normal data) ×100%

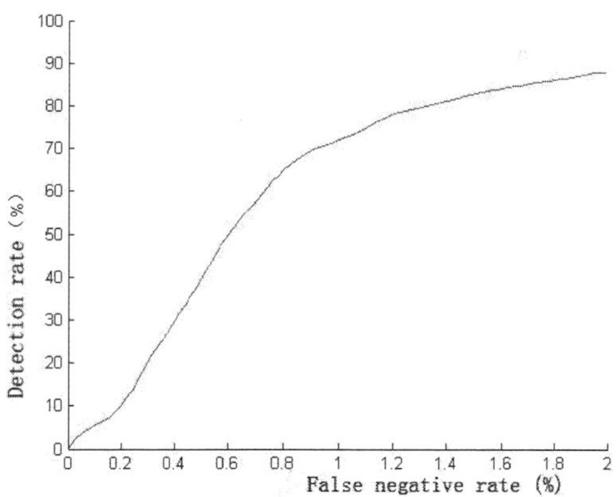

Fig. 2 ROC curve for detection rate and false positive rate

To validate the ability of Trojan horses detection algorithm, obtained D1, D2, D3 data set, the average detection rate and average false positive rate of the ROC curve shown in Fig.2: It can be seen from the figure. when the detection rate of 85%, the false detection rate was 1.6%; the traditional Trojan detection system was 50% and a false detection rate of 1.5%, Full show of the strategy for the detection of unknown Trojans behavior is feasible and effective.

In order to verify the efficiency of this strategy, it is comparison with the anti-Trojan based on Bayesian experimental strategy. First, set all the same experimental conditions, including the sample space, feature vectors and other conditions. Comparison of experimental results are shown in Table 1.

Table 1 Experimental results

	Training time(s)	Detection time (s)	Detection rate (%)	False negative rate(%)
Bayesian	5.754	0.098	88.5	11.5
SVM	1.300	0.109	99.7	2.3

Comparative analysis of above results, we can see, the use of the proposed non-linear support vector machine Trojan horse detection algorithm has higher detection rate, lower false negative rate, and higher accuracy.

5 Conclusion

In this paper, a new Trojan detection technology based on SVM is proposed. Some simulation experiments and actual data show that the proposed method is validated. Experimental results and analysis prove that this method is not only can find known attacks but also can find unknown Trojan horse, and has a high detection rate, and false alarm rate is very low. However, there are some questions need to improve, such as how to change the real-time Trojan horse attack data into the characteristics data which support vector machine can identify, and classification accuracy of support vector machines need to be improved.

References

1. Eskin, E., Amold, A., Prerau, M., et al.: A geometric framework for unsupervised anomaly detection: Detecting intrusions in unlabeled data. In: Data Mining for Security Applications (2002)
2. You, C., Hua-wei, S.: A High Performance Oriented Lightweight Intrusion Detection System Feature Selection Algorithm. Computer Journal 30, 1398–1407 (2007)
3. Hughes, L.A., DeLone, G.J.: Viruses, Worms, and Trojan Horses: Serious Crimes, Nuisance, or Both? Social Science Computer Review 25, 78–98 (2007)
4. Lee, W., Dong, X.: Information-Theoretic measures for anomaly detection. In: Proceedings of the 2001 IEEE Symposium on Security and Privacy, pp. 130–143. IEEE Computer Society Press, Oakland (2001)

High Capacity Lossless VQ Hiding with Grouping Strategy

Chin-Feng Lee*, Kuan-Ting Li, and Dushyant Goyal

Abstract. This paper proposes a reversible data hiding technique on Vector Quantization (VQ) encoding. Our scheme increases the embedding capacity of secret messages by increasing the embeddable index values. First, we count the number of index values generated through image compressed by VQ. Then, we sort the codewords in the VQ codebook in descending order according to frequencies of index values. The rearranged codebook is evenly divided into five groups which are employed to embed secret messages into an index table and then a complete stego index table with embedded secret information is created. From experimental results, our scheme is capable of increasing the embedding capacity of secret data compared to Yang et al.'s. In addition, our scheme can recover the lossless original VQ-compressed index value after extracting the embedding secret data.

Keywords: Vector Quantization, reversible data hiding, embedding capacity, visual quality.

1 Introduction

Many information-hiding schemes have been developed [1][2][5][6][8][9]. The secret message can then be covertly delivered successfully because the human eye cannot easily distinguish between the cover medium and stego-medium. With recent widespread of internet and advance in information technology, many people frequently transmit multimedia such as text, image, voice, video, etc.

Chin-Feng Lee · Kuan-Ting Li
Department of Information Management, Chaoyang University of Technology, Taiwan
No. 168, Jifeng E. Rd., Wufeng District, Taichung 41349, Taiwan, R.O.C
e-mail: amylee.cf168@gmail.com

Dushyant Goyal
LNM Institute of Information Technology, Jaipur, India
e-mail: dushyant.lnmiit@gmail.com

* Corresponding author.

F.L. Gaol (Ed.): Recent Progress in DEIT, Vol. 2, LNEE 157, pp. 203–209.
springerlink.com

The implementation cost to raise the Internet bandwidth is high, thus data compression is exploited before transmission such that the speed of transmission is increased while lowering storage space for data. VQ compression [3] really achieves significant reduction of bit rates and is quite popular to apply to the field of information hiding.

Mostly information hiding based on compression techniques uses VQ [5] or SMVQ [2][4][5] to embed secret data. The fundamental scheme of VQ creates a characteristic codebook by LBG algorithm[7]. This codebook has many vectors, and each vector is called a codeword. When an image block is encoded, the distance between each image block codebook and its corresponding vector with the Euclidean operator is calculated to find the most similar codeword. Then, the index of codeword can replace all pixel values in the block, which completes the goal of image compression.

In 2007, Chang *et al.* [1] proposed a data hiding technique based on VQ. They claimed that they first hid the secret data into compression code, the index values can be lossless recovered to VQ index values with after secret data have been extracted. However, this scheme categorizes the codebooks into three groups, the space for embedding secret data is about one-third of the codebooks. In other words, the index values of two groups are incapable of embedding any secret data. To further increase the space for embedding secret data, Yang *et al.* (2009) [8] proposed another grouping scheme for the codebooks. Through the exchange of positions for codewords, they embedded secret data in cover media, which is a rearranged index table and have better results in embedding capacity than the scheme of Chang *et al.*

The goal of our research is to increase the embedding capacity compared with the scheme of Yang *et al.* A sorted codebook according to index frequencies is evenly divided into five groups to increase the space for embedding more secret data. The scheme is carried out with the exchange of relative positions for codewords to complete the embedding of secret data. At the stage of extracting secret data, not only the secret data can be completely extracted, but the cover media can be restored to the original index values by VQ.

Our proposed scheme will be introduced in Section 2 with experimental results shown in Section 3. Finally, the conclusion is presented in Section 4.

2 Proposed Scheme

The scheme Yang *et al.* (2009) [8] can only embed the secret data in the first half of codeword of the codebook. In order to significantly increase the embedding capacity of secret data, we have divided the codebook into five groups, and the secret data is embedded in the first three groups of the codewords that have the highest frequencies of appearance. The notations below are described for later use. CB: Codebook obtained by LBG algorithm [7], $|CB|$: Size of codebook, CB': New codebook generated from sorted CB, cb_k: Group generated from even grouping of CB', $|cb_k|$: Size of group cb_k.

2.1 Grouping and Sorting of Codebooks

Before embedding secret data, VQ technique is first used to compress the image into an index table. Since the embedding capacity of secret data and the times of appearance of index value obtained by VQ coding are directly related to each other, our scheme sorts the index values of codebook according to frequencies of appearance in an index table and then generates a sorted codebook *CB′*. From the new codebook *CB′* we choose the last three index values (253, 254, 255) and define I_0, I_1, I_2 as indicator flags for image recovering and secret data extracting. Then, by Eq. 2 we divide the remainders of the new codebook *CB′* into five groups; every group has $|cb_k|$ members.

$$\left|cb_k\right| = \left\lfloor \frac{|CB'|-3}{5} \right\rfloor \tag{2}$$

2.2 Embedding Procedures of Secret Data

After obtaining five sub-codebooks, the steps of secret embedding are shown as follows:

Step 1: Assign indicators for the sorted and grouped codebook cb_k such that cb_2 is assigned to I_0, cb_3 is assigned to I_1 while cb_4 and cb_5 are assigned to I_2, as shown in Figure 1. Through the indicators, we can lossless recover the original image after VQ compression when secret data is being extracted.

Step 2: Take the original image with size $W \times H$ and partition it into non overlapping blocks with size $M \times N$.

Step 3: Proceed with VQ compression on every block to generate an index table corresponding to the original image.

Step 4: If index value Ind_i belongs to group cb_1, then according to the to-be-embedded secret data, it is replaced by the index value of relative horizontal position to complete embedding of secret data. If embedded secret data is 00_2 then it is replaced by the index value from group cb_1; if embedded secret data is 01_2 then it is replaced by index value from group cb_2; if embedded secret data is 10_2 then it is replaced by index value from group cb_3; if embedded secret data is 11_2 then it is replaced by index value from group cb_4.

Step 5: If index value Ind_i belongs to group cb_2, then according to the to-be-embedded secret data, it is replaced by the index value of relative horizontal position and indicator I_0 is added in front to complete embedding of secret data. If embedded secret data is 00_2 then it is replaced by indicator I_0 and the index value from group cb_1; if embedded secret data is 01_2 then it is replaced by indicator I_0 and index value from group cb_3; if embedded secret data is 10_2 then it is replaced by indicator I_0 index value from group cb_4; if embedded secret data is 11_2 then it is replaced by indicator I_0 and index value from group cb_5.

Step 6: If index value Ind_i belongs to group cb_3, then according to the to-be-embedded secret data, it is replaced by the index value of relative horizontal position and indicator I_1 is added in front to complete embedding of secret data. If embedded secret data is 00_2 then it is replaced by indicator I_1 and the index value from group cb_1; if embedded secret data is 01_2 then it is replaced by indicator I_1 and index value from group cb_2; if embedded secret data is 10_2 then it is replaced by indicator I_1 index value from group cb_4; if embedded secret data is 11_2 then it is replaced by indicator I_1 and index value from group cb_5.

Step 7: If index value Ind_i belongs to group cb_4 or group cb_5, then no secret data can be embedded, and indicator I_2 is added in front of the index value.

Step 8: Repeat Steps 4 to 7 until all blocks have been coded.

The following example will further exemplify the embedding procedure of our scheme. Let's assume the original codebook has a size of 33, through Section 2.2 we divided it into five groups. According to Eq. 2, each sub-codebook contains six codewords, and the remaining three index values are assigned with I_0, I_1, and I_2 as indicators. For the five sub-codebooks, cb_2 is assigned with indicator I_0, cb_3 is assigned with indicator I_1, cb_4 and cb_5 are assigned with indicator I_2. Next, we partition a cover image into non-overlapping blocks of sizes 4×4, and then we proceed with coding of the blocks from the previously sorted codebook. The index value of such codeword is used as the result of block compression to generate an index table as shown in Figure 1a. Secret data can be embedded in indices if these indices belong to the three sub-codebooks cb_1, cb_2, and cb_3 which will be coded as shown in Figure 1b.

Let's assume the to-be-embedded secret data is 1001_2, to embed a 2-bit secret data at once, we first proceed with coding of the first block. Since the index of the first block is 3, or the third index value of group cb_1, and the to-be-embedded secret data is 10_2, then the third index value 21 from group cb_4 is used for coding. The index for the second block is 29, or the fifth index value from group cb_5, and the to-be-embedded secret data is 01_2; however, no secret data can be embedded in the index values of group cb_4 and cb_5, an indicator I_2 needs to be embedded first in order to prevent inaccurate extraction of secret data and recovery of index value of VQ compression by the receiver. The previous to-be-embedded secret data 01_2 will be processed in the next block. The index for the third block is 2, or the second index value of group cb_1, when it encounters the next to-be-embedded secret data 01_2, we take the second index value, 8, from group cb_2 to proceed with coding. Then, we can completely embed all the desired secret data, as shown in Figure 1c.

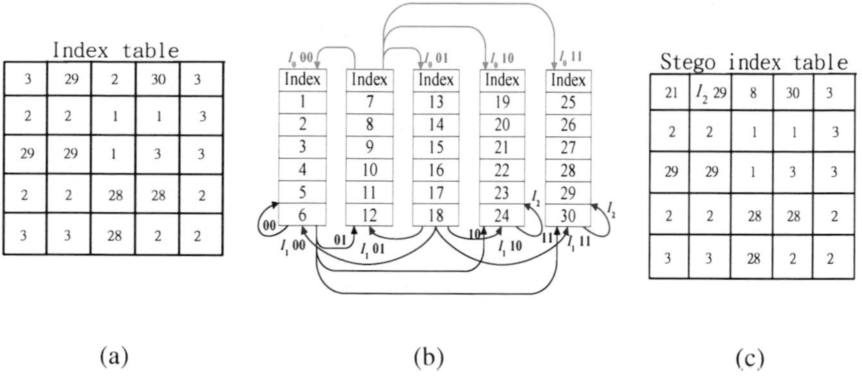

Fig. 1 Embedding of secret data

2.3 *Extraction Procedures of Secret Data*

After the receiver has obtained the index table with secret data embedded, has sorted and grouped the original codebooks, the receiver can proceed with extraction of secret data and restore the original index table according to the following cases.

Case 1: If the to-be-processed index value in the index table belongs to group cb_1 without any front indicator, then secret data 00_2 is extracted while the original index value of group cb_1 is maintained; if front indicator I_0 is present, then secret data 00_2 is extracted and the corresponding index value of group cb_2 is restored; if front indicator I_1 is present, then secret data 00_2 is extracted and the corresponding index value of group cb_3 is restored.

Case 2: If the to-be-processed index value in the index table belongs to group cb_2 without any front indicator, then secret data 01_2 is extracted while the original index value of group cb_1 is restored; if front indicator I_0 is present, then secret data 01_2 is extracted and the corresponding index value of group cb_2 is restored; if front indicator I_1 is present, then secret data 01_2 is extracted and the corresponding index value of group cb_3 is restored.

Case 3: If the to-be-processed index value in the index table belongs to group cb_3 with front indicator I_0, then secret data 01_2 is extracted and the original index value of group cb_2 is restored.

Case 4: If the to-be-processed index value in the index table belongs to group cb_4 without any front indicator, then secret data 10_2 is extracted while the original index value of group cb_1 is restored. If front indicator I_0 is present, then secret data 10_2 is extracted and the corresponding index value of group cb_2 is restored. If front indicator I_1 is present, then secret data 10_2 is extracted and the corresponding index value of group cb_3 is restored.

Case 5: If the to-be-processed index value in the index table belongs to group cb_5 without any front indicator, then secret data 11_2 is extracted while the original index value of group cb_1 is restored. If front indicator I_0 is present, then secret data 11_2 is extracted and the corresponding index value of group cb_2 is restored. If front indicator I_1 is present, then secret data 11_2 is extracted and the corresponding index value of group cb_3 is restored.

The demonstration of extraction procedures of our proposed scheme is shown in Figure 1c. All the secret data can be extracted and the VQ index values of original image can be lossless restored, as shown in Figure 1a.

3 Experimental Results

To prove the performance of our proposed scheme, we used six 8-bit grey images of size 512×512 Lena, Jet, Peppers, Toys, GoldHill, and Zelda to carry out the test. Before the experiment, we generated a codebook with a few characteristic grey images, and this codebook is sorted evenly into five groups according to the appearance times of index values in descending order.

The bit rate is being used to analyze the performances of compression. Calculation of bit rate is carried out with Eq. 3, where L represents the length of output compression code, $H \times W$ is the size of the original image. The lower the bit rate, the better the image compression effect.

$$bit\_rate = \frac{L}{H \times W} (\text{bpp}) \quad (3)$$

Related experimental results are shown in Table 1. Even though the length of compression code in our proposed scheme is 8% longer than that of Yang *et al.* [8], shown in Column 8 of Table 1, the results showed a 13% increase of the embedding capacity of secret data in some of the images (Lena, Jet, Toys, Peppers). As for GoldHill and Zelda, their secret data embedding capacities increased about 9%, as shown in Column 9 of Table 1. The results showed that although our proposed scheme slightly increases the length of compression code, the embedding capacity of secret data is significantly increased by our proposed scheme with an average increase about 11% in capacity.

Table 1 Comparison of experimental data of our proposed scheme and Yang *et al.*'s scheme

Test Images	Yang et al.			Our proposed			Compression Results	
	Embedded Capacity	Compression code size	Bit rate	Embedded Capacity	Compression code size	Bit rate	Increased Compression code size	Increased Capacity
	(A)	(C)	(E)	(B)	(D)	(F)	(D-C)/C	(B-A)/A
Lena	26,980	169,607	0.647	30,559	184,287	0.703	8.66%	13.27%
Jet	28,840	155,976	0.595	32,701	168,186	0.642	7.83%	13.39%
Toys	30,692	150,209	0.573	34,788	163,316	0.623	8.73%	13.35%
Pepers	28,818	159,646	0.609	32,521	173,277	0.661	8.54%	12.85%
GoldHill	23,594	182,714	0.697	25,726	194,253	0.741	6.32%	9.04%
Zelda	27,304	171,442	0.654	29,679	181,647	0.693	5.95%	8.70%

4 Conclusions

This paper proposed a reversible data hiding scheme based on VQ compression. Compared with Yang *et al.*'s scheme, ours first sort the codebook in descending order according to appearance of index word and then evenly divide the sorted codebook into five groups such that three fifths of the least possible values for embedding secret data are used to create a rule for data hiding. The experimental results showed that even though the final compression rate is slightly higher than that of Yang *et al.*'s, our proposed scheme has achieved a significant increase in embedding capacity of secret data.

References

1. Chang, C.C., Wu, W.C., Hu, Y.C.: Lossless recovery of a VQ index table with embedded secret data. Journal of Visual Communication and Image Representation 18(3), 207–216 (2007)
2. Chen, C.C., Chang, C.C.: High capacity SMVQ-based hiding scheme using adaptive index. Signal Processing 90(7), 2141–2149 (2010)
3. Gray, R.M.: Vector quantization. IEEE ASSP Magazine 1(2), 4–29 (1984)
4. Kim, T.: Side match and overlap match vector quantizers for images. IEEE Transactions on Image Processing 1(2), 170–185 (1992)
5. Lee, C.F., Chen, H.L., Lai, S.H.: An adaptive data hiding scheme with high embedding capacity and visual image quality based on SMVQ prediction through classification codebooks. Image and Vision Computing 28(8), 1293–1302 (2010)
6. Li, X., Orchard, M.T.: Edge-Directed prediction for lossless compression of natural Images. IEEE Transactions on Image Processing 10(6), 813–817 (2001)
7. Linde, Y., Buzo, A., Gray, R.M.: An algorithm for vector quantization design. IEEE Transactions on Communications 28, 84–95 (1980)
8. Yang, C.H., Lin, Y.C.: Reversible data hiding of a VQ index table based on referred counts. Visual Communication and Image Representation 20(6), 399–407 (2009)
9. Yang, C.H., Weng, C.Y., Wang, S.J., Sun, H.M.: Adaptive data hiding in edge areas of images with spatial LSB domain systems. IEEE Transactions on Information Forensics and Security 3(3), 488–497 (2008)

Traffic-Oriented STDMA Scheduling in Multi-hop Wireless Networks

Jung-Shian Li, Kun-Hsuan Liu, Chun-Yen Wang, and Chien-Hung Wu

Abstract. For complicated traffic-oriented transmissions (e.g., multiple multicasting) in Spatial Time Division Multiple Access (STDMA) network, scheduling algorithms are employed to determine an efficient assignment for the transmissions. In this paper, a scheduling algorithm, designated as traffic-oriented scheduling assignment (TOSA), is proposed to effectively allocate available network resource in the link-based and node-based STDMA networks. The objective of TOSA is to reduce schedule frame length by enhancing the spatial utilization within each time slot. Simulation results demonstrate that the proposed algorithms achieve a lower schedule frame length than existing STDMA scheduling algorithms.

1 Introduction

A spatial time division multiple access (STDMA) scheduling network [1] has been proposed to enhance network performance by permitting multiple concurrent transmissions within the same time slot. Specifically, the STDMA networks employ interference models to classify wireless devices into different conflict-free graphs and the devices within the same conflict-free graph can be guaranteed to simultaneously activate without collision. The graph-based or SINR-based methods are generally used as interference models to establish conflict-free graphs [2]. In graph-based methods, primary and secondary interference transmission constraints are used to determine which devices within the network are able to be grouped into the same conflict-free graph. In SINR-based methods, the conflict-free graphs are constructed by means of the signal-to-interference and noise ratio (SINR) [2]. According to the activated devices, scheduling algorithms in STDMA networks can allocate transmission rights to either nodes or links [3-6]. In addition, the schedule frame length is usually used to evaluate the performance

---

Jung-Shian Li · Kun-Hsuan Liu · Chun-Yen Wang · Chien-Hung Wu
Department of Electrical Engineering, Institute of Computer and Communication Engineering, National Cheng Kung University, 1, University Road, Tainan 710, Taiwan
e-mail: jsli@mail.ncku.edu.tw, kunhsuan@gmail.com, eejames8@msn.com, bbb123aaaaa@hotmail.com

F.L. Gaol (Ed.): Recent Progress in DEIT, Vol. 2, LNEE 157, pp. 211–216.
springerlink.com

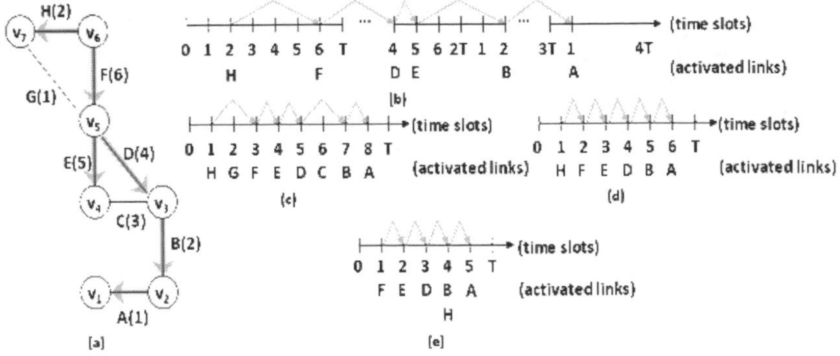

Fig. 1 Link scheduling in a given topology using different assignment strategies.

of scheduling algorithms. However, when traffic-oriented transmissions (e.g., multicasting) are implemented in STDMA networks, two factors cause whether scheduling algorithms are able to yield an effective schedule sequence for all transmissions. Firstly, the utilization of spatial reuse should be raised in each time slot. Secondly, the scheduling assignments must satisfy the traffic demands. For example, if the granted schedule sequence is different from the actual traffic flows, gaps may be induced within the schedule frame, i.e., some time slots have no devices to be activated. As a result, a scheduling delay [6] occurs and therefore the scheduler needs to allocate more time slots to complete transmissions.

Consider the network topology shown in Fig. 1(a) comprising a set of nodes v_i, a set of links X, and a set of conflict-free graphs n, where i=1~6, X=A~H and n=1~6. Assume that the network employs a link scheduling algorithm and performs multicasting traffic from node v_6. Figures 1(b)~(d) show that the transmission order within the schedule frame is specified in accordance with the order of the conflict-free graphs, the reverse link sequence and multicasting traffic flow, respectively. Note that the light lines represent actual activated devices in the time slots. However, only when the transmission order has a one-to-one correspondence with the multicasting traffic, the scheduling delay within the schedule frame can be minimized (see Fig. 1(d)). In addition, the length of the schedule frame can be further reduced by activating multiple devices within the same time slot. Figure 1(e) shows the case where the schedule assignment considers both the traffic flow and spatially-reusable property. In this case, the schedule length is reduced to five. Therefore, a good scheduling assignment should not only minimize the scheduling delay but also enhance utilization of the spatial reuse in each time slot. In this paper, the problem of allocating the time slots to all multicasting transmissions in the STDMA network is called as a multiple multicasting traffic-oriented scheduling (MMTS) problem. The remainder of this paper is summarized as follows. Section 2 proposes two schedule algorithms, designated as link-based traffic-oriented scheduling assignment (TOSA) and node-based TOSA for the MMTS problem. Section 3 compares the performance of the proposed scheduling algorithms with existing schedule algorithms. Finally, Section 4 presents the conclusions of this paper.

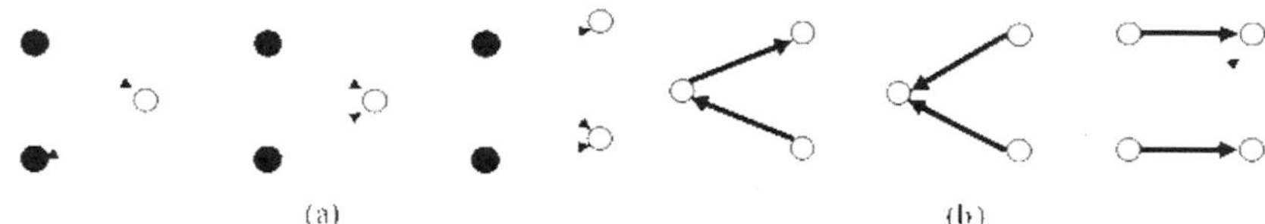

Fig. 2 Schematic illustration of node and link interference constraints in graph-based interference model.

2 Traffic Oriented STDMA Scheduling Algorithms

This paper considers that multiple multicasting transmissions are implemented in STDMA wireless networks. Assume that each node in the network uses a single half-duplex transceiver to receive or transmit information at a fixed rate. It is also assumed that each node (or link) is classified into a unique conflict-free graph by a graph-based interference model [2]. Figure 2 shows the primary and secondary interference constraints for link-based or node-based scheduling networks. Note that the solid points (or the darkened arrow lines) indicate that two nodes (or links) must be grouped into different conflict-free graphs. In this paper, the link-based and node-based traffic-oriented scheduling algorithms are developed for solving MMTS problem. The objective of the proposed algorithms is to find a schedule sequence which not only satisfies traffic demands but also maximizes all the possible spatially-reused transmissions within the schedule frame.

A greedy scheduling algorithm named as traffic-oriented scheduling assignment (TOSA) is proposed to provide an approximated solution for the MMTS problem. A three-step approach is used to implement TOSA in the link-based and node-based STDMA networks. Firstly, a multi-tiers graph is constructed from all the multicast trees. Specifically, since the multicast tree is a kind of tree structure, the sibling nodes are in the same level of the tree. As a result, all the nodes from the same level of all multicast trees are collected into a tier. For example, the tier 1 of the multi-tier graph gathers the first activated devices (links or nodes) from all the multicast trees and so on. Then, several tiers are grouped into a multi-tier graph and thus the TOSA allocates transmission rights in accordance with this graph. Moreover, the designated devices within the same tier have equal priority to assign into time slots since they belong to the same level. In other words, TOSA can arbitrary exchange their schedule positions. Secondly, the TOSA adjusts the activation order of the devices in the same tier to determine the schedule sequence based on the number of activated devices in the conflict-free graph. Thirdly, the TOSA repeats steps 1 and 2 till all the tiers are assigned to the schedule frame. Figure 3 presents an example of TOSA operations. Consider a network topology shown in Fig. 3(a), in which the graph-based interference model is used to construct conflict-free graphs. The link identifiers show which links belong to the same conflict-free graph, e.g., the conflict-free graphs c_1={1A, 1B}, c_2={2A, 2B}, c_3={3A, 3B}, c_4={4}, c_5={5} and c_6={6A, 6B}. Note that the links within the same conflict-free graph can be activated without collision. Assume that each node within the network performs multicast transmission to deliver information for all other nodes. Figure 3(b) demonstrates a multicast tree established from source v_0. As shown, the links within the tree can be divided into three different

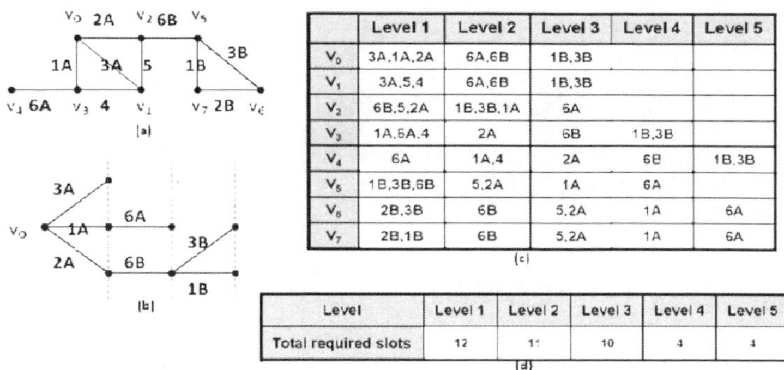

	Level 1	Level 2	Level 3	Level 4	Level 5
V_0	3A,1A,2A	6A,6B	1B,3B		
V_1	3A,5,4	6A,6B	1B,3B		
V_2	6B,5,2A	1B,3B,1A	6A		
V_3	1A,6A,4	2A	6B	1B,3B	
V_4	6A	1A,4	2A	6B	1B,3B
V_5	1B,3B,6B	5,2A	1A	6A	
V_6	2B,3B	6B	5,2A	1A	6A
V_7	2B,1B	6B	5,2A	1A	6A

(c)

Level	Level 1	Level 2	Level 3	Level 4	Level 5
Total required slots	12	11	10	4	4

(d)

Fig. 3 Illustrative example of link-based TOSA.

levels. Consequently, Figure 3(c) shows the links within the identical level collected from all the multicast trees. Then, TOSA determines the schedule sequence according to the number of active links in the conflict-free graph. In other words, the conflict-free graph which includes maximal links in the present level can be firstly allocated in the time slot, e.g., conflict-free graphs c_1 and c_6 in the level 4 have a high priority to be assigned into time slots. Note that the identical conflict-free graph may appear more than one time in the same level and such graph should be allocated several time slots to complete the transmissions. Finally, Figure 3(d) presents the total number of required time slots used for each level.

3 Performance Evaluation

In this section, two performance metrics are used to evaluate two proposed scheduling algorithms with existing scheduling algorithms. Firstly, normalized schedule frame length is defined as F/F_{TDMA}, where F_{TDMA} indicates the schedule length generated from the conventional TDMA scheduling assignment and F represents scheduling results from the STDMA scheduling algorithm. In other words, this ratio is used to show the efficiency of scheduling algorithms. Secondly, the number of reusable time slots is used to examine whether the scheduling algorithm properly uses available spatial channels. Note that the notations link-based STDMA and node-based STDMA indicate the schedule assignments without satisfying traffic-oriented demands. In the simulation, a random topology generated in a wireless multi-hop network is used to examine the performance of six different scheduling algorithms and each evaluation is repeated thirty times to show the average results. Note that the network size is set to 15. Figure 4(a) represents the simulation results of six different scheduling algorithms. In the simulation, the topology size is fixed and the network degree is used to demonstrate the effects on the scale of interferences. Note that the network degree indicates the maximal degree used in the network. As shown, node-based TOSA has the better performance than other scheduling algorithms since this algorithm is suitable for multicasting transmission and attempts to enhance the number of spatially-reusable nodes within each time slot. Moreover, when the network degree is

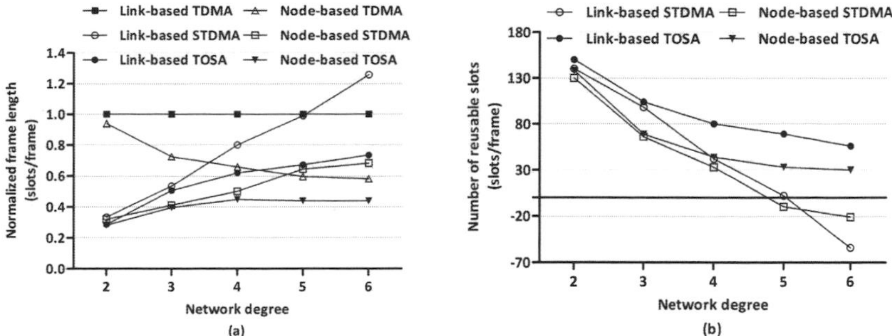

Fig. 4 Variations of normalized frame length and reusable time slots in a random topology.

increased, the scheduling algorithms allocate more time slots to complete the multicasting transmissions since the high network degree results more interference conditions. Figure 3(b) shows the variation of reusable time slots per frame with network degree for the four scheduling algorithms. As shown, link-based and node-based TOSA algorithms can effectively use the available channels in each schedule frame. When the network degree is more than four, link and node-based STDMA perform poor spatial utilization since the devices (links or nodes) in the conflict-free graphs are gradually decreased. As a result, the network degree has great effects on the number of reusable time slots in the schedule frame. Moreover, it is seen that the link-based TOSA outperforms the node-based TOSA in evaluation of reusable time slots. However, the link-based TOSA yields longer schedule length than node-based TOSA since the transmission type of node-based TOSA is more suitable for multicasting. In other words, link-based STDMA networks perform multiple-unicast transmission instead of multicasting transmission.

4 Conclusion

This paper has examined the problem of multiple multicasting traffic-oriented scheduling (MMTS) in the STDMA networks. A scheduling algorithm, designated as TOSA, has been proposed for enhancing the network performance by minimizing the scheduling delay and improving the spatial utilization within each time slot. The simulation results have demonstrated that proposed algorithm outperforms existing STDMA-based scheduling algorithms.

References

1. Nelson, R., Kleinrock, L.: Spatial TDMA: A Collision-Free Multihop Channel Access Protocol. IEEE Trans. Commun. 33(9), 934–944 (1985)
2. Behzad, A., Rubin, I.: On the Performance of Graph-based Scheduling Algorithms for Packet Radio Networks. In: Proceedings of IEEE Global Telecommunications Conference (GLOBECOM), vol. 6, pp. 3432–3436 (2003)

3. Ramanathan, S., Lloyd, E.L.: Scheduling Algorithms for Multihop Radio Networks. IEEE/ACM Trans. Netw. 1(2), 166–177 (1993)
4. Ephremides, A., Truong, T.V.: Scheduling Broadcasts in Multihop Radio Networks. IEEE Trans. Commun. 38(4), 456–460 (1990)
5. Badia, L., Erta, A., Lenzini, L., Zorzi, M.: A General Interference-Aware Framework for Joint Routing and Link Scheduling in Wireless Mesh Networks. IEEE Netw 22(1), 32–38 (2008)
6. Djukic, P., Valaee, S.: Delay Aware Link Scheduling for Multi-hop TDMA Wireless Networks. IEEE/ACM Trans. Netw. 17(3), 870–883 (2009)

Cloud Computing: Understanding the Technology before Getting "Clouded"

Sam Goundar

Abstract. The IT Industry and the wider media are abuzz with the phrase "cloud computing" and most of us have no idea what this latest terminology means. This paper starts by addressing the question – "what is cloud computing, and provides an understanding of the technology?" Cloud computing is related to technologies that has been around for ages, however, what can be done differently with the same technology when integrated with Internet is cloud computing.

Individuals who might be considering cloud computing services for personal use will also benefit by the information provided. Web sites like Flickr, FaceBook, YouTube and others are used to store and share personal photos, music and videos for free. Easy as: 1). Register, 2). Upload, 3). Share. Free and easy, there has to be a downside, and there is awareness regarding this.

There is information regarding cloud computing for businesses! Businesses, who wish to get on the "flight to the clouds" like their peers to keep their heads above the "cloud" and competition. With cloud service providers, businesses can tap into the IT services that they need, when they need, for as long as they need; without investing in any IT infrastructure. The result is a far more agile and cost-effective IT services and this paper looks at how and why.

After looking at the good of cloud computing, the bad, and the ugly is revealed. Not all our experiences on the Internet have been totally positive. A range of security and trust issues exist. Individuals and businesses ask is "how secured are their information once it is with the cloud service providers?" There is analysis on the ownership of data and information stored with the cloud service providers and taxonomy of data that you post on social networking web sites.

Finally, anecdotes about cloud computing that support our endeavour to compute in the cloud. Cloud computing is not a hype, it is a reality. Cloud computing is not a fad, it is here to stay. The question is no longer "will cloud computing happen?" but "how you are going to exploit this technology?"

Sam Goundar
Bay of Plenty Polytechnic, New Zealand
e-mail: goundar_s@hotmail.com

F.L. Gaol (Ed.): Recent Progress in DEIT, Vol. 2, LNEE 157, pp. 217–222.
springerlink.com

Keywords: cloud computing, virtual computing, cluster computing, distributed computing, pervasive computing.

1 Understanding Cloud Computing

What is cloud computing? Let's start with a simple concept. Emailing information to yourself from office and then retrieving and using that information at home is an example. Most of us are doing that, therefore we are already computing in cloud. This is a cloud user's beginners level, as cloud computing can offer a lot more than that. What else can be done with cloud computing to actually become a "cloud user" and stay in tune with the latest trend and technology of the IT Industry?

[3] Vision of 21st century is accessing Internet services from light weight portable devices, instead of accessing it from a traditional Desktop PC. Cloud computing is a technology which will facilitate companies or organisation to host their services without worrying about IT infrastructure, let alone invest in it and other supporting services.

The cloud concept draws on the existing technologies which are not new such as Virtual Computing, Cluster Computing, Utility Computing, Distributed Computing and Software-as-a-Service [SaaS]. It is new in the way it integrates all of the above and shifts them from a localised processing unit to a globalised network [11].

A simple explanation for cloud computing would be not storing data and information on local hard disk or local servers, but storing it away from our physical location. When the need arises to use that data or reference that information, access is obtained via the Internet. Since access to that data and information is via the Internet, it is available from anywhere via Internet. Access to data and information is not confined to any location, and that is the essence of cloud computing!!

How did a simple explanation acquire a fancy name as "cloud computing"? Clouds can be seen somewhere in the sky whenever we look up regardless of physical location. Whenever we connect to the Internet, data and information is up there regardless of physical location. There are claims of technical reasons behind the name. Network configuration diagrams show connections to the Internet away from the local area network [LAN] and virtual private network [VPN] connections with a fluffy cloud symbol. Any organisation or institution that uses an Internet application, even as simple as webmail applications can claim to be computing in cloud.

"Clouds are a large pool of easily usable and accessible virtualized resources (such as hardware, development platforms and/or services). These resources can be dynamically reconfigured to adjust to a variable load (scale), allowing also for an optimum resource utilization. This pool of resources is typically exploited by a pay-per-use model in which guarantees are offered by the Infrastructure Provider by means of customized SLA [Service Level Agreement]". [9]

2 Cloud Computing for Individuals

It is evident that cloud computing will change the computing environment. This section looks at how cloud computing will influence the way in which individuals compute. Cloud computing will also save them money, time, and effort if used and implemented correctly. Individuals provide and store their personal data and information with cloud service providers based on trust and promise of security, however, time and again their data and information is abused. How this happens and why this happens is investigated and analysed.

Software as a service (SaaS) refers to the paradigm where software and other solutions are delivered to the end user as a service using the Internet rather than as a product that can be installed on the user's computer. An example of SaaS for individuals is Google Apps consisting of Gmail, Google calendar and Google docs, which allow one to store and access documents online rather than on their personal computers. [1]

SaaS for individuals can be conceptualized in terms of two dimensions: cloud services and cloud storage [10]. Similar to cloud services, which refer to the online delivery of software, cloud storage refers to the online delivery of data storage, for example, Amazon EC2 - part of Amazon's web services. Cloud storage is more popularly known as Storage as a Service (STaaS).

Some common scenarios that require use of cloud storage and cloud service providers are: the hard disk on the PC is reaching its storage capacity – where can more files be stored? Hundreds of photos and videos that were taken while holidaying in Fiji needs to be shared with family and friends? Personal documents, photos, music and video clips are not allowed to be stored on the office PC – what is the alternative?

Processing power and storage capacity of large companies like Google can be utilised for free or by paying an insignificant amount. There is no need to reinvent the wheel or to start from scratch, because cloud service providers like Google, MSN, Yahoo, Amazon and others have done the hard yards. For individual users, Google, MSN, Yahoo!! and others will provide free web-based emails, and a range of web based applications, including common productivity software.

Save money by not paying for processing power and storage. Just pay for devices like $29 mobile phones, $20 Tablet PCs, and other portable devices. They are cheap because they don't have much processing power and storage capacity. However, these devices provide connection to cloud service providers. And once connected, the cloud service provider will do all the processing and store whatever is required. Time and effort is not spent on processing. It is just like using Google's Search Engine to find information on whatever is needed, as compared to going to the Library and taking the information out by browsing through several dozen books.

3 Cloud Computing for Business

Cloud computing is going to change business processes. This section looks at how businesses can use cloud computing and what cloud computing technology can offer them. We are now in what is considered the "knowledge" economy. In this knowledge economy, the currency is "information". Wealth in this economy is measured by the amount of information that can be accessed, manipulated and provided. Founders of FaceBook, Google, Yahoo, and many other cloud service providers and social networking web sites have become overnight billionaires by trading information. Their trade and trading secrets are known now. The world is full of people with a voracious appetite for information. For business considering of going into the trade of information, their link to the people of this world is through cloud computing.

Organizations are being founded with very little physical capital. For a services or knowledge intensive business, free tools and low-cost computing cycles can mostly be expensed, changing the fund-raising and organizational strategies significantly. Smartphone's, Tablet PCs and other devices built without mass storage can thrive in a cloud-centric environment, particularly if the organization is designed to be fluid and mobile. Coburn Ventures in New York, for example, is comprised of a small team of mobile knowledge workers who, for the first five years, had no corporate office; the organization operated from Wi-Fi hot spots, with only face-to-face meetings [4].

Computing as a utility has reached the mainstream. Vendors [cloud] now rent all or portions of physical machines for hourly periods for web services. The cloud computing model provides flexible support for "pay as you go" systems. In addition to no upfront investment in large clusters or supercomputers, such systems incur no maintenance costs. Furthermore, they can be expanded and reduced on-demand in real-time. The cloud computing model emphasizes the ability to scale computing resources on demand. The advantages for businesses are numerous [6].

Consider the following scenarios: businesses that are planning to start small, but do not have the capital to invest in IT infrastructure that is critical for the business to succeed; businesses thinking of backing up their critical data online as part of their data recovery contingency plan; businesses thinking of using an Application Service Provider [ASP] to supplement existing business and avail it 24/7 for their customers? These businesses can consider the following:

Small businesses can get up and running without investing in IT, office space, and other physical establishment requirements. They will be able to advertise, market, sell and communicate with their customers using services provided for free by many web sites. I know of some small business and car dealerships that have closed their offices, laid off their staff, and the owner/operator now sells on TradeMe <www.trademe.co.nz> in New Zealand. And they don't pay any tax.

4 Cloud Computing Caveats

Any technology has its downside. There are many with cloud computing. This section looks at some common implications and abuse of cloud computing infrastructure. It is not within the scope of this paper or any one paper to cover all. Privacy, Security, and Trust in cloud computing have been the buzzword of cloud computing naysayers. Many of these issues have been addressed by the pioneers of cloud computing like Google, Amazon, and others, but as we continue riding high in the clouds, new security, privacy, and trust violations time and again ground us. Regardless, cloud computing will not be stalled.

Data Security requires resistance to environmental, second-party and third-party threats to content, both in remote storage and in transit. Large organisations may be more highly professional, and are capable of investing more in security. On the other hand, they are highly visible and attractive targets. There has already been a report of Amazon's EC2 cloud service being compromised to the extent of having a "botnet" command and control module inserted [7].

The potential risks of using cloud services include loss of direct control of resources and increased liability risk due to security breaches and data leaks due to shared external resources. Additionally, reliability loss is a distinct risk since service providers may go out of business, causing business continuity and data recovery issues [5].

While it is evident that cloud providers endeavour to improve their offerings to meet clients' enterprise-grade security needs, it may not be enough in some key sectors. In sectors such as defense, aerospace and brokerage, security and compliance requirements which include the physical location of the data have made SaaS and Hardware public clouds unacceptable for a while [8]. In a recent survey, 64% of respondents in the US Federal Government say security is their topmost concern in the context of Cloud Computing [2].

Every one of the web's elite destinations has suffered from major outages at some point. In light of that history, what does a fault-tolerant cloud environment look like, require and cost? How does optimization work in a cloud? A company selling cloud services may be optimizing power consumption, say, while customer A wants stable (not necessarily fast, but predictable) transaction times for a shopping cart scenario. Customer B needs fast compute capability despite big and frequent reads and writes to disk. How can all three parties go home happy at the end of the day?

Cloud Computing's lifeline is connectivity. All the services, benefits and goodies that it is envisaged to deliver only happens when we are connected to the Internet. No connectivity = no cloud!! How consistent is connectivity. Different readers from different parts of the world will have different answers.

5 Discussion and Conclusion

Cloud Computing refers to both the applications delivered as services over the Internet and the hardware and systems software in the datacenters that provide those services. The services themselves have long been referred to as Software as a Service (SaaS). The datacenter hardware and software is what we will call a Cloud. When a Cloud is made available in a pay-as-you-go manner to the general public, we call it a Public Cloud. We use the term Private Cloud to refer to internal datacenters of a business or other organization, not made available to the general public.

There has been a lot of debate around the virtues of cloud computing. There are some that believe cloud computing is a hype, while others say cloud computing is a fad, and some are even claiming it as "old wine in a new bottle". Whether it is hype or a fad, cloud computing technology is having an effect on the IT ecosystem and its inhabitants. I believe cloud computing is here to stay and cannot be avoided, let alone sidelined. It will become the foundation of future IT infrastructure and the new IT ecosystem will be built on it. Announcements by leaders of the IT Industry on how much they are going to spend on cloud computing technology should be an indication to the rest of us. Cloud computing has the potential to do what Internet has done to the computing landscape.

References

[1] Ambrose, P., Chiravuri, A.: An Empirical Investigation of Cloud Computing for Personal Use. In: Proceedings of the Fifth Midwest Association for Information Systems Conference, Moorhead (May 2010)

[2] Chabrow, E.: Rules Make Adoption of Cloud Computing Challenge for Agencies. Survey: 64% in Government Rank Security as Top Cloud Concern (2009)

[3] Dikaiakos, M., Katsaros, D., et al.: Cloud Computing: Distributed Internet Computing for IT and Scientific Research. IEEE Internet Computing 13(5), 10–13 (2009)

[4] Jordan, J.M.: What the Cloud Really Does? Department of Supply Chain and Information Systems at Penn State University (June 04, 2010), http://www.topics.forbes.com (retrieved)

[5] Mather, T., Kumaraswamy, S., Latif, S.: Cloud Security and Privacy: An Enterprise Perspec-tive on Risks and Compliance. O'Reilly Media, Inc. (2009)

[6] Napper, J., Bientinesi, P.: Can Cloud Computing Reach the TOP500? Aachen Institute for Advanced Study in Computational Engineering Science (February 20, 2009)

[7] Prince, B.: Amazon EC2 Used as Botnet Command and Control. eWeek Security Watch (2009), http://securitywatch.eweek.com/botnets/amazon_ec2_used_as_botnet_command_and_control.html

[8] Swaminathan, K.S.: Computing in the Clouds. Outlook Journal. Computing in the Clouds, By Kishore S. Swaminathan, Chief Scientist, Accenture (2008)

[9] Vaquero, L., Rodero-Merino, et al.: A Break in the Clouds: Towards a Cloud Definition. ACM SIGCOMM Computer Communication Review 39(1), 50–55 (2008)

[10] Vogels, W.: Eventually Consistent," Association for Computing Machinery. Communications of the ACM - Rural Engineering Development 52(1), 40 (2009)

[11] Weiss, A.: Computing in the Clouds. Cloud Computing: PC Functions Move Onto The Web. Networker, 16–25 (December 2007)

Literature Review of Network Traffic Classification Using Neural Networks

Pang Bin and Li Ru*

Abstract. The management and surveillance of the network operation are vital to the managers of the network. However the traditional network management software or tools can not archive this objective. Currently there are mainly three methods for network traffic classification and recognition, one is rely on 'well known' TCP or UDP port numbers, second is deeply packet inspection and the third is based on features of traffic flow. The first two methods both have some shortcomings. while the third method can be through selecting the different pattern recognition methods, Such as linear models for classification, kernel methods, clustering methods and neural networks methods to overcome this shortcoming by identifying network applications based on per-flow statistics, derived from payload-independent features such as packet length and inter-arrival time distributions and so on. In this literature review, with the articles we get from the Google scholar, analyze the features these article use to classify the traffic by using the neural networks, and the classification accuracy of these articles announced. Finally discuss the improvement of the algorithms these articles adopted.

1 Introduction

1.1 Historical Background

With the rapid development of Internet technologies, the architecture of the Internet becomes more and more complex; we must face the challenges of the network security, management and traditional network-based applications.

Pang Bin
Centre of Network Information Inner Mongolia University Hohhot, 010021, China
e-mail: cspb@imu.edu.cn

Li Ru
College of Computer Science Inner Mongolia University Hohhot, 010021, China
e-mail: liru@imu.edu.cn

Corresponding author.

F.L. Gaol (Ed.): Recent Progress in DEIT, Vol. 2, LNEE 157, pp. 223–231.
springerlink.com

Traditional traffic monitors focus on OSI model layer 3 and 4 to detect only general network performance, recognizing application's identity by assuming that most applications consistently use 'well known' TCP or UDP port numbers (visible in the TCP or UDP headers). However, many applications are increasingly using unpredictable (or at least obscure) port numbers so it is difficult to get the details of the network traffic component, for this reason re-searchers propose a new method to identify current sophisticated traffic data generated from various newly emerging network-based applications such as streaming media, peer-to-peer and other applications. The classification which designed is based on application layer payload to inspect the application layer characteristics, especially aimed at the dynamically determining ports, pseudoing HTTP and data segmenting applications in order to escape scan and monitor. This method is named deeply packet inspection. The process of design is separate into several layers and modules to adapt different protocol's characters. And at the same we must get the every applications signature that different with each other in order to classify them. This approach is limited by the fact that classification signature must be updated whenever a new application emerged, and privacy laws and encryption can effectively make the payload inaccessible. Now more and more researchers are looking particularly closely at the application of pattern recognition and Machine Learning (ML) techniques to IP traffic classification.

1.2 Why Traffic Classification Is Important

Because Accurate identification of network applications is fundamental to numerous network activities, from security monitoring[1] to accounting, and from quality of service measurements[2] to forecasting for long-term provisioning [3]. Governments are also clarifying ISP obligations with respect to 'lawful interception' (LI) of IP data traffic [4].

2 Methods

2.1 Screening Criteria

Based on the outcome of the above work of the literature, we also need take some practical and methodological criteria for the better search result.

The First practical criterion is whether the literature that I founded is relevant to the topic, and this document can be helpful for my work. Of course this document is not necessarily identical to my study topic, for example, some documents in network security area like IDS (Intrusion Detection System) or IPS (Intrusion Prevention System) using neural networks also can be referenced in my study area.

The second practical criterion is the theory the paper or document has been used; it must relevant to the neural network and network traffic classification.

And the first methodological criterion is the impact factor of journals or conferences. If the impact factor of this paper is very low, for example below 0.4, or its ranking is not high, I think I will drop it.

The second methodological criterion is the traffic features adopted by these articles, because there are many different features can affect the accuracy and the efficiency, and we must be careful of these features selection.

There still some other criteria but not list above. As literature review is an iterative process, so I can revise the list continually.

2.2 Screening Results

After the selection of the key words and based on the criteria including practical criteria and methodological criteria ,the next step is to apply the searching process to get the papers that I wanted, this is a iterative procedure, and need to revise the key words sometimes the searching result are shown in table 1.

Table 1 Searching Result of the Key Words.

Name of search engine	Key word searches conducted	Results of search Subhead
Google Scholar	Traffic recognition with neural networks	32,700
Same to above	network traffic recognition neural	31,700
Same to above	network traffic classification recognition neural network	19,100
Same to above	supervised network traffic classification recognition neural network	4,560
Same to above	supervised network traffic classification recognition neural network after 2005	1,870

3 The Literature

3.1 Filtering

Based on the practical criteria and methodological criteria, we can filter the papers, With the fast reading of the nearly about 50 papers that I can get, I found that some of them are not relevant to my topic such as [5, 6, 7]. And the papers that only focused on neural networks but with no relationship with the network classification are also dropped [8, 9, 10, 11, 12], similarly the papers that only focused on network classification but have less relevant to neural networks are also dropped[13, 14, 15].

There are another group of paper that has relevant to the topic but not very important to my study, and these papers belongs to the network security field, while sometimes it will help for my topic [16, 17, 18, 19, 20, 21, 22, 23, 24, 25].

Finally, the rest 10 papers are fitted the wants that I needed, and accord to the practical criteria and methodological criteria."

3.2 10 Selected Papers

Based on the practical criteria and methodological criteria, we select the 10 papers for my literature review. For the first paper [3], is the document that I based on, with much relevant to my work, my further study is based on this paper's author' work. and the second and third paper [4, 1], are the review paper for network classification and Machine Learning, the fourth paper I selected, [26] improving neural network classification, and I can adopt it for my further study. In the fifth paper,[27] noted a new algorithm of neural network can improve and monitor the network flow, and its algorithm maybe useful to my study on improving the efficiency where the Moore uses. And 6, 7, 8 paper [28, 29, 30] focused on using neural networks to classify the just one type of the traffic, peer-to-peer network flows. For the ninth paper [31], it says that number of iterations required to get convergence of the system are also less with the rough neural network when compared with the pure neural network, so its method can improve the classification efficiency with considerable progress, so it can be worth to get more attention with it. And the last paper focused on the online game traffic using the neural networks, and must have some use for reference.

All the papers selected are shown in Table2.

Table 2 Selected 10 papers

No.	Title	Impact Factor	Selected Reason
1	Bayesian neural networks for internet traffic classification	3.726	This is the article that I based on ,with much relationship with my work
2	Internet traffic classification demystified: myths, caveats, and the best practices		seven traces with payload collected in Japan, Korea, and the US ,compare with other method of traffic classification
3	A survey of techniques for internet traffic classification using machine learning	2.071	review 18 significant works that cover the dominant period from 2004 to early 2007
4	Improving Neural Network Classification Using Further Division of Recognition Space	0.402	Improving Neural Network Classification, and I can adopt it
5	Neural projection techniques for the visual inspection of network traffic	0.86	A new algorithm of Neural Network can improve and monitor the netflow

Table 2 (*continued*)

6	Learning on class imbalanced data to classify peer-to-peer applications in IP traffic using resampling techniques		I can learn some classify technology of peer-to-peer applications in IP traffic
7	Exploiting unlabeled data to improve peer-to-peer traffic classification using incremental tri-training method		And How to improve peer-to-peer traffic classification
8	Online hybrid traffic classifier for Peer-to-Peer systems based on network processors	1.909	Same to above.
9	Rough Set Approach to Unsupervised Neural Network based Pattern Classifier		This paper says that number of iterations required to get convergence of the system are also less with the rough neural network when compared with the pure neural network,
10	Evaluating Machine Learning Methods for Online Game Traffic Identification		This paper focus on online game traffic, and have some use for reference

4 Discussions

From this admittedly small review, we can get some conclusion from compare. And at the same time we can get some interesting tentative conclusions, questions, new issues, or challenges that draw from the literature reviewed above. Next we can discuss these discover from several aspect.

4.1 Objective

The Selected 10 papers in Table 2 objectives are divided into two aspects. Paper 1, 2, 3, 6, 7, 8, 10 focus on the classification while paper4, 5, 9 mainly focus on the algorithm of improvements of neural networks. The method and the type of the traffic these papers aimed are also different. paper 1 using Bayesian neural networks to classify many type of the traffic, while paper 2 and 3 discuss different methods to classify the traffic, including port-based classification and deeply packet inspection and Machine Learning method and compare their advantages and disadvantages. Additional, in paper 6, 7, 8 and 10 using networks to classify only one type of the traffic just as peer-to-peer flows or online game flows from the all traffic captured from the Internet. Based on these findings, I can make out my own objective such as using the improvement algorithm to promote the efficiency of the classification or apply the classification to another specific type of traffic such as pseudoing HTTP flow and so on. Or even can be propose a new algorithm to expand the scope that the classification applied.

4.2 Theory

All of these papers using neural networks for the base method, but paper 1 using Bayesian neural networks to accelerate the speed of the calculation progress. Paper 2 and 3 just compare the neural networks technology with other methods just like port-based classification and deeply packet inspection and Machine Learning. paper 4 improve the algorithm by using further division of recognition space, can improve the classification performance and accuracy by expanding and further dividing the recognition space. While paper 9 use the rough set approach on neural network to improve the classification accuracy, which is calculated as the number of correct recognitions by the total number of samples while testing, shows that the efficiency of the system increases considerably when rough neural network is applied. Paper 5 using neural projection models network traffic monitoring and classification. While paper 7 focuses on the training method neural networks adopted. Especially, paper 8, proposed a network processor (NPs) based online hybrid traffic classifier. The designed hardware classifier is able to classify P2P traffic based on the static characteristic namely on line speed, and the Flexible Neural Tree(FNT) based software classifier helps learning and selecting P2P traffic attributes from the statistical characteristics of the P2P traffic. Other papers not mention new technologies for neural networks, just using it to classify the different type of the traffic, only.

4.3 Traffic Features

In paper 1, they take a summary of the 246 per-flow features that are available to them. A full description of these features is provided in [33].The computational overhead for the total feature group has led them to use an off-line technique at this time. And to my feel this features are too many to be adopted, my further work can study how to reduce this features to a small set. Paper 2 and 10 adopt Correlation-based Filter (CFS), which is computationally practical and outperforms the other filter method (Consistency based Filter) in terms of classification accuracy and efficiency [2].

The other papers as 7 and 9 are not focus on features based traffic classification so not provide the features that they select in the paper, or even provided, like paper 9, total 31 features are extracted and are mainly of either structural or statistical type and help in obtaining local characteristics than the global ones, note for image preprocessing, not for network flow.

The features selection can influence the efficiency of the classification algorithm, so with the beginning of the classification, we must select the best features subset in order to avoid the sudden drop of the classification efficiency, As we have noted before, adopt Correlation-based Filter (CFS), or other papers' suggest maybe a shortcut to the success, but I think there will need to take much more experiments to verify them.

The features relationship with each other is also can be dividing into a single area. If possible we can do some develop in this field.

5 Contributions, Limitations

5.1 Contributions

By all of the works that we completed, the contribution to the question that started it all mainly is to construct a framework of the topic that I studied, using neural networks for network traffic classification. Although this framework is not comprehensive and have to be add much articles and theories to support. Additionally through the comparison of the literature, we can find the shortcoming of the articles and the gaps among them. I think this is the fundamental of my next step of research and study. Based on the analysis of the articles, and with the support of the framework, we can narrow the field; focus on the one or two point to reach my own achievement. Another contribution will be this review take a summary of the front of the technologies and findings in this field, can make the readers have a overview of researcher's field, and make followers or beginners of this field grasp the comprehensive knowledge and theories ,even to develop new technologies or algorithms to get the further progress in this field.

5.2 Limitations

Of course there are many limitations in this literature review. Because the time for doing the job is not enough, the preparation of the search and analysis are not arrangement. And the database to searching for only in Google Scholar, also make the searching result maybe limited.

Acknowledgments. This work was supported by Natural Science Foundation of Inner Mongolia Autonomous Region Project" Research based on machine learning network traffic analysis and management " (No.2010MS0917), and Inner Mongolia Science and Technology Department Social Development Project "Research on heterogeneous network system based on emergency disaster "(No.20090512).

References

1. Nguyen, Armitage, G.: A survey of techniques for internet traffic classification using machine learning. IEEE Communications Surveys & Tutorials 10, 56–76 (2008)
2. Williams, Zander, S., Armitage, G.: Evaluating machine learning algorithms for automated network application identification. Center for Advanced Internet Architectures, CAIA, Technical Report 060410B (2006)
3. Auld, T., Moore, A., Gull, S.: Bayesian neural networks for internet traffic classification. IEEE Transactions on Neural Networks 18(1), 223–239 (2007)
4. Kim, H., Claffy, K., Fomenkov, M., Barman, D., Faloutsos, M., Lee, K.: Internet traffic classification demystified: myths, caveats, and the best practices. In: Internet Traffic Classification Demystified: Myths, Caveats, and the Best Practices. Internet traffic classification demystified: myths, caveats, and the best practices, pp. 1–12. ACM, City (2008)

5. Brans, E., Brombach, B., Zeidler, T., Bimber, O.: Enabling mobile phones to support large-scale museum guidance. IEEE Multimedia 14(2), 16–25 (2007)
6. Dietterich, T., Langley, P.: 1 Machine Learning for Cognitive Networks: Technology Assessment and Research Challenges (2008)
7. Chitaliya, N., Trivedi, A.: Feature Extraction Using Wavelet-PCA and Neural Network for Application of Object Classification & Face Recognition. In: Feature Extraction Using Wavelet-PCA and Neural Network for Application of Object Classification & Face Recognition. Feature Extraction Using Wavelet-PCA and Neural Network for Application of Object Classification & Face Recognition, pp. 510–514. IEEE, City (2010)
8. Lippmann, R.: Pattern classification using neural networks. IEEE Communications Magazine 27(11), 47–50 (1989)
9. Jing, X., Yanghong, T.: Handwritten digital recognition method based on new feature extraction method and quantum neural network. Electronic Measurement Technology (2009)
10. Xu, J., Wang, G.: Cellular Neural Network in Gray-scale Video Traffic Image Processing (2009)
11. Yan-Ping, L., Zhen-Min, T., Hui, D., Yan, Z.: Speaker Identification Algorithm of Semi-Supervised Learning Mechanism. Computer Engineering 14 (2009)
12. Chitaliya, N., Trivedi, A.: Feature Extraction Using Wavelet-PCA and Neural Network for Application of Object Classification & Face Recognition. In: Feature Extraction Using Wavelet-PCA and Neural Network for Application of Object Classification & Face Recognition. Feature Extraction Using Wavelet-PCA and Neural Network for Application of Object Classification & Face Recognition, pp. 510–514. IEEE, City (2010)
13. Sporns, O., Honey, C., Kotter, R.: Identification and classification of hubs in brain networks. PLoS One 2(10) (2007)
14. Zhong, S., Khoshgoftaar, T., Seliya, N.: Clustering-based network intrusion detection. International Journal of Reliability Quality and Safety Engineering 14(2), 169 (2007)
15. Shrivastav, A., Tiwari, A.: Network Traffic Classification Using Semi-Supervised Approach. In: Network Traffic Classification Using Semi-Supervised Approach. Network Traffic Classification Using Semi-Supervised Approach, pp. 345–349. IEEE, City (2010)
16. Locasto, M., Wang, K., Keromytis, A., Stolfo, S.: Flips: Hybrid adaptive intrusion prevention. In: Flips: Hybrid Adaptive Intrusion Prevention. Flips: Hybrid adaptive intrusion prevention, pp. 82–101. Springer, City
17. Moradi, M., Zulkernine, M.: A neural network based system for intrusion detection and classification of attacks. In: A Neural Network based System for Intrusion Detection and Classification of Attacks. A neural network based system for intrusion detection and classification of attacks, City
18. Balagani, K., Phoha, V., Kuchimanchi, G.: A Divergence-measure Based Classification Method for Detecting Anomalies in Network Traffic. In: A Divergence-measure Based Classification Method for Detecting Anomalies in Network Traffic. A Divergence-measure Based Classification Method for Detecting Anomalies in Network Traffic, City, pp. 374–379 (2007)
19. Cordella, L., Sansone, C.: A multi-stage classification system for detecting intrusions in computer networks. Pattern Analysis & Applications 10(2), 83–100 (2007)

20. Golovko, V., Vaitsekhovich, L., Kochurko, P., Rubanau, U.: Dimensionality reduction and attack recognition using neural network approaches. In: Dimensionality Reduction and Attack Recognition using Neural Network Approaches. Dimensionality reduction and attack recognition using neural network approaches, City, pp. 2734–2739 (2007)
21. Jin, S., Yeung, D., Wang, X.: Network intrusion detection in covariance feature space. Pattern Recognition 40(8), 2185–2197 (2007)
22. Li, Y., Guo, L.: An active learning based TCM-KNN algorithm for supervised network intrusion detection. Computers & Security 26(7-8), 459–467 (2007)
23. Oke, G., Loukas, G.: A denial of service detector based on maximum likelihood detection and the random neural network. The Computer Journal 50(6), 717 (2007)
24. Shon, T., Moon, J.: A hybrid machine learning approach to network anomaly detection. Information Sciences 177(18), 3799–3821 (2007)
25. Beghdad, R.: Critical study of neural networks in detecting intrusions. Computers & Security 27(5-6), 168–175 (2008)
26. Chen, Z., Wang, H., Abraham, A., Grosan, C., Yang, B., Chen, Y., Wang, L.: Improving Neural Network Classification Using Further Division of Recognition Space. International Journal of Innovative, Computing, Information and Control 5(2) (2009)
27. Herrero, Corchado, E., Gastaldo, P., Zunino, R.: Neural projection techniques for the visual inspection of network traffic. Neurocomputing (2009)
28. Raahemi, B., Kouznetsov, A., Hayajneh, A., Rabinovitch, P.: Classification of Peer-to-Peer traffic using incremental neural networks (Fuzzy ARTMAP). In: Classification of Peer-to-Peer traffic using incremental neural networks (Fuzzy ARTMAP). Classification of Peer-to-Peer traffic using incremental neural networks (Fuzzy ARTMAP), City, pp. 000719–000724 (2008)
29. Raahemi, B., Zhong, W., Liu, J.: Exploiting unlabeled data to improve peer-to-peer traffic classification using incremental tri-training method. Peer-to-Peer Networking and Applications 2(2), 87–97 (2009)
30. Zhong, W., Raahemi, B., Liu, J.: Learning on class imbalanced data to classify peer-to-peer applications in IP traffic using resampling techniques. In: Learning on Class Imbalanced Data to Classify Peer-to-Peer Applications in IP Traffic using Resampling Techniques. Learning on class imbalanced data to classify peer-to-peer applications in IP traffic using resampling techniques, pp. 1573–1579. Institute of Electrical and Electronics Engineers Inc., City (2009)
31. Kothari, A., Keskar, A.: Rough Set Approaches to Unsupervised Neural Network Based Pattern Classifier. In: Advances in Machine Learning and Data Analysis, p. 151 (2010)
32. Ridley, D.: The literature review: a step-by-step guide for students. Sage (2008)
33. Moore, A., Zuev, D., Crogan, M.: Discriminators for use in flow-based classification. RR-05.13 Department of Computer Science, University of London (2005)

AUS: A Scalable IPv6 Multicast Solution

Bizhen Fu, Kun Yu, Congxiao Bao, and Xing Li

Abstract. While the deployment of IPv4 network-layer multicast has been mostly limited to LAN level networks due to scalability and management issues, the application of a large-scale, secure and controllable IPv6 multicast is of great interest in the current high-bandwidth transmission networks. In this paper we (1) define multicast application requirements, (2) propose a scalable IPv6 multicast framework solution (AUS) and (3) provide the evaluation and deployment of the AUS framework in CERNET2. Given the deployment of this scalable multicast framework, we believe that our solution offer the advantages of both a bandwidth-saving delivery network and strong controllability.

1 Introduction

Multicast is widely believed efficient for high-bandwidth communication applications such as network conferencing, IPTV and remote 1-n/n-n group video applications, e.g. education and medical consultation. In the past years, academic and enterprise networks have adopted small-scale multicast networks; however, a scalable design, implementation and deployment of multicast network is still far away from success from an ISP's perspective.

According to our evaluation of current IPv4 multicast models and solutions, we believe that there are several difficulties in the deployment of global IPv4 multicast network. However, within current IPv6 multicast models and protocols, achieving a large scale IPv6 multicast infrastructure with strong security and management is still possible.

In the past 10 years, many multicast solutions have been proposed to solve the underlying deficiency in current IPv4 multicast model (see section 3). [1][2][3][5] They either tried to introduce an overlay application layer solution, to use a hybrid application/network layer solution, or even to modify multicast protocols. However, none of these approaches has been proved to be success. The scalability is always the limitation in these solutions which confine them from large scale

Bizhen Fu · Kun Yu · Congxiao Bao · Xing Li
Network Research Center, Tsinghua University, Beijing 100084, P.R. China
e-mail: {fbzisme,yukun2005}@gmail.com
{congxiao,xing}@cernet.edu.cn

F.L. Gaol (Ed.): Recent Progress in DEIT, Vol. 2, LNEE 157, pp. 233–243.
springerlink.com

deployment. Moreover, the security and management issues haven't been considered carefully so that the service providers will not adopt them as commercial solution.

In this paper, we clarify the requirements of a scalable IPv6 multicast framework in consideration of sender/receiver and traffic management. With these requirements, we can easily evaluate the advantages and disadvantages of any multicast solution and find out which aspect to improve. Then we propose ASM/Unicast/SSM network (AUS) within this framework. AUS is a complete design and implementation of scalable IPv6 multicast framework. It adapts to customers with any network support, e.g. ASM, SSM or even simply unicast only. Experimental results on the AUS system reveal that AUS system can achieve no delay stretch upon the native IP multicast and scale well with increasing customer nodes. Finally, we describe an application scenario of the system – the IPTV live streaming of 2010 FIFA world cup in CERNET2, China.

The rest of this paper is organized as follows. In Section 2 and 3 we analyze the requirements of a scalable IPv6 multicast framework and review related works. The design of our multicast solution within this framework is provided in Section 4. In Section 5, we discuss the implementation of the framework and the current AUS system details. In Section 6 we describe the AUS system deployment in CERNET2 and its application in live streaming. We finally conclude in Section 7.

2 Requirements

Before we dive into the scalable IPv6 multicast framework, we will analyze current IPv4 and IPv6 multicast models and existing frameworks to figure out the requirements of a well-defined and appropriate multicast framework.

Scalability
Scalability is the most important issue in multicast networks. Multicast routing is far more complicated than unicast routing since the routers need to map the group address to a set of destination addresses which results in frequent maintenance of the states on routers as the nodes dynamically joins/leaves.

Reachability
Several network support scenarios exist in current Internet. Some customers may have SSM and/or ASM support while others can only be reached via unicast. The IPv6 multicast framework should be able to serve customers in any of these scenarios. Meanwhile, customers' operating systems restraint our reachability consideration. Most customers' operating systems are Windows XP or Windows 7. Windows XP only supports IPv6 ASM multicast and Windows 7 supports both IPv6 SSM and ASM multicast. Besides, Windows Media Player can't play multicast streaming.

Management

Multicast models involve several senders and receivers. The IPv6 multicast framework should enable the management of both such as accounting and security for future requirements from service providers. Moreover, the bandwidth of each group should be controlled by the framework due to the same reason.

3 Related Works

In IPv4 multicast networks, several group communication approaches have been proposed. However, most of these researches focus mainly on the reachability of multicast services. Overcast [2] introduces a unicast application-layer network simulating the multicast tree, but it does not take the advantage of network layer multicast. Host Multicast [3] combines the IP-layer and application-layer multicast under an assumption that only the customer's local networks support multicast, while currently some AS such as CERNET2 does have network-layer multicast support in the backbone. CAN[5] tries to reduce and optimize the state maintenance in multicast routers by distributed hash table (DHT), but utilized only unicast for data transmission. Scalability is only partially concerned in these works, and management is seldom carefully considered.

4 Framework Design

4.1 Infrastructure

The scalable IPv6 multicast framework is shown in Figure 1. Two components are involved in this general framework, i.e. transmission component and management component, marked as circle and hexagon respectively.

Routers are the transmission components in the framework, with access to SSM/ASM networks and/or unicast only networks. Routers may forward contents from some address/port pair to any destination set, composed of either multicast group or unicast address/port pairs. Also, routers should be able to forward packets selectively and set the bandwidth limit upon any transmission channel. An important characteristic of the transmission components is that they only forward UDP packets, which is appropriate for high-bandwidth and real-time streaming applications.

Each router supports a universal operation interface that consists of several possible router operations, and we define the router operation strategy as the *state maintenance problem*, i.e. how to operate all the routers so that the framework will satisfy the receiver/sender and bandwidth limitations of all channels specified by the framework administrators. The solution to *state maintenance problem* is the state updates sent to all routers. Finally routers will take on the corresponding operations according to these state updates.

Routers are controlled by the management component, which maintains the group states (e.g. multicast tree), handles group mapping, source/receiver joining/leaving and additional states such as bandwidth allocation and customer authentication. Management component is constituted with a state storage (of mapping state, either in the memory or in a database) and a *management algorithm*. The management component periodically applies the management algorithm to solve the state maintenance problem.

In current IP-layer multicast model, the transmission components are under the assumption that the multicast routers (an example of our *Routers*) are all in a single multicast network, either SSM or ASM. The routers only forward packets from a multicast group to unicast destinations and do not support any bandwidth limitation. Router also functions as the management component, and both the state storage and management algorithm are distributed in multicast routers and RP (Rendezvous Point) router. [10]

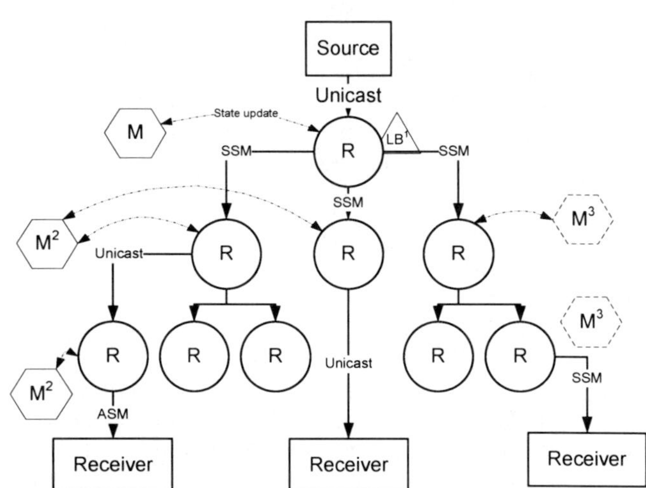

Fig. 1 Scalable IPv6 Multicast Framework. (1) Limit bandwidth in routers; (2) management component hierarchy; (3) distributed management component

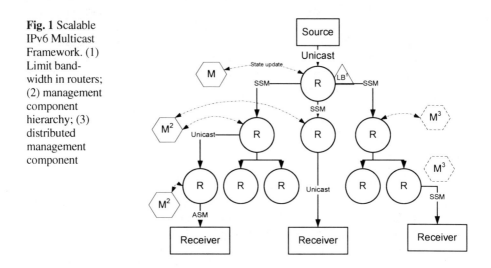

4.2 Router Operation Interface

Router operation interface defines a set of operations supported by multicast router, listed in Table 1.

Router operation interface can be accessed by either a negotiated message (e.g. the join message in the PIM-SSM [10] protocol) or a packet composed of both message and the data payload (e.g. the registering message in the source registration process between the source and RP in the PIM-SSM protocol).

Table 1 Router Operation Interface

Name	Function	Operation
Prepare Receive (PR)	rsocket = PR(LRA, LRP)	Create a socket rsocket listening to router's local address:port LRA[a]:LRP[a]
Prepare Send (PS)	wsocket = PS(LSA, LSP)	Create a socket wsocket for sending out packets on router's local address:port LSA[a]:LSP[a]
Forward (F)	fchannel = F(SA, SP, {(DA_i, DP_i)})	Create a fchannel to forward packet from source address:port SA[a]:SP[a] on rsocket to address pair set {(DA_i[a]:DP_i[a])}
Join ASM (JA)	JA(G, rsocket)	Join ASM group G for a rsocket
Join SSM (JS)	JS(S, G, rsocket)	Join ASM group <S,G> for a rsocket
Limit Bandwidth (LB)	LB(blimit, fchannel)	Set[b] bandwidth limit blimit for fchannel

a. Local Receiving Address, Local Receiving Port, Local Sending Address, Local Sending Port, Source Address, Source Port, Destination Address, Destination Port, respectively. b. The bandwidth can be limited on either the ingress or the egress stream.

4.3 Scalability Consideration

When the system scales, the state storage capacity and the management algorithm's execution time will increase remarkably. Basically, the best practice to relieve a framework's scalability issue is to reduce the states in the system, which is our choice too. However, states introduced in multicast could not be reduced to none due to the group address to receiver addresses mapping. Instead, we would reduce the states in the network's backbone and relocate them to the network's edge.

The technique is to split the multicast streams into two categories, static multicast streams and dynamic multicast streams. Static multicasts are the sustaining multicast streams with very limited sources and almost fixed bandwidth, e.g. IPTV. For static multicasts, we define their state as static, which does not require the multicasts and dynamic multicasts. The dynamic multicasts are streams with dynamically joining/leaving sources/receivers and variable bandwidth. For this category, the management algorithm has to be solved periodically.

Another technique is adopting distributed state storage and management algorithm. The flat structure of current multicast model introduces numerous churns (group joining/leaving and tree construction) when the multicast tree scales with more routers, causing the scalability issue. The alternative solution is to separate the multicast framework into several sub frameworks. We can treat each framework as a virtual router, which also supports the router operation interface in the framework level, forwarding packets and limiting bandwidth. Hence all the sub frameworks constitute a larger framework, and this hierarchy will delegate the state storage and management algorithm into far smaller scope and thus the *state maintenance problem* will be solved faster.

4.4 Management Consideration

As with the requirements, access control (sender/receiver) and bandwidth should be applied in the framework. The access control is done by the routers at the edge of the framework reacting to router operations such as Forward, JA and JS; bandwidth is filtered by the LB operation. Management Algorithm will take the access control and bandwidth control into account and thus inform the routers via calling the router operation interface.

For each group, the management algorithm will figure out the customers in it and thus send out signals to the routers telling them only to join these ASM/SSM groups, listen to the packets sent from the source set and forward to the receiver set of addresses. The first router in the forwarding tree will also set a traffic fence on the stream so the bandwidth in the framework will be strictly limited.

4.5 Reachability Consideration

Routers should adapt to their network support and the customers'. Typically when customers require a multicast streaming, a router will be selected as the designated router which works in the same multicast island with the customer, called a matched service. In this scenario, the router will forward the multicast stream to customer's preferred multicast type, e.g. SSM or ASM.

Besides, for mismatched routers and customers, unicast is the last and feasible solution to provide the multicast service.

5 AUS Implementation

We implemented AUS (ASM-Unicast-SSM) system based on the scalable IPv6 multicast framework within the CERNET2 IPv6 controllable large-scale multicast project. CERNET2 has 25 backbone nodes which enable IPv6 SSM multicast and 100 campus nodes. The difficulty lies in that the campus networks may have no multicast support at all.

5.1 Transmission Components

We deploy 25 backbone AUS gateways and 100 campus gateways in CERNET2, each running our implementation of the transmission router that provides Router Operation Interface. Three measures are suitable for the forwarding task, as

- Application-layer forwarding, which can be implemented trivially and easily controlled by the operation interface; however the latency and resource consumption, requires careful consideration.
- UDP Tunnel, which encapsulates the multicast packets into a UDP packet and sends them out via unicast. The receiver unpacks the encapsulated packets, pushes them onto the packet queue and applications will virtually receive the multicast packet. This method requires additional modules on customers' desktops.

- Packet hacking in OS kernel, which modifies the packet addresses in the OS (e.g. Linux) protocol stack. However, whether the hacked packets are visible to IGMP/MLD protocol depends on the LAN switches connecting to the routers.

We have a limited number of 100 IPTV channels (1Mbps each) in AUS system; hence we adopt application-layer forwarding for the router implementation. However, the other two measures are trivial to implement too.

In application-layer gateways, we support ASM, SSM and unicast receiving in the ingress side, including multicast joining operation. In the egress side, forwarding the packets to ASM, SSM or unicast address shares the same logic – it's just a destination address. Hence we have an identical API to support ASM/SSM/unicast, e.g.

$$
\begin{aligned}
&Forward(s_{remote}, r_{local}, s_{local}, \{r_{remote}\}_i) \\
&\quad where \quad s_{remote}, r_{local}, s_{local}, (r_{remote})_i \\
&\qquad = (source, group, port)
\end{aligned} \tag{1}
$$

For example, when the gateway needs to receive a SSM stream and forward to ASM group($(r_{remote})_1$) and a unicast destination($(r_{remote})_2$), it will join in s_{remote} 's source/group pair, listen on r_{local} 's source:port, bind to s_{local} 's source:port, and then send the packets to $(r_{remote})_1$'s group:port and $(r_{remote})_2$'s source:port.

We evaluate the transmission components' performance by comparing the router-forwarded streams (Figure 2) with direct-routed unicast streams. Dvping/dvmcast [11] is used to calculate the packet loss, bandwidth and RTT between two nodes. We forward packets via AUS gateway, evaluate the performance from end to end and as a contrast, the end-to-end performance of direct unicast is tested in dvping and dvmcast.

The results are shown in Figure 3. The packet loss, bandwidth and latency are almost the same as a direct unicast transmission. The result is reasonable because we decoupled the signals from the forwarding channel, so AUS gateways' performance is hardly abated.

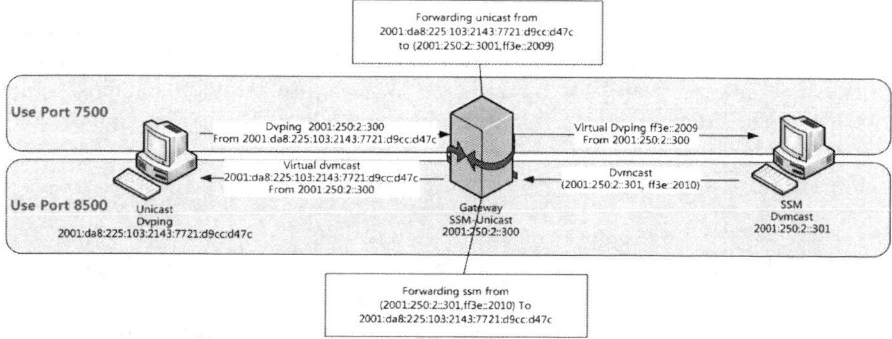

Fig. 2 AUS gateway performance evaluation topology

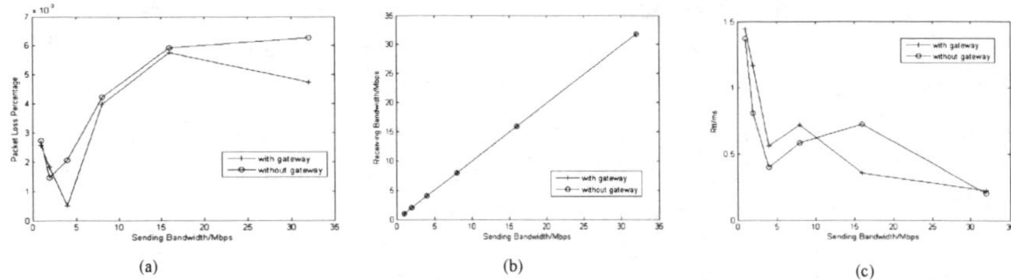

Fig. 3 Performance of AUS gateway in forwarding stream compared to direct unicast stream. (a) packet loss (b) bandwidth (c) RTT

5.2 Management Components

We implemented a centralized management server, which keeps the AUS gateways' topology information, receives their state report (traffic history, channel availability, etc.) and solves the state maintenance problem.

The AUS gateways' topology is very structured where 25 backbone gateways are reachable via SSM multicast and the 100 campus gateways only accept unicast stream from the backbone and then forward it to the ASM network in campuses. All the customers are served by one and only one gateway, for the customer must be in any campus network of CERNET2. To set up multicast tree for each channel, the management server will look up the state storage for the source's corresponding gateway, and build the forwarding tree rooted from the source.

We improve the system's scalability by static joining SSM group in the backbone gateways. The AUS system is responsible for the 100 IPTV channels, so there's no state change after system initialization. The dynamic part of the system is the users' access network, where ASM streams are sent out by campus gateways. However, the states are limited in the campus network, so the backbone remains clean and simple with a very small and static state table.

We control the multicast source by operations on the campus gateways, forwarding only the streams from specific addresses. Simultaneity the bandwidth of the a channel is rigorously bounded to some level (100kbps, 1Mbps, 10Mbps and 100Mbps) by the ingress traffic control on the source gateway.

Due to the very limited gateways, a centralized algorithm is sufficient and efficient (unlike a distributed algorithm which takes more time to reach convergence) to finish the management task. Nevertheless, when more and more campuses join in CERNET2, a distributed AUS system hierarchy will come in. For example, the system will be divided to 2 layers. Level 1 will operate in the backbone communicating with Level 2 systems such as Beijing system, Shanghai system, etc.

6 AUS Deployment

As an example of the AUS system application, we deployed the system to broadcast the 2010 FIFA World Cup. The architecture of the broadcasting instruments are shown in Figure 4 and the topology we deployed in three universities of Beijing and Shenyang is shown in Figure 5. On the ingress side, as we described above, the AUS system only transmits UDP streams. If we employ VideoLan Media Player[6] as the streaming server, we can easily satisfy the requirements of the system. However, some widely-used commercial media server (e.g. Microsoft Windows Media Service [7]) only supports MMS protocol[8], which is based on RTSP over TCP. Thus we have to import a conversion module outside the system to generate an UDP streaming. VideoLan Media Player can also accomplish this task.

On the egress side, pushing the multicast streaming to customers is not trivial due to the operating system support described in Section 2. So we have two options, (1) recommend our customers to install a VideoLan Media Player to receive the broadcasting or (2) convert the streaming from UDP back to MMS in the campus network before they arrive at customers' desktop. In most CERNET2 campus networks, multicast is only partially or not deployed, so we apply the latter approach as customer solution.

Before we deploy AUS system campus students can only access MMS service from the sender university, which results in huge redundancy in CERNET2 backbone and the sender's egress traffic, and thus severely affecting the video quality and bringing in high latency due to the media server's limited CPU and memory resources. We successfully reduce traffic with the AUS system during the World Cup broadcasting, for there are only no more than 25 duplicated streams in CERNET2 backbone. Moreover, the duplication occurs only in some links for the streams are transmitted via IPv6 SSM multicast. At the same time, the sender reduces lots of egress traffic.

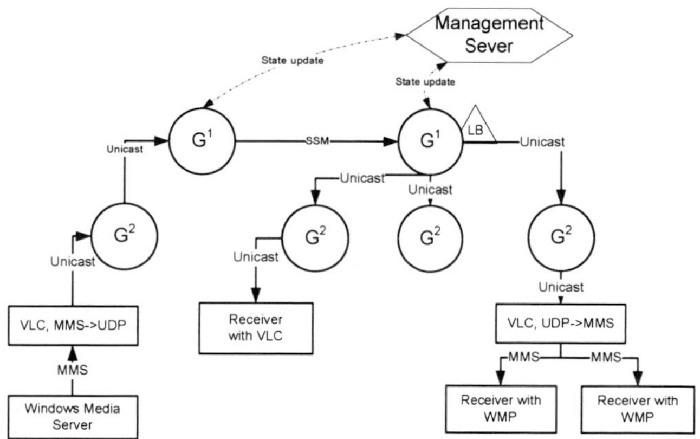

Fig. 4 AUS Deployment in World Cup IPTV (1) AUS gateway in CERNET2 backbone (2) AUS gateway in CERNET2 campus.

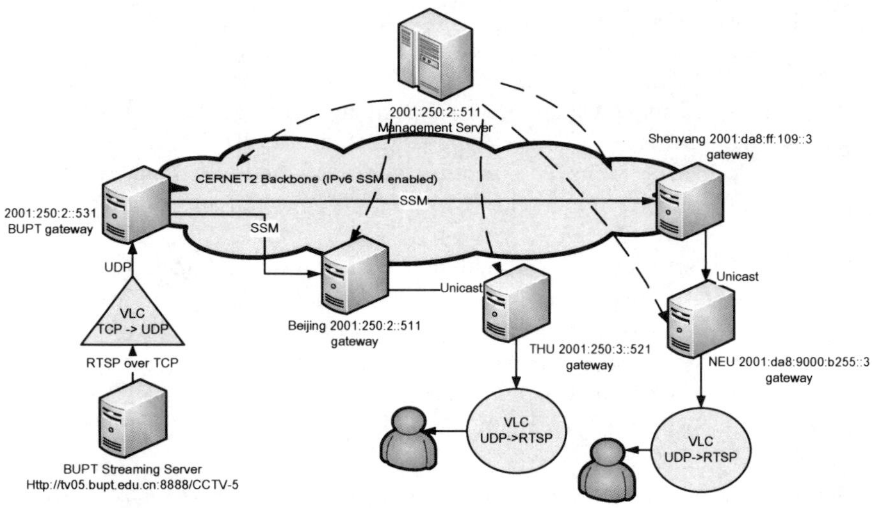

Fig. 5 AUS system topology in World Cup.

7 Conclusion and Future Work

In this paper, we introduce AUS, a scalable IPv6 multicast framework which utilized both network and application-layer multicast to achieve high reachability to end customers in various network situations. With our implemented AUS system based on the framework, we strongly believe that this framework is an appropriate solution for high-bandwidth applications such as IPTV and video conference in IPv6 network, with very few limitations and high controllability. In our deployment of the AUS system, we demonstrated a scenario that the framework can be applied for large scale IPTV application while reducing both data redundancy in network and media server's load.

References

1. Banerjee, S., Bhattacharjee, B., Kommareddy, C.: Scalable Application Layer Multicast. In: Proceedings of ACM Sigcomm, Pittsburgh, Pennsylvania (August 2002)
2. Jannotti, J., Gifford, D.K., Johnson, K.L.: Overcast: Reliable Multicasting With An Overlay Network. In: USENIX Symposium on Operating System Design and Implementation, San Diego, CA (October 2000)
3. Zhang, B., Jamin, S., Zhang, L.: Host Multicast: A Framework for Delivering Multicast To End Users. In: IEEE INFOCOM 2002 (2002)
4. Boivie, R., Feldman, N., Metz, C.: Small Group Multicast: A New Solution for Multicasting on the Internet. IEEE Internet Computing 4(3), 75–79 (2000)
5. Ratnasamy, S., Handley, M., Karp, R., Shenker, S.: Application-level Multicast using Content-Addressable Networks. In: Proceedings of NGC (November 2001)
6. Cellerier, A., et al.: VideoLAN - VLC media player, http://www.videolan.org

7. Microsoft, Windows Media Services, http://www.microsoft.com/windows/windowsmedia/forpros/server/server.aspx
8. Microsoft, Windows Media HTTP Streaming Protocol Specification, http://msdn.microsoft.com/en-us/library/cc251059.aspx
9. Protocol Independent Multicast-Sparse Mode: Protocol Specification, IETF RFC 2362, http://www.ietf.org/rfc/rfc2362.txt
10. Bhattacharyya, S.: An Overview of Source-Specific Multicast Deployment. Internet draft (January 2002), http://www.ietf.org/internet-drafts/draft-ietf-ssm-overview-02.txt
11. Li, X., Bao, C., Jiang, J.: dvping/dvmcast, http://sasm3.net/DVPING/

Indonesian Shallow Stemmer for Text Reading Support System

Hajime Mochizuki, Yuhei Nakamura, and Kohji Shibano

Abstract. Our project involves the construction of a web-based system to facilitate the reading and comprehension of Indonesian text. The system will help users to understand difficult words in a text by displaying dictionary information about the words in a window. A large number of words in the Indonesian language are formed by combining root words with affixes and other combining forms. To search for the related dictionary entry, we need a stemming program to extract these root words. We develop an Indonesian stemming program for ourselves. Our stemmer does not need to be perfect because our application is limited to that of a text reading system. In this paper, we describe such a stemmer and present the results of preliminary examinations to evaluate it. We also describe a design for the text reading support system that uses the developed stemming program.

1 Introduction

Reading comprehension is one of the most important skills in language learning. Its importance and necessity are demonstrated by the fact that the large majority of information now circulating in the world takes the form of text media. Due to the spread of the Internet, people have an increased opportunity to come in contact with a variety of languages. Today, if someone has an interest in Indonesian culture,

Hajime Mochizuki
Institute of Global Studies, Tokyo University of Foreign Studies, 3-11-1 Asahi-cho, Fuchu-shi, Tokyo 183-8534, Japan
e-mail: motizuki@tufs.ac.jp

Yuhei Nakamura
Mitsubishi UFJ Lease & Finance Co. Ltd.

Kohji Shibano
Research Institute for Languages and Cultures of Asia and Africa,
Tokyo University of Foreign Studies, 3-11-1 Asahi-cho, Fuchu-shi, Tokyo 183-8534, Japan
e-mail: sibano@aa.tufs.ac.jp

F.L. Gaol (Ed.): Recent Progress in DEIT, Vol. 2, LNEE 157, pp. 245–255.
springerlink.com

he or she can find much relevant information by accessing Indonesian Web sites. Certainly, it is preferable for a person who wants to learn about another country to learn the language of that country. It can be said that the spread of the Internet has increased people's need and motivation for language learning. In our project, we are constructing a system to support reading and comprehension of Indonesian text. The system will help the user to understand difficult words in a text by displaying dictionary information about them in a window. In such a system, it is important for each word in a text to be automatically linked with its correct entry in a dictionary.

A large number of Indonesian words are formed by combining root words with affixes and other combining forms. On the other hand, typical Indonesian dictionaries have only root words as their entry words; therefore, to know the meaning of a derived word, one has to know its root word. Just as in human cognition, an automated system needs a word stemmer in order to link each word in the text with its entry in the dictionary. Many stemming programs and morphological analyzers have already been developed for many languages such as Japanese and English. However, there has not been much research on stemming in Indonesian[1][2][3].

In this research, we first develop an Indonesian stemming program. Our stemmer does not have to be perfect because its function is limited to use in our text reading system. In this paper, we describe a shallow stemmer and present the results of a preliminary examination to evaluate it. In the definition of our stemming rules, we refer to the rules for forming Indonesian words in TruAlfa and IndoDic.com[4].

We also describe a design for a text reading support system that uses the developed stemming program. In this research, we use the CICC Indonesian Basic Dictionary[5] as a dictionary of the Indonesian language.

2 Shallow Stemming Algorithm

2.1 Forming Indonesian Words

Indonesian words are composed of root words and their derived words. The derived words are formed by combining the root words with one or more affixes and other combining forms. The number of derived words in Indonesian is large. The relationships among a root word and affixes and other forms can be defined as follows.

(affix| other form)* root word + (affix| other form)*

Here, '|' is equivalent to the logical symbol 'OR,' and '*' means 0 or more repetitions of the preceding character. Stemming an Indonesian word involves dividing the derived word into the affixes or other forms and the root word. Therefore, stemming can basically be achieved by taking the following steps.

1. Prepare a list of affixes and other combining forms.
2. Remove the matching string on the list from the input word.
3. Search for the remaining string in a dictionary.
4. Store the remaining string as a candidate of the root word, if the string is in the dictionary.

The above processes are examined recursively because there is a possibility that multiple affixes and other combining forms on the list will match the input word. After all these steps are taken, the stored strings with their affixes and other combining forms are shown as the candidates of the root word and stemming result. Here, each entry of the string list in the first step must include a description of the characteristics of a root word to which they can be connected. The kind of attached affixes and other forms depends on the kind of root words. To understand and formulate how affixes and other combining forms are used is the key to the development of an Indonesian stemming program. Fortunately, TruAlfa and IndoDic.com[4] offer brief explanations about the formation of Indonesian words for learners of the Indonesian language. We created the rules for our Indonesian stemming program by referring to their explanations.

2.2 *Affixes and Other Forms*

We classify affixes and other combining forms into the following five categories.

(1) prefix: an affix attached at the beginning of a root word that creates a new word. (2) infix: an affix that is inserted within a root word and which creates a new word. (3) suffix: an affix attached at the end of a root word that creates a new word. (4) confix: a prefix and a suffix that are both attached to a root word to create a new word. (5) combining form: a linguistic unit whose function is similar to that of an affix, but which cannot be technically classified as an affix. The affixes and the combining form that belongs to each classification are listed in Table 1.

Table 1 Affixes and other forms

category	strings
prefix	ber-, di-, ke-, *me-group* (me-, mem-, men-, meng-, menge-, meny-), *pe-group* (pe-, pem-, pen-, peng-, penge-, peny-), per-, se-, ter-
infix	-el-, -em-, -er-
suffix	-an, -kan, -i, -lah, -kah, -nya
confix	ber-an, ke-an, per-an, se-nya, *pe-an group* (pe-an, pem-an, pen-an, peng-an, penge-an, peny-an)
combining form (a)	antar-, para-, eka-, kau-, ku-, oto-
combining form (b)	-pun, -ku, -mu, -nya

In these affixes, we exclude infixes from the scope of our stemming program because they are rarely used in Indonesian.[1]

[1] According to [4], there are only about 20 examples of infixes in common usage in Indonesian.

2.3 Spelling Changes of Root Words

The prefixes, confixes, and combining forms (a) in Table 1 can be classified into the following two types.

a This type of affix and combining form can be attached to any root word and other affixes and combining forms. The strings that belong to this type can be matched with di-, ke-, se-, antar-, para-, eka-, oto-, kau- and ku-.
b hese types of affixes and combining forms can only be attached to certain root words. The strings that belong to this type can be matched with *me-group*, *pe-group*, ber-, per- and ter-.

Furthermore, affixes and other combining forms in type b are classified into the following three sub-types.

b1 The spelling of root words that attach to them are not changed.
b2 The first letters of root words that attach to them are deleted.
b3 The spelling of root words that attach to them are partly changed.

We next describe the details of the spelling change rules for type b strings, which are pe-group, *me-group*, ber-, per- and ter-.

- *pe-group*
  - 'pe' from *pe-group* can be attached to root words whose first letter is l, m, n, r, w, y or z. (b1)
  - 'pem' from *pe-group* can be attached to root words whose first letter is b, f or v. (b1)
  - 'peng' from *pe-group* can be attached to root words whose first letter is a, e, g, h, i, o or u.
  - 'pen' from *pe-group* can be attached to root words whose first letter is k, and the k is deleted.
  - 'peny' from *pe-group* can be attached to root words whose first letter is s, and the s is deleted.
  - 'penge' from *pe-group* can be attached to root words that are monosyllabic.
- *me-group*
  - 'me, mem, memg, men, meny and menge' from *me-group* are the same as in the case of 'pe, pem, pemg, pen, peny and penge' from *pe-group*.
- *'ber'*
  - 'ber' can be attached to root words whose first letter is r, and the r is deleted.
  - 'ber' can be attached to root words whose first syllable includes er, and the r is deleted.
  - 'ber' can be attached to the other root words.
- *'per'*; 'per' is the same as in the case of 'ber.'
- *'ter'*; 'ter' is the same as in the cases of 'ber' and 'per.'

2.4 Rules of Attachment for Affixes and Combining Forms

In this subsection, we define the rules for constructing candidate strings of root words from input words. The format of the rules is defined as follows.

rule id, prefix form, suffix form, delete letter, first letter, connecting condition code

In order to construct candidate strings of the root word from the input word, we subject the input to testing by all the rules in the following steps.

First, we take one rule and check whether the strings of the 'prefix form' and 'suffix form' of that rule match the input word. If these strings match the word, we construct a candidate of the root word according to the test procedure, which is defined by the 'connecting condition code.'

We prepare seven kinds of connecting condition codes-*none*, *first letter*, *drop letter*, *syllable*, *no double r*, *no double r 2* and *no double r 3*-and define the test procedures for each code.

- *none*: The remaining string becomes a candidate of the root word.
- *first letter*: Compare the first letter of the remaining string to *first letter* in the rule. If they match each other, the remaining string becomes a candidate of the root word.
- *drop letter*: Attach a letter in *drop letter* of the rule to the front of the remaining string. The new supplemented string becomes a root word candidate.
- *syllable*: Count the number of syllables in the remaining string. If it is composed of one syllable, it becomes a root word candidate.
- *no double r*: If the first letter of the remaining string is not r, the string becomes a root word candidate.
- *no double r 2*: Attach the letter r to the front of the remaining string. The new supplemented string becomes a root word candidate.
- *no double r 3*: If the first syllable includes er, the remaining string becomes a root word candidate.

Examples of the rules for constructing candidate strings of root words are listed in Table 2.

Table 2 Example of Rules for Constructing Candidate Strings

rule id	1600	1900	2100	560	600	601	602
prefix form	ke	pe	pem	penge	per	per	pe
suffix form	an	an	an				
drop letter			p				
first letter		l	p				
connecting condition	None	first letter	Syllable	drop letter	no dbl r	no dbl r 2	no dbl r 3

For example, using the input word *keadaan* when the rule id 1600 is tested, the prefix form *ke* and the suffix form *an* are deleted from the input word. The remaining string *ada* becomes a candidate string of the root word because the connecting condition code is *none*. When the input word is *pelajaran* and the rule id 1900 is tested, the prefix form *pe* and the suffix form *an* are deleted from the input word. The remaining string *lajar* becomes a candidate string because the connecting condition code is *first letter* and *l* is matched to the first letter of the remaining string. When the same rule is applied to *pegangan*, the remaining string *gang* does not become a root word candidate because the first letter is not *l*.

We have to count the number of syllables when the connecting condition code is *syllable*. A component of one syllable in an Indonesian word can be defined as C^*VC^*, where C indicates a consonant and V indicates a vowel. The sign $*$ indicates a repetition of 0 times or more. If the remaining string matches this syllable pattern exactly, we can infer that the string has a single syllable.

3 Shallow Stemming Algorithm

We next describe our shallow stemming technique in detail.

1. Take one rule from the rules for attached affixes and combining forms sequentially. If all rules were tested, this step is finished.
   a. Check whether the prefix form and suffix form of the rule match the string; if there is no match, return to Step 1.
   b. Apply the connecting condition code of the rule to the string; if a candidate string of the root word is extracted, check the dictionary for the candidate.
   c. If the lookup succeeds, the candidate string is considered to be one of the stemming results and is stored in memory.
   d. Make a recursive call of Step 1 with the candidate string as new input.
2. At the end of processing, show all results stored in memory.
3. If no data is stored in memory, the algorithm shows the original unstemmed word.

Our stemming system does not disambiguate the stemming results and all stemming results are shown. Our intention is to use the stemming results in a text reading comprehension support system. We believe that it is better for language learners if the system offers choices rather than shows the answer.

The stemming result for the word *keadaan*, which means *existence*, is shown in Fig. 1. The derived word *keadaan* is decomposed into three candidates: *keada* + *an*, *ke* + *ada* + *an*, and *ke* + *adaan* in this sample. The correct stemming is *ke* + *ada* + *an* and the correct root word ada means *there is*.

```
keadaan
keadaan      [keada]   an (900)
keada   ke    [ada]    an (900+300)
keadaan ke    [adaan]     (300)
```

Fig. 1 The sample of stemming result

4 Experimental Evaluation for Shallow Stemming

To determine the effectiveness of our shallow stemming technique, we performed two experiments. First, we examined how well it performs on Indonesian derived words. Second, we examined how well it performs, not on derived words, but on base words that have the same forms as the root words.

For these experiments, we required collections of Indonesian words and their correct root words as answers. Unfortunately, there was no suitable existing testbed for us; therefore, we created our own test data for our experiments. We describe these in the following subsections.

4.1 Data Collections

We formed a collection of words to be stemmed by extracting every word from a set of news stories published in the online edition of the newspaper Kompas on July 25 and 26, 2009. First, we extracted a total of 60,146 words from 128 news articles. These are the 4,590 words in the types. Next, we counted the frequencies of these 4,590 words and selected the 1,000 most frequent words, from which we deleted words that include capital letters, symbols, Arabic numerals, and English words. We also deleted plurals and composite words that are hyphenated. Finally, the remaining 779 words were classified into the two categories listed below, and the correct root words and their affixes were attached to them.

- The collection of derived words: 279 words were put into this classification. One of the authors determined the correct root words and affixes.
- he collection of base words: 500 words were put into this classification. The correct root words are the same as the words themselves and they have no affixes.

We used the collection of derived words in the first experiment and the collection of base words in the second experiment.

4.2 Experiment with Derived Words

The first experiment we report assesses the effectiveness of our stemmer with derived words.

4.2.1 Evaluation Criteria

We used the following two criteria for evaluation:

1. Correct answer rate. This criterion indicates the ratio of correct answers. If the correct root word is included in candidates of the root words, it is considered correct.

$$\frac{num.\_of\_words\_which\_include\_correct\_answer}{num.\_of\_words} \qquad (1)$$

2. Average number of candidates in the results. This criterion indicates how many candidates of the root word are created per word.

$$\frac{num.\_of\_candidate\_words\_of\_the\_base\_words}{num.\_of\_words} \tag{2}$$

As mentioned in the previous section, our system shows all candidates rather than one answer. Therefore, when evaluating our stemming algorithm, if a correct stemming result is included in all stemming results for a word, we consider that our system can output the correct stemming result for the word.

4.2.2 Result of Experiment 1

Table 3 lists the results of the first experiment.

Table 3 Results of First Experiment

	Correct	average number of candidates
without dicitonary	97.5% (272)	4.1 (1141 candidates)
with dictionary	96.8% (270)	1.1 (320)

In Table 3, 'with dictionary' shows the results of our algorithm and 'without dictionary' shows those for a baseline system for comparison. The baseline system does not check the dictionary at step 1-(b) of the algorithm, and all lookups are treated as in the case of success in step 1-(c) in the previous section. Therefore, the correct answer ratio would become a maximum, though the average number of candidates would become worse in the baseline system.

Our system produces an average of 1.1 candidates per word and produces correct stemming for 96.8% of words.

Comparing the baseline and our system with regard to the correct answer ratio, our system scored a little lower than the baseline, but the difference between the two was not very large.

Comparing the two approaches with regard to the average number of candidates, our system achieved scores that were about four times better.

The results show that our system is able to stem most derived words in the collection. The number of candidates can be reduced sufficiently using the dictionary.

4.3 Experiment with Base Words

The second experiment we report upon assesses the effectiveness of our stemmer with base words. The base words are not stemmed in the correct stemming process. Thus, the effectiveness of the system is lowered, if the system does stem root words in Experiment 2. For our evaluation, we used the same two criteria as in Exp. 1.

4.3.1 Result of Experiment 2

Table 4 lists the results of the second experiment.

Table 4 Results of Second Experiment

	Correct	average number of candidates
without dictionary	100% (500)	1.5 (759 candidates)
with dictionary	95.0% (475)	1.1 (525)

In Table 4, as with Experiment 1, 'with dictionary' represents our algorithm and 'without dictionary' represents a baseline system for comparison.

Our system produces an average of 1.1 candidates per word, and produces correct stemming for 95.0% of words.

Comparing the baseline and our system with regard to the ratio of correct answers, our system scored a little lower than the baseline, but the difference between the two was not very large.

Comparing the two approaches with regard to the average number of candidates, our system achieved scores that were about 1.5 times better.

The results show that our system can produce correct results for most base words. The number of candidates can be reduced by using the dictionary.

5 Text Reading Comprehension Support System

The text reading system consists of a client part that works for a user as an interface and a server part that can attach a variety of information to an original text.

First, a learner inputs or pastes a text that he or she wants to read and understand. Next, the text is sent to the main program on a WWW server via CGI. The main program then filters the input text to remove unnecessary spaces and characters from a sentence and joins sentence fragments. Next, the filtered text is decomposed into word components by our Indonesian shallow stemmer so that the following processes can extract information and attach it. Finally, a learner can read the text with the information given to facilitate his/her understanding of the words. A screenshot of the current system is shown in Fig. 2.

In the current version, if the learner puts the mouse cursor on a word in a text, a small window in which information is displayed will come up beneath the word.

We use CICC[5], the Indonesian Basic Dictionary, as a source of root word information. This dictionary is the fruit of the project of the Multilingual Machine Translation System for Asian Languages organized by the Center for International Cooperation in Computerization (CICC). It is divided into 5 files: basic terms (Indonesian Master Dictionary, IMD), special terminologies (TECH), idioms (IDIOMS), acronyms (ACRONYMS), and other words (OTHER). In these files, we use the IMD. The IMD contains a total of about 17,000 headwords, and each entry has a head concept, corresponding word entries containing derivation words with their conjugation codes, the part of speech, grammatical properties, conceptual descriptions in English, and more.

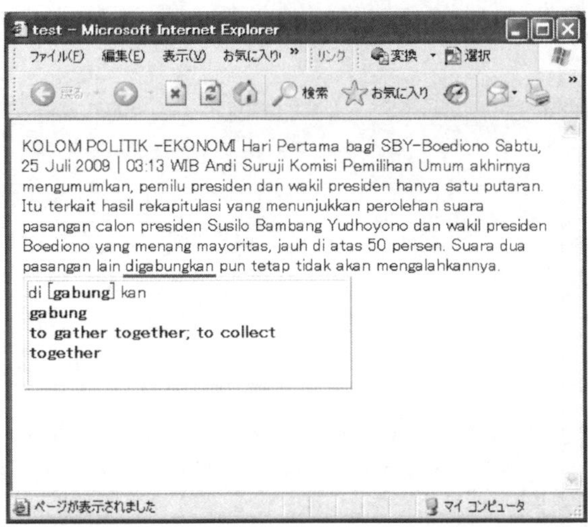

Fig. 2 A Screenshot of text reading support system

Using IMD, the following information can be shown by our current prototype system.

- A list of possible stemming forms. The learner is urged to choose the correct answer from this list during the reading process.
- Meanings of the root words. Conceptual descriptions of the root words are shown in English.

6 Conclusion and Future Work

In this paper, we described an Indonesian shallow stemming algorithm and performed experiments to assess its effectiveness. The results showed that our algorithm can stem Indonesian words well enough for use in the text reading support system. We also described a design for our system for Indonesian text reading support and reported the current status of the system.

We plan the following future projects to further our work with stemming.

- Extend the stemming algorithm to enable it to treat plurals and composite words that are hyphenated, which the current program is unable to treat.
- Extend the stemming algorithm to enable it to perform light disambiguation. It is better if an obviously unsuitable candidate is excluded, though we do not consider full disambiguation to be necessary for our system.

We think the text reading support system also needs to be extended. In this regard, the following points should be considered in future works.

- Show the part of speech information: If the part of speech information is attached to the stemming result, it is helpful to the learner who is trying to understand the meaning of the word and the sentence.

- Show hints for understanding word composition: Knowing how affixes and combining forms are used is the key to understanding the meanings of derived words and to learning to read Indonesian. Our system should therefore be able to show grammatical explanations of the affixes, combining forms, and the root words.

Acknowledgements. This study was supported by 'Advanced Language Teacher Education Based on CEFR and ICT' (head: Prof. Koji Shibano) selected for the 'Support Program for Improving Graduate School Education.'

References

1. Yusuf, H.R.: An analysis of indonesian language for interlingual machine-ranslation system. In: Proceedings of the 15th International Conference on Computational Linguistics, pp. 1228–1232 (1992)
2. Nazief, B.: Panel: Development of computational linguistics research: A challenge for indonesia. In: Proceedings of the 38th Annual Meeting of the Association for Computational Linguistics, pp. 1–2. Association for Computational Linguistics, Hong Kong (2000), http://www.aclweb.org/anthology/P00-1075
3. Adriani, M., Asian, J., Nazief, B., Tahaghoghi, S.M.M., Williams, H.E.: Stemming indonesian: A confix-stripping approach. ACM Transactions on Asian Language Information Processing 6(4), 1–33 (2007)
4. TruAlfa and IndoDic.com. Forming Indonesian Words & using Indonesian Affixes, http://indodic.com/index.html
5. CICC, Indonesian basic dictionary, Center of the International Cooperation for Computerization Technical Report. Tech. Rep. 6-CICC-MT 53 (1995)

Intellectual Property Online Education System Based on Moodle

Yutaro Ikeda and Yasuhiko Higaki

Abstract. A serious issue being raised currently in Japan is the shortage of human resources and the decreasing recognition of Intellectual Property Rights, which support the economy of Japan but lag behind the global trend. However, certain changes in education are now providing opportunities for creating and acquiring the rights to Intellectual Property. To help develop education on Intellectual Property Rights, the Intellectual Property Online Education System (IPOES) is proposed in this research. IPOES utilizes the features of Moodle, a system for online classes. For now, IPOES targets science and engineering students at University. In this paper, an outline is presented for IPOES development on Moodle and the ideal infrastructure for education in collaboration with the Technology Licensing Organization (TLO).

1 Introduction

For students studying science and engineering, the research objectives include developing beneficial technologies for humans, inventing products and manufacturing methods, and using scientific knowledge. Not only departments of engineering but also universities themselves are expected to return innovative products, manufacturing methods, and Intellectual Property back to society.

Unfortunately, too many students, teachers, and staff do not recognize the importance of Intellectual Property. Although they are in the position of creating, using,

Yutaro Ikeda
Department of Master of Intellectual Property(MIP), Tokyo University of Science Graduate School of Innovation Studies, Iidabashi Central Plaza 2F 4-25-1-12 Iidabashi, Chiyoda-ku, Tokyo 102-0072, Japan
e-mail: jm311003@ed.tus.ac.jp

Yasuhiko Higaki
Graduate School of Engineering, Chiba University, 1-33, Yayoi-cho, Inage-ku, Chiba, 263-8522, Japan
e-mail: higaki@tu.chiba-u.ac.jp

F.L. Gaol (Ed.): Recent Progress in DEIT, Vol. 2, LNEE 157, pp. 257–262.
springerlink.com

and protecting Intellectual Property, they have a remarkably low awareness of this, and it is unlikely that society will automatically develop into one that puts a greater importance on Intellectual Property.

Since the Strategic Intellectual Property Conference took place in February 2002 in Japan, a project for creation, protection and use of Intellectual Property, a strategy for enhancing its competitiveness, and human resource development strategies have progressed, as listed in the following initiatives:

- Intellectual Property Strategic Program[1]
- Strategic Council on Intellectual Property[2]
- The Comprehensive Strategy for the Development of Human Resources[3].

However, even though the government of Japan has raised the importance of Intellectual Property, a shortage of human resources having the ability and the knowledge to secure Intellectual Property is still predicted. Therefore, the nation's recognition of Intellectual Property will continue to be very low unless definite actions are taken.

At University, the use of a free and easy-to-introduce online education support system called "Moodle" is becoming more and more popular. However, we have not yet fully utilized the functions of that system. Moreover, even though Japan has a superior Internet infrastructure in comparison with other countries, online education does not have a clear direction, and so introduction of a technology education that meets Japan's economic standards is an urgent need.

The purpose of this research is to develop a wide variety of professional human resources, to work on Intellectual Property education that advances Intellectual Property, and to establish an Intellectual Property Online Education System (IPOES) that is provided using Moodle. If we can create an environment that improves people's awareness of Intellectual Property, Japan's economics and technologies will also improve.

This research is based on "The Comprehensive Strategy for the Development of Human Resources" which was organized to specifically develop the policies proposed by the "Strategic Council on Intellectual Property". This research concentrates on engineers especially because they have multiple opportunities to take part in the creation, the protection, and the use of Intellectual properties.

The system proposed in this research will be used in lectures, seminars, and symposiums related to Intellectual Property at University. The primary target audience of this system will be graduate and underground students in science and engineering, teachers, and the staff of the Organization for Academic-industrial Collaboration and Intellectual Property sponsored by the Technology Licensing Organization (TLO).

2 Moodle

Moodle[4, 5, 6] is a Course Management System (CMS), also called a Learning Management System (LMS), which is required in the operation of e-learning. Also, since Moodle is a free web application, it supports online education by allowing all users to conduct their own course development.

The main feature of Moodle is the setting up of courses. The Admin User has full control over all courses and adds input to the Resource Module and Activity Module in Moodle.

The Activity Module includes a Quiz Module that makes it possible to create questions in a variety of forms. Consequently, it is possible to provide questions that meet the academic levels of the targeted learners and to provide appropriate feedback for the answers given. The Forums Module generally has the same functions as a BBS or bulletin, and so it activates discussions. The Assignment Module is for submitting quiz answers through written texts and uploaded files. The Assessment Module manages grades and allows the visualization of the grades of individual assignments as well as the overall grades.

In addition to the functions mentioned above, Moodle offers other useful functions, such as chatting, sending messages, lessons, polls, glossaries, and Wiki, as standard features. Many of the available plug-in modules were originally constructed by developers around the world, but all users can freely install and use these functions.

Therefore, the advantages of Moodle are as follows: it is free, it flexibly responds to developers' needs, and it centralizes the information of educational sites. Its education infrastructure can serve as a class-exclusive site that allows functions such as the distribution of teaching materials, communications, tests, and submission of assignments.

3 TLO

The Technology Licensing Organization (TLO) [7, 8] is a corporation that awards patents to university researchers and transfers the patented technologies to companies, thereby acting as a conduit between the academic side and the production side. The profits earned by the universities by creating new industries will return to the researchers as research capital, and this will further stimulate research. TLO acts as a core, known as the "Intellectual creation cycle," of the Organization for Academic-Industrial Collaboration and Intellectual Property. Although some universities in Japan have many research resources and some are likely to act as "seeds" in new industries, they are not fully utilized in all industries. Therefore, TLO serves a useful purpose in Japan.

By developing TLO, researchers are able to focus on their research. Moreover, along with their original duties, going into partnership with the proposed system would make it possible to support Intellectual Property in terms of education and to support Intellectual Property related to universities.

4 The Proposed System

Registration, administration, and user management are performed by centralizing all user IDs and information. Universities provide exclusive e-mail addresses to students, teachers, and staff. The exclusive addresses come from three types of

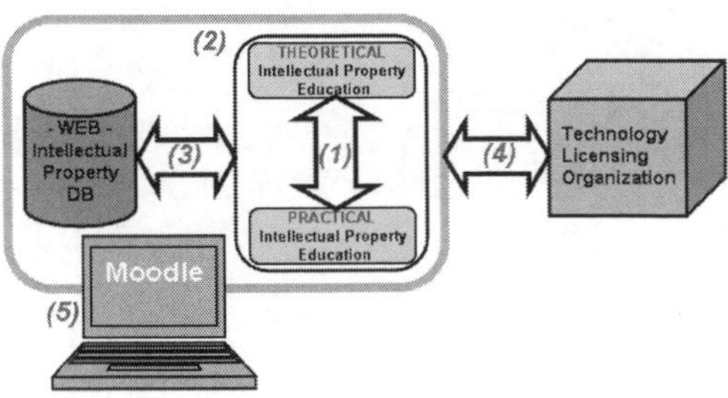

Fig. 1 General image of IPOES.

domains. The principle of "one user ID, one e-mail address" would prevent unauthorized access and at the same time make it easier for staff and teachers lacking computer knowledge to manage. This will help increase use of this system.

Fig.1 shows an image of the IPOES. The flow of text implies this would follow Fig.1.

(1)

1. Carrying out "Practical" Intellectual Property education that instructs users on how to use knowledge along with conventional "Theoretical" Intellectual Property education that focuses on obtaining knowledge.
2. Providing the education discussed in 1 based on the *Industrial Property Standard Textbook*[9], which was published by the National Center for Industrial Property Information and Training (INPIT)[10]. The purpose of this text is to help students in "higher education" to obtain proper knowledge and practice of Intellectual Property rights.
3. Providing many examples of how Intellectual properties are used and what they should be like in the future. (It is assumed that basic knowledge and skills have already been obtained.)

(2) IPOES is divided into two categories: lectures and seminars. The lectures are further divided into different levels that determine the education provided. The seminars are divided and provided according to themes (see Table 1).

- The "Primary" level targets users who have no knowledge or skills regarding Intellectual properties, but who are highly motivated to learn.
- The "Secondary" level targets those who have basic knowledge of Intellectual properties.
- The "Superior" level targets those who intend to obtain the qualifications related to Intellectual properties, or who will need knowledge of Intellectual properties in the future.
- Last, the "Specialist" level targets those who already have the qualifications related to Intellectual properties, who have jobs that pertain to Intellectual Property, or who research Intellectual Property.

Table 1 Educational Level of IP

Category	Level
Lecture	1. Primary
	2. Secondary
	3. Superior
	4. Specialist
Seminar	1st grade Certified Skilled Worker of IP Management
	2nd grade Certified Skilled Worker of IP Management
	3rd grade Certified Skilled Worker of IP Managament
	Job hunting

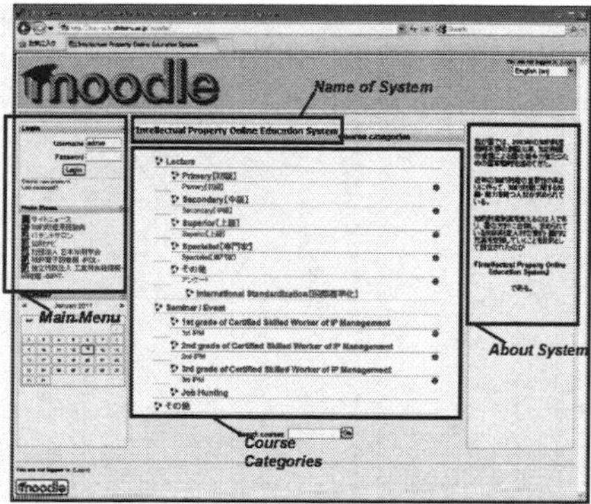

Fig. 2 The Top Page of IPOES

(3) IPOES is always in direct link with dozens of useful web sites, such as the Industrial Property Digital Library (IPDL)[11] and J-STORE[12], providing an ideal environment for searching and obtaining information on Intellectual Properties. In addition, using IPOES as a teaching material will lead to continually updated Intellectual Property information.

(4) It is possible to form a technical environment of human support by forming a partnership with the university's Organization for Academic-industrial Collaboration and Intellectual Property. This support can help nurture technicians at an early stage. Based on this partnership, Intellectual Property education would propagate an Intellectual Property Human Resource Cycle and an Intellectual Creation Cycle.

(5) Moodle can centralize and manage the ideas discussed in *(1)* to *(4)*, thereby systematizing the Intellectual Property education curriculum at Universities. The top page of IPOES using Moodle shows on Fig.2.

5 Conclusion and Future Work

In the future, we will construct a database based on the Industrial Property Standard Textbook series, which is published with the cooperation of companies and organizations in Japan and implement it as teaching material. Also, we will cooperate with the closely related subject of information education and technology to distribute the knowledge gained in the process of establishing the IPOES.

Furthermore, we will support teachers, staff, researchers and those who wish to obtain the qualifications related to Intellectual Property.

By using the module functions of Moodle, which were developed to reflect the input of both teachers and students, it is possible to create teaching materials, manage the submission of assignments, administer quizzes, and obtain data on student attendance and grades. Also, a future objective of Moodle is to provide more accurate and careful guidance to students by data mining. Teachers could receive various kinds of data from students, and in turn they could give back the results of the data analysis as guidance.

Moodle, which itself is an Intellectual Property and an Imaginary Classroom, is a necessary tool for Japan's economic growth and will play a substantial role in developing Japan's educational system and technologies, which are currently lagging behind the global trend. Thus, by targeting technicians and students studying science and engineering, who will be the future leaders of Japan, the essential objectives can be achieved.

References

1. Intellectual Property Strategy Headquarters, Japan, Intellectual Property Strategic Program (2003-2010)
2. 2003–2010 Strategic Council on Intellectual Property, Japan, Intellectual Property Policy Outline (2002)
3. Intellectual Property Strategy Headquarters, Japan, Comprehensive Strategy for the Development of Human Resources (2006)
4. moodle.org, `http://moodle.org/` (cited December 28, 2010)
5. moodle.com, `http://moodle.com/?moodlead=moodle.general` (cited December 28, 2010)
6. William, H., Rice, I.V.: Moodle 1.9:E-Learning Course Development. Packt Publishing (2008)
7. Organization of MEXT, `http://www.mext.go.jp/a_menu/shinkou/sangaku/sangakub/sangakub5.htm` (cited December 28, 2010)
8. University Technology Transfer Association, Japan, `http://unitt.jp/en` (cited December 28, 2010)
9. National Center for Industrial Property Information and Training, Japan, Industrial Property Standard Textbook (2010)
10. National Center for Industrial Property Information and Training, Japan, `http://www.inpit.go.jp/english/` (cited December 28, 2010)
11. National Center for Industrial Property Information and Training, Japan, Industrial Property Digital Library, `http://www.ipdl.inpit.go.jp/homepg_e.ipdl` (cited December 28, 2010)
12. JST Science and Technology Research Result Database for Enterprise Development, J-STORE, `http://jstore.jst.go.jp/index.html?lang=en` (cited December 28, 2010)

Knowledge Sharing in Virtual Teams a Perspective View of Social Capital

Ying Chieh Liu

Abstract. Although research on virtual teams is becoming more popular, there is a gap in the understanding of how social capital affects the processes of knowledge sharing and creating, and their impacts on virtual team performance. To fill in this gap, this study formed a framework by incorporating social capital and SECI model [7] and examined it by an experiment with 65 virtual teams collaborating in a Wiki platform. The results showed that social capital was positively related to the four SECI modes (socialization, internalization, combination and externalization) and three SECI modes (internalization, combination and externalization) were found to be positively related to virtual team performance. The contributions of this study were twofold. Firstly the framework brought a broader view of researching knowledge management in a virtual team context. Secondly leaders and managers of virtual teams should be made aware of enhancing the effects of social capital to encourage internalization, combination and externalization to substitute the role of socialization.

1 Introduction

Little has been known how social capital affects the knowledge sharing and creating, and its influence on virtual team performance. To address this gap, we drew on the notion of social capital to examine its impact on knowledge sharing and virtual team performance.

Social capital refers to the connections within and between social networks and it resides in the relations among the nodes and facilitates team productivity [10]. Information provided via CMC was felt to be less reliable, inadequate and difficult to interpret [11]. This hindered the sharing of knowledge between workers and prevented coworkers from completing their jobs and inhibited the development of social relationships [6].

Knowledge sharing has been identified as a major issue for knowledge management and an important factor for evaluating the performance of organizations [4]. SECI model by Nonaka [7] is the most outstanding research work in knowledge management, As SECI stands for "socialization", "externalization",

F.L. Gaol (Ed.): Recent Progress in DEIT, Vol. 2, LNEE 157, pp. 263–268.
springerlink.com

"combination" and "internalization", knowledge is transferred from individual level to group level and vice versa. SECI model regards the process of generating knowledge a spiral which was known as "knowledge spiral". Although Nonaka [7] argued that the four modes could not be clearly separated as single isolated steps progressing in a sequential arrangement, however in order to analyze and understand the complicated processes of how virtual team members share and create knowledge upon the view of social capital, a conceptual distinction between the four modes of SECI is compulsory and provides the basis of the hypotheses and the empirical examination.

The purpose of this study is to provide a more detailed understanding of the relationships between social capital and SECI model and their impacts on virtual team performance. Thus, this study incorporated social capital and SECI model to build a framework and examined their relationships on virtual team performance by conducting a survey after an experiment which had 65 virtual teams formed in a Wiki platform.

2 Theoretical Framework and Hypotheses Development

Social capital incorporates three dimensions: structural capital, cognitive capital and relational capital. For fitting the context of virtual teams, network ties, shared vision and trust were selected to represent the three dimensions correspondingly. Moreover, four components of SECI model (socialization, internalization, combination and externalization) were regarded as four individual behaviors of knowledge sharing in this study. We assumed that three social capital factors (network ties, shared vision and trust) are positively related to the four behaviors and the four behaviors are positively related to performance of virtual teams. Thus, 16 hypotheses are described below.

Hypothesis 1: Network ties will be positively related to the disseminated of tacit knowledge within virtual teams (socialization).

Hypothesis 2: Network ties will be positively related to the transformation of explicit knowledge into tacit knowledge within virtual teams (internalization).

Hypothesis 3: Network ties will be positively related to the transformation of explicit knowledge into explicit knowledge within virtual teams (combination).

Hypothesis 4: Network ties will be positively related to the transformation of tacit knowledge into explicit knowledge (externalization).

Hypothesis 5: Shared vision will be positively related to the transformation of tacit knowledge into tacit knowledge (socialization).

Hypothesis 6: Shared vision will be positively related to the transformation of explicit knowledge into explicit knowledge (combination).

Hypothesis 7: Shared vision will be positively related to the transformation of tacit knowledge into explicit knowledge (externalization).

Hypothesis 8: Shared vision will be positively related to the transformation of explicit knowledge into tacit knowledge (internalization).

Hypothesis 9: Trust will be positively related to the disseminated of tacit knowledge within virtual teams (socialization).
Hypothesis 10: Trust will be positively related to the transformation of explicit knowledge into tacit knowledge within virtual teams (internalization).
Hypothesis 11: Trust will be positively related to the transformation of explicit knowledge into explicit knowledge within virtual teams (combination).
Hypothesis 12: Trust will be positively related to the transformation of tacit knowledge into explicit knowledge within virtual teams (externalization).
Hypothesis 13: Socialization will be positively related to virtual team performance.
Hypothesis 14: Internalization will be positively related to virtual team performance.
Hypothesis 15: Combination will be positively related to virtual team performance.
Hypothesis 16: Externalization will be positively related to virtual team performance.

3 Research Design

Students who took "Management Information Systems" from seven classes in three universities which were located in three dispersed counties in Taiwan were required to complete a given case study as their final report which was a group writing assignment. It describes a restaurant facing some serious problems and challenges such as the inefficiency of managing orders, the disorganization of stock management, and difficulties in calculating payroll and tax. Four questions related to these issues were given and teams were asked to propose the solutions for the restaurant. Team members could only communicate each other via a Wiki platform, and any other methods of communication were prohibited (such as MSN, email and phone call).

The project operated over five weeks. Students were chosen randomly from each class and were put into a group with five members. Some groups decreased to 4 members because some students dropped the course after the project started. This brought 65 groups comprised of 302 students in total. Each group was pre-assigned to their Wiki working space and had to access their discussion boards to discuss and exchange information to complete the assignments which were posted on the Wiki platform. To ensure their disability to contact others by other prohibited methods (such as email, msn) or even meet face-to-face, three assistants checked discussion boards two times per day in order to remove any personal contact message posted by members and left warnings. After completing the assignments, hard copy questionnaires were distributed in the lectures. Participants were asked to use a 7-point Likert scale in answering the questions related to the framework. 287 validated questionnaires were collected, giving a return rate of 95%.

The measurement items were derived from past studies. The questionnaire was composed of five parts. The measurement items of social ties, shared vision and trust referred to the research by Tsai and Ghoshal [9] and Chiu et al. [1] while those of socialization, internalization, combination and externalization referred to the study by Schulze & Hoegl [8]. The measurement items for performance came from a study by Lurey & Raisinghani [5].

4 Data Analysis and Result

Appendix 1 presents the descriptive statistics and covariance for the constructs. Convergent validity was assessed by examining composite reliability (CR) and average variance extracted (AVE) from the measures. The CR values of the eight constructs were between 0.86 and 0.98 and all are above the suggested minimum of 0.70 [3]. Their AVE values were all above 0.50 and these values provided further evidence of convergent validity and discriminant validity [2]. The reliability of these constructs was evaluated using Cronbach's coefficient alpha (α) and their α values were all above 0.70, which indicates a reliable measurement instrument.

SEM (Structural Equation Modeling) was applied to estimate the path diagram of the framework using LISREL 8.72. The overall fit of the proposed structural model was quite satisfactory (X2=796.40, df=368, X2/df = 2.16, Root mean square error of approximation (RMSEA)=0.06, Comparative fit index (CFI)= 0.98, Normed fit index (NFI)=0.98, Goodness-of-fit index (GFI)=0.84, and Adjusted goodness of fit index (AGFI)=0.67).

Appendix 2 showed the results of applying Structural Equation Modeling (SEM) and the only insignificant path was "Socialization Performance", which brought the rejection of hypothesis 13. Other hypotheses were supported. The results showed that network ties, shared vision and trust positively were related to the socialization, internalization, combination and externalization. Internalization, combination and externalization were positively related to virtual team performance. Only socialization had no significant impact on virtual team performance.

5 Conclusions

This study examined the relationships between social capital and SECI mode and the results showed that all paths were significant except the one between socialization and performance. There are two contributions of this study. In a theoretical view, this study took the lead in combining social capital and SECI mode to frame a model to examine virtual team performance. As most research addressed the issues of knowledge sharing in an organizational setting, this study depicted how social capital affected knowledge sharing and creating conducted by the processes of SECI in a virtual team context, which brings a broader thought of studying knowledge management in a team level.

In a practical view, it can be concluded that individuals contributed knowledge to the teams when they were structurally embedded in the network, when they enjoyed helping each other and when they shared the same visions and goals. Based on this, leaders and managers of virtual teams should be made aware of the effects of social capital on the smooth processes of knowledge sharing and creating.

References

1. Chiu, C.-M., Hsu, M.-H., Wang, E.T.G.: Understanding knowledge sharing in virtual communities: An integration of social capital and social cognitive theories. Decision Support Systems 42(3), 1872–1888 (2006)
2. Fornell, C., Larcker, D.F.: Evaluating structural equation models with unobservable variables and measurement error. Journal of Marketing Research in Organizational Behavior 18, 39–50 (1981)
3. Hair, J.F., Anderson, R.E., Tatham, R.L., Black, W.C.: Multivariate data analysis, 5th edn. Prentice Hall, New York (1998)
4. Hendriks, P.: Why share knowledge? The influence of ICT on the motivation for knowledge sharing. Knowledge and Process Management 6(2), 91–100 (1999)
5. Lurey, J.S., Raisinghani, M.S.: An empirical study of best practices in virtual teams. Information and Management 38(8), 523–544 (2001)
6. Nandhakumar, J., Baskerville, R.: Durability of online teamworking: Patterns of trust. Information Technology and People 19(4), 371–389 (2006)
7. Nonaka, I.: A dynamic theory of organizational knowledge creation. Organization Science 5(1), 14–37 (1994)
8. Schulze, A., Hoegl, M.: Organizational knowledge creation and the generation of new product ideas: A behavioral approach. Research Policy 37(10), 1742–1750 (2008)
9. Tsai, W., Ghoshal, S.: Social capital and value creation: The role of intrafirm networks. Academy of Management Journal 41(4), 464–476 (1998)
10. White, L.: Connection matters: Exploring the implications of social capital and social networks for social policy. Systems Research and Behavioral Science 19(3), 255–296 (2002)
11. Zolin, R., Hinds, P.J., Fruchter, R., Levitt, R.E.: Interpersonal trust in cross-functional, geographically distributed work: A longitudinal study. Information and Organization 14, 1–26 (2004)

Appendix 1. Correlation matrix, Cronbach alpha, CR, AVE

	Mean	S.D.	C1	C2	C3	C4	C5	C6	C7	C8	Alpha[a]	CR[b]
C1 Socialization	4.44	2.00	*.87*[c]								.90	.95
C2 Externalization	4.30	2.23	.86	*.94*							.80	.98
C3 Combination	4.64	2.18	.79	.73	*.83*						.73	.94
C4 Internalization	4.26	2.00	.82	.82	.70	*.91*					.83	.97
C5 Performance	4.58	2.42	.66	.75	.71	.79	*.79*				.90	.86
C6 Social ties	3.15	2.00	.68	.51	.66	.49	.42	*.89*			.76	.95
C7 Trust	4.59	1.70	.77	.83	.66	.81	.69	.30	*.92*		.88	.97
C8 Shared Value	4.27	2.04	.89	.89	.73	.83	.70	.47	.77	*.96*	.90	.98

[a] Internal Consistency Reliability (Cronbach's coefficient alpha).
[b] Composite Reliability
[c] The diagonal (in italics) shows the average variance extracted

Appendix 2. Structural Equation Model Results (** Significant at 0.05 level)

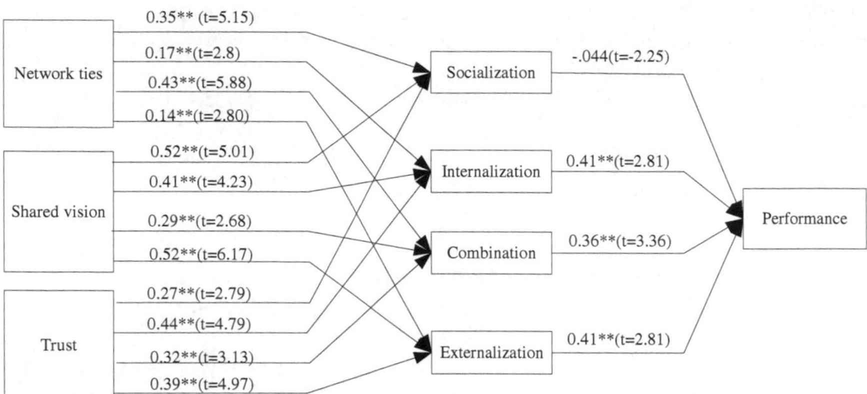

A Study on Correlation between Web Search Data and CPI

Chong Zhang, Benfu Lv, Geng Peng, Ying Liu, and Qingyu Yuan

Abstract. The web search data, which recorded hundreds of Millions of searchers concerns and interests, reflected the trends of their behavior and provided essential data basis for the study of macro-economic issues. This paper established a concept frame based on commodity market and equilibrium price theory, revealed a certain correlation and lead-lag relationship between web search data and consumer price index (CPI). Empirical results indicated that there is a co-integration relationship between web search data and CPI. The model was able to obtain a good fit with CPI. Model fitting is 0.978.

1 Introduction

CPI (consumer price index) is an important basis for national macroeconomic policies and the economic regulation. Therefore it has great significance to enabling real-time monitoring and forecasting CPI in advance.

CPI reflects the price level of a specific group of specification commodity and and services, the essence of CPI is price. According to the equilibrium price theory, the price of commodities is determined by supply and demand relation. Supply and demand are generated by the producers and consumers of the market. Both the production decision-making and purchase decision-making need Massive information, such as the price , quality and performance of goods and raw materials, the scale, credit of the producers, the macroeconomic policies and situation, etc. With the development of the Internet , Internet has transcended the traditional media and become the most important way for more than 300 million Web

Chong Zhang · Benfu Lv
Management School, Graduate University of Chinese Academy of Sciences, Beijing, China
e-mail: zczoln@hotmail.com, lubf@gucas.ac.cn

F.L. Gaol (Ed.): Recent Progress in DEIT, Vol. 2, LNEE 157, pp. 269–274.
springerlink.com

users access to information. In many internet services, search engine is the most popular tool to get information. When producers and consumers get information through the web search , the search engine has also recorded their searching and browsing behaviors. These data can reflect the producers and consumers' attention and interest, mirrors the change of their behavior trend of manufacturing and purchasing. Therefore the behavior of producers and consumers will have signs both in the commodities market and the internet. The signs in the commodity market are the changes of price. And the signs in the internet are the changes of the Web browsing and the Web search data. They are both response of the same thing, so that they should have certain relationship.

There had some research achievements abroad using the Web search data for monitoring and predicting economic and social activities. Based on the Google search data of influenza prediction [1] and unemployment monitoring [2] has achieved high accuracy. Comparing with the traditional method ,web search data can be good for real-time monitoring because of its instant. But using the Web search data for the prediction research of comprehensive index as the CPI is still a blank . Therefore we make CPI and Web search data as the research object. Firstly, we discuss theoretically the relationship between them and build the conceptual framework; secondly, we make empirical analysis and inspection on the relationship between them; Finally, we conclude the whole paper and give the future research direction of extensible.

2 The Theoretical Analysis and the Conceptual Framework

This section we set a conceptual framework from the angle of the relation between supply and demand of the commodity market , to analyze the relationship between the search data and CPI. The basic thoughts of the conceptual framework are as follows: Macroeconomic factors influence the relation between supply and demand which determined price; as the main body of microeconomic, producers and consumers' behavior are reflected in the commodities market and the Web, in the commodity market for trading the price reflects the change, while the Web is embodied in the Web search, browse, etc; Because of the behaviors of market reflected in the changes of price are lagging behind and the reflects on the Web search data are immediate, there must be a chronological order between them.

According to the theory of consumer behavior, we divide buying behavior into consumer demand, information search and purchase decisions three links; Likewise the producers will be divided into the investment demand, information search and production decision-making three links. The decision-making process of consumers and producers requires a lot of information. The information searched by producers and consumers in the Web can be divided into the micro and macro level. Micro-level information includes a specific commodity prices, performance, quality, size of business, credit, raw materials prices, the search for such information produced many keywords such as "laptop offer," "Which brand of digital camera good", "Nokia N95 "," cement price " etc. Macroeconomic information includes national macroeconomic policies, economic trends and market conditions, etc. On the macro information search on the Web produced many keywords such as "exchange rates", "deposits", "Chinese Price "etc. We select the Web search words, using the search changes of relevant related words to the movements of the price level of commodity to having judgment or prediction.

3 Empirical Studies

A. Data description and Preprocessing

CPI data in this paper originate from National Bureau of Statistics Web site. We use 58 months data from January 2004 to August 2009. and the web search data was take from Google insights, this was a product from google which launched in 2008, we can get the weekly search volume of a given keyword from January 2004 to the current week, the search volume we got is not the absolute search volume, but a relative data, that is "In the past period of time, how many times that your keywords are searched as opposed to the total search volume on google,", so the search volume data reflect the level of popularity to a particular keyword . In order to better fit with CPI, we do some preprocess on search volume data from Google :

(1) merging the Weekly data to monthly data;
(2) using moving average method to smooth the search data, eliminate the short-term fluctuations and highlighting their long-term features
(3) converting the search data to year-on-year data in keeping with the CPI data.

B. Keywords Selection

According to the composition of CPI and its influencing factors, we firstly selected about 300 candidate-related keywords. then we calculate the Pearson correlation coefficient between every candidate search data word and CPI data. The coefficient presents the similarity between the two variables. Finally we Just select the key words whose correlation coefficient with CPI are greater than 0.4, and totally 44 keywords were selected.

According to the different economic implications, the 44 keywords will be divided into two categories of indices: the macroeconomic search index (X1) and the supply-demand search index (X2).

C. Search Index and CPI Lags

In order to discover the lag period between the search index and the CPI, we set CPI(t) as dependent variable and X1(t) ,X1(t-n), X1(t+n), X2(t), X2(t-n), X2(t+n) as the independent variable to establish several regression models. We determine the lag period based on R2 of the model. Table 1 shows the R2 value of each model ,the result shows the variables which have the best fitting of the CPI(t) is the X1(t-5) , X1(t-6) and X2(t-1), R2 equals 0.92/0.76 respectively. This means the lags from the changes in the macroeconomic policy and changes in supply and demand on the CPI are different. Macroeconomic policy changes on the CPI of the conduction delay for about the six months, however, the supply and demand changes in the transmission delay on the CPI for one month.

Table 1 Search index and CPI lags

	t+2	t+1	t	t-1	t-2	t-3	t-4	t-5	t-6	t-7
(X1)	0.16	0.28	0.40	0.54	0.66	0.76	0.86	0.92	0.92	0.85
(X2)	0.58	0.68	0.72	0.76	0.72	0.66	0.59	0.47	0.35	0.22

D. Co-integration Test of Search Index and CPI

Co-integration test is used in order to verify whether there is a long-term stability relations between search index and CPI. We choose CPI (Y) as the dependent variable, the macroeconomic search index (X1) and supply-demand search index (X2) as independent variables, firstly, we test the stationary of each variable (unit root) using the extended Dickey - Fuller test (the ADF test). The original hypothesis is the sequence has at least one unit root. Test results as shown in Table 2, the original series are non-stationary series, but the test results by the first-order difference are at 1% level of significance rejected the null hypothesis, indicating that they are integrated of order one differnce.

Table 2. Variable stability test

Variables	ADF t-Stat	MacKinnon threshold 1%	5%	10%	ADF Result
Y	-2.11	-3.56	2.92	-2.60	non-stationary
X1	-1.73	-3.56	2.92	-2.60	non-stationary
X2	-1.24	-3.56	2.92	-2.60	non-stationary
ΔY	-6.62	-3.56	2.92	-2.60	stationary
ΔX1	-4.56	-3.56	2.92	-2.60	stationary
ΔX2	-5.35	-3.56	2.92	-2.60	stationary

Then further tests between dependent variables and independent variables are long-term stable relationship, we use Engle and Granger (1987) proposed two-step co-integration (Co-integration) relationship test, the first step is to establish the regression equation, according to search index and the CPI, respectively, X1 (t-5), X1 (t-6) and X2 (t-1), X2 (t-2) as the independent variables. In this way, we get the four regression equations. The second step is unit root test on residuals. If there exists a co-integration relationship between the variables and the dependent variable, the regression residuals should have stability. The results show in Table 3: the residuals of the four regression equation are stability, so the co-integration of long-term stability are existing between search index and CPI.

Table 3 Regression results and co-integration test

		Model 1	Model 2	Model 3	Model 4	Model 5
C		67.890***	67.254***	69.338***	69.440***	49.784***
LOGX1(t-5)		4.420***	4.424***			2.559***
LOGX1(t-6)				4.728***	4.793***	
LOGX2(t-1)		2.945***		2.324**		
LOGX2(t-2)			3.076***		2.239***	1.102**
Y(t-1)						0.464***
Y(t-12)						-0.116***
Adj-R2		0.938	0.946	0.924	0.925	0.978
F-stati		390.25	471.28	304.21	323.98	491.86
DW		0.888	1.027	0.787	0.862	2.47
Residual Statio-narity	ADF	-4.70	-5.30	-4.18	-4.49	-8.22
	1%	-3.57	-3.56	-3.57	-3.57	-3.58
	5%	-2.92	-2.92	-2.92	-2.92	-2.93
	10%	-2.60	-2.60	-2.60	-2.60	-2.60
	results	stationary	Stationary	stationary	stationary	stationary

Table 3shows, among the four models above, the model with LOGX1 (t-5) and LOGX2 (t-2) as the independent variable model fit with the highest R2, adjusted R2 is 0.946; the DW value of the model is 1.027, indicating there exist serial correlation issues. In order to eliminate this problem, we add lagged dependent variable item Y (t-1) and Y (t-12) to Model 5, adj-R2 is improved to 0.978, and all variables are significant at 5% level.

The regression equation of Model 5 shows that change of CPI are determined by the macroeconomic situation, supply- demand relation and the historical price level.

4 Conclusions

In this paper we first theoretically analyze the relationship between the search data and CPI, reveal the internal mechanism of CPI forecast based on search data. Then we have an empirical study to testing their relationship and build models to predict the CPI. The main conclusions and innovation are as follows:

(1). There is a co-integration of long-term stability relation between the web search data and CPI. The model of CPI forecast based on the web search data is fitting to 0.978, the absolute forecast error is 0.48. The increase of search index will cause the CPI increase, in which the macroeconomic search index and supply-demand search index changes 1%, the corresponding rate of CPI change is 2.559 percent and 1.102 percent; two types of search index relative to the CPI of the early period are approximately five months and two months.

(2). Compared with traditional monitoring methods, the model based on the web search data monitors the CPI timely. This method can make up the traditional monitoring methods which have information dissemination lag up to two weeks. And also it can release about a month ahead of the national Bureau of Statistics. In this way, real-time monitoring of CPI is effective, The CPI search data can be used as indirect indicators for monitoring. And our method provides an effective supplement for traditional monitoring methods.

References

[1] Kahneman, D.: Attention and Effort. Prentice-Hall, Englewood Cliffs (1973)
[2] Barber, B.M., Odean, T.: All That Glitters: The Effect of Attention and News on the Buying Behavior of Individual and Institutional Investors. Review of Financial Studies 21(2), 785–818 (2008)
[3] Hou, K., Lin, P., Wei, X.: A tale of two anomalies: The implications of investor attention for price and earnings momentum. Working Paper. Ohio State University and Princeton University (2008)
[4] Ginsberg, Mohebbi, Patel, Brammer, Smolinski, Brilliant: Detecting influenza epidemics using search engine query data. Nature 457, 1012–1014 (2009)
[5] Doornik, J.A.: Improving the Timeliness of Data on Influenza-like Illnesses using Google Search Data. Working paper (2009)
[6] Goel, S., Hofman, J.M., Lahaie, S., Pennock, D.M., Watts, D.J.: What Can Search Predict. Workingpaper (2009)
[7] Tierney, H.L.R., Pan, B.: A Poisson Regression Examination of the Relationship between Website Traffic and Search Engine Queries. Workingpaper (2010)
[8] Choi, H., Varian, H.: Predicting the Present with Google Trends. workingpaper.Technical Report, Google Inc. (2009)
[9] Choi, H., Varian, H.: Predicting Initial Claims for Unemployment Benefits (2009)

Improve the User Experience with Quick-Browser on Netbook

Fangli Xiang

Abstract. In this paper, we discuss how to improve user experience with web browser on Netbook. Internet browser is the most important application for Netbook. The start up speed of browsers is too slow on Atom processor, such as Firefox, IE, etc... So this paper proposal a quick-browser to overcome this defect, and improve the cold booting user experience.

1 Introduction

Netbook is a new type of laptop computer, defined by size, price, and horsepower. They are small, cheap, under-powered, and run either an old or unfamiliar operating system. DisplaySearch forecasted that 33M Netbooks shipped in 2009 (see Table 1). This market grows to 20% worldwide.

Table 1 Mini-Note and Notebook Shipments and Growth by Region[1].

Brand	2008 Mini-Note	2008 Notebook PC	2009 Mini-Note PC	2009 Notebook PC	Mini-Note Y/Y Growth	Notebook PC Y/Y Growth	Total Y/Y Growth
Japan	1.5M	8.5M	1.9M	7.4M	29.1%	-13.0%	-6.8%
North America	3.7M	40.3M	8.8M	39.9M	136.9%	-1.1%	10.6%
EMEA	7.3M	51.4M	13.3M	46.7M	80.6%	-9.1%	2.2%
Greater China	1.1M	13.0M	3.9M	16.0M	260.3%	22.3%	40.4%
Latin America	1.0M	4.5M	1.9M	5.4M	88.1%	19.8%	32.4%
Asia Pacific	1.8M	11.9M	3.0M	14.1M	64.6%	18.7%	24.9%
Total	16.4M	129.6M	32.7M	129.5M	99.1%	-0.1%	11.1%

Fangli Xiang
Department of Computer, Zhengjiang Changzheng Technical and Vocational College
e-mail: susanxfl@163.com

F.L. Gaol (Ed.): Recent Progress in DEIT, Vol. 2, LNEE 157, pp. 275–279.
springerlink.com

One of the big differences between Netbook and Notebook is the CPU. Atom processor has so far dominated the Netbook category. Atom is lower power consumption than any Notebook processor, is therefore slower.

Netbook as the laptop is focusing on internet connection. The light-weight feature of Netbook decides it was originally designed for mobile Internet, will certainly need the networks of telecom operators. In addition, the arriving of 3G and 3.5G also has laid the solid foundation for its applications. In Japan (Data from Yahoo Japan and Splashtop Blog), 40% use Netbooks in cafes (compared to 4% for notebook users), and 28% use Netbooks while commuting (compared to 1% for notebooks). Remember, this is a country with good public transport. Internet browser will be the most important application for Netbook.

Because of the slow Atom processor, if you want use a normal and powerful browser such as IE or Firefox, you should waiting more than 7 seconds after you click the browser button in your system[2]. It's too awful to wait so long for an application launch. To improve user experiments, a Netbook needs a quick and full-featured internet browser.

2 Design Overview

In this project we want to control the Firefox by an external app. The framework of this project is shown in the Fig.1. In this project, Firefox is just used to read web content, and render it on user's screen. And all the menus and tool bar are designed and controlled by the external app. In order to control Firefox we need a client embedded in to the Firefox, we use a Firefox extension as a client. And we use socket to sends the control command to Firefox and accept feedback.[3]

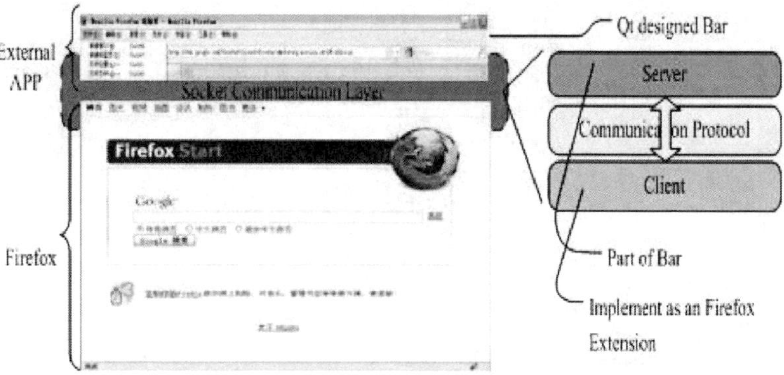

Fig. 1 Project Framework

3 Implementation Details

3.1 Server Part

We creates an embedded web server on a listening (i.e. -server) socket (For example localhost::8010). This web server will only accept connections from the local host, and implements a small set of requests used by Firefox to communicate with the server (see feedback protocol part).

3.2 Client Parts

In this project the client is an extension for Firefox. So it can use XMLHttpRequest for socket communication (WebSocket is an alternative, but it was non-normative). And we need bidirectional communications with server-side processes. For server to client, it can send any type of messages through socket, and we called this message as command. But for client to server, we could just use the XMLHttpRequest's[4] GET method to posting messages. We call this message as feedback.

We need two XMLHttpRequest in client, one for waiting commands, and another for posting feedback. When the client started, it will send an XmlHttpReqeust to the server. This is the command waiting request. The server keeps this connection alive and waiting for user command. When user launches a command, the server will send the command to the client through this connection, and close this socket. When the client receives the command, it will parse the command and fire a specified action. At the end of the action, client will send another XMLHttpRequest to server and waiting for command input. Another request is the feedback. When the client needs feedback to server, it will setup an XMLHttpRequest to the server with the feedback info. When the server received it, it will parse the info and fire some action and close the socket at the end. This part seems awful, but I have no choice.

3.3 Communication Protocol

In the follow parts, we will define the command protocol and feedback protocol.

Command protocol: We divide the command protocol into three parts: menu command, toolbar command and tabbar command. We use XML to define commands [5]. All the commands start with tag <command>, and end with </command>.The different among menu commands, toolbar commands, and tabbar commands is the second tag. For menu command is start with <menu> and end with </menu>, for toolbar command is <toolbar> and </toolbar>, and for tabbar command is <tabbar> end with </tabbar>.

For example toolbar command back, it can be represented as:
<command><toolbar>BACK</toolbar></command>
A menu command new window under the menu file can be represented as:
<command><menu>FILE_NEW_WINDOW</menu></command>

And a tabbar command focus tab number X can be represented as:

<command><tabbar>FOCUS</tabbar><tabindex>X</tabindex></command>

If you have any attached message, you can add a tag. For example attach a URL after the menu command new:

<command>

<menu>FILE_NEW</menu><url>www.splashtop.com.cn</url>

<command>

Feedback protocol: The server part needs to update the tab bar when a clicking on a Hyperlink in a tab happened and a tab closed.

When a New tab made, the client need inform the server that a new tab made with a URL. Then client also should inform the server loading start. After loading complete, the client also needs to inform the server. So the server can made a new tab on the tabbar, and updating tab information.

When a new page loading in the current tab happened, the client needs inform the server that this tab is loading another URL, and also needs to inform server when loading complete.

The flow of a hyperlink click is shown in the Fig.2.

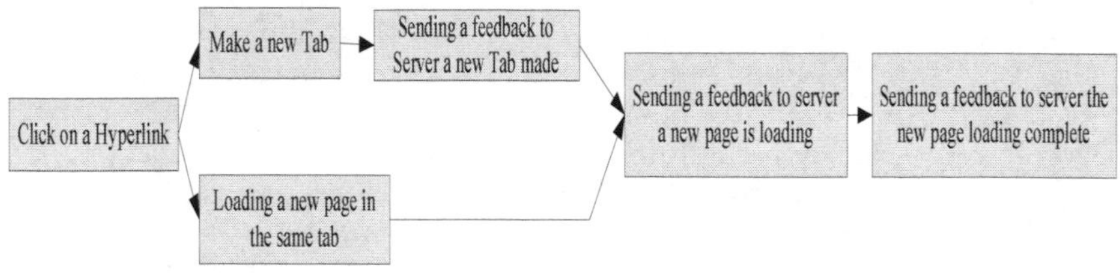

Fig. 2 Working flow of hyperlink clicking happened.

So the client should feed the following information back to the server:

- If a new tab made, the new tab URL, and index which was created by a click on Hyperlink should be feedback;
- When a loading happened the tab index and loading URL should be feedback;
- When a loading is complete, the tab index and tab title should be feedback.

When a tab is closed, we need to send a back a feed back to the server. Then server can close a tab correctly.

Because of XMLHttpRequest is a high level TCP/IP protocol, so we can only use the Http protocol in our client. The Http protocol we used is the GET. And we encapsulate the feedback information in the request URL with different tags.

4 Protocol Specification

4.1 Command Protocol

Let's start to define all the protocols. We start Command protocol first. This command protocol definition takes reference to the Firefox menu.

4.2 Feedback Protocol

All the request made by client using XMLHttpRequest.open("GET", URL, true);This protocol is using the parameter URL in open method as a communication carrying.

5 Conclusion

After the text edit has been completed, the paper is ready for the template. Duplicate the template file by using the Save As command, and use the naming convention prescribed by your conference for the name of your paper. In this newly created file, highlight all of the contents and import your prepared text file. You are now ready to style your paper.

References

1. http://www.devicevm.com
2. Mozilla Chrome, http://www.mozilla.org/xpfe/ConfigChromeSpec.html
3. Mozilla XUL, http://www.mozilla.org/projects/xul/
4. XMLHttpRequest, http://en.wikipedia.org/wiki/XMLHttpRequest
5. Extensible Markup Language (XML) 1.0, http://www.w3.org/TR/1998/REC-xml-19980210

A Study on Sharing Grid of Instructional Resources Management System

Bin Li, Fei Li, and Shuqiang Song

Abstract. With changed requirements, the old management system for the Audio & Video Resources Library of Tsinghua University had not met needs any more and should be improved. But there is a question: how to design it? In past decade, many universities in China had constructed resource libraries by themselves, and it is rewarding to sharing the resources in them with each other. But there still exist some barriers coming from technology and management etc, for instance, how exchange and transfer the resources? Thus, this article will discuss how to design a sharing grid of the instructional resources management system satisfying current demands.

Keywords: ICT, Resource Library, grid technology, web service.

1 Introduction

1.1 Background

In 2004, there built the Audio & Video Resource Library of Tsinghua University (AVRLTU), which source of resources come from a traditional video magnetic tape library. With these digitized video clips, it is very convenient to provide services for teaching and learning through Internet.

At first, AVRLTU was designed especially for storing video format contents. With rapid development and application of information and communication technology (ICT), some other format resources have accumulated and need to be put into the library so as to serve for the university, because they are also an important source for instructional resources, i.e. e-Book[1] etc.

In fact, supporting different media formats is not a real challenge. The tough problem is that newly updated resources have been decreasing year after year due

Bin Li · Fei Li · Shuqiang Song
Educational Technology Institute, Tsinghua University, Beijing 100084, P.R. China
e-mail: binli@tsinghua.edu.cn, lifei@tsinghua.edu.cn

F.L. Gaol (Ed.): Recent Progress in DEIT, Vol. 2, LNEE 157, pp. 281–286.
springerlink.com

to insufficient budget and ascending cost. As a result the users of the library have been cutting down accordingly. In past decade, many universities had constructed resource libraries for their own students and faculty, which contents are complementary with each other in a certain extent. If the resources in libraries could be mutually shared, the quantity of resources for their own schools would grow up more easily. But there still exist some barriers coming from technology and management etc, for instance, how exchange and transfer the resources?

The emergence and development of grid technology and open source software (OSS) provides a revolutionary way to share scattered sources. Grid technology can help to carry out scattered resource sharing, information management, and information service. This article is trying to describe and design an open platform based on open source software by taking use of the notion and technology of grid.

1.2 Related Work

DSpace is an open source platform for academic, non-profit, and commercial organizations to build digital repositories, and there are over 800 organizations that are currently using it in a production or project environment now [2]. Different DSpace deployments can be loosely coupled for sharing by the application of some technologies, for instance, DSpace uses the CNRI Handle System [3] as a means of assigning globally unique identifiers to resource objects; in addition, DSpace offers Storage Resource Broker (SRB) Support, and SRB provides a means to organize information stored on multiple heterogeneous systems into logical collections for ease of use [4].

1.3 Our Work

At present, it is well known that there exist large numbers of projects and technology standards for realizing resource sharing. Why do this research? Our work is not trying to develop new technologies or standards, but gives a combined design and application of current technologies especially for open source and grid technologies so as to contrive a sharing grid of instructional resource management system that suits the situation of resource construction and management in China. For example, DSpace is an excellent resource management system which a few designs of the system refer to, but it is short of file transfer and unfit for manage gigabytes-scale file based on Web. In addition, it does not have a flexible architecture to develop own web application etc.

2 The Topology Structure

The whole system hopes to aggregate the resources distributing in different universities and institutes, but there still lays two impediments making against the purpose: Firstly, instructional resources are a valuable part of universities or institutes, and it is almost impossible to take them away for forming a more large scale library. Secondly, the management mode and routine of some schools are different from others, and this will make for diversities of function and interface of resource management system. For this reason, a uniform portal is not feasible yet, such as CersGrid.

Therefore, as depicted in Fig. 1, the topology structure of the grid should be open, distributed, and flexible, and each node is equal with one another. In the system, node is an independent deployment of Instructional Resources Management System (IRMS). According to the demands of university or institute, one node might mutually share and exchange its resources with another node.

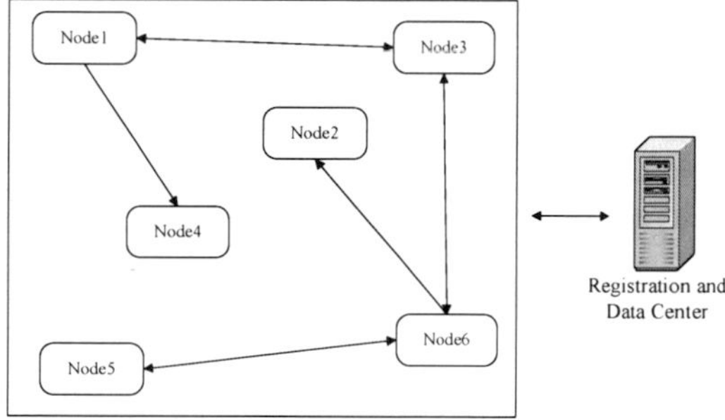

Fig. 1 The topology structure of the whole system.

In order to be found, as well as find other nodes and their resources, a new deployment of node must register itself in Registration and Data Center (RDC). RDC is just like a server providing directory services, through which one node is transparent from another with the exclusion of resource files.

3 Instructional Resource Management System Architecture

Instructional Resource Management System (IRMS) realizes the core functions of querying, managing and transferring resources. Every deployment of IRMS in university or institute could independently provide service for its students and faulty, as well as the deployment could become a fundamental unit of the sharing grid, i.e. a node, if it would like to. The work is very simple, and all what to do are to register the relevant information of the node in RDC.

The architecture of IRMS, shown in Fig. 2, is composed of Data Storage, Fundamental Service, Business Components, Web and GridFTP. In deed, GridFTP is not a layer, but a module, which spans two layers and is the basis for weaving the resource file sharing gird.

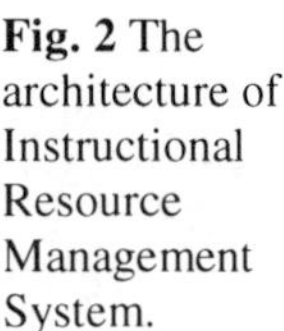

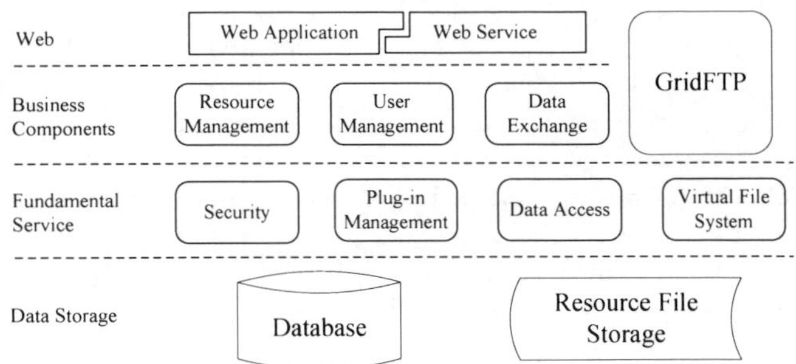

Fig. 2 The architecture of Instructional Resource Management System.

3.1 Data Storage

Data Storage layer concerns data storage. The system needs save two kinds of data: relational data and resource files. Database, such as MySQL or Oracle, is used for maintaining these relational data like user info and resource metadata etc. On the other hand, resource files are preserved as operating system file system into large-capacity disks.

3.2 Fundamental Service

The layer, providing the fundamental parts of the framework on which the Business Components build, consists of Security, Plug-in Management, Data Access, and Virtual File System modules.

The Plug-in Management module

A plug-in is indeed a module, and the system is made up of many modules. Here, the Plug-in Management uses Spring Framework as plug-in container. In this way, developers might insert a plug-in (module) more easily in Business Components layer.

The Virtual File System module

It is the only entrance for GridFTP to download and upload resource files, and its purpose is to allow GridFTP to access different types of concrete file systems in a uniform way. In addition, the module provides storage and backup management, and each node can customize its storage strategy, for example, adding a new disk if there is no free space; saving different kinds of resources into different disks.

3.3 Business Components

The Business Components is a more flexible and extensible layer, and users are able to develop new modules satisfying their own demands and insert them into the layer like a plug-in. Basically, the system provides 3 modules: Resource Management, User Management and Data Exchange.

The Resource Management module

It is concerned with resource metadata management such as adding, removing and modifying the metadata of the resource item, batchly importing resource items using an XML metadata document with some content files, realizing a workflow process that allows one or more human reviewers to check over the submission and ensure it is suitable for inclusion in the library, and providing an interface working together with GridFTP, which helps to treat metadata and file as a whole.

The Data Exchange module

The module exchanges data with other nodes and RDC by the means of web service. In practice, there are differences in metadata schemas on account of application situation, media format, etc. As a result, the framework should have the ability to add or change any field to customize the metadata schemas.

3.4 Web

Users can develop their own web application on the basis of web service and modules in Business Components layer. If there deploy more than two web applications, it is constructive to use single sign-on service JA-SIG CAS provides.

3.5 GridFTP

The GridFTP based on FTP, the highly-popular Internet file transfer protocol, is in charge of the file transfer of the grid system. It supports sending and receiving more than two gigabytes size file.

There are a few additional features in GridFTP to improve bandwidth capability. From above, it is known that each node could exchange resources with another voluntarily. So a resource item might have multiple replicas. With the help of RDC and UUID, it is very easy to find them. Thus GridFTP is able to carry out multi-channel parallel transfer so as to accelerate download speed (see Fig. 3).

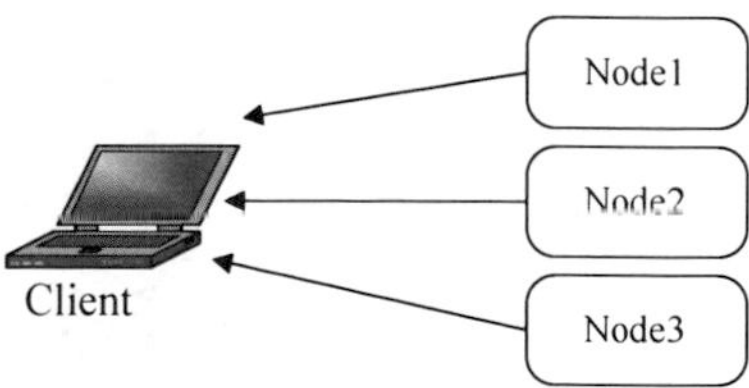

Fig. 3 The multi-channel parallel transfer.

4 Conclusion

Technology just offer a possibility of sharing resources distributed in different locations, and its function and design should accord with the current actuality of constructing and managing instructional resources.

The system is in the process of developing. In future, it will be published as an open source software, and any university or institute can make use of it for setting up own instructional resource library freely. At the same time, with the help of the system, it is hoped that the users with close relationship could share and exchange their resources. Then more users would do like this gradually. In this process, the collection of user requirements is very important, and it will help us to improve the system.

References

1. Green, K.C.: The 2010 National Survey of Information Technology in U.S. Higher Education, USA (2010), http://www.campuscomputing.net
2. DSpace project, http://www.dspace.org/
3. The Handle System, http://www.handle.net/
4. Rajasekar, A.K., Moore, R.W.: Data and Metadata Collections for Scientific Applications. In: High Performance Computing and Networking (HPCN 2001), Amsterdam, NL (June 2001)

Learning with Serious Games: The SELEAG Approach

Carlos Vaz de Carvalho

Abstract. Games can be instantiated for learning as they involve mental and physical stimulation and they develop practical skills – they force the player to decide, to choose, to define priorities, to solve problems, etc. Serious Games are specifically designed to change behaviors and impart knowledge. The objective of the project SELEAG is to evaluate the use of Serious Games for learning history, culture and social relations. An extensible, online, multi-language, multi-player, collaborative and social game platform for sharing and acquiring knowledge of the history of European regions was developed for this purpose and was tested with students.

1 Introduction

A Game is a particular context where users (players) have structured or semi-structured goals that they try to achieve by overcoming challenges. Games can involve one player acting alone, two or more players acting cooperatively, and, more frequently, players or teams of players competing between themselves [1].

Computer games are highly interactive products. These technologies, together with the different modes of interaction and communication, allow different participants, who have similar interests, to play together through ICT tools [2].

According to Mark Riyis, the use of games for learning is effective due to the following characteristics: they are motivational, they are cooperative and they meet educational objectives, they allow the resolution of problematic situations, they allow the application of concepts in practical situations, they are interdisciplinary, they favor oral expression and cultural awareness, respect for others, teamwork and cooperative learning [3].

Carlos Vaz de Carvalho
GILT-Instituto Superior de Engenharia do Porto, Rua Dr. António Bernardino de Almeida, 472 P4200-072 Porto, Portugal
e-mail: cmc@isep.ipp.pt

F.L. Gaol (Ed.): Recent Progress in DEIT, Vol. 2, LNEE 157, pp. 287–292.
springerlink.com

The pursuit of knowledge through interaction and cooperation among players is enhanced, according to [4], by the very same structure of the game when played in groups. It is an exercise in dialogue, group decision and consensus [5], supported by environments of simulations and practices, experiences and creativity enhancers, participation, research and integration [6]. To Papert, children develop better cognition by discovering themselves the specific knowledge, becoming thus the active constructor of their knowledge and that is promoted by games [7].

Serious Games can be defined as "a mental contest, played with a computer in accordance with specific rules, that uses entertainment to further government or corporate training, education, health, public policy, and strategic communication objectives" [8] or as "Games that do not have entertainment, enjoyment or fun as their primary purpose." [9]

But games may not be an effective teaching tool for all the students. Squire found that about 25% of students withdrew from his study, which used Civilization (an historical simulation game) to teach geography and history, as they found it too hard, complicated and uninteresting [10] while another 25% of the students (particularly academic underachievers) loved playing the game, they thought it was a "perfect" way to learn history.

2 SELEAG

Three game scenarios were chosen and are based on cultural and historic realities of different European eras: the Maritime Discoveries, the Industrial Revolution and the Second World War. The scenarios provide an historical context that allows students to better understand the process of how Europe has been shaped and restructured as a result of history and will help students understand the formation and evolution of the countries and regions. Scenarios emphasize the cooperation between European countries to strengthen the notion of European citizenship. This understanding is expected to assist students to identify and respect cultural diversity. SELEAG will also evaluate if the game does in fact contribute to any of the planned objectives. Therefore, the main Learning Outcomes for the game are:

- Understanding and using appropriately dates, vocabulary and conventions that describe historical periods and the passing of time.
- Developing a sense of period through describing and analyzing the relationships between the characteristic features of periods and societies.
- Building a chronological framework of periods and using this to place new knowledge in its historical context.
- Understanding and respecting the diverse experiences and ideas, beliefs and attitudes of men, women and children in past societies and how these have shaped the world.
- Analyzing and explaining the reasons for, and results of, historical events, situations and changes.
- Considering the significance of events, people and developments in their historical context and in the present day.

2.1 Pedagogical Approach

The project is based on a different pedagogical approach, where games become important educational tools. They promote the collaboration between children and their knowledge of history, geography, culture and social relations. But they will also be integrated in a larger community of learning that includes all the schools belonging to the project.

2.2 Game Concept

The main game concept is that the player notices something is wrong with their current time period; E.g. Newspapers are in a different language. This is the trigger event for a specific scenario action. The player then learns of a past event or sequence of events in history that somehow happened differently and changed history. This event or sequence of events can be explained by a short narrative and is the basis for a game scenario.

2.2.1 World War II Scenario

In September, 1940, Hitler was preparing England's invasion which finally aborted. He postponed it till next Spring and it never happen. In our story he was successful because his plans were never intercepted - the German code was never broken because the ENIGMA machine was not available - England was successfully invaded. The player must go back and make sure that the Enigma machine codes were delivered as historically factual. The player knows that this was the fundamental event because the cleverness of the German code engineers and the unbreakability of their cryptographic system is pointed out. For this scenario a set of Learning Outcomes was established:

- Knowledge about the second world war in general. Focus on events at the start of the war. Focus on religious and other persecutions. Focus on main characters (Hitler, Churchill, etc.)
- Understand what a war really means in terms of loss of lives and impact on a "normal", everyday life.
- Understand basic cryptographic mechanisms
- Realize that people can behave strangely under pressure
- Understand co-operation between the different countries (passing information, etc.), movement between countries, life in wartime, and the importance of technology advancements to the war effort.

2.2.2 Maritime History / Discoveries

The concept of this scenario lies on a change in History that caused Madeira not being colonized by Portugal. When the first navigators were coming back from the Island discovery, some maps were stolen from Portugal and Madeira became a pirates/corsair base. Due to the balance of forces in Europe, it later became an independent country which has now nuclear power and is threatening Europe. Players must go back and restore historical correctness. The learning outcomes for this scenario combine Historical aspects with notions related to politics, governance and maritime technology:

- Understand the role of Discoveries and Maritime trade in the development of Europe
- Become aware of major maritime navigation instruments and associated sciences (Maths, Astronomy, etc.)
- Understand political balancing between neighboring countries
- Become aware of life in the 15th and 16th centuries

2.2.3 Industrial Revolution

The basic History twist in this scenario is related to the fact that the Mines Act (law that limited the age and hours that kids could work, therefore heavily ruling against child labour) was not approved by the English Parliament therefore kids are still working today. Player must go back and make sure it is approved. The Learning Outcomes for this scenario are related to aspects in technology, social awareness and parliamentary systems:

- Become aware of life in the Industrial Revolution time. In particular, realize what was like being a child worker
- Understand the role of the Industrial Revolution in the development of Europe
- Understand principles of Parliamentary democracies

3 Assessment of Learning

A crucial aspect to measure is the actual, effective impact that the game will have on learning. Therefore, this process will be thoroughly evaluated and assessed to ensure the best use and the production of guidelines for its replication. Several steps were defined in the evaluation methodology.

3.1 Alpha Testing

The objective of an Alpha Testing procedure is to simulate an actual operational testing and is performed by members of the development team that have not been involved in the development of the particular features to be tested. Alpha Testing allows anticipating internally problems that would only be detected by external testers in the Beta Testing phase. This stage is more rewarding if qualitative data collection is used because it provides richer information. A protocol was established for the implementation of the Alpha Testing:

- Identify participants which have not been involved in the development
- Participants play the scenario and try to finish it

- Participants measure how much time it took to finish each scene
- Participants identify learning outcomes addressed.
- Participants identify other learning outcomes that should be addressed
- Participants assess if the game is motivating and identify problems
- Participants assess if the graphical environment / usability is adequate. If not, identify the issues
- Report

3.2 Beta Testing

Beta testing comes after alpha testing and it is planned to release the game to a limited audience outside of the consortium. The target group is composed of teachers familiar with the subject and with the technology that can assess pedagogically and technically the prototype so that further testing can ensure the game has no faults or bugs. In this stage a mix of qualitative and quantitative information will be collected, because there is still interest in receiving rich comments but there is already a relatively important number of people evaluating the game that can provide statistically valid results. The protocol for this stage is the following:

- Identify participants (teachers)
- Participants play the game and try to finish it
- Participants measure how much time it took to finish each scene
- Participants answer the questionnaire
- Interview the participants using the semi-structured interview guide
- Report the results

The questionnaire is meant to provide a quick data filling tool for the beta testers. This way they are able to report immediately their impressions, just after finishing the scenario. It is based on a mixture of adapting heuristics for evaluating playability of games, heuristics for usability evaluation for history educational games and educational effectiveness factors.

3.3 Final Stage

For the final evaluation, the quality space aggregates the dimensions – Functionality; Efficiency and Adaptability. The Functionality dimension reflects the characteristics related to its operational aspects. It aggregates two factors: ease of use and content's quality. The Efficiency dimension aggregates five factors: audiovisual quality, technical and static elements, navigation and interaction, originality and use of advanced technology. Through this dimension we measure the system's ability for presenting different views on its content with minimum effort. The Adaptability dimension is the aggregation of five factors: versatility, pedagogical aspects, didactical resources, stimulates the initiative and self-learning and cognitive effort of the activities.

4 Conclusions

The proposed game learning methodology can introduce a new relation between young learners and learning and promote their interest in their education. By doing so, the children themselves will become more motivated to learn and to go to school as they will be able to relate their game experience with curricular contents and subjects.

The first development stage corresponded to a thorough needs analysis, specification and design for the whole game approach. The final report of this stage provides info on the state of the art of serious and learning games, an analysis of game development tools and game engines, the design of the scenarios and the evaluation methodology. It is an extensive report that answers to the requirements for a steady project development.

The development stage already started and two scenarios are fully developed, going through the beta testing phase. The other scenario will be ready by nest January. The extensive implementation stage in schools will last until next June but will allow for implementation in different countries. The game can easily be replicated around the world because it is delivered through the Internet. New scenarios can be developed and integrated in the platform.

Acknowledgment. The SELEAG Project (503900-LLP-1-2009-1-PT-COMENIUS-CMP) has been funded by the European Commission, under the Lifelong Programme, Comenius action.

References

1. Batista, R., de Vaz Carvalho, C.: Learning Through Role Play Games. In: Proceedings of FIE 2008 - 38th IEEE Annual Frontiers in Education Conference (October 2008)
2. Batista, R., Vaz de Carvalho, C.: Funchal 500 Years: Learning Through Role Play Games. In: Proceedings of the 2nd European Conference on Games Based Learning (October 2008)
3. Riyis, M.T.: RPG e Educação, `http://www.jogodeaprender.com.br/artigos_1.html`
4. Andrade, F.: Possibilidades de uso do RPG, `http://www.historias.interativas.nom.br/educ/rpgtese.htm`
5. Marcatto, A.: O que é o RPG?, `http://www.alfmarc.psc.br/avent_edu_o_que.asp`
6. Nunes, H.: O jogo RPG e a socialização do conhecimento. Palestra nos Encontros Bibli. Universidade Federal de Santa Catarina, Florianoplis, `http://www.encontros-bibli.ufsc.br/bibesp/esp_02/5_nunes.pdf`
7. Papert, S.: A Máquina das Crianças: Repensando a Escola na Era da Informática. Trad. Sandra Costa. Artes Médicas, Porto Alegre, `http://www.psicolatina.org/revista/index.php?option=com_content&task=view&id=26&Itemid=20`
8. Chen, S., Michael, D.: Serious Games: Games that Educate, Train and Inform. Thomson Course Technology, USA (2005)
9. Zyda, M.: From visual simulation to virtual reality to games. Computer 38(9), 25–32 (2005)
10. Squire, K.D.: Replaying history: Learning world history through playing Civilization III. Indiana University, Bloomington, IN. 200

A Framework for e-Content Generation, Management and Integration in MYREN Network

Vala Ali Rohani and Siew Hock Ow

Abstract. Malaysian Research and Education Network (MYREN) aims to provide accessible broadband to the Malaysian researchers to achieve the country's knowledge economy aspiration. This networking super highway enables researchers to run data-intensive applications, share computing equipments and run advanced applications within Malaysia as well as overseas. In this paper we provide a social network based framework for enabling all universities and higher education colleges in Malaysia to generate their academic content using MYREN infrastructure and also share them with each other and even with the several other international research communities in Asia Pacific, Europe and North America.

Keywords: Social Network, e-Content, Integration, MYREN.

1 Introduction

A huge number of Internet users have visited thousands of social networking sites. They have taken advantage of the free services of such sites in order to stay connected online with their offline friends and new online acquaintances, or to share user-created contents, such as photos, videos, bookmarks, blogs, etc.

Most of these social networks are developed for public purposes which allow a wide range of internet users in different ages, interests, social and academic levels to join them and make their own friendship network [8,9].

Universities, Colleges and higher education institutions are constantly being refreshed with new members while being connected with graduates seems to be vital too. In most cases, Students use the systems only out of necessity rather than choice, preferring to self create study groups on Facebook [2].

Vala Ali Rohani · Siew Hock Ow
University of Malaya, Department of Software Engineering,
Faculty of Computer Science and Information Technology
e-mail: V.Rohani@siswa.um.edu.my, Show@um.edu.my

F.L. Gaol (Ed.): Recent Progress in DEIT, Vol. 2, LNEE 157, pp. 293–298.
springerlink.com

But is obvious that the academic society and definitely their members, need some special facilities to leverage the regarding communication between students, graduates, instructors and all other members of academic environments [1].

As a government-funded Project, MYREN provides an accessible broadband infrastructure for connecting all universities in Malaysia. Now, it is necessary to have a social network-based solution to help all universities to generate their academic content autonomously and also making them enable to share some common web based facilities with each other for achieving an integrated academic environments which includes all universities and higher education colleges in Malaysia.

2 Social Network-Based Framework

The institutions of higher learning in Malaysia play an important role in broadening the knowledge horizons in Malaysia. This research aims to design an infrastructure which can be a platform for each Malaysian university to create its own social network. Also, the supervision features can be included to gather valuable information for the use of the Malaysian institutions of higher learning.

In this section, we will provide taxonomy of essential features of academic social web sites as a framework for generating, managing and integrating e-content by all universities and higher education colleges in Malaysia that are the members of MYREN network [3].

These features are separated in 4 categories:

a) Administrative Features
b) Collaborative Features
c) Reporting Features
d) Integrating Features

As mentioned in Table 1, the administrative features of academic social networks are categorized to 6 groups. The first 6 capabilities are common between all types of users are considered as General. The assignment uploading is considered for students who use this model and join this academic social network. Lecturers also need some special feature for handling their tasks in this model. Four features are considered for lecturers. There are other types of users which we can them as Researchers. They are not necessarily the lecturers. So we consider Research Field Following Management and My Publications Management for this kind of users. There are two level of administration in this model which will be handled by university and social network administrators. The last two groups of features are dedicated to this type of users.

For persuading the social network users to check their profiles continuously, we should pay special attention to the collaborative features. Table 2 shows the list of suggested collaborative features in academic social networks. Personal billboards are amazing capability for announcing the personal feelings while the general billboard can be helpful in informing the last news of an organization (e.g. a university). All of the social network users need peer-to- peer message passing system and friendship network to be connected to their friends. Discussion forums and online groups can play a significant role in mechanizing the organizations in academic environments. We cannot forget the online chat as one of the most favorite features of social networks.

Reporting services in social network can update the user information about his/her current position among the other users. Table 3 shows the list of necessary reporting features for our model in academic social networks.

Table 1 Administrative Features of Academic Social Networks

A) Administrative Features					
General	1.Create and Editing profile 2.Editing Privacy Settings 3.Uploading Publications 4.Billboard management 5.File Repository Management 6.Freinds Network Management 7.Membership 8. Forum activities	**For University Administrators**	1.Editing University Information 2.Managing Faculties Information 3.Abuse Report Management 4.News Letter Management	**For Lecturers**	1.Uploading assignments 2.Building Courses 3.Managing Course Student Forums 4.Course Membership Management
For Researchers	1.Research Field Following management 2.My Publications management	**For Students**	1. Assignment Uploading	**For Social Network Admin**	1.Confirming a University 2.Abuse Report Management

Pointing and ranking mechanism can create an enthusiasm for users to increase their rank and enhance their position in the social network of their university or colleges. We can consider different parameters for using in pointing mechanisms. Usually the number of invites, the count of scientific posts and also rate of user activities in answering to other members' questions are considered as basic parameters for this purpose.

Sending daily scientific news to members email is one of the most effective alternatives for encouraging users to login to their profile periodically.

The most important feature group of this model is Integrating Features (Table 4)

Table 2 Collaborative Features of Academic Social Networks

B) Collaborative Feature	
1. Personal and General Billboard	5. Discussion Forums
2. Peer-to-Peer Message Passing	6. Online Groups
3. General Announcement	7. Online Chat
4. Friendship Network	

According to survey in academic environment, we face to the need of integration between academic organizations social networks. Hence, we categorize these types of features in five groups according to its user domains.

Some features like Common Billboard and Calendar are common among all social network members. Also, this model should prepare a special service to enable the its users to have several profiles regarding different social networks while having just a single Username and password. We named this service the Single Signup Mechanism.

Table 3 Reporting Features of Academic Social Networks

C) Reporting Features
1. Points & Ranks Management and Reporting
2. Daily Scientific News Letter Sending
3. Different Transaction Notifications

Regarding the students view point, this model provides some features for building scientific groups and also some services for doing teamwork in group assignments.

This kind of features helps lecturers managing the shared course contents as well as handling the common Question and Answers services. Usually, the lecturers and researchers needs to cooperate with each others in some scholarly tasks such as paper reviewing. They can use the prepared service in this manner to interact with each other as well as possible [4].

Table 4 Integrating Features of Academic Social Networks

D) Integrating Features					
General	1.Common Billboard 2.Single Signup Mechanism 3.Common Calendar	**For Students**	1.Group Building 2.Common Assignments	**For Lecturers**	1.Common Course Management 2.Common Q&A Service 3.Common Paper Reviewing Service
For Researchers	1.Common Research Following UP 2.Common Paper Reviewing Service	**For University Administrator**	1.Common News Letter 2.Common Discussion Forum Topics 3.Common Survey		

Maybe it would be too amazing for autonomous universities to share some services of their separated social networks with each other. In these cases, the integrating features which considered for University Administrators level can enable them to have inter-organization newsletters and discussion forums.

3 Malaysian Experts Academic Social Network

As the first working prototype, the Malaysian Experts academic social network has been created by the author to gather some experimental data from 10 universities in Malaysia. In this first try, more than 200 members has joint this web site to create their own profile which most of them are postgraduate students.

In this working version, some basic facilities like profile, personal bill board, offline message passing, academic news broadcasting , university-based search services and ranking mechanism has been created to motivate the university students and graduates to join this academic social network [10].

4 Conclusion

This paper presented a social network-based framework for enabling the universities and higher education colleges in Malaysia to generate, manage and integrate academic e-contents. This solution is applicable on MYREN network as a broadband infrastructure which will connect all universities in Malaysia to each other.

In this framework, we categorized the most important and necessary features of a model for developing academic social networks in four categories includes Administrative, Collaborative, Reporting and Integrating features.

This set of requirements can be used to design the architecture for autonomous and integrated social networks to enhance the interaction and teamwork between the students, graduates, lecturers, official staffs and researchers.

Acknowledgment. The authors wish to acknowledge the collaborative funding support from the Fundamental Research Grant Scheme (FRGS) Under grant no. FP061/2010A.

References

1. James, L.: JISC Academic Social Networking Final Report (April 22, 2010)
2. Boston, I.: Racing towards academic social networks. On the Horizon 17(3), 218–225 (2009)
3. Rohani, V.A., Ow, S.H.: Eliciting Essential Requirements for Social Networks in Academic Environments. In: Proc. 2011 IEEE Symposium on Computers & Informatics (2011) (accepted for publication)
4. Rohani, V.A., Ow, S.H.: A Social Network Based Solution for Integrating the Universities in Malaysia. In: Proc. 4th International Conference in Postgarduate Education, ICPE4 (2010) (accepted for publication)
5. `http://www.myren.net.my/about/what-myren` (accessed October 18, 2010)
6. `http://www.myren.net.my/network/domestic-network` (accessed October 23, 2010)
7. `http://www.myren.net.my/network/international-linkages` (accessed October 23, 2010)
8. Kim, W., Jeong, O.-R., Lee, S.-W.: On social Web sites. Information Systems 35, 215–236 (2010)
9. Bhattacharyya, P., Garg, A., Wu, S.F.: Social Network Model based on Keyword categorization. In: Proc. Of International Conference on Advances in Social Network Analysis and Mining, ASONAM 2009. IEEE Computer Society (2009)
10. `http://www.MalaysianExperts.com.my` (accessed October 28, 2010) (last upadted October 20, 2010)

A Case of Study of Investigating Users' Acceptance toward Mobile Learning

Shu-Sheng Liaw and Hsiu-Mei Huang

Abstract. Mobile learning (m-learning) is a relatively new learning tool in the pedagogical arsenal to support students and teachers as they navigate the options available in the expanding world of distance learning. Regarding to wide application of mobile learning, investigating learners' acceptance toward it is an essential issue. This research investigates learners' acceptance toward m-learning. An m-learning questionnaire survey that based on 168 university students' attitudes was investigated. From statistical results, learner autonomy of using m-learning, perceived interaction of using m-learning, quality of m-learning functions, and perceived satisfaction of using m-learning are positive predictors on m-learning acceptance.

1 Introduction

Mobile learning (m-learning) is a relatively new tool in the pedagogical arsenal to support students and teachers as they navigate the options available in the expanding world of distance learning. M-learning is the learning accomplished with the use of small, portable computing devices. These computing devices may include: smart phones, PDAs and similar handheld devices. This type of learning is facilitated by the convergence of the Internet, wireless networks, mobile devices and e-learning systems (McConatha & Praul, 2008). With a mobile or handheld

---

Shu-Sheng Liaw
General Education Center, China Medical University, Taiwan
e-mail: ssliaw@mail.cmu.edu.tw

Hsiu-Mei Huang
Graduate School of Computer Science and Information Technology,
National Taichung Institute of Technology, Taiwan
e-mail: hmhuang@ntit.edu.tw

F.L. Gaol (Ed.): Recent Progress in DEIT, Vol. 2, LNEE 157, pp. 299–305.
springerlink.com

device, the relationship between the device and its owner becomes one-to-one interaction. Mobile devices have the potential to change the way students behave, the way students interact each other and the attitude toward learning (Motiwalla, 2007). The key features of using mobile devices for m-learning, such as one-to-one interaction and place and time independence; are personalization capability and extended reach (Zurita & Nussbaum, 2004).

Even though m-learning provides useful overviews of different applications in education, there is an emerging need for a more applicable framework to provide teachers, educational policy-makers and researchers with a better representation of educational affordances of m-learning. Regarding to wide application of mobile learning, investigating learners' attitudes toward it is an essential issue. This research will investigate learners' attitudes toward m-learning. At first, we will present an m-learning framework. Furthermore, m-learning questionnaire survey that based on 168 university students will be investigated. In this research, the users' acceptance is based on factors such as m-learning interaction, m-learning system's functions, learners' autonomy, perceived satisfaction, and perceived usefulness. The last section is discussion.

2 Mobile Learning Framework

With the trend of the educational media becoming more portable and individualized, the form of learning is being changed dramatically. This work aims to synthesize the cognition and technology domains to establish a new learning model. Therefore, the mobile learning environment posses many unique characteristics, they are (Chen, Kao & Mobility of learning setting, (d), Interactivity of the learning process, (e), Situating of Sheu, 2003): (a), Urgency of learning need, (b), Initiative of knowledge acquisition, (c),instructional activity, (f), Integration of instructional content.

From the viewpoint of Chen, et al. (2003); indeed, the m-learning environment provides a flexible and powerful learning opportunity. Additionally, Liu, et al. (2003) point out that ubiquitous computing that integrates mobile devices, wireless communication, and network technology to construct a mobile learning environment has the following features: (a), reducing the time for tedious activities, (b), engaging students in learning activities, (c), empowering the teacher to monitor students' learning statues, (d), facilitating group collaborative learning, (e), implementing technology-supported activities smoothly.

3 Research Hypotheses

In this research, we will apply this conceptual model to investigate learners' attitudes toward the acceptance of m-learning.

As Liaw, Hatala, and Huang (2010) stated, based on the activity theory approach, system satisfaction, system activities, learners' autonomy, and system functions have positive influence on system acceptance toward m-learning. Based on their researching findings, they have proposed a theoretical conceptual model when applying m-learning. Four affordances will improve the acceptance of m-learning systems: enhance learners' satisfaction, encourage learners' autonomy, empower system functions, and enrich interaction and communication activities. Fig. 1 presents learners' acceptance toward m-learning systems.

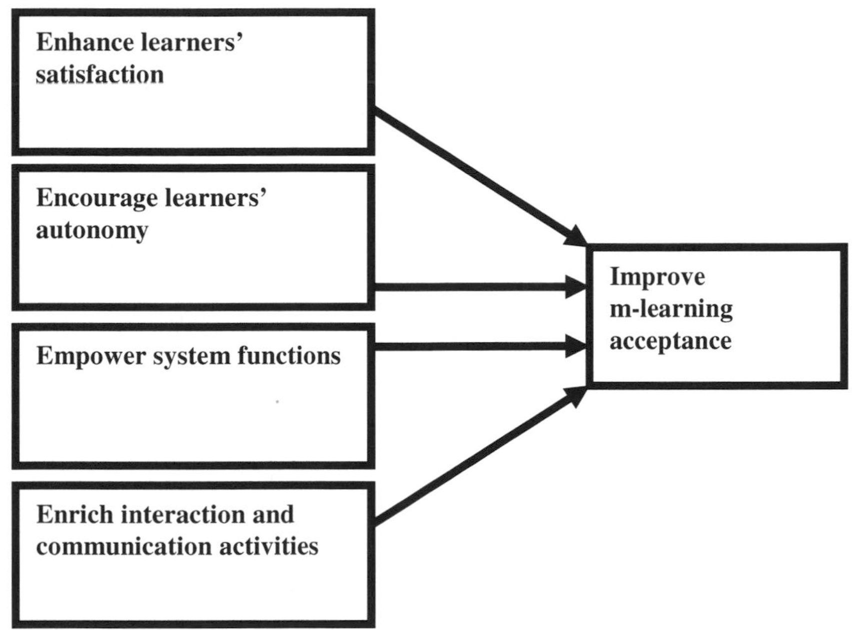

Fig. 1 The acceptance conceptual model of m-learning

Based on the Figure 1, this research proposes the following hypotheses:

H1 With the increase perceived interaction by using m-learning provides, the acceptance of the m-learning system increases.

H2 With the increase perceived satisfaction of using m-learning provides, the acceptance of the m-learning system increases.

H3 With the increase learners' autonomy of using m-learning provides, the acceptance of the m-learning system increases.

H4 With the increase quality of m-learning functions provides, the acceptance of the m-learning system increases.

4 Research Design

4.1 Participants

This study conducts a survey for understanding learner attitudes toward the m-leaning environment. A total of 190 university students were taught on how to use the system. Students were allowed to use the system anytime for a period of one month. After that, a questionnaire survey for m-learning was distributed to participants during class. Participants were invited to complete the questionnaire. All subjects were asked to respond to the questionnaire and their responses were guaranteed to be confidential. All 190 students participated the questionnaire survey. However, 22 missing responses for m-learning questionnaire were eliminated. The study group comprised of 168 students which includes 71 male students and 97 female students.

4.2 Measurement

The questionnaire included three major components: (a) demographic information, (b) computer experience, and (c) attitudes towards m-learning. The following shows the content of the questionnaire.

Demographic information: The demographic component covered gender and the field of study.

Computer experience: In this component, participants were asked to indicate their experience with computers, the Internet, PDA, and m-learning.

Attitudes toward m-learning: Participants were asked to indicate their attitudes towards m-learning. These 19 questions were adopting a 7-point Likert scale (ranging from 1 which means ''strongly disagree'' to 7 which means ''strongly agree''). The attitudes toward m-learning includes five factors, quality of m-learning functions, perceived interaction of using m-learning, perceived satisfaction of using m-learning, learners' autonomy of using m-learning, and perceived acceptance toward using m-learning.

5 Conclusion

The internal consistency reliability was assessed by computing Cronbach's αs. The alpha reliability was highly accepted (α=0.94) and items' coefficients are presented in Table 1. The values ranged from 0.40 to 0.77. Given the exploratory nature of the study, reliability of the scales was deemed adequate.

Table 1 The Mean, Standard Deviation, item-total correlations of m-learning from 1 which means "strongly disagree" to 7 which means "strongly agree")

Items	M	SD.	r*
Quality of m-learning functions:			
I know how to operate PDA for m-learning	2.94	1.57	0.39
I know how to navigate web pages with PDA	3.15	1.72	0.44
It is convenient to use m-learning to read course materials.	5.45	1.28	0.73
It is convenient to use PDA for m-learning.	4.70	1.49	0.72
Perceived interaction of using m-learning:			
The m-learning provides interactive opportunities.	5.67	1.38	0.56
The m-learning provides opportunities for communication among learners.	5.61	1.39	0.59
The m-learning provides opportunities for navigating and downloading instruction.	5.04	1.35	0.70
Learners' autonomy of using m-learning:			
I am active in finding Internet resources.	5.65	1.61	0.58
I am active in finding m-learning resources.	3.73	1.68	0.72
I am active in learning m-learning instruction.	3.71	1.67	0.68
Perceived acceptance toward using m-learning:			
The system is useful to find Internet resources.	5.84	1.19	0.59
The system is useful to retrieve online learning instructions.	5.61	1.28	0.68
PDA is a useful tool for m-learning.	4.94	1.38	0.75
The system is acceptable to improve my learning capacity.	4.13	1.68	0.76

Table 1 *(continued)*

The system is acceptable to improve my problem-solving capacity.	4.12	1.65	0.74
Perceived satisfaction of using m-learning:			
I am satisfied with using a PDA to find Internet resources.	5.73	1.46	0.57
I am satisfied with using the m-learning to interact with others.	5.46	1.46	0.55
I am satisfied with m-learning to learn web-based instruction.	4.01	1.58	0.71
I am satisfied with m-learningto retrieve online learning resources.	4.00	1.61	0.66

r*: Corrected Item-total correlation

For verifying hypotheses H1 to H2, the result of stepwise multiple regression for the path associated with the variables are presented in Table 2. To investigate H1 to H4 , a regression analysis was performed to check the effects of quality of m-learning functions, perceived interaction of using m-learning, perceived satisfaction of using m-learning, and learners' autonomy of using m-learning on perceived acceptance toward using m-learning. The result showed that all four factors were predictors and learners' autonomy of using m-learning had more contributions than other three other factors ($F(4, 163)=42.19$, $p<0.001$, $R2=0.77$).

Table 2 Regression results of m-learning

Dependent variable	Independent variables	ß	R^2	P
Perceived acceptance	Learners' autonomy	0.42	0.67	<0.001
	Perceived interaction	0.19	0.06	<0.001
	Quality of functions	0.21	0.02	<0.001
	Perceived satisfaction	0.20	0.02	0.002

6 Discussion and Conclusion

From the statistical results, this research shows that five factors have significant correlations among them. Furthermore, this research also provides that four factors (learner autonomy of using m-learning, perceived interaction of using m-learning, quality of m-learning functions, and perceived satisfaction of using m-learning) are all predictors on m-learning acceptance. In other words, this research supports the conceptual model proposed in the research which conducted by Liaw, et al. (2010). This research also proves that encouraging learners' autonomy, enriching interaction activities, empowering system's functions, and enhancing learners' satisfaction will directly influence learners' acceptance toward m-learning.

In summary, although the mobile devices have a limitation in screen size, this study confirms that mobile devices are valuable tools for mobile learning. As mobile learning systems have become more individualized, learner-centered, situated, and ubiquitous, understanding learners acceptance toward those mobile systems are more crucial to enhance learning performance.

Acknowledgment. This study was supported by NSC99-2511-S-039-001-MY2, and CMU98-S-050. The full version will be submitted to as a book chapter.

References

1. Chen, Y.S., Kao, T.C., Sheu, J.P.: A mobile learning system for scaffolding bird watching learning. Journal of Computer Assisted Learning 19, 347–359 (2003)
2. Liu, T.C., Wang, H.Y., Liang, J.K., Chan, T.W., Ko, H.W., Yang, J.C.: Wireless and mobile technologies to enhance teaching and learning. Journal of Computer Assisted Learning 19, 371–382 (2003)
3. Liaw, S.S., Hatala, M., Huang, H.M.: Investigating acceptance of mobile learning to assist individual knowledge management: Based on Activity Theory approach. Computers & Education 54, 446–454 (2010)
4. McConatha, D., Praul, M.: Mobile learning in higher education: An empirical assessment of a new educational tool. The Turkish Online Journal of Educational Technology 7(3), 15–21 (2008)
5. Motiwalla, L.F.: Mobile learning: a framework and evaluation. Computers & Education 49(3), 581–596 (2007)
6. Zurita, G., Nussbaum, M.: A constructivist mobile learning environment supported by a wireless handheld network. Journal of Computer Assisted Learning 20, 235–243 (2004)

A Layered Service-Oriented Architecture for Exercise Prescriptions Generation and Follow-Up

Wen-Shin Hsu and Jiann-I Pan*

Abstract. Recently, the services that integrating hospital care and home care from the patient-centered concept are more and more important to modern healthcare. The main challenge faced is the integration of heterogeneous platforms. For example, generating and follow-up one exercise prescription is one of such typical services. Service-Oriented Architecture (SOA) has been proposed as a key technology to overcome various problems involved in the integration of heterogeneous platforms and reuse legacy systems. In this paper, we present a layered SOA model for generating exercise prescriptions and follow-up that facilitates interoperability between intranet (hospital) and internet (homecare). In hospital layer, the system automatically produces an exercise prescription for one particular patient, and follows the effectiveness of the exercise prescription in homecare layer. Exercise prescriptions are produced and followed by diversely web services which interoperable over inter and intra networks. The implementation of exercise prescription and monitoring system (named EPMS) is demonstrated by a brief scenario for heart disease patient in this paper.

1 Introduction

Recently, the services that integrating hospital care and home care from the patient-centered concept are more and more important to modern healthcare. A typical example is the exercise prescription generation and follow-up. As many researches show that adequate exercises can be very helpful for maintaining

Wen-Shin Hsu
Institute of Medical Science, Tzu Chi University, Hualien, Taiwan, ROC
e-mail: 95351114@stmail.tcu.edu.tw

Jiann-I Pan
Department of Medical Informatics, Tzu Chi University, Hualien, Taiwan, ROC
e-mail: jipan@mail.tcu.edu.tw

* Corresponding author.

F.L. Gaol (Ed.): Recent Progress in DEIT, Vol. 2, LNEE 157, pp. 307–312.
springerlink.com

and promoting human health [1, 2]. More and more physicians and psychiatrists are now developing exercise prescriptions or guidelines for patients and users as regular therapy process. Exercise prescription programs include a variety of exercise items and a balanced routine to build core strength, endurance, flexibility, and based fitness, which must target patient's health status, goals, abilities, and interests [3, 4]. In order to formulate a good exercise prescription, the program must reference to various types of information, such as medical records, medical history, physical examination records, and physical fitness testing records. Furthermore, according to every disease has particular characteristics, exercise prescription must be case sensitive, and integrating both physical and psychological viewpoints. On the other hand, tracking whether the exercise prescriptions have dispensed by patients well or not is also important. If a patient did not follow the prescription, the medication would fail to produce positive results. In this example, services for generating and tracking exercise prescriptions are allocated on different platforms over intranet and internet.

Service-Oriented Architecture (SOA) is an important system development methodology in many applications [5, 6] for facilitating the integration of heterogeneous systems, and enhancing the reusability of programs. In this paper, we present a layered SOA model for generating exercise prescriptions and follow-up that facilitates interoperability between intranet (hospital) and internet (homecare). In hospital layer, the system automatically produces an exercise prescription for one particular patient, and follows the effectiveness of the exercise prescription in homecare layer. Exercise prescriptions are produced and followed by diversely web services which interoperable over inter and intra networks. This framework integrates internal hospital functions and patient participation, which emerging a number of security concerns such as how to identify users and whether users are authorized to access the mechanism.

This paper is organized as follows. Section 2 introduces the design of the system and analysis of the protocol. Section 3 presents the implementation of the EPMS environment. Finally, a brief conclusion is provided in Section 4.

2 A Layered SOA Model for Exercise Prescription and Monitoring

2.1 Layered Service-Oriented Architecture

Based on the service-oriented architecture [7, 8], the two topics of this study are included in two-level structure (see Fig. 1). The first level includes web services for generating exercises prescriptions around the intranet environment. All available web services, such as service for accessing the patient's records and service for calculating appropriate exercises, are communicated in Simple Object Access Protocol (SOAP). Each service is described in Web Service Description Language (WSDL), and registered in Universal Description, Discovery, and Integration (UDDI) protocols. The bottom layer is focus on the existed healthcare services who are interacted by different professionals. The next layer focus on organizing the web services and providing interfaces for the professionals. The top layer in this level is concerned on the quality of service, includes access authorization model, authentication model. When a physician is prescribing an exercise prescription, the services for requesting medical records, assessing exercises items and strength, may interoperable under these layers.

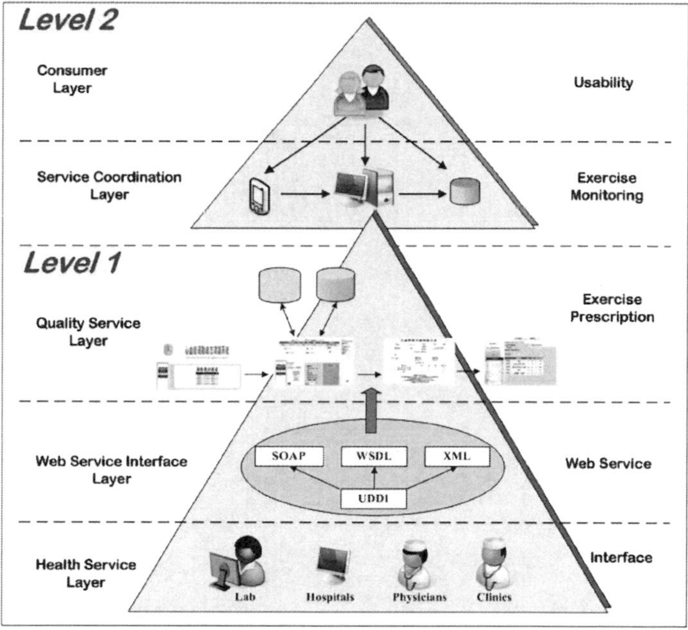

Fig. 1 A layered SOA model for exercise prescription.

A patient may access the exercise prescription through the web service. As the meanwhile, the physician may monitor the exercise effectiveness of that patient, even may update exercise items or durations via web services. The second level concentrates on the services coordination for supporting patients, Healthcare Service Center (HSC), and physicians around the internet environment. The top layer in this level focus on the consumer interface to operate the exercise prescription and design a self-exercise schedule. The next layer is focus on the coordination process of services who works for patients, physicians, and monitoring process.

2.2 *Exercise Prescriptions Generation and Monitoring*

In this research, the guideline for exercise testing and prescription from American College of Sports Medicine (ACSM)[1] is adopted as the exercise prescription base. The developed equations allow users to estimate oxygen consumption with reasonable accuracy regardless of the age, sex, weight, blood pressure, and medicine history. Predictable modes of exercise are those which easily measure the workload and for which the physiatrist is able to maintain a steady intensity of time-running, swimming, walking, and cycling.

A detecting system with intellectual mobile sensors is developed to estimate and record exercise routines and processes. The exercise records may be transferred to and stored in healthcare service center. Physicians or physiatrists can thus have a clear picture of the medication patients taken from the information these systems deliver to the hospital.

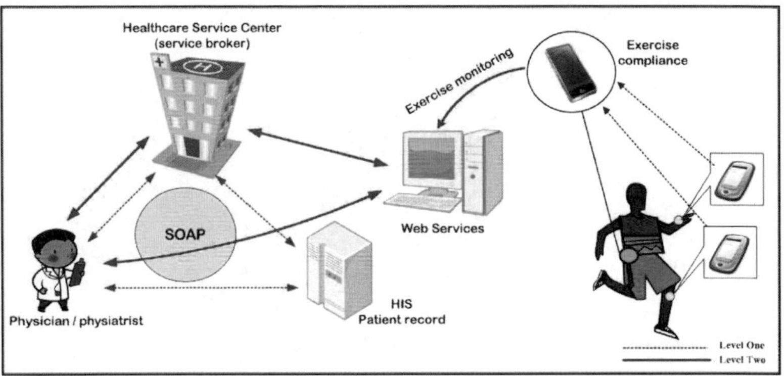

Fig. 2 An overview of Exercise prescription and monitoring system.

The main components of the proposed exercise prescription and monitoring system (EPMS, as shown in Fig. 2) are described as follows:

1. Physician/Physiatrist: to review and assess the patient's medical record and calculate an exercise prescription.
2. Healthcare Service Center (HSC): to implement the ACSM exercise prescription guidelines as the web service. On the other hands, served as service broker to coordinate the web services around the intranet and internet.
3. HIS Patient record: provides a web service to access HIS data and expert system under an authorized and authenticated access controlled.
4. Exercise compliance: use exercise detection system to records patient's daily movements to monitor whether the patient's exercises to the required intensity or for a sufficient period of time.
5. Exercise Monitoring: Vital to understanding the appropriateness of an exercise prescription by designing a system to assist a physician in prescribing an exercise prescription and tracking whether the patient follows the prescription.

3 Implementation

An exercise prescription considers human health needs, physical conditions, and clinical status. Assessing and customizing each individual case constitutes the ultimate goal for the exercise prescription.

The following shows a scenario of case study: Mr. Wang is 56 years old. He is 166cm tall and weights 70kg. He quit smoking 20 years ago. His blood pressure is 120/90 mmHg. He has history of heart disease that includes two myocardial infarctions, each followed by a bypass operation. The most recent bypass was two years ago. He has decided to begin an exercise program. The physician ordered a cardiopulmonary exercise stress test and an echocardiography test to evaluate Mr. Wang's heart and exercise capacity.

The physiatrist can utilize exercise prescriptions via EPMS by connecting with the HIS using internet web browsers. The physiatrist applies to HIS as well as EPMS to regulate the exercise prescription, or to create and modify defined exercise prescriptions. The results of the EPMS were as follows:

Maximal heart rate: 164 beats per minute; Low-exercise intensity: 66 beats per minute; High-exercise intensity: 131 beats per minute

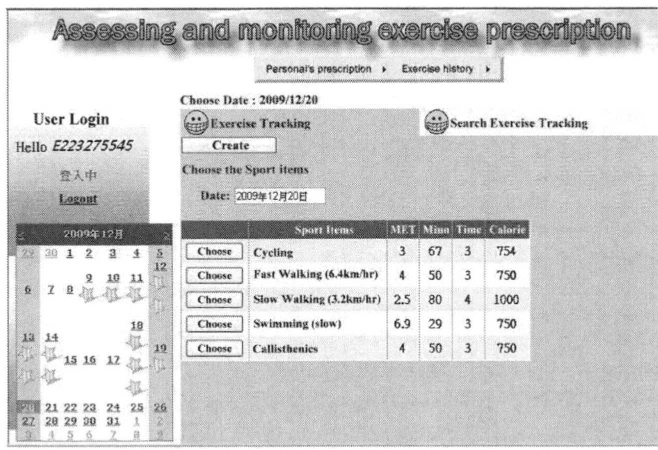

Fig. 3 Clinic interface of EPMS.

The physiatrist indicates that Mr. Wang should begin his session with at least 15 minutes of exercise prescription, progressing to 60 minutes. During the exercise prescription, the sport items have slow-walking (3.2/hr), fast-walking (6.4/hr), time-running, swimming (slow), cycling, and calisthenics. The interface to the clinic module is shown as Fig. 3.

Mr. Wang should do exercise prescription four to six days per week. Encourage Mr. Wang enrolled in rehabilitation programs that meet only three days per week to exercise on off days. He can perform range-of-motion and resistance training two or three days per week. Patients with diabetes should avoid high-intensity exercises to prevent acute hyperglycemia. Patients having a history of osteoporosis, high-impact exercises should be avoided. A revisit to the doctor is recommended twelve weeks after the prescription, so that the doctor can reassess the suitability of the exercise prescription.

4 Conclusion

In this paper, we introduced a layered service-oriented architecture for automatically generate the exercise prescription and follow-up the compliance of such prescription. The web service and service-oriented architecture approach to integrating the hospital care services and home care services can reduce the complexity and cost of system development.

Acknowledgments This project is supported by the National Science Council, Taiwan (grant number NSC99-2221-E-320-005).

References

1. Whaley, M.H., et al. (eds.): ACSM's Guidelines for exercise testing and prescription, 7th edn. Lippincott Williams and Wilkins, Baltimore (2006)
2. Kinetics, H.: Exercise Prescription: a case study approach to the ACSM, guidelines, 2nd edn., Swain DP, Leutholtz BC (2007)
3. Anderson, R.N., Smith, B.L.: Deaths: Leading causes for 2002. CDC National Vital Statistics Reports 53(17) (2005)
4. Petersen, S., Peto, V., Scarborough, P., Rayner, M.: Coronary heart disease statistics 2005 edition, British heart foundation health promotion research group. Department of Public Health. University of Oxford (2005)
5. Kart, F., Moser, L.E., Melliar-Smith, M.: Building a distributed e-healthcare system using SOA. IEEE IT Professional 10(2), 24–30 (2008)
6. Omar, W.M., Taleb-Bendiab, A.: E-health support services based on service-oriented architecture. IEEE IT Professional 8, 35–41 (2006)
7. Krafzig, D., Banke, K., Slama, D.: Enterprise SOA: Service-oriented architecture best practices. Prentice-Hall, Englewood Cliffs (2005)
8. Papazoglou, M.P., Heuvel, W.: Service oriented architectures: approaches, technologies and research issues. The VLDB Journal, 389–415 (2007)

Web GIS-Based Regional Social Networking Service as Participatory GIS

Satoshi Kubota, Kazuya Soga, Yuki Sasaki, and Akihiro Abe

Abstract. Participatory GIS (PGIS) is used in regional urban development and in solving regional problems that affect local communities. Accordingly, the widespread adoption of PGIS is important. In this paper, the development of a Web GIS-based regional social networking service (SNS) is presented. The SNS focuses on PGIS and takes into account user activity support. The system has three functions: information extraction by user activity area, GPS location referencing, and location based information posting function. The proposed SNS is evaluated in Takizawa Village, Japan. Furthermore, a PGIS usage model is proposed for actively utilizing PGIS. Local governments, nonprofit organizations, and universities each have a role in this model.

1 Introduction

In urban development at the regional level, cooperation through public and private partnerships is important. Participatory Geographic Information System (PGIS) is used in regional urban development and in solving regional problems that affect local communities [1]–[3]. In this study PGIS is defined as an information system with Web GIS that is used for urban development, local development, and support of community activities.

With the rapid development of mobile devices, wireless networks, and Web 2.0 technology, a number of location based social networking services (SNSs) have emerged in recent years [4]. These location-based SNSs allow users to connect with friends; explore places such as restaurants, stores, and cinemas; share their locations; and upload photos, video, and blog posts [5]. As city and neighborhood exploration is a main theme in many location-based SNSs, providing location recommendations to users is highly desirable for such services. Moreover, because the numbers of users and locations in location-based SNSs are rapidly growing, it is essential that SNS adopt efficient techniques for making location recommendations.

Satoshi Kubota · Kazuya Soga · Yuki Sasaki · Akihiro Abe
Iwate Prefectural University
e-mail: s-kubota@iwate-pu.ac.jp

F.L. Gaol (Ed.): Recent Progress in DEIT, Vol. 2, LNEE 157, pp. 313–321.
springerlink.com

In Japan, an increasing number of local governments and organizations have been operating SNSs since 2005 [6]. These SNSs are called "regional SNSs" and are established for the participation of the local community. These services allow citizens to disseminate and exchange information freely over the Internet. Also, there is a growing trend of local governments using this exchange of information to promote civic participation and interaction. Existing regional SNSs are compatible with Web GIS. However, they have problems regarding their utility and operation. Location data in regional SNSs can be used to share journal entries, events, and information through the network. Location information can be referenced through journal functions and by members of the community, but is not used for active communication among the participants of the SNS. In operation, regional SNSs have the problem that separate functions must be used to display maps and post location information.

In this study, Web GIS-based regional SNS is proposed as PGIS and an SNS is presented that considers user and community activity support. The system is evaluated in Takizawa Village, Japan, on the basis of usability and availability. Also, this paper proposes a usage model of regional SNS.

2 Regional Social Networking Service

2.1 Participatory GIS

Participatory activity occurs when users collect and transmit information in the field, organize, and examine their information, and transmit the results using Web GIS. Examples of PGIS are presented in references [1]–[3], and [7]. In existing PGIS, it can be difficult for users to manage and to organize information from field activities and to transmit information to the public. Web GIS is necessary for organizing and examining information gathered by users in field activities.

2.2 Regional SNS

An increasing number of local governments and organizations have been operating regional SNSs since 2005 in Japan. The aim of regional SNSs is to increase the participation of residents in activities such as local government decision making, regional activities, sightseeing, and other events [8]. Regional SNSs have various functions that are similar to other SNSs such as Facebook [9] in the United States and Mixi [10] in Japan, including journals, community communications, and messaging. The main features of regional SNSs target people living in the community. Regional SNSs are managed by local governments, nonprofit organizations (NPOs), and private companies. Shoji [6] classified regional SNSs into four categories based on the size of the area targeted by the SNS and the forms of online and interpersonal communications.

2.3 Regional SNS Using Web GIS

Existing regional SNSs are compatible with Web GIS. However, performance problems arise when manipulating and running applications. Location-based data in regional SNSs can be used to share journal entries, events, and information through the network. Location data is referenced by journal functions and members of the community, but is not used for active communication among the participants of the SNS. In operation, regional SNSs have the problem that separate functions must be used to display maps and post location data. Because regional SNSs target people living in a community, persons who are active in the field will transmit the information on maps using Web GIS.

3 Design and Development of System

3.1 Design Principles

In this paper, three design principles are proposed. (1) The proposed regional SNS has an extraction function based on the user activity area. Users can collect information in this area. This idea is used for promoting the attention and interests of the local community. Personal computers and mobile phones are used to access the regional SNS. (2) Web GIS facilitates the use of maps. When users post location-based information, they can connect it to journals and community messages, and transmit the information easily. Google Maps is used for Web GIS because of its versatility. (3) Mobile phones with GPS (Global Positioning System) are used to obtain users' positions, to transmit local information based on their present position, and to extract local information from postings.

3.2 System Architecture

The proposed Web GIS-based regional SNS is accessed via PCs and mobile phones. Open source software such as open-gorotto, OpenSNP, and OpenPNE is used for developing regional SNSs. The functions of the proposed regional SNS include extracting information by user activity area, posting location-based information, and referring to locations using GPS. The proposed SNS was developed by adding these functionalities to an OpenPNE-based SNS. The system was developed using a Linux server, Apache, MySQL, and PHP. Google Maps API for PC and Google Static Maps API for mobile phone were used for Web GIS. The system architecture is shown in Figure 1.

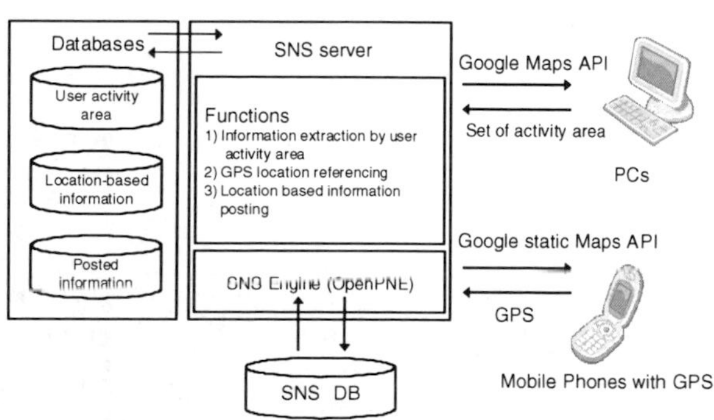

Fig. 1 System architecture of the proposed regional SNS.

3.3 System Development

The Web API (Application Program Interface) of the regional SNS was developed to operate on OpenPNE version 2.12.0.

1. Information extraction by user activity area: This function extracts information from the user activity area and supports users and communities activities through the use of PCs and mobile phones. A user sets a location (such as the user's home or business, frequently accessed roads, or an arbitrary location) on a two-dimensional map in the regional SNS. A block-triangular model is used to calculate the user activity area. Figure 2 and 3 show the user activity area map.
2. GPS location referencing: This function uses GPS on mobile phones to reference the location of users with mobile phones. This function extracts local information from the areas in which users are located. On Google Maps on a mobile phone, markers are displayed and information is filtered by category.
3. Location based information posting function: This function can be used to post location-based information by persons involved in local community activities. The posted information is displayed in user's main window on the regional SNS, which is in the user activity area. Users can submit information by retrieving a location, using GPS, and writing comments for the location using their mobile phones. Geocoding is used to retrieve locations based on words.

Fig. 2 User activity area map in PC.

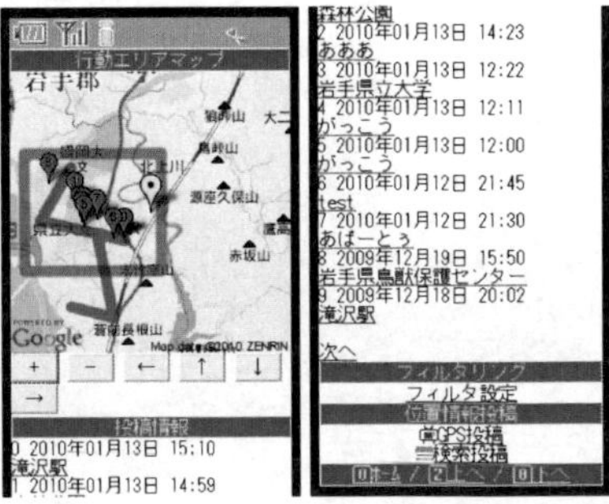

Fig. 3 User activity area map in mobile phone.

4 Evaluation

The proposed regional SNS was evaluated to determine performance when local participants are manipulating data and running applications. The evaluation was conducted using PCs and mobile phones.

4.1 PC Use

The developed system was operated in Takizawa Village, Japan, and evaluated considering user activity. To evaluate the applications, a questionnaire shown in Figure 4 was administered to 42 users from local government and NPOs to determine whether the system could be used in urban development and in communications during local participatory activities. The system was operated from November to December 2008. In about one month of operation, 67 items of location-based information were posted.

17 users responded to the questionnaire on the performance of data manipulation and applications in urban development and local activity. The questionnaire was made refering universal design practical guidelines [11]. Because all system functions operated without any incident, the functionality of the system was verified. The evaluation results are shown in Figure 4. Based upon these results, we conclude that the developed regional SNS in Takizawa Village was a practical system.

The system could possibly be used to create policy for local government and support NPO activities by sharing community information among participants. However, it is not possible to verify this application of the system conclusively because information was posted in the SNS only for a short period. Therefore, it is necessary to operate the developed regional SNS over a longer time frame for further study.

The results of the evaluation show that the performance of the system for manipulation of data by users was average. Users responded that they could operate the system without difficulty. However, maintaining an operating manual and teaching data manipulation in training is required to facilitate the use of the system.

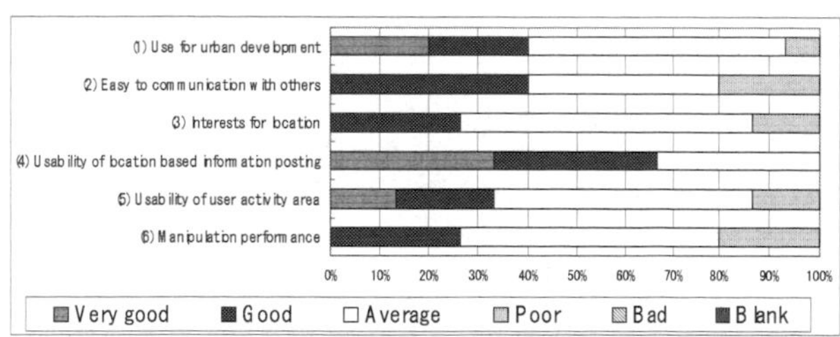

Fig. 4 Evaluation results.

4.2 Mobile Phone Use

The developed mobile phone system was installed on the Takizawa Village regional SNS. To evaluate its functionality, 28 university students used the developed functions and answered a questionnaire about the system application and manipulation performance.

The system was used easily and without difficulty by participants. More than 60% of users responded that the system can be used easily for each of the data manipulation functions. For applications, more than 60% of users also responded that they could refer to interesting information in the user activity area by using the functions for information extraction and GPS location referral.

5 Practical Use and Operation

5.1 PGIS Usage Model

It is necessary to construct the management systems for stakeholders to operate regional SNS-centered PGIS. To use and to operate a regional SNS continuously, this paper proposes PGIS as a SNS usage model, as shown in Figure 5. In PGIS, cooperation among local governments, private companies, NPOs, and universities is important, and thus, these actors are included in the usage model. PGIS will be used to collect and to post ideas in a positive manner by defining the participants' role in operating the regional SNS and motivating participation. As shown in Figure 5, we propose that NPOs steer regional SNSs. NPO can operate flexibly and construct an atmosphere of participation.

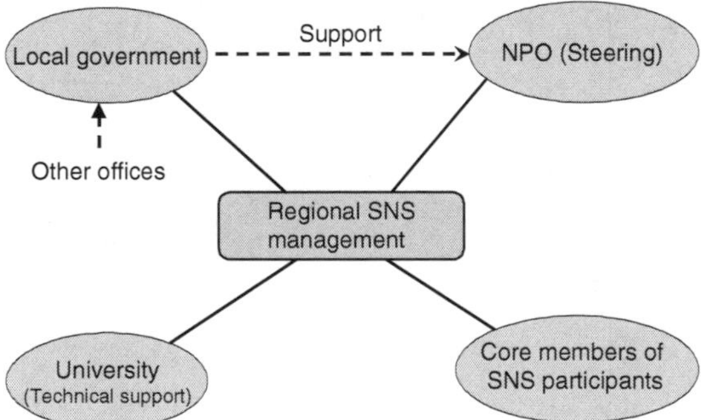

Fig. 5 Usage model of regional SNS as PGIS.

5.1.1 Takizawa Village Regional SNS

It is clarified that regional SNS-centered PGIS is used for urban development by local government participation. NPOs can take responsibility for the installation and maintenance of the system, respond to participants' inquiries and plan events. Also, NPOs should use the GIS functions of the regional SNS for urban development and local community activities, and demonstrate its utilization methods and effects. Local governments can provide a feeling of security, recruit participants, transmit information, and provide ongoing management support. Universities can loan and install system software, conduct training, and ensure that the system is simple to use. Core members of SNS participants can support new participants, plan events, and share knowledge and experience regarding PGIS.

5.1.2 Elements of Ongoing Regional SNS Operations

This section presents the elements for ongoing management of regional SNS-centered PGIS.

- Financial resources of regional SNS management

  It is important to acquire enough financial resources to manage the regional SNS independently and continuously. For NPOs, an administrator is necessary to acquire financial resources to operate independently and to maintain the support of local governments. In cases where the local governments provide support, the NPO must keep track of the funding provided.

- GIS education

  Human resources must be developed that can use the spatial analysis functions of GIS. In urban development and local community activities, there are large amounts of temporal and spatial information. This information is used to understand the targeted community. Knowledge of GIS and manipulation of location data should be taught.

- Mobile Web GIS environment

  Urban development and local community activities often are outdoors. Regional SNS-centered PGIS is used outdoors to refer to the collected and organized information. Therefore, the GIS functions of mobile phones with GPS will be used.

- System installation and maintenance

  To install and to maintain the system, professional expertise and technology are necessary. Universities can support NPOs by providing these skills and knowledge.

5.2 Practical Use in Local Community

The developed Web GIS-based regional SNS was used on the Takizawa Village SNS, where the PC version has been in operation since 24 October 2009. The mobile system began operations in January 2010. Figure 6 shows the development and operation of the Takizawa Village regional SNS, which is based on the proposed PGIS usage model. In the system interface, users can post journal entries and messages to the community and use the activity area map. As of December 2010, there are 158 participants.

Fig. 6 Takizawa village regional SNS.

6 Conclusion

In this paper, Web GIS-based regional SNS was proposed as PGIS that takes into account user behavior support for PC and mobile phone users. The proposed regional SNS was evaluated in Takizawa Village, Japan. The results indicated validity of system design principles. The system has been in operation. This system has general versatility and can be transported for a big city like Tokyo by using open source SNS engine and Google Maps API as Web GIS. We can make regional communication using Mixi. However, Mixi is too huge to communicate with a sense of security and interpersonal communication in local community activity. The proposed Web GIS-based regional SNS will be used in regional urban development and in solving regional problems that affect local communities. And, a PGIS usage model is proposed for actively utilizing PGIS in partnership with local governments, nonprofit organizations, and universities.

References

1. Craig, W.J., Harris, T.M., Weiner, D.: Community participation and geographic information systems, p. 383. Taylor & Francis (2002)
2. Minang, P.A., Mccall, M.K.: Participatory GIS and local knowledge enhancement for community carbon forestry planning: an example from Cameroon. Participatory Learning and Action 54, 85–91 (2006)
3. Zhu, J., Qiao, Q., Zhang, T.: High Performance Participatory GIS - Application in Emergency Evacuation. In: 2009 First International Workshop on Database Technology and Applications. IEEE (2009)

4. Weaver, A.C., Morrison, B.B.: Social Networking, pp. 97–100. IEEE Computer Society (2008)
5. Ye, M., Yin, P., Lee, W.: Location Recommendation for Location-based Social Networks. In: ACM GIS 2010, pp. 458–461 (2010)
6. Shoji, M., Miura, S., Sugo, Y., Wasaki, H.: Forefront of Regional Social Networking Service. In: ASCII, 241 p (2007)
7. Imai, O.: The Study of GIS Use in Participation Activity of Solving Regional Problem. Papers and Procedings of the Geographic Information Systems Association, vol. 17 (2008)
8. Soga, K., Kubota, S., Ichikawa, H., Abe, A.: A prototype of Regional SNS Cooperation Map Considering User Behavior Support. Papers and Procedings of the Geographic Information Systems Association, vol. 17 (2008)
9. Facebook, `http://www.facebook.com/`
10. Mixi, `http://mixi.jp/`
11. Japan Ergonomics Society, The Universal Design Practical Guidelines. Kyoritsu Publishing (2003)

Network and Data Management Techniques for Converged IP Networks

Sangita Rajankar and Abhay Gandhi

Abstract. In this paper an attempt has been made to design and develop a prototype of Web Based Server of Servers (WBSOS) which will automatically configure some network parameters for various servers such as DHCP, NETCONF, etc, connected in a network and maintain the data in a centralized manner in an RDBMS, through a browser based application.

In WBSOS, various servers can be managed using a single browser based application. Testing is being done using two servers' viz. Dynamic host Configuration Protocol (DHCP) and Internet Engineering Task Force's (IETF), Network Configuration Protocol (NETCONF) implemented in the same browser based application. The system is being tested on windows platform. As NETCONF based Network Management System (NMS) is being implemented, constraints in existing SNMP based NMS have been removed.

NETCONF is being implemented using Web Services Description Language (WSDL) that imports the definitions of XML types and elements from the base NETCONF schema. Using the conversion tools, basic JAVA APIs are generated. These APIs are modified to access the parameters from the graphical user interface (GUI) and generate a request message. This message is encapsulated with SOAP envelope and HTTP and then encrypted with SSL agent and finally sent to the agent. At the agent, after decryption by SOAP & HTTP, the message is parsed by the XML parser and sends to the operation processor to get back the operation commands such as get, edit, etc. The command is executed at the agent and reply is send back to the server. On receiving the reply, the manager saves the data in the centralized data store. Here, data is stored in My SQL database.

Sangita Rajankar
Maharashtra Remote Sensing Applications Centre, Nagpur, Maharashtra
e-mail: sangitarajankar@gmail.com

Abhay Gandhi
Visveswaraya National Institute of Technology
e-mail: asgandhi@ece.vnit.ac.in

F.L. Gaol (Ed.): Recent Progress in DEIT, Vol. 2, LNEE 157, pp. 323–331.
springerlink.com

The WBSOS will share the data of the network thus reducing the overall burden on the network. As the data is shared, collective and relative analysis can also be carried out. Use of NETCONF finds many benefits in this environment: from the reuse of existing standards, to ease of software development, to integration with deployed systems. Thus WBSOS will be capable of handling a wide range of devices in converged network system.

1 Introduction

With the growing technology crowd, the network is converging and becoming more and more complex. The network is not only increasing in size but also increasing in variety of devices. Today's problem lies in integration of such network systems and service management [1]. Large network generally run several management platforms in parallel, each dedicated to a management plane or a technology (e. g one for network management, another for system management, one for service management, one for VPN management, etc). Each NMS works separately and maintain its own database. Thus creating redundant databases. These databases are also not shared. For getting the basic configuration, each server sends the same command separately thus blocking the network for the same output. Internet Engineering Task Force (IETF) is encouraging various techno-persons to provide solutions to the various challenges in network environment from time to time.

The technology is moving out of the closed proprietary environment and entering a new world of open source and interfaces. 4open sources use the concept of Object Oriented Technology (OOT). Implementation of OOT in network management lower downs the burden on network administrator by introducing "Anywhere Anytime" administration. The paper demonstrates the use of new web based technologies for Network and data management.

1.1 Future Network and Its Management Issues

The convergence of voice, video and data networks has been evolving and gaining momentum since last few years. The irresistible logic is that digitized voice is just another kind of data, so why not carry it on the same data links that handle all ones and zeros? The economies of convergence can be considerable—there is no need to build and support separate voice and data infrastructures when you can have just one.

Organizations are increasingly looking at network management which will converge the management systems also. Secondly, organizations are considering remote management, through web browser. This increase in the mesh has lead to shortage of IPv4 address which is going to be replaced by IPv6, the next generation protocol.

Advanced client/server services also allow VoIP systems and devices to be provisioned and managed remotely. Remote management reduces costs, including expenditures that are associated with user Moves, Adds, and Changes (MACs) and costs that are related to updating edge devices with the latest customer applications and services.

SNMP has been widely used as an industrial standard [Reference of SNMP] in the past recent years. However, SNMP has been used for mostly in monitoring for fault and performance management, but very less used for configuration management [2]. Also SNMP uses UDP to transfer messages, which results in unreliability and limited size. Secondly the Command Line Interface (CLI) was also used for configuration management. But due to the efforts required to operate different commands for different devices and due to lack of formal description language to define all properties of the programmatic interface, this technique too was not much popular.

To overcome the shortcomings of SNMP and to provide integration of managing various systems, IETF is working on standardization of configuration management protocol for network devices using Extensible Markup Language (XML) technologies like XML schema, XML Path Language (XPath), Extensible Style-Sheet Language (XSL), Simple object access protocol (SOAP), and Web Services Description Language (WSDL)[3].

Based on the above, NETCONF was proposed in 2006 which used XML to send and receive management information between manager and agent [4].

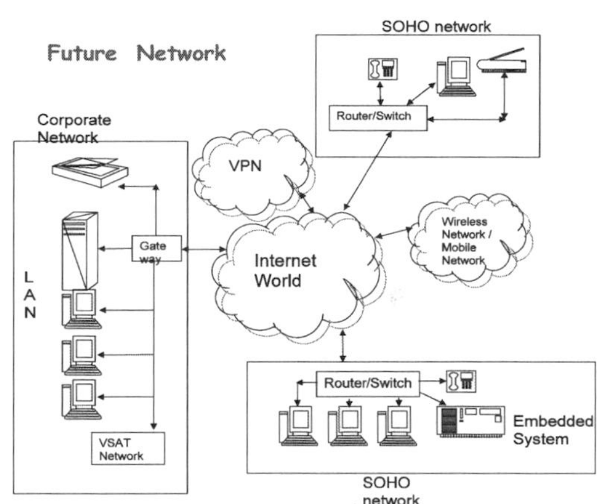

Fig. 1 Integrated Network Management System

2 NETCONF Protocol

NETCONF was proposed in 2006 which has many advantages in NMS scenario. Netconf protocol [4] defines a simple mechanism by which a network device can be managed, configuration information can be retrieved and set new configuration data. NETCONF separates the configuration data and state data. In addition, NETCONF not only defines the basic operations but also puts forward several high level configuration management operations. It uses RPC paradigm to send and receive request and responses as described in the XML schema file. It adopts the connection- oriented Simple Object Access Protocol (SOAP) to ensure the security of the transmission [5]. Netconf allows client to discover the "capabilities" of the server and act accordingly. The capability is identified by a uniform resource identifier (URI). NETCONF defines the existence of one or more configuration datastores and allows configuration operations on them. A configuration datastore is defined as the complete set of configuration data that is required to get a device from its initial default state into a desired operational state. The configuration datastore does not include state data. Only the <running> configuration datastore is present in the base model. Additional configuration datastores may be defined by capabilities. Such configuration datastores are available only on devices that advertise the capabilities. The netconf supports operations like <get>, <get-config>, <edit-config>, <copy-config>, <delete-config>, <lock>, <unlock>, <close-session> and <kill-session>.

Thus, there is a great need to build a NETCONF-based network configuration management system to solve the problems encountered using SNMP and to cater the needs of future converged network.

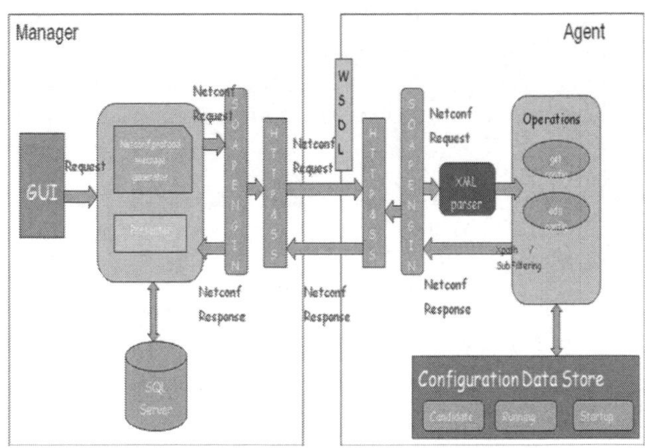

Fig. 2 Communication between Netconf manager and agent

The fig 2 here explains the communication between the basic Netconf manager and agent:

- Using GUI, the manager will select the parameters he wants to configure or monitor for the agent.
- The NETCONF message generator receives the command from the GUI to make certain operations.
- The message is encapsulated with SOAP envelope and HTTP package and then encrypted with SSL agent and finally sent to the agent.
- The manager refers the information from SQL datastore
- At the agent end, the reverse process of decryption by SOAP, HTTP & SSL is done
- The messge is further parsed by the XML parser & send to the Operation Processor to get the operation commands (get, edit, etc)
- Accordingly, the commands will be executed by referring the configuration data store

- Agent uses xpath/subfiltering mechanism to validate & examine the message
- On execution of the command, agent generates reply message for the manager

3 Related Work

J.P Martin-Flatin presented an idea to use XML for integrated management in his research on web-based integrated network management architecture (WIMA) [6]. WIMA provides a way to exchange management information between a manager and an agent through HTTP. WIMA implemented push-based network management prototype to transfer SNMP on JAVA platform.

Mi-Jung Choi, James W. Hong and Hong-Taek Ju found that XML manager and XML agent is the most ideal framework for gaining the maximum advantages of XML-based network management [7].

Yanan Chang, et al, implemented a prototype including a NETCONF manager and a NETCONF agent and then the performance analysis between NETCONF agent and SNMP agent were discussed[8]. In his study, he has mentioned that very few prototypes have been developed, most of them based on XML rather than Netconf.

4 Web Based Server of Servers (WBSOS)

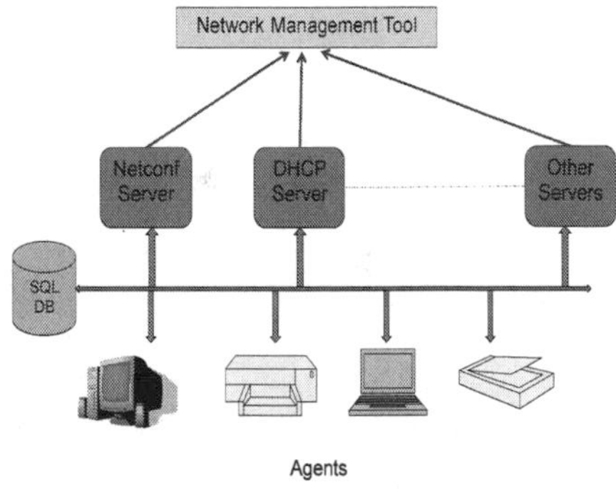

Fig. 3 Server of Servers

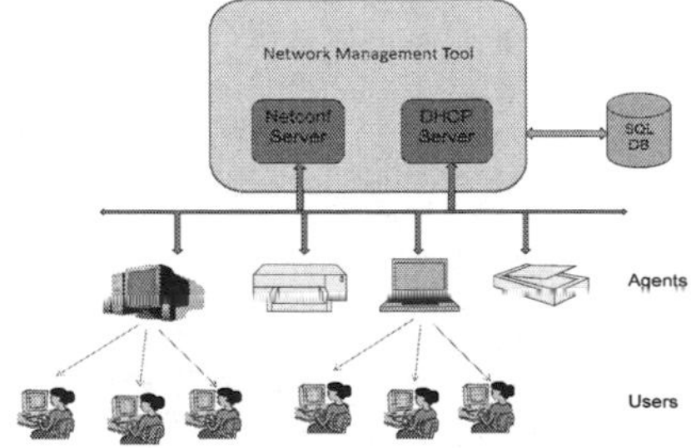

Fig. 4 Implemented Web Based Server of Servers (WBSOS)

Fig. 3 shows the block diagram of what a Web Based server of servers really should be. However, prototype of WBSOS which has been implemented is shown in fig 4. The WBSOS implements two servers namely DHCP server and NETCONF server. The parameters of both the servers can be configured using a single browser based application called as a "Network Management Tool". The DHCP server mainly configures the basic network parameters for a system that brings it in the network such as : allowed IP address range, not-allowed IP address range and lease time. The network device acts as DHCP client and get its IP address from the WBSOS. The DHCP server saves the information of all the network devices in the centralized data repository of MySQL.

The Network management tool has been set in "auto" mode which helps netconf to gather information stored in the centralized data store and thus detect number of devices and its basic configuration. The other configuration information such as its netconf capabilities can be viewed by clicking on the properties button in the GUI of Network Management tool. The basic operations such as <get>, <get-config>, <edit-config>, <copy-config>, <close-session>, <kill-session>, <lock> and <unlock> sessions has been implemented in this prototype.

4.1 Operation Flow Diagram

The following shows the flow of operation messages in the form of request and response between the Network Management Tool and the Netconf agent implemented in WBSOS.

1) Hello Request/Response (Open Session)

hello
Network Management tool ------→ NETCONF Agent
←-----
Capabilities & Session Id

2) get Request/Response (Get Config)

rpc request (msg Id, rpc operation)
Network Management tool ------→ NETCONF Agent
←-----
rpc reply (data)

For other sessions such as "getconfig", "closeSessionElement", "killSessionElement", "lockSessionElement", "unlockSessionElement", "deleteConfigElement", "copyConfigElement" and "editConfigElement" the rpc reply will return "OK" message.

MySQL database is being used to store the information by DHCP as well as Netconf server. They share the same database. Various tables such as list of IP addresses, lease time, users information, department list, etc has been designed.

The server stores the collected information in this database in respective tables. Apart from the XSD and WSDL file, configuration files config.xsd and stats.xsd were designed to implement the user defined data tags and interface data tags respectively.

4.2 Components Used for Implementation of WBSOS Using Web Services Technology

As discussed earlier, the WBSOS implements two servers, DHCP and Netconf. DHCP server has been implemented using J2EE, Javascript, html and struts framework. Netconf uses the standard XSD and WSDL file format from the IETF's location. WSDL is an XML-based language for describing web services. A WSDL file contains the following types of information [9]:

- Information about the functionality of the web service
- Information about how to access the web service

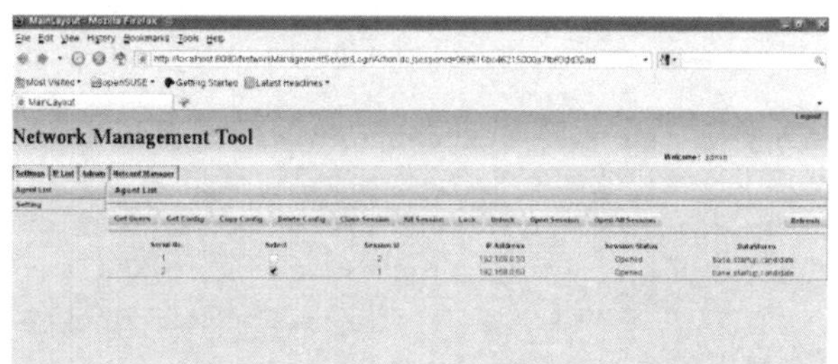

Fig. 5 Operational GUI of WBSOS

XSD defines the schema of the data elements used in the WSDL file. However, it was noticed that the elements in XSD and WSDL file had syntactical errors which were further rectified. Using the modified WSDL, web services were created for the manager and the client. Web services are distributed application components that conform to standards that make them externally available. They solve the problem of integrating diverse computer applications that have been developed independently and run on a variety of software and hardware platforms. The web services architecture is to allow you to connect applications that were developed on different platforms and in different programming languages.

The web service created in WBSOS, is based on the model "Java API for XML Web Services" (JAX-WS). JAX-WS simplifies the task of developing web services using Java technology. It addresses some of the issues in JAX-RPC 1.1 by providing, for example, annotations to simplify web service development and reduce the size of runtime JAR files. The mechanism by which a web service is called and how data is returned is defined by Simple Object Access Protocol (SOAP).

GlassFish application server is a Java EE compatible application server that supports JavaServer PagesTM (JSPTM), Java Servlet, and Enterprise JavaBeansTM (EJBTM) component-based applications. Libraries included with the GlassFish application server can be used to compile web services and clients, verify that your application implements the Java EE specification correctly, create a new web application with JavaServer Faces (JSF) support, or add JSF support to an existing web application.

Thus the components of WBSOS included: SOAP, WSDL, XSD, JAVA API and JAX-WS. The entire application is developed using Netbeans 6.8 editor.

5 Results and Discussions

WBSOS has demonstrated a methodology and given a simple solution of mixed environment which can be used in converged integrated network with ease. There are various advantages such as centralized datastore, manage various servers using one application from anywhere, ease of management operations. It can be noticed that Netconf can be used to configure network as well as system configuration parameters. However, it does not mean that DHCP should no longer be considered for implementation. Since DHCP is at such a matured level, many devices are easier to be set in this environment. DHCP can thus be used to conduct basic network configuration, and use NETCONF for advanced network and system configuration. Thus an attempt has been made to contribute to the standardization and stability of Netconf protocol.

There is a lot of scope in improving the management system. IN WBSOS, the basic capabilities of the Netconf have been used. No extra capabilities are defined in the network. Using more capabilities a more stable NMS can be designed. Apart from network configuration many new characteristics of fault, access control and security managements can also be included.

References

[1] Pras, A., Martin-Flatin, J.P.: What Can Web Services Bring To Integrated Management? In: Handbook of Network and System Administration. Elsevier, Amsterdam (2007) ISBN 978-0-444-52198-9

[2] Schonwalder, J., Pras, A., Martin-Flatin, J.P.: On the Future of Internet Management Technologies. IEEE Communication Magazine 41, 90–97 (2003)

[3] Choi, M.-J., Choi, H.-M., Hong, J.W., Ju, H.-T.: XML-Based Configuration Management for IP Network Devices. IEEE Communication Magazine (July 2004)

[4] RFC 4741, NETCONF Configuration Protocol (December 2006)

[5] RFC 4743, Using NETCONF over the Simple Object Access Protocol (SOAP) (December 2006)

[6] Martin-Flatin, J.P.: Web-Based Management or IP Networks and Systems. Ph. D. Thesis. Swiss Federal Institute of Technology, Lausanne, EPFL (October 2000)

[7] Choi, M.-J., Hong, J.W., Ju, H.-T.: XML-Based Network Management for IP Networks. ETRI Journal 25(6) (December 2003)

[8] Chang, Y., Xiao, D.: Design and Implementation of NETCONF-Based Network Management System. In: IEEE, Second International Conference on Future Generation Communication and Networking (2008)

[9] Netbeans 6.8 IDE help

[10] Web Service Description Language (WSDL) 1.1., http://www.w3.org/TR/wsdl
[11] NETCONF.xsd, http://www.w3.org/2001/XMLSchema
[12] Yoo, S.-M.: Web Services Based Configuration Management for IP Network Devices. IFIP International Federation for Information Processing (2005)
[13] Lee, J.-O.: Enabling Network Management Using Java Technologies. WarePlus Inc./Korea Telecom (January 2000); Issue of IEEE Communications
[14] Experience of Implementing NETCONF over SOAP (October 2008)
[15] Xu, H., Xiao, D.: Evaluation on Data Modeling Languages for NETCONF-Base Network Management (2008)
[16] Kotsakis, E., Bohm, K.: XML Schema Directory: A Data Structure for XML Data Processing. IEEE (2000)
[17] Schildt, H.: JAVA 2 The complete Reference, 5th edn. Tata McGraw Hill Publication
[18] [JDK] Java SE, http://java.sun.com/javase/index.jsp
[19] [NetBeans] NetBeans, http://www.netbeans.org/index.html

Developing WikiBOK: A Wiki-Based BOK Formulation-Aid System

Yoshifumi Masunaga, Masaki Chiba, Nobutaka Fukuda, Hiroyuki Ishida, Kazunari Ito, Mamoru Ito, Toshiyuki Masamura, Hiroyasu Nagata, Yasushi Shimizu, Yoshiyuki Shoji, Toru Takahashi, and Taro Yabuki

Abstract. The design and implementation of WikiBOK, a Wiki-based body of knowledge (BOK) formulation-aid system, is investigated in this paper. In contrast to formulating a BOK for a matured discipline such as computer science, BOK formulation for a new discipline such as social informatics needs a "bottom-up" approach because academics in a new discipline cannot draw its entire figure par avance. Therefore, an open collaboration approach based on the collective intelligence concept seems promising. WikiBOK is under development as part of our project based on BOK+, which is a novel BOK formulation principle for new disciplines. It uses Semantic MediaWiki (SMW) to facilitate its fundamental functions. To support a rich graphical user interface for WikiBOKers, a a graph visualization software, Graphviz, is adopted. SMW is enhanced to work in conjunction with Graphviz. Because edit conflicts occur when WikiBOKers collaborate, a resolution principle is investigated to resolve BOK tree edit conflicts.

Yoshifumi Masunaga · Nobutaka Fukuda · Hiroyuki Ishida ·
Kazunari Ito · Yasushi Shimizu
School of Social Informatics, Aoyama Gakuin University, Kanagawa, Japan
e-mail: masunaga@si.aoyama.ac.jp

Masaki Chiba · Hiroyasu Nagata · Toru Takahashi
Faculty of Social Information, Sapporo Gakuin University, Hokkaido, Japan

Mamoru Ito
Faculty of Education and Integrated Arts and Sciences, Waseda University, Tokyo, Japan

Toshiyuki Masamura
Graduate School of Arts and Letters, Tohoku University, Miyagi, Japan

Yoshiyuki Shoji
Graduate School of Informatics, Kyoto University, Kyoto, Japan

Taro Yabuki
College of Science and Engineering, Aoyama Gakuin University, Kanagawa, Japan

F.L. Gaol (Ed.): Recent Progress in DEIT, Vol. 2, LNEE 157, pp. 333–341.
springerlink.com

1 Introduction

BOK (body of knowledge) is a term used for representing a complete set of concepts, terms, and activities that make up a professional domain, as defined by the relevant professional association. CC2001 (computing curricula 2001) [1] is a well-known BOK for the academic field of computer science; we call this BOK as CSBOK (computer science body of knowledge). In contrast to the formulation of a BOK for a matured discipline such as computer science, it is very difficult to formulate a BOK of a new discipline because of the discipline's novelty, i.e., academics in a new discipline cannot draw its entire figure par avance. Therefore, it seems promising to introduce a "bottom-up" open collaboration approach to formulate a BOK for a new field. Note here that a traditional BOK formulation is done top-down.

In order to validate the effectiveness of the bottom-up approach, we have been developing an open collaborative environment, which may assist BOK formulators in formulating a BOK using "Social Informatics" (SI) as an example of a new discipline. It is expected that if an SIBOK (social informatics body of knowledge) is formulated, we can compare it with other BOKs such as CSBOK to realize the difference between the two disciplines.

In order to materialize our idea, we have proposed a novel BOK formulation principle named BOK+ [2], which is innovative in the following sense: (1) In contrast to a traditional BOK formulation scheme, it is bottom-up. (2) It handles not only a BOK tree but also descriptions and materials that may help a BOK formulator's work. (3) It is based on the collective intelligence approach [3].

In order to develop a system based on our approach, MediaWiki (http://www.mediawiki.org/wiki/MediaWiki) is adopted, which is an open source wiki [4] package written in PHP and was developed for building Wikipedia (http://www.wikipedia.org). It is known that Wikipedia editors, called Wikipedians, from all over the world work collaboratively to construct Wikipedia [5]. Our approach is similar to the Wikipedia approach in the sense that, for open collaboration, our system also uses a wiki clone named Semantic MediaWiki (SMW) as the kernel. However, it differs from Wikipedia in that our system manages not only articles but also BOK trees. The editing tree is absolutely required for formulating a BOK because a BOK is characterized as a conceptual tree. An SMW-based prototype named "WikiBOK" is currently under development as part of our project. To provide BOK formulators using WikiBOK, whom we call WikiBOKers, with a user-friendly, rich interface, a graph visualization software called Graphviz (http://www.graphviz.org/) is used. By enhancing the SMW to work in conjunction with Graphviz, it has become possible for WikiBOKers to directly manipulate a BOK tree. WikiBOK consists of a facility to upload academic materials, a facility to edit wiki pages named "descriptions" to describe the academic topics that a WikiBOKer regards meaningful, and a facility to edit a BOK tree. Note that all such facilities are developed as a "front-end" of the SMW.

The rest of this paper is organized as follows: In section 2, we will provide an overview of BOK+. The prototype development is described in section 3. The BOK tree edit conflict issues are discussed in section 4. Section 5 concludes this paper.

2 BOK+: A Novel BOK Formulation Principle

In order to formulate a BOK in a bottom-up manner, a novel BOK formulation principle BOK+ was introduced [2]. It is the general term for the three types of resources that the WikiBOK manages, and it is a term to refer to a novel BOK formulation principle for new disciplines as well. The resources are materials, descriptions, and the BOK tree, which are linked. Fig. 1 presents an overview of BOK+. A set of all materials or material files such as text and PPT files constitutes a wiki namespace "Material." By referring to the materials, the WikiBOKers of a targeted discipline can select an academic topic to describe its content. This is done using a wiki page, and it is termed "description." Descriptions can be linked together under semantic links that yield a semantic network. The set of all descriptions constitutes a wiki namespace "Description." A BOK is represented by a BOK tree, and it is collaboratively formulated stepwise by WikiBOKers. By interacting with WikiBOK, WikiBOKers formulate a BOK of a targeted discipline from scratch. It is noted that the nodes and the edges of the BOK tree are written using wiki pages so that the entire functions of the WikiBOK are realized as a front-end of the SMW: There is a one-to-one correspondence from each node of the BOK tree to a description. However, the correspondence between Description and Material is many to many.

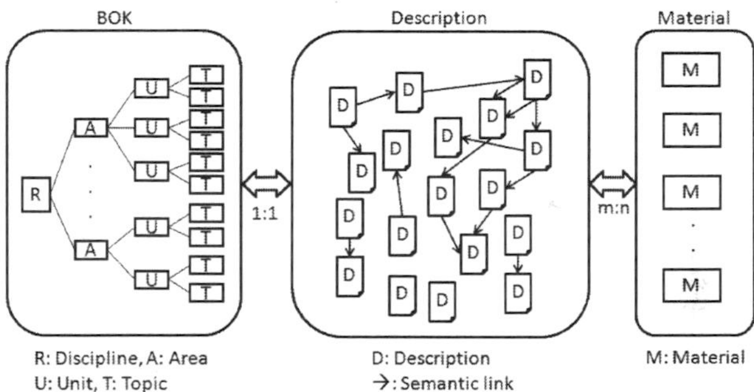

Fig. 1 Overview of BOK+, a novel BOK formulation principle.

3 WikiBOK

3.1 Overview of WikiBOK

A WikiBOK is designed and implemented to support WikiBOKers' open collaborative work. It manages BOK+ objects described in the previous section and provides a rich graphical user interface (GUI) for the direct manipulation of the BOK tree. The SMW is enhanced so that it can store all the WikiBOK objects, i.e., the nodes and the edges of the BOK tree, descriptions, and wiki pages that refer to the actual data files storing the materials. Note here that the WikiBOK is implemented as a "front-end" of the SMW in the sense that all functions are developed using the SMW. In other words, the WikiBOK is defined as a special page on the SMW named Wiki Editor. Since the BOK tree nodes and edges are described as wiki pages, the editing history of the BOK tree is implemented using the page history management capability of MediaWiki. Fig. 2 presents an overview of the WikiBOK architecture currently under development. As it is shown, its architecture is very simple in the sense that the WikiBOK is an enhanced SMW with Graphviz. We call it the ESMW. Its current version consists of five modules: BOK Editor, Description Editor, Uploader, Authorization & Privilege module, and Edit Conflict Resolver. Among these modules, the BOK Editor plays an essential role supporting WikiBOKers in the sense that it provides a rich graphical user interface (GUI) for editing the BOK tree, descriptions, and materials. The graph visualization software Graphviz is used for developing the BOK Editor. A rich GUI implementation using Graphviz is described in section 3 B. The editing functions are described in section 3 C. The Description Editor is introduced to edit descriptions. The Uploader is introduced to upload materials. The Authorization & Privilege module is introduced to ensure that only authorized people can edit WikiBOK objects. Since WikiBOKers work together in an open collaboration environment, edit conflicts may occur. This issue is investigated in section 4. MySQL is used for storing all WikiBOK objects.

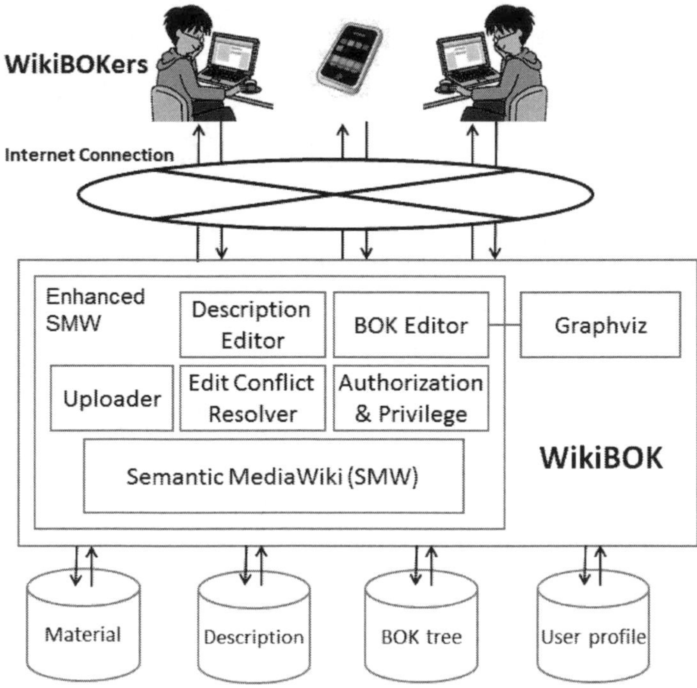

Fig. 2 Overview of WikiBOK architecture.

3.2 Rich GUI Implementation Using Graphviz

The BOK Editor provides a rich GUI for WIkiBOKers. Graphviz, the open-source graph visualization software, is used for implementing the direct manipulation of BOK. The BOK Editor includes the following three modules to realize the aim: (1) a dot file generation function, (2) an SVG file generation using a dot file, and (3) embedding inline an SVG file into a PHP file (xhtml file). Fig. 3 shows how the SMW was enhanced for Graphviz cooperation. As is seen, according to the WikiBOKer's action, the necessary BOK data are retrieved from the database, and they are fed to the dot file generation module. Then, the SVG file generation module works with Graphviz to generate a PHP file. Note that the dot file in the BOK Editor describes all the necessary information that Graphviz needs to draw the BOK tree on a screen. Safari is adopted as the Web browser because it supports SVG files, and it is a Web browser used by iPad. In order to overcome the inflexibility of the traditional computer screen when a large number of objects are displayed, the multi-touch functionality of iPad seems attractive for resolving this problem.

Fig. 4 shows a screenshot of a BOK tree editing result. As is shown, the root node represents the targeted discipline, i.e., "social informatics" in our case. The first-, the second-, and the third-level nodes represent "areas," "units," and "topics" of the targeted discipline in terms of CC2001. We presume that SIBOK may constitute 1000 to 2000 nodes.

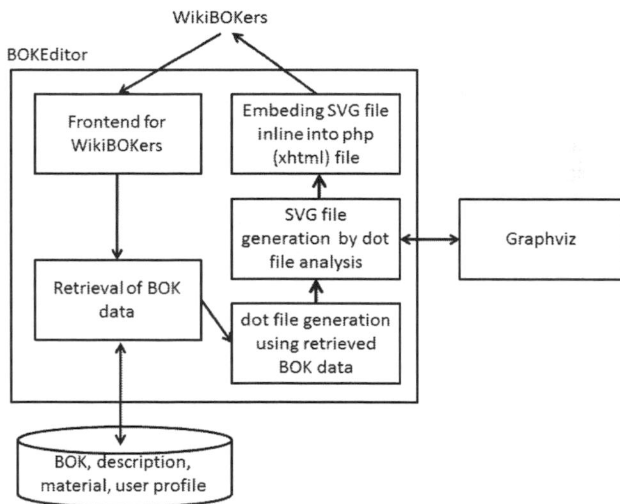

Fig. 3 Overview of BOK Editor configuration.

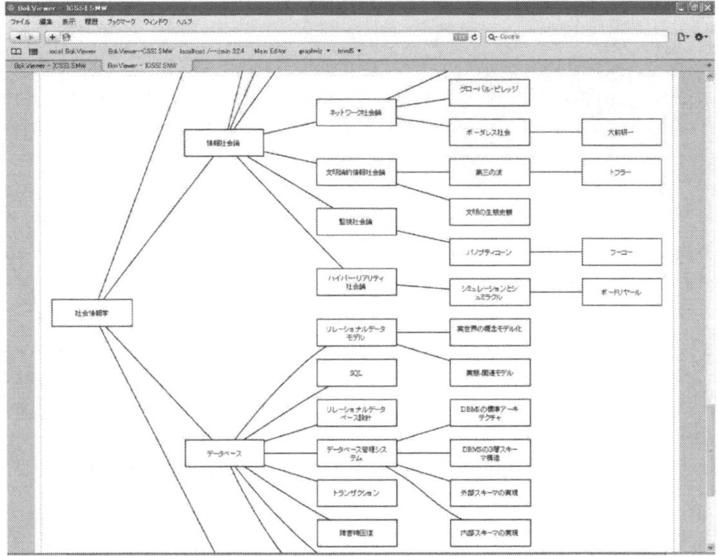

Fig. 4 Screenshot of BOK tree editing result.

3.3 BOK Manipulation

WikiBOKers can edit a BOK tree and descriptions directly through a rich graphical user interface provided by the WikiBOK. For example, if a WikiBOKer wants to create a node, he/she right-clicks the mouse button on any place of the screen supported by the BOK Editor. Such BOK manipulation operations are classified into the following four categories: (1) node creation, deletion, and modification (re-naming) of the BOK tree; (2) edge creation, deletion, and modification of the BOK tree; (3) description creation, deletion, and modification; and (4) undo, redo, and commit instructions of the editing work. Notice that if a WikiBOKer clicks a node to delete, then its sub-tree is also deleted. This is because the relationship between the higher node and the lower node of the BOK tree is a subsumptive containment hierarchy, i.e., a lower-level object "is a" member of the higher class; we delete the entire sub-tree if its root node is deleted. If a WikiBOKer clicks an edge for the purpose of deletion, then the edge is deleted and the sub-tree connected to the edge becomes free in the sense that it is listed in the "pool" area on the screen for subsequent use.

4 Edit Conflict Issues

4.1 Resloving Edit Conflict

An edit conflict is an unavoidable phenomenon when we support open collaboration. Therefore, a resolving policy and mechanism is to be investigated. Thus far, "diff" has been a well-known file comparison utility that is used for showing the changes between one version of a file and the former version of the same file. There are utilities such as "tdiff" for the purpose of a directory comparison and "xmldiff" for the purpose of a comparison of XML documents. Version control in software development also closely relates to this issue. Apache Subversion (SVN) is a software versioning and revision control system that does not have file locking. The "check-out" and "check-in" concepts found in the version control are also used in the realization of the concurrency control mechanism of object-oriented database management systems.

An edit conflict frequently occurs in Wikipedia editing, and in order to resolve the edit conflict among Wikipedians, it obeys the edit conflict resolution rule introduced by "wiki," which states that "If two or more people try to edit a page simultaneously, the first one to save wins and the rest are notified that they can't save. Should you be so notified, try opening a new edit window in a new browser and copying your changes to it." (http://c2.com/cgi/wiki?EditConflictResolution)

We follow the same principle. However, we need to manage not only the description edit conflict but also the BOK tree edit conflict. This is essentially different from the Wikipedia situation and is discussed below:

4.2 Edit Conflict Resolving Principle of BOK Tree

In order for WikiBOKers to not be disturbed by handling a wiki page edit conflict and a BOK tree editing conflict differently, we adopt the same resolution principle to handle both. Since an SMW is used in a WikiBOK, we adopted the resolution principle proposed by wiki, which has been stated in the previous section. A sample of this principle is illustrated as follows:

Suppose that WikiBOKers Alice, Bob, and Carol checked out a BOK tree with version V0 at around the same time. Suppose that Alice completed her work and checked it in at time t1. Since there is no edit conflict, her update is accepted, and the tree version becomes V1. Then, suppose Bob finished his work and issued the check-in command to make the BOK tree of V0 to V2. This request is not accepted because the version of the tree had has already changed from V0 to V1. Then, the system asks Bob to merge his version with that of Alice, i.e., V1, which might yield a BOK tree with version V3. Such a situation is shown in Fig. 5.

Of course, to help Bob in the merging process, the difference in the BOK trees edited by Alice and Bob should be shown to him. In general, calculating the difference between the two trees seems a time-consuming task. However, since a WikiBOK uses a dot file to represent a BOK tree, the tree difference can be easily obtained by taking the "diff" of the two dot files that represent the BOK trees saved by Alice and Bob, respectively.

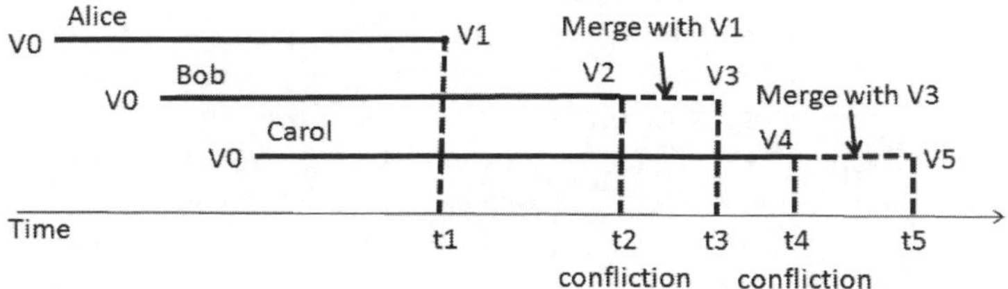

Fig. 5 Resolution principle of edit conflicts in wiki.

4.3 Resolving BOK Tree Edit Conflict in WikiBOK

BOK tree editing operations are classified into three categories: (1) insertion of a sub-tree into the BOK tree, (2) deletion of a sub-tree from the BOK tree, and (3) deletion of an edge from the BOK tree. Note that after operation (2), the deleted sub-tree is erased. In contrast, after operation (3), the deleted sub-tree is "pooled" for future use and is displayed on the same screen. As mentioned before, since the relationship between the higher node and the lower node of the BOK tree is a subsumptive containment hierarchy, we delete the entire sub-tree if its root node is deleted.

Now suppose that Alice and Bob checked out the BOK tree with the same version number and Alice checked in before Bob did. As mentioned earlier, since the BOK tree edit functions are classified into three categories, we discuss how WikiBOKers behave when a BOK tree edit conflict occur. Table 1 summarizes their actions. The rows represent three cases taken by Alice, and the three columns represent how Bob behaves corresponding to the Alice's edit. In the matrix, P represents a sub-tree that Alice has inserted or deleted, while S represents a sub-tree that Bob has intended to insert or delete. Since P and S are sub-trees, we can calculate whether the intersection of P and S is empty or not. By "light merge" and "heavy merge," we mean that the merging process may be simple and complicated, respectively. For example, suppose that Alice inserted sub-tree P and this update was accepted so that a new version number is given to the BOK tree. After a while, Bob issued an update request to insert sub-tree S. Since Bob was using the same version tree as Alice used before, the WikiBOK asked Bob to merge his update request with the BOK tree updated by Alice. In this case, the merge task seems "light" because P and S are disjoint. However, in the case where P and S are not disjoint, it is presumed that the merge task seems "heavy." Table 1 summarizes the BOK tree edit conflict resolution scheme of the WikiBOK, where Alice and Bob represent WikiBOKers who issue the BOK tree update earlier and later, respectively. P and S represent sub-trees that Alice and Bob wanted to update. "Delete Edge to Sub-tree P" means the deletion of an edge to which sub-tree P is connected.

Table 1 BOK Tree Resolution Principle in WikiBOK

Bob / Alice	Insert Sub-tree S		Delete Sub-tree S		Delete Edge to Sub-tree S	
Insert Sub-tree P	$P \cap S = \varphi$	Light merge	$P \cap S = \varphi$	Light merge	$P \cap S = \varphi$	Light merge
	$P \cap S \neq \phi$	Heavy merge	$P \cap S \neq \phi$	Heavy merge	$P \cap S \neq \phi$	Heavy merge
Delete Sub-tree P	$P \cap S = \varphi$	Light merge	$P \cap S = \varphi$	Light merge	$P \cap S = \varphi$	Light merge
	$P \cap S \neq \phi$	Heavy merge	$P \cap S \neq \phi$	Heavy merge	$P \cap S \neq \phi$	Heavy merge
Delete Edge to Sub-tree P	$P \cap S = \varphi$	Light merge	$P \cap S = \varphi$	Light merge	$P \cap S = \varphi$	Light merge
	$P \cap S \neq \phi$	Heavy merge	$P \cap S \neq \phi$	Heavy merge	$P \cap S \neq \phi$	Heavy merge

5 Conclusions

In this paper, we reported a WikiBOK that is a Wiki-based BOK (body of knowledge) formulation-aid system. We explained how the WikiBOK was designed and implemented using an enhanced SMW (Semantic MediaWiki) with Graphviz. Because we cannot avoid the edit conflict issue, the resolution of the BOK tree edit conflicts was investigated. The WikiBOK is currently under development under our project, and it is expected that the entire figure of the SIBOK (social informatics body of knowledge) will appear as a collective work result of academics with various related fields of social informatics.

Acknowledgement. This research is partly supported by the Grant-in-Aid from the Aoyama Gakuin University Research Institute and the Grant-in-Aid for Scientific Research (B) 22300036 from the MEXT of Japan. Thanks also to Mr. Yuho Ogashiwa and Ms. Saya Tokuyama for their contributions to system implementation.

References

1. The Joint Task Force on Computing Curricula of IEEE Computer Society and Association for Computing Machinery. Computing Curricula 2001 Computer Science — Final Report –, 236 p (December 15, 2001)
2. Masunaga, Y., Shoji, Y., Ito, K.: A Wiki-based Collective Intelligence Approach to Formulate a Body of Knowledge (BOK) for a New Discipline. In: Proceedings of WikiSym 2010, Gdansk, Poland, article no 11 (July 2010)
3. Surowiecki, J.: The Wisdom of Crowds: Why the Many Are Smarter Than the Few and How Collective Wisdom Shapes Business, Economies, Societies and Nations (2004)
4. Leuf, B., Cunningham, W.: The Wiki Way: Quick Collaboration on the Web. Addison Wesley (2001)
5. Lih, A.: The Wikipedia Revolution: How a Bunch of Nobodies Created the World's Greatest Encyclopedia, Hyperion (2009)

Secure Spaces and Spatio-temporal Weblog Sensors with Temporal Shift and Propagation

Shun Hattori

Abstract. This paper defines three kinds of Weblog Sensors to mine the Web, especially CGM such as Weblog documents for spatio-temporal data about a target phenomenon in the physical world, and tries to validate the potential and reliability of these Weblog Sensors' spatio-temporal data by measuring the correlation with weather statistics of Japan Meteorological Agency as real-world data.

1 Introduction

Many researches on mining the Web, especially CGM (Consumer Generated Media) such as Weblogs, for knowledge about various phenomena and events in the real world have been done very actively [2,3,4,5,6,7]. Meanwhile, Web services with the Web-mined knowledge have begun to be developed for the public, and more and more ordinary people actually utilize them as information for choosing better products, services, and actions in the real world. However, there is no detailed investigation on how accurately Web-mined data about a phenomenon or event held in the real world reflect real-world data. Therefore, while choosing better products, services, and actions in the real world, it must be problematic to idolatrously utilize the Web-mined data in public Web services without ensuring their accuracy sufficiently.

This paper defines three kinds of Weblog Sensors [8] to mine the Web, especially CGM such as Weblog documents for spatio-temporal data about a target phenomenon (e.g., precipitation and temperature) in the physical world. And then this paper tries to validate the potential and reliability of these Weblog Sensors' spatio-temporal data by measuring the correlation coefficient with precipitation and temperature statistics of JMA (Japan Meteorological Agency)[1] as real-world data.

Shun Hattori
School of Computer Science, Tokyo University of Technology, 1404–1 Katakura-machi, Hachioji, Tokyo 192–0982, Japan
e-mail: hattori@cs.teu.ac.jp

[1] http://www.jma.go.jp/jma/indexe.html

F.L. Gaol (Ed.): Recent Progress in DEIT, Vol. 2, LNEE 157, pp. 343–349.
springerlink.com

2 Method

This section constructs three kinds of Weblog Sensors to mine the Web, especially Weblog documents for spatio-temporal data about a target phenomenon in the physical world (e.g., "precipitation" and "temperature"). First, I define the basic Weblog Sensor with a geographical space s (e.g., 47 prefectural capitals in Japan such as "Tokyo" and "Kyoto"), a time period t (e.g., a day such as "2007/1/1" and "2009/12/31"), and a keyword kw (e.g., "ame" and "atsui" in Japanese that mean "rain" and "hot" in English respectively) for a target phenomenon. Next, I define the temporal-shifted Weblog Sensors with a temporal shift parameter d (days). Last, I define the temporal-propagated Weblog Sensors with a temporal propagation parameter σ or with temporal propagation parameters σ_b and σ_a.

First, the value $\mathrm{ws}(s,t,kw)$ indicated by the basic Weblog Sensor with a geographical space s, a time period t, and a keyword kw for a target phenomenon is defined as follows:

$$\mathrm{ws}(s,t,kw) := \mathrm{bf}_t(\texttt{[}s\ \texttt{AND}\ \ kw\texttt{]}),$$

where $\mathrm{bf}_t(\texttt{[}q\texttt{]})$ stands for the Frequency of weBlog documents searched by submitting the query q with the custom time range t to Google Blog Search.

Next, the value $\mathrm{ws}_d(s,t,kw)$ indicated by the temporal-shifted Weblog Sensor with a temporal shift parameter d (days), a geographical space s, a time period t, and a keyword kw for a target phenomenon is defined as follows:

$$\mathrm{ws}_d(s,t,kw) := \mathrm{ws}(s,t+d,kw).$$

Last, the value $\mathrm{ws}^{\sigma}(s,t,kw)$ indicated by the temporal-propagated Weblog Sensor with a temporal propagation parameter σ, a geographical space s, a time period t, and a keyword kw for a target phenomenon is defined as follows:

$$\begin{aligned}
\mathrm{ws}^{\sigma}(s,t,kw) &:= \sum_{\forall d} \mathrm{ws}_d(s,t,kw)\cdot p^{\sigma}(d),\\
p^{\sigma}(d) &:= N(0,\sigma,d),\\
N(\mu,\sigma,d) &:= \frac{1}{\sqrt{2\pi}\sigma}\exp\left(-\frac{(d-\mu)^2}{2\sigma^2}\right),
\end{aligned}$$

where $N(\mu,\sigma,d)$ stands for the Normal Distribution with an average μ and a standard deviation σ. In the below-mentioned experiment, $\forall d$ is restricted to $[-30,30]$.

The another value $\mathrm{ws}^{\sigma_b,\sigma_a}(s,t,kw)$ indicated by the temporal-propagated Weblog Sensor with a temporal propagation parameter σ_b before a time period t, the other parameter σ_a after the time period t, a geographical space s, and a keyword kw for a target phenomenon is defined as follows:

$$\begin{aligned}
\mathrm{ws}^{\sigma_b,\sigma_a}(s,t,kw) := &\sum_{\forall d<0} \mathrm{ws}_d(s,t,kw)\cdot p^{\sigma_b}(d)\\
&+\ \mathrm{ws}_0(s,t,kw)\cdot\frac{p^{\sigma_b}(0)+p^{\sigma_a}(0)}{2} + \sum_{\forall d>0} \mathrm{ws}_d(s,t,kw)\cdot p^{\sigma_a}(d).
\end{aligned}$$

3 Experiment

This section shows several experimental results to validate the basic Weblog Sensor, the temporal-shifted Weblog Sensors, and the temporal-propagated Weblog Sensors in the case of the target phenomenon "precipitation" or "temperature" in the real world. The whole experiments evaluate the correlation coefficient between real statistics by JMA (Japan Meteorological Agency) as a physical sensor and spatio-temporal data mined by the proposed Weblog Sensors.

Fig. 1 shows the average of correlation coefficient between the JMA's precipitation and the temporal-shifted Weblog Sensor $ws_d(s,t,kw)$ with a temporal shift parameter d (days) using a Japanese noun "ame" ("rain" in English) as a positive correlated keyword to the target phenomenon "precipitation", a Japanese noun "hare" ("shine" in English) as a negative correlated keyword to the target phenomenon "precipitation", and Japanese adjectives "samui" and "atsui" ("chill" and "hot" in English) as non-correlated keywords to the target phenomenon "precipitation" for each space $s \in 47$ prefectural capitals in Japan and each day $t \in$ January 1st, 2007 to December 31st, 2009. In only Fig. 1(a) from among these four figures, the temporal-shifted Weblog Sensor with only $d = 0$ which is equivalent to the basic Weblog Sensor shows positive correlation with JMA's precipitation to some extent, the temporal-shifted Weblog Sensors with $d \in \{-1, 1\}$ show a little positive correlation, i.e., the temporal shift parameter d is ineffective to improve the correlation of the Weblog Sensor at least in the case of the target phenomenon "precipitation".

Fig. 2 shows the average of correlation coefficient between the JMA's temperature and the temporal-shifted Weblog Sensor $ws_d(s,t,kw)$ with a temporal shift d (days) using a Japanese adjective "atsui" ("hot" in English) as a positive correlated keyword to the target phenomenon "temperature", a Japanese adjective "samui" ("chill" in English) as a negative correlated keyword to the target phenomenon "temperature", and "ame" and "hare" ("rain" and "shine" in English) as non-correlated keywords to the target phenomenon "temperature" for each space $s \in 47$ prefectural capitals in Japan and each day $t \in$ January 1st, 2007 to December 31st, 2009. In Fig. 2(a) among these four figures, the temporal-shifted Weblog Sensor with any d shows much more positive correlation with JMA's temperature than in the case of the target phenomenon "precipitation", but gives the best performance when $d = 0$. Meanwhile, in Fig. 2(b) among these four figures, the temporal-shifted Weblog Sensor with any d shows negative correlation with JMA's temperature to some extent, and gives the best performance when $d = -15$. So, the temporal shift parameter d has a possibility to improve the correlation of the Weblog Sensor at least in the case of the target phenomenon "temperature".

Fig. 3(a) shows the average of correlation coefficient between the JMA's precipitation and the temporal-propagated Weblog Sensor $ws^{\sigma}(s,t,kw)$ with a temporal propagation parameter σ using a Japanese noun "ame" ("rain" in English) as a keyword representing the target phenomenon "precipitation" for each space $s \in 47$ prefectural capitals in Japan and each day $t \in$ January 1st, 2007 to December 31st, 2009. Fig. 3(b) shows the average of correlation coefficient between the JMA's precipitation and the temporal-propagated Weblog Sensor $ws^{\sigma_b,\sigma_a}(s,t,kw)$ with temporal

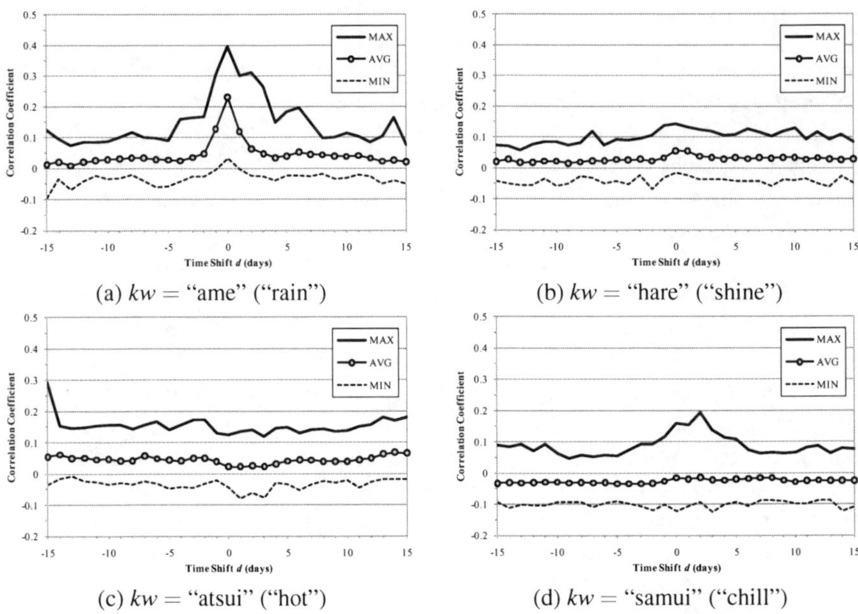

(a) kw = "ame" ("rain")   (b) kw = "hare" ("shine")

(c) kw = "atsui" ("hot")   (d) kw = "samui" ("chill")

Fig. 1 Correlation between JMA's precipitation & temporal-shifted Weblog Sensor $\mathrm{ws}_d(s,t,kw)$.

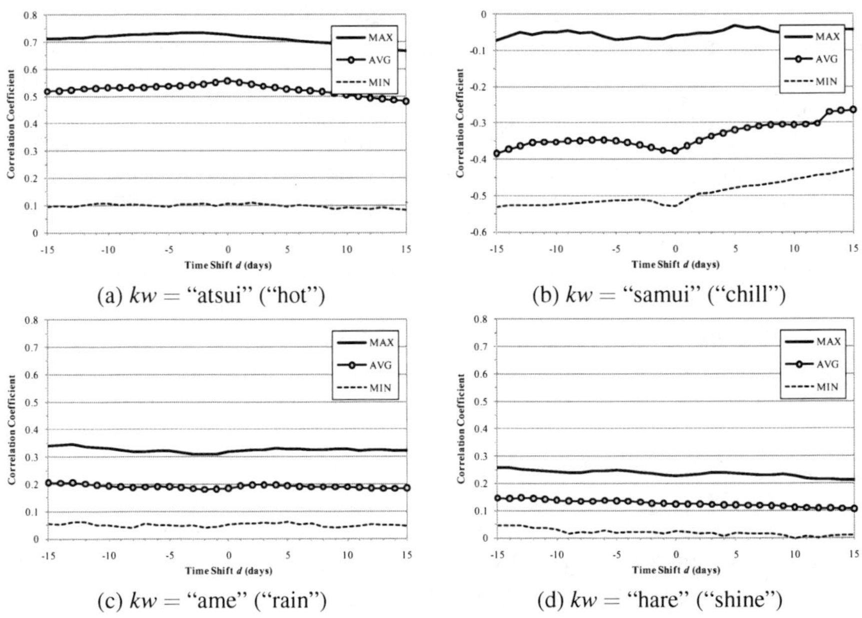

(a) kw = "atsui" ("hot")   (b) kw = "samui" ("chill")

(c) kw = "ame" ("rain")   (d) kw = "hare" ("shine")

Fig. 2 Correlation between JMA's temperature & temporal-shifted Weblog Sensor $\mathrm{ws}_d(s,t,kw)$.

pr-opagation parameters σ_b before a time period t and σ_a after the time period t using a Japanese noun "ame" ("rain" in English) as a keyword representing the target phenomenon "precipitation" for each space $s \in 47$ prefectural capitals in Japan and each day $t \in$ January 1st, 2007 to December 31st, 2009. These figures show that the basic Weblog Sensor $\mathrm{ws}(s,t,kw)$ is superior to any temporal-propagated Weblog Sensor $\mathrm{ws}^{\sigma_b,\sigma_a}(s,t,kw)$ in the case of the target phenomenon "precipitation" and that the basic Weblog Sensor is followed by the temporal-propagated Weblog Sensor $\mathrm{ws}^{0.4,0.5}(s,t,kw)$, $\mathrm{ws}^{0.3,0.4}(s,t,kw)$, and $\mathrm{ws}^{0.5,0.6}(s,t,kw)$, i.e., the performance of the temporal-propagated Weblog Sensors is not monotonic decreasing dependent on a temporal propagation parameter σ, σ_b, or σ_a.

Fig. 4(a) shows the average of correlation coefficient between the JMA's temperature and the temporal-propagated Weblog Sensor $\mathrm{ws}^{\sigma}(s,t,kw)$ with a temporal propagation parameter σ using a Japanese adjective "atsui" ("hot" in English) as a representation of the target phenomenon "temperature" for each space $s \in 47$ prefectural capitals in Japan and each day $t \in$ January 1st, 2007 to December 31st, 2009. Fig. 4(b) shows the average of correlation coefficient between the JMA's temperature and the temporal-propagated Weblog Sensor $\mathrm{ws}^{\sigma_b,\sigma_a}(s,t,kw)$ with temporal propagation parameters σ_b before a time period t and σ_a after the time period t using a Japanese adjective "atsui" ("hot" in English) as a representation of the target phenomenon "temperature" for each space $s \in 47$ prefectural capitals in Japan and each day $t \in$ January 1st, 2007 to December 31st, 2009. These figures show that any temporal-propagated Weblog Sensor $\mathrm{ws}^{\sigma_b,\sigma_a}(s,t,kw)$ is superior to the basic Weblog Sensor $\mathrm{ws}(s,t,kw)$ in the case of the target phenomenon "temperature" unlike the case of the target phenomenon "precipitation", that the temporal-propagated Weblog Sensor $\mathrm{ws}^{20,20}(s,t,kw) = \mathrm{ws}^{20}(s,t,kw)$ gives the best performance and gains about 25% of the basic Weblog Sensor, and that the performance of the temporal-propagated Weblog Sensors is monotonic increasing dependent on σ but not always monotonic increasing dependent on a temporal propagation parameter σ_b or σ_a.

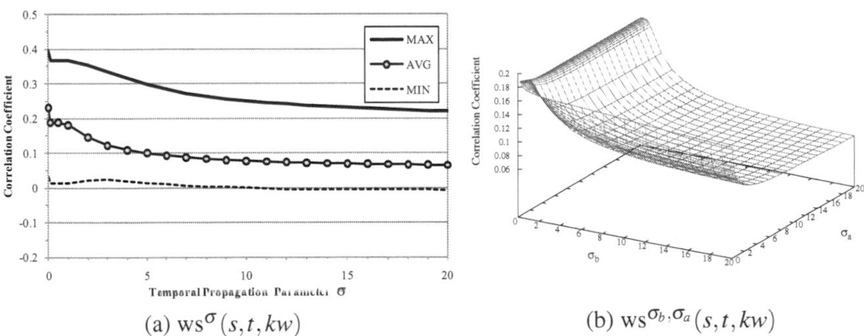

(a) $\mathrm{ws}^{\sigma}(s,t,kw)$ (b) $\mathrm{ws}^{\sigma_b,\sigma_a}(s,t,kw)$

Fig. 3 Correlation between JMA's precipitation & temporal-propagated Web Sensors, *kw*="ame".

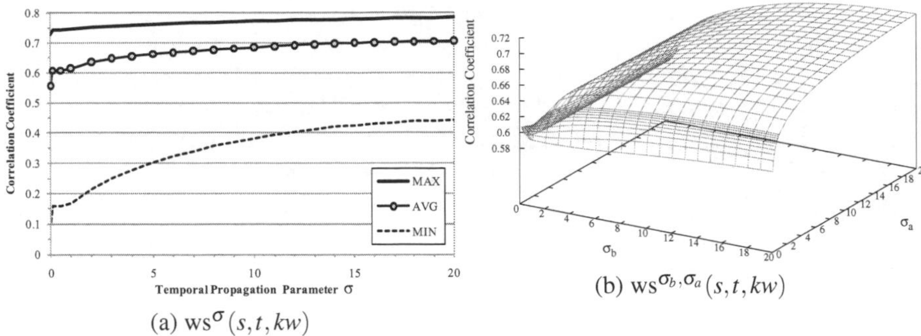

(a) $ws^{\sigma}(s,t,kw)$

(b) $ws^{\sigma_b,\sigma_a}(s,t,kw)$

Fig. 4 Correlation between JMA's temperature & temporal-propagated Web Sensor, *kw*="atsui".

4 Conclusion

This paper has defined three kinds of Weblog Sensors to mine the Weblog documents as CGM for spatio-temporal data about a target phenomenon (e.g., precipitation and temperature) in the real world: the basic Weblog Sensor, the temporal-shifted Weblog Sensors, and the temporal-propagated Weblog Sensors. And also this paper has validated some potential and reliability of these Weblog Sensors' spatio-temporal data by measuring the correlation coefficient with weather (precipitation and temperature) statistics of Japan Meteorological Agency as real-world data.

Acknowledgements. This work was supported in part by JSPS Grant-in-Aid for Young Scientists (B) "A research on Web Sensors to extract spatio-temporal data from the Web" (23700129).

References

1. Hattori, S., Tanaka, K.: Towards building secure smart spaces for information security in the physical world. Journal of Advanced Computational Intelligence and Intelligent Informatics (JACIII) 11(8), 1023–1029 (2007)
2. Dave, K., Lawrence, S., Pennock, D.M.: Mining the peanut gallery: Opinion extraction and semantic classification of product reviews. In: Proc. of WWW 2003, pp. 519–528 (2003)
3. Tezuka, T., Kurashima, T., Tanaka, K.: Toward tighter integration of Web search with a geographic information system. In: Proceedings of WWW 2006, pp. 277–286 (2006)
4. Hattori, S., Ohshima, H., Oyama, S., Tanaka, K.: Mining the Web for Hyponymy Relations Based on Property Inheritance. In: Zhang, Y., Yu, G., Bertino, E., Xu, G. (eds.) APWeb 2008. LNCS, vol. 4976, pp. 99–110. Springer, Heidelberg (2008)
5. Hattori, S., Tanaka, K.: Extracting concept hierarchy knowledge from the Web based on property inheritance and aggregation. In: Proceedings of WI 2008, pp. 432–437 (2008)

6. Hattori, S., Tezuka, T., Tanaka, K.: Mining the Web for Appearance Description. In: Wagner, R., Revell, N., Pernul, G. (eds.) DEXA 2007. LNCS, vol. 4653, pp. 790–800. Springer, Heidelberg (2007)
7. Hattori, S.: Peculiar image search by Web-extracted appearance descriptions. In: Proceedings of SoCPaR 2010, pp. 127–132 (2010)
8. Hattori, S., Tanaka, K.: Mining the Web for access decision-making in secure spaces. In: Proceedings of SCIS&ISIS, TH-G3-4 (2008)

Neighbor Information Table-Based Handoff Research for Mobile IPv6

Zhang Lin, Qiu Shu-Wei, Zhou Jian, and Huang Jian

Abstract. To analysis the standard mobile IPv6 (MIPv6)fast handoff scheme, an neighbor information table-based fast handoff scheme for Mobile IPv6 was proposed. Through pre-configured and regularly updated way to let MN get ahead of target limited areas of information table-- neighbor information table, Switching process with a neighbor information table to shorten the uniqueness of the care-of address verification and testing of time delay. The simulation results show that this mechanism does not take up additional network resources, Reduce the switching process of ping-pong exchange and cut down the switching process of switching delay time, this scheme is a kind of superior performance of mobile IPv6 fast handoff scheme.

Keywords: Mobile IPv6, Neighbor Information Table, Fast Handover.

1 Introduction

Along with mobile communication technology rapidly expand, It is hoped that the process of moving can maintain a continuous Internet access and communications. Mobile IPv6 protocol (MIPv6, the Mobile IPv6)[1] is a next generation network technology to achieve a seamless roaming. It enables the mobile node (MN, the Mobile node) in heterogeneous networks and roaming freely. In November 1996, IETF published the first draft of a mobile IPv6 protocol, in June 2004, the mobile IPv6 protocol (MIPv6) to submit a standard[1]. Mobile IPv6 technology has become one of the hot spots, but there is existing switching delay large groups and packet loss rate of defects in mobile IPv6 protocol, it can not meet the large delay-sensitive real-time applications and applications sensitive to data loss [2]. In response to these problems, a neighbor information table based fast handoff

Zhang Lin · Qiu Shu-wei
Anhui Institute of Architecture and Industry Information & Network Center, Hefei, China, 230022
e-mail: zl@aiai.edu.cn, qiusw@aiai.edu.cn

F.L. Gaol (Ed.): Recent Progress in DEIT, Vol. 2, LNEE 157, pp. 351–356.
springerlink.com

mechanism for mobile IPv6 was proposed. Neighbor information table with this mechanism reduces the care-of address sole verification time and detection delay time, to reduce packet loss and reduce the delay effect.

2 Analysis of Mobile IPv6 Fast Handover Mechanism

In order to remains in the transport layer connection when the mobile node at move time, mobile nodes must always maintain a fixed IP address. This address is home address. Home address is used to identify a static address to connect end to end, no matter where the mobile node moves to. On the other hand, the mobility of mobile nodes, in order to communicate smoothly, the mobile node must also be bound to another IP address, that is, care-of address. In basically mobile IPv6 protocol, the mobile node handoff between different networks need movement detection, new forwarded address configuration, duplicate Address Detection, binding registration process which resulted in large switching delay[3][4]. When the mobile node left the home network to handoff the outside areas, Mobile node detects that its network has been switched to the new position, Configure the current network of care-of address, Duplicate Address Detection verify the address uniqueness, send binding update message to the home agent [5][6]. Home agent binding the mobile node's home address and care-of address,send a binding acknowledgment message to mobile node.The mobile node send binding update message to communication node in order to directly establish a connection with the communication nodes. The basic Mobile IPv6 protocol exist long switching delay time, high data loss rate, large signaling load such problems.

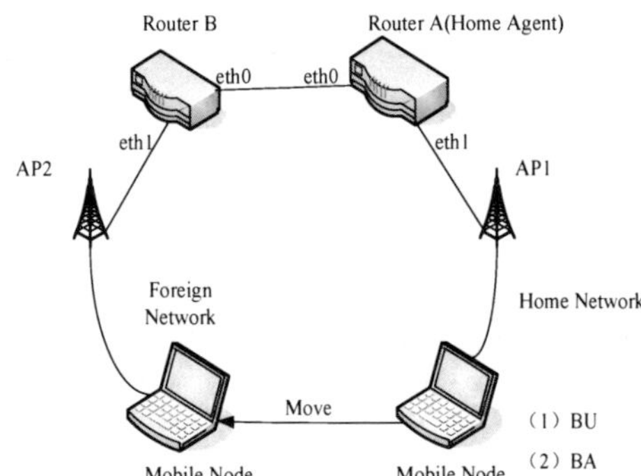

Fig. 1 Basic Mobile IPv6 protocol switching process

Basic mobile IPv6 handover delay, including mobile detection delay, care-of address configuration and duplicate address detection register delay and binding registration delay mobile detection delay Use T_{Detect} to represent. Care-of address configuration latency refers to the mobile node configuring a new network of care-of address, do duplicate address detection to ensure that the time required to address unique, Use T_{Coa} to represent. Binding registration latency refers to mobile node to switch to the new network, the home agent sends binding update message, registered its current care-of address, and wait for the home agent sends back confirmation that the process of binding the time, Use T_{BU} to represent. Network layer switching delay including T_{Detect}, T_{Coa} and T_{BU}. Use T_{L_3} to represent. That is $T_{L3}=T_{Detect}+T_{Coa}+T_{BU}$. Link layer handover latency Use T_{L2} to represent. The total switching delay of the mobile node including T_{L2} and T_{L3}. The total delay time $T=T_{L2}+T_{L3}$. Network layer handoff latency which is longer, is the impact of mobile IPv6 handover performance of the main factors.

3 Based on Neighbor Information Table Switching Mechanism for Mobile IPv6

According to the analysis can be seen above, the network layer handoff latency is a key factor to constrain mobile IPv6 handover performance. When the MN from one AP coverage area to another, AP through radio messages between the MN to confirm its Regional ownership. This process will produce a large number of broadcast messages, take up a lot of the communication channel, Because the MN's mobility are not sure what it might lead to reach the AP region, so this process will take up a lot of time, causing delays. Although the MN moving target coverage area is uncertain, but it is only likely to move to its neighboring AP coverage area(this is called a neighbor). If it can not be moved to the neighbor AP coverage area, then communication interruption occurs, the so-called dead zone of communication. When the MN in a mobile AP coverage area, it may move to the target AP coverage area is limited kinds of potential. The MN's mobile regional, and may switch to enter the area, is a finite set, this set is defined as MN Mobile area set.

According to the analysis above, Establish a mathematical model. MN Mobile area set to represent the finite set Φ. Φ's elements, including the current AP, Use a to represent, as well as all the neighbors AP, to express with Φ1, It is a subset of Φ. Set of relations between them can be expressed as: Φ={a,Φ1}. For unknown, Including the temporary or new AP, A finite set of elements in MN Mobile Regional, At this point Φ1 elements increase with the number of new AP. Next, describe the elements Φ, MN Mobile area for a limited element, Information can be abstracted as a neighbor node. This finite set can be described as a finite number of neighbor node information to form a neighbor information table.

Table 1 Structure of a neighbor node information

<table>
<tr><td colspan="2">ver</td><td colspan="2">len</td><td>AP</td><td colspan="2">Reserve</td></tr>
<tr><td>IP1</td><td>flag</td><td>IP2</td><td>flag</td><td>……</td><td>IPn</td><td>Flag</td></tr>
</table>

ver is the IPv4 or IPv6 recognition sites, len is the length of the node; AP is the AP's basic information structure; Neighbor information table based on mobile

IPv6 switching process (hereinafter referred to as NFMIPv6). Whether through broadcast messages for MN to move to the critical area. When switching, the link layer of communication continue to use the home address, The network layer information directly from the neighbor table to find a care-of address, With hash algorithms to find a neighbor information table in the care-of address, can guarantee the uniqueness of agency address, Reduce the standard Mobile IPv6 proxy address unique verification time. MN to obtain proxy address, then send binding and registration information to the home network, mobile switching completed.

Mobile switching algorithm described as follows:

```
NFMIPv6HandOver()
{Initiate Handover = false;
 Read configure file;
 Create neighbor information by Hash function;
 Initiate Update Timer;// Initialize timer
 Initiate Channel scanning;
 Initiate MN;
 While (MN Scanning Channel)
     {Read probe responses;
           If   (new AR)
           {HANDOVER = true;}
           Else
           {HANDOVER = false;}
     }
     If (true == HANDOVER)
     { Read Neighbour Information Table;
         Search Care-of address by HASH;/* HASH algorithms to find care-of
address information in the table of neighbors*/
   Reset Care-of-address of MN;
     }// Reset the MN's care-of address
     Else
     {   If (TIME is over)
         { Update Neighbor Information table;         }
     }
}
```

4 Simulation and Analysis

In NS-2 simulation environment, Add FHMIPv6 module [7] [8] to the NS-2.31 version, Modify the protocol to adapt this simulation requirements.

4.1 Simulation Scenarios

Network simulation scenario shown in Figure 2, a total of 11 nodes, each node bandwidth and latency indicated in the figure. NS from the 0.0s running, and ends in 80.0003s. CN send data, MN to receive data. Map region's wired link transmission delay is 2 ms, the total simulation time is 80 s. MN with 1m / s speed to move and through the adjacent AR. CN in the simulation 10 ms intervals beginning with 256-byte UDP packet sent to the MN, until the simulation ended.

Fig. 2 Network simulation topology

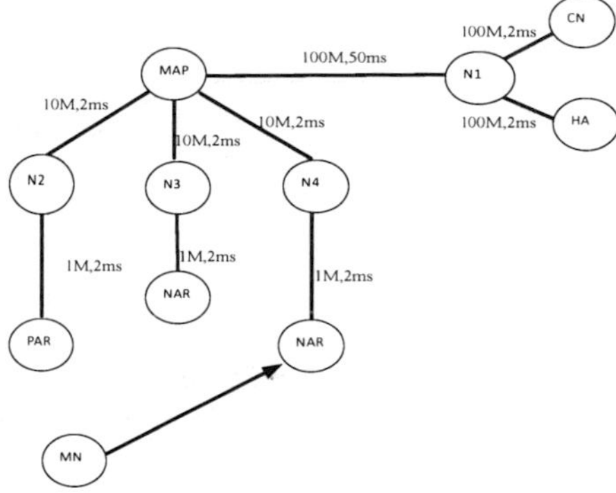

4.2 Simulation Performance

As can be seen from Figure 3, with the increase in the number of mobile nodes, The switching delay is different of two kinks of switching mechanism. When the number of mobile nodes increased to 25 above, the channel has already reached saturation point, and radio link switching delay become serious. In the case of the optimal mechanism for switching NFMIPv6 delay of about 20% better than FMIPv6. the switching delay time reduced 43ms. This is because the switching mechanism NFMIPv6 according to a neighbor information table to obtain the MN's care-of address in advance to shorten the address of the only validation time. In the worst case performance, NFMIPv6 switching mechanism will degenerate into FMIPv6 handoff mechanism.

Fig. 3 The number of mobile nodes and the relationship between the transmission delay

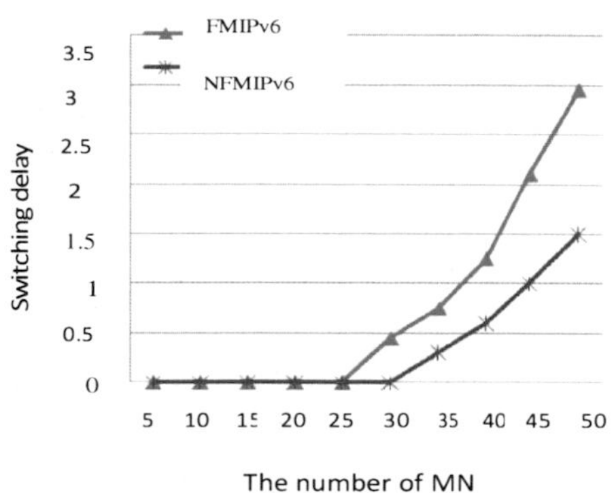

We can see from Figure 4, with the increase of network transmission delay, Two kinds of handoff packet loss rate varies, FMIPv6 handoff compared with NFMIPv6 handoff, the former has a larger packet loss rate. This is because the NFMIPv6 handoff reduce the Ping-pong exchange.

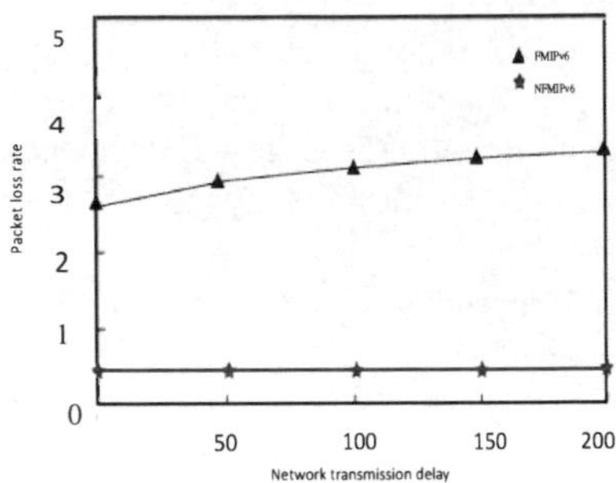

Fig. 4 The network transmission delay and packet loss rate relationship

5 Conclusion

By a small amount of information to ensure the burden does not increase network traffic, With auxiliary information to improve the mobile IPv6 handoff, and make the mobile node switching smoothly. Neighbor information table will not take up too much resources, This article designed the smallest neighbor information table. Shorten the care-of address verification time and detection delay time. Using NS-2 simulation software to evaluate the performance of the NFMIPv6 handoff. According to the simulation results, NFMIPv6 mechanism for switching latency and packet loss rate lower than the FMIPv6, Switching latency and packet loss rate is affected by the impact of network transmission delay is very small, Can achieve the smooth switching requirements, so this mechanism is feasible.

References

1. Johnson, D., Perkins, C., Arkko, J.: Mobility Support in IPv6. RFC3775. Internet Engineering Task Force (June 2004)
2. IETF RFC 4140, Hierarchical mobile IPv6 mobility management
3. Fast handovers for mobile IPv6. Internet. Draft, `http://www.packetizer.com/rfc/rfc4068/`
4. Hierarchical Mobile IPv6 mobility management (HMIPv6), `http://datatracker.ietf.org/doc/rfc5380/`
5. Fast handover support in Hierarchical mobile IPv6. In: International Conference on Advanced Communication Technology (ICACT), Phoenix, Park, Korea, vol. (2), pp. 551–554 (February 2004)
6. Park, B., Lee, S., Latchman, H.: A fast neighbor discovery and DAD scheme for fast handover in mobile IPv6 networks. In: Proceedings of International Conference on Systems and International Conference on Mobile Communications and Learning Technologies, Washington, pp. 201–201 (2006)
7. Natalizio, E., Scicchitano, A., Marano, S.: Mobility anchor point selection based on user mobility in HMIPv6 integrated with fast handover mechanism. In: IEEE Wireless Communications and Networking Conference, New Orleans, LA, USA, March 12-17, vol. 3, pp. 1434–1439 (2005)
8. Hsieh, R., Seneviratne, A.: A compareson of mechanisms for improving Mobile IP handoff latency for end-to-end TCP. In: Proceedings of the 9th Annual Internationl Conference on Mobile Computing and Networking, San Diego, California, USA, September 14-19, pp. 29–41 (2003)

Channel Assignment in Wireless Networks Using Simple Genetic Algorithms

Li Ming Li

Abstract. Channel assignments in wireless telecommunication and mobile computing have been studied using a simple genetic algorithm. In genetic algorithm, each cell populates all channels and the fitness function is set a function as channel interferences. Co-channels constraints, adjacent channels constraints, and co-site channel constraints are considered and represented as an energy function. The minimum energy for each available channel is reported as the results of genetic algorithm. The channel with minimum energy is assigned to the incoming call. The simulation results show that genetic algorithm is better than fixed channel assignment in charging calls blocking problem.

Keywords: Channel Assignment, Wireless Networks, Genetic Algorithms.

1 Introduction

Recently, the wireless telecommunication and mobile computing have been developed rapidly, thus making the growth of population of mobile users tremendous. On the other hand the bandwidth of radio spectrum that can be assigned to is limited. In order to use available frequency spectrum efficiently, optimal channel assignment becomes one of the important issues in wireless networks [1-5].

Although the available resource is limited, the re-use technique can make it enough to cover a large service by using the same frequencies in spatially cells. The re-use technique must satisfied three constraints, Co-channel constraint (CCC), adjacent channel constraint (ACC) and co-site constraint (CSC) [1]. The channel assignment is the assignment of the required number of frequency channels to each cell such that the above constraints are satisfied. Neural networks and simulated annealing have been considered for solutions, which are based on the behavior of the neurons and physical process, respectively. But neural network only can provide local optimal value rather than global optimal value, and it must depend on the initial values [2]. For simulated annealing method, the major disadvantage is the long time to find optimal values.

Li Ming Li
Library of Henan Polytechnic University, Jiaozuo, China

F.L. Gaol (Ed.): Recent Progress in DEIT, Vol. 2, LNEE 157, pp. 357–362.
springerlink.com

While Genetic Algorithms can find a good solution quickly and it is an adaptive search technique that can find the global optimal solutions by manipulating and generating recursively a new population of solutions from an initial population of sample solutions [1]. GA's operators such as crossover and mutation are developed by a natural selection. The project simulated the channel assignment by using GA, which display a great advantage than previous technique.

2 Genetic Algorithms

Genetic algorithms were inspired by Darwin's theory about evolution, which is based on an analogy with nature [6]. It begins working with a set of solutions (represented by chromosomes) called population. Solutions from one population are taken and used to form a new population.

New population will be better than the old one.

Solutions selected to form new solutions (offspring) are selected according to their fitness - the more suitable they are the more chances they have to reproduce. GA is an adaptive method that can be used to solve search and optimization, and it is robust and can deal successfully with a wide range of problem areas such as image processing, routing problems.

The basic idea of GA is combining good characters from different ancestors to produce super fit offspring. The offspring will become better and better through the process. The algorithms are described below:

[Start] Generate random population of n chromosomes (suitable solutions for the problem)

[Loop] begin

[Fitness] Evaluate the fitness f(x) of each chromosome x in the population

[New population] Create a new population by repeating following steps until the new population is complete

[Selection] Select two parent chromosomes from a population according to their fitness (the better fitness, the bigger chance to be selected)

[Crossover] With a crossover probability, offspring crosses over the parents to form a new offspring (children). If no crossover was performed, offspring is an exact copy of parents.

[Mutation] with a mutation probability, mutate new offspring at each locus (position in chromosome).

[Accepting] Place new offspring in a new population

[Replace] Use new generated population for a further run of algorithm

[Test] if the end condition is satisfied, stops, and returns the best solution in current population

[Loop] Go to step 2

The two basic operations in GA are crossover and mutation, and we can get better offspring through them.

Crossover selects genes from parent chromo-somas and creates a new offspring. The simplest is to choose randomly some crossover point and everything before this point copy from a first parent and then everything after a crossover point copy from the second parent.

Crossover can then look like this (| is the crossover point) in Figure 1:

Chromosome 1	Chromosome 2
11011 | 00100110110	11011 | 11000011110
Offspring 1	Offspring 2
11011 | 11000011110	11011 | 00100110110

Fig. 1 Chromosome example

Mutation is to prevent falling all solutions in population into a local optimum of solved problems. It can Change the new offspring's randomly, shown in Figure 2:

Original offspring 1	Original offspring 2
1101111000011110	1101100100110110
Mutated offspring 1	Mutated offspring 2
1100111000011110	1101101100110100

Fig. 2 Mutation example

3 Channel Assignment

Firstly I would introduce the cellular architecture: There is a low power transmitter in each cell. Because the spectrum is limited, the frequency can be reuse, and cell can be split to increase capacity, and Handoff - moving to other cells. But there exist co-channel interference if transmission on the same frequency.

Reuse distance is the minimum distance between two cells using same channel without interference, this situation likes coloring graph. One frequency can be (re)used in all cells of the same color, and channel assignment is to minimize number of frequencies, such as colors in graphs.

There are three kinds of channel assignments: Fixed channel assignment, in which channels are permanently allocated to each cell for its exclusive use. While in dynamic channel assignment, channels are temporarily allocated to cells. Hybrid cannel assignment is the combination of these two methods.

The following three constraints we must consider when we assign channel to cells:

Co-channel constraint (CCC):

For a pair of cells, the same frequency can not be used simultaneously;

Adjacent channel constraint (ACC):

The adjacent frequencies can not be assigned to adjacent cells simultaneously

Co-site constraint (CSC):

Any pair of frequencies assigned to a cell should have a minimal distance between frequencies.

We can use the compatibility matrix C = (cij) to describe the situation in a mobile network. Where cij is the minimum frequency separation between a frequencies in cell i and cell j. If cij = 0, cell i and cell j can use the same channel, while 1, 2, 3 indicate CCC, ACC and CSC, respectively. The number of required channels for each cell i is represented by the demand vector M = (mi), and fik is the assigned frequency for the kth call in cell i. The Condition for the compatibility constraint is:

| fik-fik |>= cij

The channel assignment problem is to find the value of fik, which satisfies the constraint conditions when given the number of cells in the network N.

4 Simulation

A simple genetic algorithm (GA) has been implemented to solve the channel assignment problem. Generally, GA has two steps in the algorithm. First, the initial population is needed. Second, evaluating function is used to operate the solutions.

A population consists of a number of possible candidate solutions. In our simulation, the number of variables is set to the number of cells. The population for each cell is the number of total channels. We assign each cell with randomized integer value which should be smaller than the number of total channels.

The evaluation function is defined by energy functions which are represented by three constraint conditions mentioned above. CCC and ACC are considered together since both constraints can be represented by the value of cij when $i \neq j$. Energy function for ACC and CCC for each cell, EAC is given by

$$E_{AC} = \sum_{i=1}^{N} \sum_{k=1}^{mi} \sum_{j=1}^{N} \sum_{l=1}^{mj} V_{ikjl} \tag{1}$$

Where

Vikjl = 1 if |fik – fjl| < cij

$$V_{ikjl} = 1 if \mid fik - fjl \mid < cij \tag{2}$$

0 otherwise

i and j indicate the cell numbers and k and l are call numbers in each cell. Here Vikjl represents the satisfaction state for ACC and CCC. IF ACC or CCC is not satisfied, then Vikjl =1. On the other hand, if ACC or CCC is satisfied, then Vikjl=0.

For CSC, the energy function is given by

$$E_{CSC} = \sum_{i=1}^{N} \sum_{k=1}^{mi} V_{ik} \tag{3}$$

where

Vik = 1 if |fik-fi(k+1)| < cii

$$V_{ikjl} = 1 if \mid fik - fjl \mid < cij ,$$

or |fik-fi(k-1)| < cii (4)

0 otherwise

The total energy function E = ECSC + EAC is considered as evaluation function. So the problem to find the minimum number of channels can be used for a certain number of incoming calls is transferred to the one to find the minimum energy. When a call comes into a cell, the energy for each available channel is compared and the channel with minimum energy is chosen as the solution.

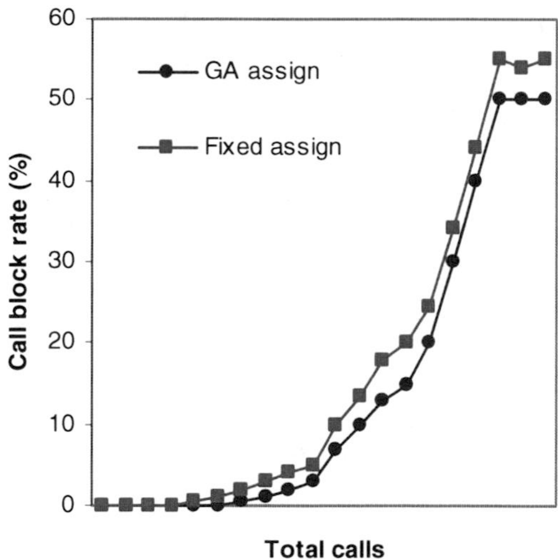

Fig. 3 Calls blocked rates

In Figure 3, the call block rates are shown as a function of total calls. In this example, calls arrived possibilities are simulated as a Poisson distribution. The total cells are 6 x 6, and total channels for each cell is 49. When no many calls arrive, both fixed channel assignment and genetic algorithm based channel assignment are satisfied with coming in calls, and no calls are blocked. Due to limited channels and co-channel constraints, fixed channel assignment has calls blocked problem when more calls arrive. The figure shows that GA based channels assignment is better than fixed channel assignment in charging calls blocked problem.

5 Conclusion

In this project, a simple genetic algorithm has been used to study channel assignment problem in wireless networks. In genetic algorithm, each cell populates all channels and the fitness function is set a function as channel interferences. Co-channels constraints, adjacent channels constraints, and co-site channel constraints are considered and represented as an energy function. The minimum energy for each available channel is reported as the results of genetic algorithm. The channel with minimum energy is assigned to the incoming call. The simulation results show that genetic algorithm is better than fixed channel assignment in charging calls blocked problem.

References

Kim, J.S., Park, S., Dowd, P., Nasrabadi, N.: Channel assignment in cellular radio using genetic algorithms. Wireless Personal Communication (1995) (under review)

Tissainayagam, D., Palaniswami, M., Everett, D.: Dynamic Channel Assignment Schemes Using Neural Networks. In: Computational Intelligence in Telecommunications Networks, ch. 17, p. 433 (2001)

Kassotakis, I., Markaki, M., Vasilako, A.: A hybrid genetic approach for channel reuse in multiple access telecommunication networks. In: Computational Intelligence in Telecommunications Networks, ch. 15, p. 383 (2001)

Makaya, C., Pierre, S.: Enhanced fast handoff scheme for heterogeneous wireless networks. Computer Communications 31(10), 2016–2029 (2008)

Lim, K.Y., Kumar, M., Das, S.K.: Message ring-based channel reallocation for cellular wireless networks. Computer Communications 23(5-6), 483–498 (2000)

Hefny, H.A., Bahnasawi, A.A., Abdel Wahab, A.H.: Logical radial basis function networks a hybrid intelligent model for function approximation. Advances in Engineering Software 30(6), 407–417 (1999)

Application Study of Brand Niche Theory in B2C Brands Overlap and Breadth Measurement

Xiuting Li, Geng Peng, Hong Zhao, and Fan Liu

Abstract. This paper uses theories in ecology for resource competition to study the brand competition in business. It provides a quantitative method to measure brand breadth and brand overlap and examines their relationships with brand competition. Four Chinese online bookstores are used as examples to illustrate the utility of the proposed method in evaluating their degree of competition. This study can help companies identify main competitors in the industry and the main aspects that they are competing in.

1 Introduction

Brand homogeneity has become a reality [4]. About 86% brands possess the same key properties [8]. Trends of increasing brand proliferation make positioning brands an increasingly difficult and important task for enterprises. As many brand experts put, brands can be taken as beings with life characteristics [6]. Brands consumed and competed for resources in brand ecosystem defined as a complex, dynamic, and constantly changing organic system by Winkler [12]. Wang also proposed the concept of "brand ecosystem" in China [10, 11]. Zhang further discussed internal diversity of eco-brands in his book "brand ecology" [13, 14]. Other domestic scholars like C. F. Huang, X. Wang, J. Lu, F. Chen, etc. also studied the ecological characteristics of brands. Marketers have long advocated the use of ecological theory to study brands, but the abstract definitions of key constructs

Xiuting Li
Management School, Graduate University of Chinese Academy of Sciences,
NO.80 Zhongguancun East Road Haidian District, Beijing, China
e-mail: lindaall@163.com

Geng Peng
Management School, Graduate University of Chinese Academy of Sciences,
NO.80 Zhongguancun East Road Haidian District, Beijing, China
e-mail: Penggeng@gucas.ac.cn

F.L. Gaol (Ed.): Recent Progress in DEIT, Vol. 2, LNEE 157, pp. 363–369.
springerlink.com

have hampered further studies and empirical applications. This paper takes a first step in applying ecological theory to brand research.

2 Construct and Measurement of Brand Niche

"Niche" was firstly proposed by Johnson, and he put that different species occupy different niche in the environment. Grinnel then defined niche as "the final distribution of units occupied by a species [1]. Hutchinson held that the niche is the n-dimensional space which describes the characteristics of the resources a species needs for survival [3]. There are also other similar definitions by different ecologists. Van Valen defined the "niche breadth" as "the percentage of scarce resources used by a species or biome in a multi-dimensional eco-space". Niche breadth indicated the extent that species utilize different types of resources. And niche overlap was defined as the degree of the sharing utilization of resources by different species [2]. Niche overlap is the essence of competition. When the environment is very rich in resources, different species can share resources without causing harm to each other, and contrarily lack of resources may bring about fierce competition. Larger the overlap lead more intense competition. The development of Enterprise Bionics has promoted the introduction of ecological theory into business and brand research. Just after M. E. Porter proposed the enterprise communities theory [9], J. Moore proposed the concept of business ecosystem and established the commercial eco-system theory [7]. Then in the field of brand management, Lynn, a well-known U.S. brand expert, described the brand as "a complex species" which compete with other brand species for resources in the market ecological environment. Form an ecological view, market can be taken as an ecological environment where brands, as species, compete for scarce resources. And competition can be understood from an ecological perspective. In this paper, brand niche is proposed as "the resources a brand needs for survival in an n-dimensional market ecological environment". Brand niche breadth measures the range of resource characteristics across which a brand exists. Brand niche overlap indicates the extent to which two brands sharing the same resources. The intensity of competition is determined by the extent to which two brands compete for the same scarce resources.

2.1 *Measuring Brand Niche Breadth and Brand Niche Overlap*

Brand niche breadth indicates part of the ability the brand competing for market resources, which is measured as formula 1. B_i measures brand niche breadth, p_{ir} stands for respectively the amount of resource r utilized by species i and $R_{ir} = P_{ir}/R_i$ stands for the proportion of resource r to the total resources utilized by species i in the resource space. Lots of formulas have been put forward to measure the niche overlap, among which the most typical one is the asymmetric αmethod (Levins formula [5]) as formula 2. p_{ir} and p_{jr} stands for respectively the amount of resource α utilized by species i and j in an n-dimensional resource space. And a_{ij} measures niche overlap of species i compared with species j, which is unequal to a_{ji}. The formula is commonly used, because it takes the unequal position of different species competing for resources into account, which is obviously closer to real world.

$$B_i = \frac{1}{\sum_{r=1}^{n} R^2_{ir}} = (\sum_{r=1}^{n} P_{ir})^2 \Big/ \sum_{r=1}^{n} P^2_{ir} \cdots \tag{1}$$

$$\tilde{\alpha}_{ij} = \frac{\sum_{r=1}^{n} (p_{ir} p_{jr})}{\sum_{r=1}^{n} p_{ir}^2} \cdots\cdots \tag{2}$$

2.2 *Brand Niche and Brand Competition*

Brands compete for resources in the market to survive, and the resources in the market are limited, which inevitably leads to competition. Brand niche breadth and overlap can jointly determine the living conditions of the brands. Accordingly, the inter-brand competition can be divided into eight cases showed in Figure 1. **Case 1:** brand A and B both have narrow niche breadth, but with large degree of niche overlap, which implies that A and B are heavily dependent on the same resources. If such resources are not enough for their survival, they would face fierce competition. **Case 2:** niche breadth of A is narrower than B's, and niche overlap between the two is large. They can co-exist, though there is a certain competition. But if A cannot maintain its ability to occupy the key resources, it may be squeezed out of the market by B. **Case 3** is the same as Case 2. **Case 4:** both of niche breadth and niche overlap of A and B are large, so there is certain degree of competition between the two for different resources. The competition is intense but not fatal. **Case 5:** A and B have relatively smaller niche breadth, but there is no niche overlap within them, which means that the two utilize different resources for survival. Little competition exists. **Case 6:** B has a larger niche breadth than A and niche overlap between the two is small. The two utilize different resources for survival. B utilizes more types of resources than A. Little competition exists. **Case 7** is the same as Case 6. **Case 8:** niche breadth of A and B is large, while the overlap is relatively small. No competition exists, which happens little in the market with similar products.

3 Empirical Studies on B2C Website Brands

B2C websites, like Dangdang and Joyo, have gradually become favorable and fashionable shopping choice for the internet users for their convenience and low costs. On the basis of brand niche theory, this paper tries to analyze the current status of Chinese B2C market through an empirical research on four main B2C website brands. According to B2C Websites Rank by Analysis in 2007, four top websites mainly selling books are selected, including Dangdang, Joyo, China-pub and 99 Read, which hold about 40% of the market share. As the market concentration is low, the four websites can represent the whole market situation to a great extent. As a 'species' in the market ecological environment, the brands consume various kinds of resources and most of them are hard to measure, which thus brings much inconvenience to relevant research. Therefore, only resources key to B2C website brands' survival are taken into consideration, including users and partners in the industrial chain. Users, as the main revenue source, are necessities to B2C website brands' survival. Also good relationships with partners are foundation for B2C website brands to provide services for the users, and therefore are crucial for the brands. Accordingly, the key resources are concluded and showed in table 1. The explanation and data source are also provided.

The performance of brands in competing for resources is measured with the scores (1, 3, 5, and 7) according to their ranks. The higher score suggests the higher utilization of such resources and ranking is based on data collected and some calculation showed table 1. Taking the indicator 'Traffic' as an example, we find that Dangdang ranks 463, Joyo ranks 435, China-pub ranks 5695 and 99Read ranks 15742 in the Alexa traffic rank, then the scores are 7 for the top 1, and 1 for 99Read. Likewise, scores of four B2C brands in other resources indicators can be gained as illustrated in table 2. The brand breadth of the four B2C brands can then be calculated from the brand niche formula, as Dangdang 7.15, Joyo 7.11, China-pub 6.55 and 99Read 6.05. All the four B2C brands have large niche breadth, which indicates that the ecological resources in the B2C market are fully used by the brands. The results are generally consistent with the actual situation. The niche overlap figure of B2C website brands can also be drawn according to the scores of the resource indicators, in which the curve and area below form the resource space of B2C website brands as displayed in Figure 4. The part overlap roughly shows the extent of brand niche overlap among B2C brands. As the figure tells, there is great degree of niche overlap among the brands, with characteristics of stratification. Dangdang and Joyo are relatively more competitive than the other two, both of which are easier to gain the resources and competition between the two is fierce. However, China-pub and 99Read hold some advantage in partner resources. It can be more vivid when we separate the resource space as two three-dimensional spaces, that is, users space S (popularity, traffic, market share), and partner space S (logistic, payment, commodities). Four B2C website brands form four dot points in the space, the father the dot away from the origin, the larger breadth the brand holds. As showed in Figure 3 Dangdang occupies the largest users space, with Joyo second and 99Read the last. The shadowed cube S3 reflects the overlap between Dangdang and Joyo, and S4 represents the overlap space among all the four B2C website brands. Likewise, the cubes in Figure 4 also represents the overlap situation among brands in partner space, with S1 as the overlap between Dangdang and Joyo, S2 as the one among four B2C website brands. Conclusively, the competition among four B2C websites brands falls into Case 4 showed in Fig 1, that is, it is intense but not fatal to any of the brands. Because of the large brand niche breadth, all the B2C website brands are hard to achieve particular prominence. It helps each brand perform excellent in some certain field, which is the result of competition, but also the way of avoiding fierce competition.

Table 1 Recourses consumed by B2C website brands

Indicators	Explanation	Data Source
Registered Users	Memberships	Analysys
Market Share	Average share in 05-07	Analysys
Loyalty	Average growth rate in 05-07	Analysys
Traffic	Cumulative traffic	Alexa traffic rank
Popularity	Search Volume in Google	Google Insight
Commodities	Number of category	B2C Websites
Logistics	Geographic coverage, relationships	B2C Websites
Payment	Categories of payment	B2C Websites

Table 2 Resource list and Scores of B2C Websites

Resources	Dangdang	Joyo	China-pub	99Read
Registered Users	7	5	3	1
Market Share	7	5	3	1
Loyalty	5	7	1	3
Traffic	5	7	3	1
Popularity	7	1	5	3
Commodities	7	5	1	3
Logistics	7	5	5	5
Payment	1	7	3	5

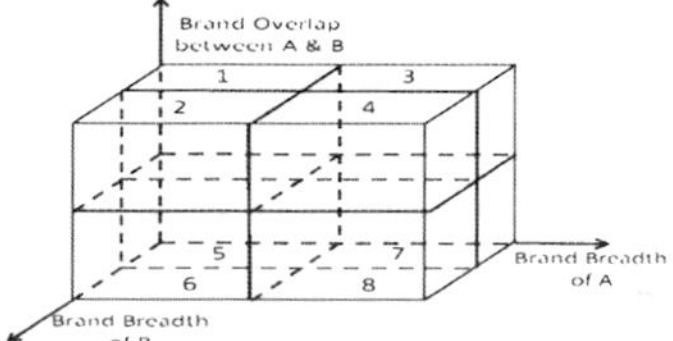

Fig. 1 Brand niche and brand competition

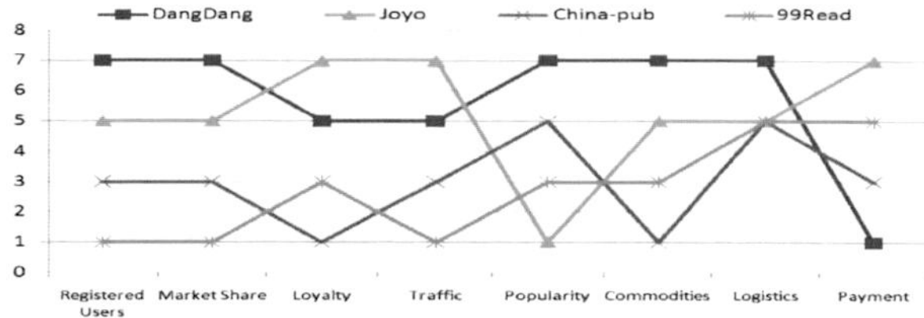

Fig. 2 The niche overlap graph of B2C websites brands

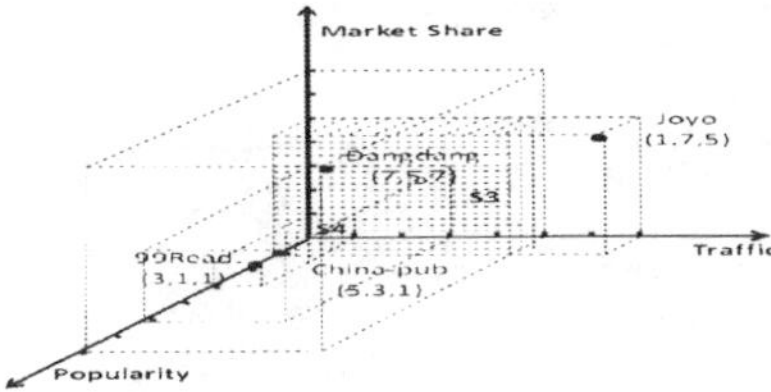

Fig. 3 Brand overlap in user resources space

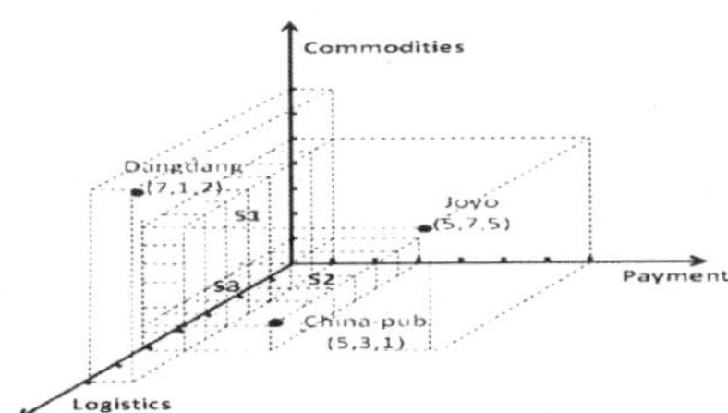

Fig. 4 Brand overlap in partner resources space

Acknowledgments. This research was supported by the National Natural Science Foundation of China under Grant 70772103, and Beijing Natural Science Foundation under Grant 9083017.

References

1. Grinnel, J.: Geography and Evolution. Ecology (1924), doi:10.2307/1929447
2. Hurlbert, S.H.: The measurement of niche overlap and some relatives. Ecology (1978), doi:10.2307/1936632
3. Hutchinson, G.E.: Concluding remarks. Cold Spring Harbor Symposia on Quantitative Biology 22(3), 415–427 (1957)
4. Kevin, J.C., Trout, J.: Brand confusion. Harvard Business Review (2002), `http://hbr.org/2002/03/brand-confusion/ar/1` (cited March 1, 2002)
5. Levins, R.: Evolution in changing environment. Princeton University Press, Princeton (1968)
6. Moon, M., Millison, D.: Torch brand——brand building in internet economy. China Machine Press, Beijing (2002)
7. Moore, J.F.: The death of competition: Leadership and Strategy in the Age of Business Ecosystems. Arts & Licensing International, Inc., New York (1999)
8. Ollé, R., Riu, D.: The new brand management: lessons from brand indifferentiation. Brand Management and Communication Lecturers at ESADE (2003), `http://www.verajordan.com/adrianajordan/pdf/BM_eng.pdf`
9. Porter, M.E.: Clusters and the New Economics of Competition. Harvard Business Review (1998), `http://hbr.org/1998/11/clusters-and-the-new-economics-of-competition/ar/1`

10. Wang, X.Y.: Study on competition and cooperation of brand ecosystem. Nankai Management Review (2000)
11. Wang, X.Y.: Background and research framework of brand ecology. Science & Technology Progress And Policy 7 (2004)
12. Winkler, A.: Build a brand in a quick pace: brand strategy in new economy era. China Machine Press, Beijing (1999)
13. Zhang, Y., Zhang, R.: Brand ecology——a new trend of brand theory evolution. Foreign Economics & Management 8 (2000)
14. Zhang, Y., Zhang, R.: Literature review of brand nature theory in China and abroad. Journal of Beijing Technology and Business University: Social Science 1 (2004)

Influenza Epidemics Detection Based on Google Search Queries

Fan Liu, Benfu Lv, Geng Peng, and Xiuting Li

Abstract. Some new researches demonstrated that the search data can be used to detect public health trends and short-term syndrome surveillance. In this paper, we study the problem of influenza epidemics surveillance using Google search data. A hybrid model with dynamic search query set is developed, which was more accurate in influenza forecast than Google flu trends, especially for the irregular new influenza strain forecasts. This research is valuable for improving the timeliness of syndrome surveillance.

1 Introduction

The Internet search engine has become an important channel for people to seek life information. Meanwhile, the Internet search data record what the searchers are concerned about, and reflect peoples' activity trends in a certain extent. Some new studies present that the search data can help to detect public health trends and syndrome surveillance [5, 6, 8, 9,]. A recent fundamental study is using Google search data to detect influenza epidemics, and their new method can improve the timeliness compared to the traditional surveillance method [7]. The model was also used to Google Flu Trend. Recently, there are also some other researches using this methodology. In the economic field, some study is about predicting unemployment and product sales [1, 4, 11].

Ginsberg et al built a simple linear regression model to predict influenza-like illness (ILI) in United States based on Google search engine query data, and it can be two weeks in advance of the release of CDC's flu report [7]. The model selects 45 best fitting ones from 50 million of most common search queries, and reached

Fan Liu · Benfu Lv
Management School, Graduate University of Chinese Academy of Sciences,
NO.80 Zhongguancun East Road Haidian District, Beijing, China
e-mail: Jeffery.van@gmail.com, lubf@gucas.ac.cn

F.L. Gaol (Ed.): Recent Progress in DEIT, Vol. 2, LNEE 157, pp. 371–376.
springerlink.com

a mean correlation of 0.97. This indicates that the Google search query data indeed has predictability of influenza epidemics. But the Google Flu Trends model has a serious deficiency. The model shows severe insensitiveness to the non-seasonal influenza outbreaks, because the selected 45 can't be amended with characteristic queries of new influenza symptoms. As a consequence, it is helpless in the new influenza forecast. An example is that at the end of April 2009 when a new swine flu began to outbreak, the model made a long period of forecast failure, and completely omitted the trend of this influenza peak.

In order to robustify the Google Flu Trends model, J. A. Doornik extended it to an auto-regression model with calendar effects [5]. In his model, the lagged dependent variables were used to capture the long-term cycle, and the calendar dummy variables were used to capture the seasonal and asymmetric effects. The new model not only improved the prediction accuracy, but also has self-correct capability that it needs two periods or more to revise when the swine flu suddenly arose in 2009-04-26 (week 17). However, we find that the robustified model still couldn't capture the turning point when the sudden and unprecedented fluctuation occurred. So the new model yet does not solve the key issue of turning point detection.

2 The Integration of Time Series Data and Search Query Data

Influenza is a seasonal infectious disease. In United States, Centers for Disease Control and Prevention (CDC) defines the flu season as the duration between the 40th week each year and the 20th week next year (about early October to mid May next year). In flu season, CDC releases weekly influenza surveillance report [3]. The main indicator of influenza surveillance is the percentage of visits for influenza-like illness (ILI%) in hospital outpatients. We can draw out the trends of historic influenza activity with the CDC reported ILI% weekly data. In Figure 1, the curves of ILI% 2006-07 and ILI% 2007-08 are only in the flu season. By comparing the three curves, the influenza activity is highly seasonal. In October of each year, the flu season began, and between January and February the peak appeared, and then gradually fell. Importantly, the trend of each year is extremely similar. Based on this characteristic, using time series modeling for influenza surveillance would be an appropriate method [2, 10].

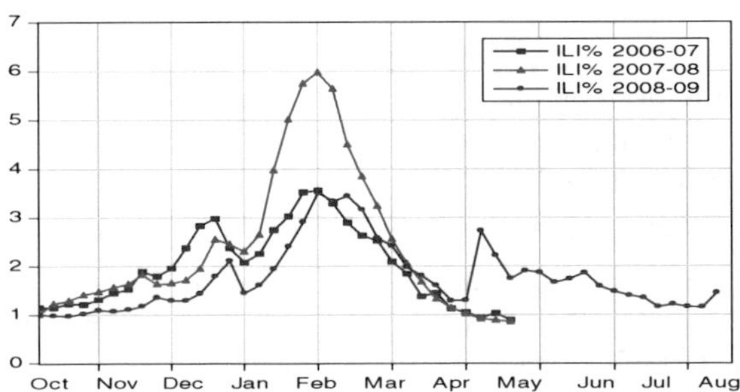

Fig. 1 The percentage of visits for influenza-like illness (ILI%) released by CDC.

In history, the influenza activity had dropped to very low level after April, and even down to the level of non-flu season after May, while Figure 1 shows that in April and May 2009 there is a significant peak of ILI% compared with the same time of past years, which is due to the outbreak of the H1N1 flu. The time series methods are unable to detect the abnormal fluctuation, which can only be adjusted after the outbreak [5]. In 2009, by massively calculating the correlation between the time series of 50 million queries and the influenza (ILI% series), Ginsberg et al finally selected 45 best fitting ones to monitor influenza activity. The 45 queries are all related to influenza. They got fairly good results using simple linear regression. However, their model also can't capture the above-mentioned volatility.

3 Model

3.1 The Hybrid Model with Fixed Query Set

We use the above-mentioned 45 queries' search volume and weekly ILI% data from CDC to build a static model. The data collection duration is the flu season from 2003 to 2009 (week 40 in 2003 to week 29 in 2009). Here we split the data into two segments: Segment 1 is from Week 40 in 2003 to Week 20 in 2007, and segment 2 is from Week 40 in 2007 to Week 29 in 2009. We use the segment 1 data to estimate our static model, while the segment 2 is for forecasting.

In order to reduce the data variation and instability, we use the logit transformation to preprocess the variable ILI%:

$$\log\text{it}(ILI\%) = \log\left(\frac{ILI\%}{100 - ILI\%}\right)$$

The log-transformation is used to convert Q, the sum of the 45 queries' weekly search volume, to lnQ. The dependent variable is logit(ILI%)t , and explanatory variables are the lagged 1-order dependent variable logit(ILI%)t-1 and current search volume lnQt. As the query set is fixed, we call it static model:

$$\log\text{it}(ILI\%)_t = \underset{(0.371)}{-4.525} + \underset{(0.030)}{0.649}\log\text{it}(ILI\%)_{t-1} + \underset{(0.059)}{0.695}\ln Q_t$$

$$\overline{R}^2 = 0.942, \hat{\sigma} = 12\%$$

The estimated coefficients are all highly significant, and their standard errors are in parentheses. The standard error of this static model regression is 12%, Doornik' auto-regression model with up to the 53rd lag is also about 12%, while the Google Flu Trends model is up to 65%.

In forecasting, since the Google Flu Trends model used real-time search data, it could detect influenza activity about two weeks in advance of CDC, but don't have predictability to the future trends of influenza. Therefore, in practice, the Google Flu Trends model can be used for influenza surveillance two weeks ahead. In Figure 2, the static model forecasts have been transformed by the anti-logit transformation. Before Week 16 in 2009 (April 26), the forecasts from static model is mainly consistent with the actual value. Although there are some minor fluctuations in the first quarter of 2009, it should belong to stochastic perturbation of prediction. But after that the prediction was significantly underestimated. Obviously, the static model did not detect the influenza abnormal fluctuation in May 2009.

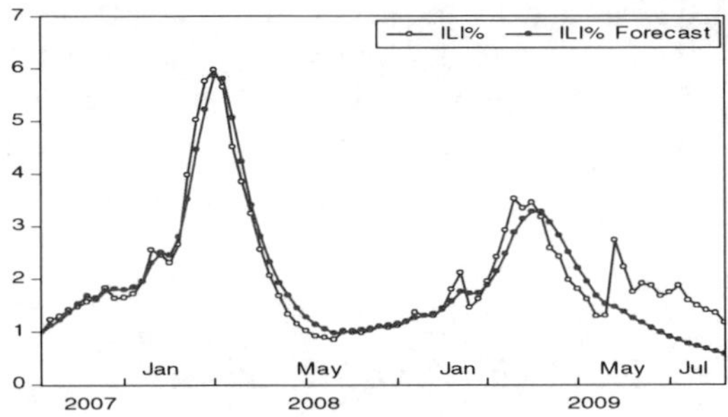

Fig. 2 The forecasts from static model and the actual ILI%

In May 2009, the H1N1 flu outbreak appeared in United States, which led to the higher level of flu activity than the same period of history. The 45 queries used in the static model are all concerning ordinary flu which were build in 2008, so they didn't reflect the new influenza such as H1N1 and swine flu. In order to monitor the unusual volatility, we must add queries about new influenza into our model.

3.2 The Hybrid Model with Dynamic Query Set

For the influenza outbreak in May 2009, pure time series model couldn't predict it, but only adjusted to a reasonable level in a few weeks after the abnormity occurred. To detect similar influenza outbreak, we try to use Google Trends (http://www.google.com/trends) to supplement our query set with queries about H1N1 and swine flu. With Google Trends, people can inquiry the terms' search volume in any topics, based on a subset of Google's search database.

According to the characteristics of new influenza and the actual situation of Google Trends data, we mainly consider the H1N1 and swine flu related queries weekly search volume after Week 13 in 2009 (April 4). All the added queries are listed in table 1. The search volumes of all queries are very small before April 2009, but there is a sudden surge in the end of April, and then dropped to a lower level quickly after May. These features are consistent with the development of new influenza.

Table 1 The added queries about H1N1 and swine flu

H1N1 related queries	Swine related queries
H1N1	Swine Flu
H1N1 Flu	Swine Flu Symptoms
Symptoms H1N1	Swine Flu Vaccine
H1N1 Virus	Swine Flu Cases
H1N1 Flu Symptoms	Swine Flu H1N1
H1N1 CDC	Swine Flu CDC

The dynamic model added new queries is estimated as follows:

$$\log it(ILI\%)_t = \underset{(0.393)}{-6.072} + \underset{(0.030)}{0.554} \log it(ILI\%)_{t-1} + \underset{(0.063)}{0.95} \ln Q_t$$

$$\overline{R}^2 = 0.965, \hat{\sigma} = 10.6\%$$

The dynamic model is estimated better than the static model in various indicators. The standard error of regression is only 10.6%, and the residual also meets the requirements. Similarly, we split the data into the same two segments for predictive test.

In table 2, the forecast statistics of dynamic model is very satisfactory. The root mean squared error (RMSE) is 0.15, the mean absolute percentage error (MAPE) is 2.87, and the covariance proportion is up to 0.93. Our hybrid models are superior to the models of Google Flu Trends and the autoregressive with calendar effects [5], and the dynamic hybrid model is far better than the static.

As shown in Figure 3, by adding new influenza-related queries, the dynamic model monitored the outbreak of influenza in May 2009, and the forecasts are highly in line with the actual in the subsequent phases. This demonstrates that the adding queries basically reflect the influence of new flu in 2009.

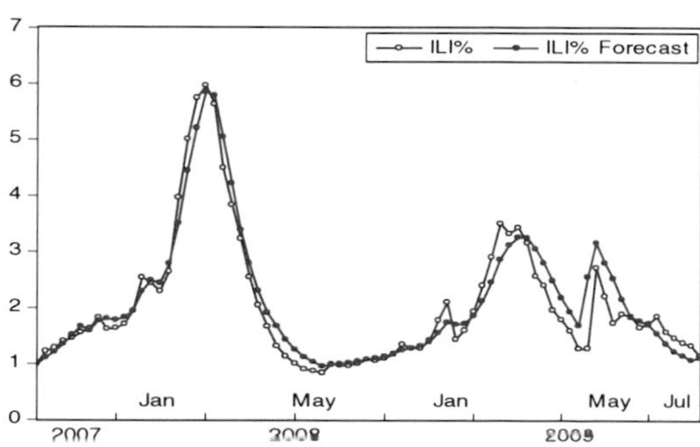

Fig. 3 The forecasts from dynamic model and the actual ILI%

Table 2 Forecast statistics for ILI% of two hybrid models, Google Flu Trends model and autoregressive with calendar effects

Models	ROME	MAPE
The Static Model	0.29	4.78
The Dynamic Model	0.15	2.87
Google Flu Trends	0.29	12
Autoregressive with calendar effects	0.41	12

Acknowledgments. This research was supported by the National Natural Science Foundation of China under Grant 70772103 and 70972104, and Beijing Natural Science Foundation under Grant 9083017.

References

1. Askitas, N., Zimmermann, K.F.: Google Econometrics and Unemployment Forecasting. Applied Economics Quarterly (2009), doi:10.3790/aeq.55.2.107
2. Batal, H.: Predicting patient visits to an urgent care clinic using calendar variables. Acad. Emerg. Med. 8(1), 48–53 (2001)
3. Centers of Disease Control and Prevention. Flu Activity & Surveillance. Reports & Surveillance Methods in the United States (2010), `http://www.cdc.gov/flu/weekly/Fluactivity.htm`
4. Varian, H., Choi, H.: Predicting the Present with Google Trends. Google Research Blog (2009); Available at SSRN, `http://googleresearch.blogspot.com/2009/04/predicting-present-with-google-trends.html`, `http://ssrn.com/abstract=1659302`
5. Doornik, J.A.: Improving the Timeliness of Data on Influenza-like Illnesses using Google Search Data. In: 8th OxMetrics User Conference. George Washington University, Washington, D.C. (2010)
6. Eysenbach, G.: Infodemiology: tracking flu-related searches on the web for syndromic surveillance. In: AMIA Annu. Symp. Proc., pp. 244–248 (2006); PMCID: PMC1839505
7. Ginsberg, J., Mohebbi, M.H., Patel, R.S., Brammer, L., Smolinski, M.S., Brilliant, L.: Detecting influenza epidemics using search engine query data. Nature (2009), doi:10.1038/nature07634
8. Hulth, A., Rydevik, G., Linde, A.: Web Queries as a Source for Syndromic Surveillance. PLoS ONE 4(2), e4378 (2009)
9. Polgreen, P.M., Chen, Y., Pennock, D.M., Nelson, F.D.: Using Internet Searches for Influenza Surveillance. Healthcare Epidemiology 47(11), 1443–1448 (2008)
10. Reis, B.Y., Mandl, K.D.: Time series modeling for syndromic surveillance. BMC Medical Informatics and Decision Making (2003), doi:10.1186/1472-6947-3-2
11. Sano, Y., Takayasu, M.: Macroscopic and microscopic statistical properties observed in blog entries. Journal of Economic Interaction and Coordination (2010), doi:10.1007/s11403-010-0065-7

A Message Prioritization Scheme for Virtual Collaboration

Dewan Tanvir Ahmed and Shervin Shirmohammadi

Abstract. Assuring quality for online games is a big challenge. The client-server architecture, currently though a widely running mode supporting hundreds of thousands of players in a regular basis, faces many difficulties. Network limitations like latency and jitter can leave timely interaction at jeopardy. In online games, the magnitude of interaction among players varies largely which generally depends on virtual distance and other complex variables. It is imperative that the frequency of interaction among players is uneven as a closer player is interacted with more often than a distant one. Similarly, the significance of interaction follows the same drift as the virtual distance increases the significance of relative interaction decreases. A message prioritization scheme can be formulated considering these ingredients for large scale collaboration where the volume of message is high. For this purpose, we exploit the importance of players temporal bond of interaction as well as a segmentation scheme for virtual space around a players area of interest where each segment has a different level of significance in terms of game states. The above techniques set a platform to devise a message passing plan for online games that can offer better game service.

1 Introduction

Over the years Internet-based multiplayer online games such as Role Playing Games (RPG) and First Person Shooter (FPS) have become increasingly popular. A Massively Multi-User Virtual Environment (MMVE) is a kind of synthetic world where numerous participants share the same virtual environment where they interact using avatars. Online games usually function in a client-server mode where players control game clients generally running on a personal computer (PC) that

---

Dewan Tanvir Ahmed · Shervin Shirmohammadi
Distributed and Collaborative Virtual Environments Research Laboratory
School of Information Technology and Engineering, University of Ottawa, Canada
e-mail: {dahmed, shervin}@discover.uottawa.ca

F.L. Gaol (Ed.): Recent Progress in DEIT, Vol. 2, LNEE 157, pp. 377–385.
springerlink.com

communicate with game servers hosting individual games. Because of resource requirement, game publishers rely on Internet Service Providers (ISPs), dedicated game hosting companies, and private individuals to host game servers. A cluster of high-end servers deployed for this purpose that can offer a lot of computational power and data centers but eventually produce a large amount of network traffic.

To manage a large number of players, the virtual world is typically divided into realms or kingdoms which are the clones of the same virtual world each hosting several thousand registered players. To be precise, realms are geographically distributed across the Internet. So, players from one particular region generally participate in the same realm. Realms are further divided into separate areas. Each area is considered as a zone. Zones can have different themes and different level of difficulties in order to hold inexperienced players advancing into the next hard level. Normally there is a server for each zone managing game traffic. The communication strategy within a zone best resembles to a multicast structure because of players common interest (i.e. regular interaction) in game logic. IP multicast is not widely used because of many reasons like deployment difficulty [7]. Current practices depend on centralized architecture that has scalability bottlenecks, which is also costly to adopt and install. New designs and proposals are introduced combining client and server side resources in a seamless manner to take advantage of the P2P design [1].

In order to provide a consistent game space, each player must keep a clone of relevant game states on his computer. When a player performs an action or generates an event affecting the virtual space, the update must be shared with other players around. The amount of data required to exchange roughly depends on population size of the interested area. However, the capacity is bounded by at least two realistic limitations network bandwidth and processing power. Latency tolerance usually varies from game to game and is typically limited to a value between 100 ms to 1000 ms depending on many factor such as game perspective (i.e. First-person or Third-person), game genres (i.e. racing or role playing game), and the sensitivity of the actions [2].

In this article, we develop a game state prioritization scheme that can improve game service. In online games, virtual position defines a players location in the game space which plays an important role for the design. We believe that even in the same interaction space, the importance of interaction among players is uneven. Simply, a closer player has higher importance than a distant one, which is intuitive in some sense. On the other hand, symmetric and asymmetric relationships among players are temporary and can change over time based on players relative position and orientation. So a message ordering procedure can take these temporal symmetric and asymmetric relationships into account while defining rules to forward games states. Considering the importance of interaction, relative orientation, and virtual locations, we devise a message forwarding plan that can work in server and client machine standalone with the local information available.

The outline of this paper is as follows. After describing an overview of the related work in the following section, we present our message prioritization scheme in Section 3. The evaluation and other relevant issues are briefly discussed in Section 4. Finally the paper is concluded in Section 5.

2 Related Work

The early proposals of online games suggested decades ago were client-server in nature where one player sends an update message to the server who then relays it to all other players. While peer-to-peer (P2P) approaches have recently been proposed to either move away from a client-server approach or to complement it, it is apparent that a pure P2P architecture is not a practical solution for online games due to business, security, and quality concerns. At present, different hybrid architectures are being proposed using peer resources, but practical deployment hurdles are not yet fully overcome.

Consistency, responsiveness, reliability, security, and persistency are fundamental issues of MMOGs [9]. The general phenomenon is that when a player interacts with other players the updated information must be sent to all the participants. Because of networking limitations and traffic conditions, some of the updates may be lost or delayed. Much research has been conducted to overcome these limitations and to build more reliable systems [8, 4]. Fritsch et al. show that *Everquest2* can run with latency up to 1250 ms [5]. For this purpose authors monitor time span in which players kill a certain number of monsters, outstanding health, and magic points that players have after the encounter. Dick et al. conduct player survey and analyze how latency affects player performance in different games [3]. According to them, a *Role Playing Game* like *Diablo II* performs well when latency is around 80 ms which can be stated as the optimal, and a latency of 120 ms as a maximum tolerable value. In this article, we present a state sharing procedure that allows players to exchange game state directly when the experienced latency via intermediate server exceeds the threshold.

The model proposed by Hampel et al. reuses architectures capable of exploiting the flexibility and scalability of peer-to-peer networks [6]. One of the main drawbacks of P2P networks for games is the lack of a central authority that can regulate access and prevent cheating. The model presented in [6] overcomes that by using a set of controller peers that can supervise each other. This kind of redundancy can prevent cheating. The model is based on the existing distributed hash table Pastry, which has been extended into SCRIBE. The key issue is unbounded end-to-end delay, which could be a problem for synchronization.

In this article, we consider the importance of local interaction and introduce a game state forwarding scheme that will improve game service. It should be noted that the procedure is presented assuming that players can communicate directly with the approval of moderators. The procedure can also work in pure client-server mode. Both are presented in this paper.

3 The Message Prioritization Scheme

The magnitude of interaction among players based on virtual distance varies largely. It is quite common that the frequency of interaction among players is different as a closer player is interacted with more actively than a distant one. In addition, the significance of interaction follows the same trend i.e., as the virtual distance increases between two players the significances of *relative interaction* decreases. Thus a message prioritization scheme can be formulated for large scale collaboration where the volume of generated message is high. In an area of interest, each player knows the virtual position of all other players. This fact makes it feasible devising a message ordering procedure in real-time.

3.1 Interest Management

Before discussing the messaging procedure, we will briefly explain *space-based* interest management model which is quite effective for MMOGs. This works based on proximity which can be realized in terms of *aura-nimbus* information model. *Aura* is the area that bounds the existence of an avatar in space, while *nimbus*, i.e., area of interest, is the space in which an object can realize other objects. In the simplest form, both aura and nimbus are usually represented by circles around the avatar. This model is more appropriate when a server maintains a connection with each client.

3.2 Space Prioritization

Let us consider the interaction space of a player as a disk of radius. In order to put different importance on players according to their visibility scope and current focus, we partition the disk into several wedges namely front wedge (w_f), two side wedges (w_s) and one back wedge (w_b) as shown in Figure 1. Front wedge represents a players current direction. For a player, the front space is more valuable than other wedges in terms of interactions. Similarly the side wedges are valuable compared to the back wedge. So the importance of wedges can be $w_f > w_s > w_b$.

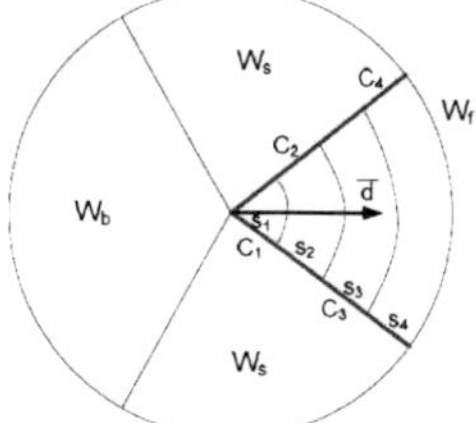

Fig. 1 The decomposition of virtual space w.r.t. the significance of interaction

We further make subdivision of the front wedge, and for this purpose concentric circles are used named coronas obtained as follows. Consider k concentric circles of radii $0 < r_1 < r_2 < ... < r_k = R$ centered at 0. Now, for every i, $1 \leq i \leq k$, the corona C_i is the sub-area of D delimited by the circles of radii r_{i-1} and r_i. The width of each corona is R/k. So, all coronas have the same width as shown in Figure 1. For $k = 4$ the area is partitioned into four coronas C_1, C_2, C_3 and C_4. As a result, the front wedge w_f is partitioned into k segments s_1, s_2, s_3 and s_4 by its intersection with k concentric circles. According to our design, each segment of the front wedge has different level of importance which is $s_1 > s_2 > s_3 > s_4$.

3.3 Inter-player Relationship

On the other hand, symmetric and asymmetric relationships can be defined in terms of message importance. In symmetric relationships, game states have equal importance for two players in both ways i.e. an update about player A for player B is as important as an update about player B for player A. When two players interact face-to-face, a symmetrical relationship is encountered. On the other hand, in asymmetric relationships, game states have different significance for two players. Here an update about player A for player B is not as important as an update about player B for player A. For example, two players A and B running towards west, player B is behind player A. So player B can see player A, but not vice versa. So the position update of player A is important for player B but the position update of player B is not so important for player A. This is called asymmetric relationship. According to the above definitions, symmetric and asymmetric relationships among players are not fixed for online games which change over time base on players relative position and orientation. So message prioritization should take temporal symmetric and asymmetric relationships into account while formulating rules to exchange game states.

3.4 Game State Forwarding Plan

Having the details of previous discussion and accessible solutions, we are now presenting a new-fangled game state forwarding procedure. The approach takes following factors into account:

1. Virtual positions of players within an area of interest
2. Symmetric and asymmetric relationships among players at the time of sharing game state
3. Players available resource

The goal of this procedure is to regulate message forwarding so that player can improve gaming experience. Active participation among players within the client-server architecture is considered for this purpose. All factors can be known and realized at runtime. Factors 1 and 2 can be realized through the game states regularly. Factor 3 is an application specific parameter which can be set at the beginning of a game session. The precedence will be given according to the order shown in Table 1.

The virtual space is segmented using the procedure defined earlier in subsection 3.2. This segmentation gives an ordering of regions in terms of relative importance which implies that two players from two different segments will be treated differently according to the order of virtual space. The ordering of regions according to

Table 1 The relative precedence of game states

Precedence	Region	Relationship
1	s_1	x
2	s_2	x
3	s_3	symmetric
4	s_3	asymmetric
5	s_4	symmetric
6	s_4	asymmetric
7	w_s	symmetric
8	w_s	asymmetric
9	w_b	x

Figure 1 is $s_1 > s_2 > s_3 > s_4 > w_s > w_b$. Let p_1 and p_2 are two players from regions s_3 and s_4, respectively. As $s_3 > s_4$ in terms of region importance, the message for player p_1 will be processed earlier than that of p_2. The method also gives importance to inter-player relationship presented in subsection caxcxz3.3. Considering the above two attributes, the message ordering plan for players is shown in Table 1. In this procedure, players can communicate either directly (if P2P feature is assumed) or via server (C/S mode) as usual, considering the importance of interaction, relative orientation, and virtual location

The virtual position of the players changes regularly in a game space. Similarly, the relative orientation of players within an area of interaction is unpredictable which also changes frequently. A current message forwarding plan has little value after some time. We cannot generate a stable rule set for message exchange because of these facts. The server and client side steps for message forwarding are given below in Figure 2 and Figure 3, respectively.

Three computationally low-cost auxiliary functions namely Virtual-Region (p_c, p_i), Inter-Player-Relationship (p_c, p_i) and Precedence-Level (r, j) are used. The descriptions of these functions are mentioned below. It is important to mention that the algorithms run with an assumption of resource availability. It tries the next option if the partner at the other hand has adequate bandwidth or stop otherwise.

`Virtual-Region` (p_c, p_i): This function determines the region of p_i with respect to the area of interest of p_c. The return value is an integer between 1–9.

`Inter-Player-Relationship` (p_c, p_i): To determine association between two players in terms of orientation this function is used. If there is a symmetric relation between two given players, it returns TRUE, otherwise FALSE.

`Precedence-Level` (r, j): This function decides precedence level for p_i with respect to p_c. Using the return values of other two functions, it looks up a table and sets the precedence for p_i with respect to p_c.

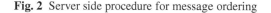

```
Algorithm Message-Ordering-at-Server
Input
  P: The set of player in the area of interest
 pc: The server running module for player pc
begin
  for each player pi ∈ P do
    r ← Virtual-Region(pc, pi)
    j ← Inter-Player-Relationship (pc, pi)
    k ← Precedence-Level (r, j)
    msg-ordering(k) ← pi
  end
  The server relays message according to msg-ordering when pc
  is a source
end
```

Fig. 2 Server side procedure for message ordering

```
Algorithm Message-Ordering-at-Client-Machine
Input
  P: The set of player the area of interest
 pc: The player running this module
begin
  for each player pi ∈ P do
    r ← Virtual-Region(pc, pi)
    j ← Inter-Player-Relationship (pc, pi)
    k ← Precedence-Level (r, j)
    msg-ordering(k) ← pi
  end
  if resource allows player pc relays message according to msg-
  ordering otherwise the server does the remaining forwarding
end
```

Fig. 3 Client side procedure for message ordering

4 Evaluation

Usually, the size of an AoI is below 100 players (approximately 30 50). For example, World of Warcraft supports Raid-groups of up to 40 people. In this simulation, the players were placed randomly and the server was approximated at the middle of the players. Some parameters of the evaluation method are given in Table 2. We tested the presented method for random movement pattern.

We also tested the significance of space segmentation while forming an ordered list of players to forward game states. For a group of 50 players, we checked how many players are important and can be treated differently when relaying or forwarding game information to other players. Such a scenario for 50 runs is shown in Figure 4. From this figure, it is revealed that the number of players in an interested area is few compared to other regions. There are relatively more players in insignificant segments like and this implies giving an equal importance to all players in different segments is impractical as some segments are more important than others. By simulation, it is discovered that the consideration of temporal inter-player relationship is useful while sharing game states.

Table 2 The parameters of the evaluation

Parameter	Value
θ_f	60^0
θ_s	45^0
θ_b	210^0
N	50
Movement pattern	random

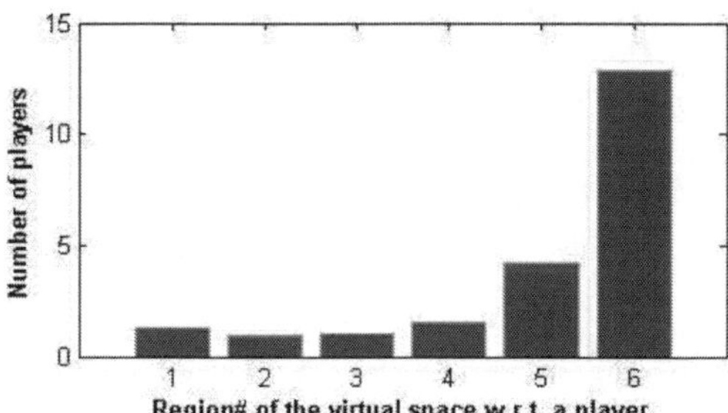

Fig. 4 Quality improvement while conisdeing symmetric relationships

5 Conclusion

We have presented a message prioritization scheme to forward game states from one player to others either in a client-server mode or a P2P fashion. For this purpose, it exploits the importance of players temporal bond of interaction. In addition, a segmentation of virtual space is carried out around a players area of interest where each segment has a different level of significance in terms of game states. The above techniques set a platform to devise a message passing plan for online games that can offer better game service. The simulation result shows that the method can be advantageous when the volume of interaction is high.

References

1. Ahmed, D.T., Shirmohammadi, S., Kazem, I.: Zone based messaging in collaborative virtual environments. In: IEEE International Workshop on Haptic Audio Visual Environments and their Applications (HAVE), pp. 165–170 (2006)

2. Claypool, M., Claypool, K.: Latency and player actions in online games. Entertainment Networking Special Issue: Entertainment Networking, 40–45 (2006)
3. Dick, M., Wellnitz, O., Wolf, L.: Analysis of factors affecting players' performance and perception in multiplayer games. In: Proceedings of 4th ACM SIGCOMM Workshop on Network and System Support for Games, NetGames 2005, pp. 1–7 (2005)
4. IEEE standard for distributed interactive simulation - application protocols (1998)
5. Fritsch, T., Ritter, H., Schiller, J.: The effect of latency and network limitations on MMORPGs: a field study of everquest2. In: Proceedings of 4th ACM SIGCOMM Workshop on Network and System Support for Games, NetGames 2005, pp. 1–9 (2005)
6. Hampel, T., Bopp, T., Hinn, R.: A peer-to-peer architecture for massive multiplayer online games. In: Proceedings of ACM SIGCOMM Workshop on Network and System Support for Games, p. 48 (2006)
7. Hosseini, M., Ahmed, D.T., Shirmohammadi, S., Georganas, N.D.: A Survey ofApplication-Layer Multicast Protocols. IEEE Communications Surveys & Tutorials 9(3), 58–74 (2007)
8. Pullen, J.M.: Reliable multicast network transport for distributed virtual simulation. In: International Workshop on Distributed Interactive Simulation and Real-Time Applications, p. 59 (1999)
9. Schiele, G., et al.: Requirements of Peer-to-Peer-based Massively Multiplayer Online Gaming. In: Proceedings of the IEEE International Symposium on Cluster Computing and the Grid, pp. 773–782 (2007)

An Online Virtual Classroom Using SIP

Vajirasak Vanijja and Budsakol Supadetvivat

Abstract. Online Virtual Classroom using SIP is another alternative for the users who have the network constrain. The system composes of two user agents, which are the teacher's user agent, and the student's user agents. Both user agents contain three main functions: slides control, live text chat, and 2-ways real-time voice communication. The user agents are compatible with an Open source SIP server called "Asterisk". No modification is required on the SIP server. The purposed system allows the teacher, and the student to communicate via text chat and real-time voice with a low bandwidth connection. PowerPoint files would be loaded into the students' computers before the class start. The slide control module will synchronize the slides between the teacher and all students. The system also supports voice communication using SIP and RTP. The network efficiency analysis is done to evaluate the performance of the purposed system compare to a commercial web conference system.

1 Introduction

e-Learning [1] is all forms of electronic supported learning and teaching, which is procedural in character and aim to effect the construction of knowledge with reference to individual experience, practice and knowledge of the learner. Information and communication systems, whether networked or not, serve as specific media to implement the learning process. In this research, e-Learning may categories into two types; on demand e-Learning and the Virtual Classroom.

The authors proposed a new system which using SIP user agents for both the teacher and the students. The online virtual classroom system in this research apply the push technology. Push technology [2,3] is automatically broadcasting data technology via the Internet and user must install program or software for receiving and translating that data. The purposed system broadcasts the classroom using two types of user agents, which are the teacher's and student's user agents. The teacher's user agent sent the control message to all students' user agent to dominate the slide in the student's machines. A new set of SIP user agent (soft-phone)

Vajirasak Vanijja · Budsakol Supadetvivat
King Mongkut's University of Technology Thonburi, Bangkok, Thailand
e-mail: vajirasak@sit.kmutt.ac.th

F.L. Gaol (Ed.): Recent Progress in DEIT, Vol. 2, LNEE 157, pp. 387–393.
springerlink.com

is developed in this research to deliver not only voice but also text message and the control message to synchronize the slides between the teacher's and the students' machine. The background is described in the section two.

2 Session Initiation Protocol (SIP)

Session Initial Protocol or SIP Protocol [5,6] is a signal protocol for controlling multimedia communication session such as voice and VDO over Internet Protocol (IP). SIP sessions can be used in unicast or multicast communication. The SIP is an Application Layer protocol. SIP works on Transmission Control Protocol (TCP), User Datagram Protocol (UDP) and Stream Control Transmission Protocol (SCTP). The SIP working group within the Internet Engineering Task Force (IETF) develops SIP. The protocol is published as IETF RFC 2543.

SIP User Agent (UA) is a logical network endpoint entity use to initiate and terminate sessions by exchanging requests and responses. RFC 2543 defines the user agent as an application which contains both a user agent client (UAC) and user agent server (UAS). UAC sends SIP request and UAS receives the request and return a SIP response. The devices, which can have UA's function in a SIP network, are: workstations, IP-phones, telephony gateways, call agents, automated answering services. For purposed virtual classroom system, teacher's and student's systems are user agent type.

SIP Server is an intermediary entity, which acts as both a server and a client for making requests on behalf of other clients. In the purposed system, the Asterisk 7 is used as the SIP server.

The SIP conference architecture is described by J. Ni and J. Luo [8]. E. H.K. Wu and et.al [9] have analyzed the characteristic of the SIP conference in their work. The SIP conference is a key component of the purposed system. The purposed system use Asterisk [7] as a SIP server and the conference server. The conference is a basic feature of the Asterisk. The Meetme [7] is a conference module in the Asterisk.

3 The Online Virsual Classroom System

The Online virtual classroom system structure is shown in Fig. 1. The system is composed of a teacher, students and the SIP server (Asterisk server).

The online virtual classroom using SIP system starts with the teacher and the students register on SIP server with their predefined SIP accounts. The SIP conference number is assigned to a conference room or the virtual classroom. This virtual classroom brings Meetme [7] function of Asterisk for the voice conference channel.

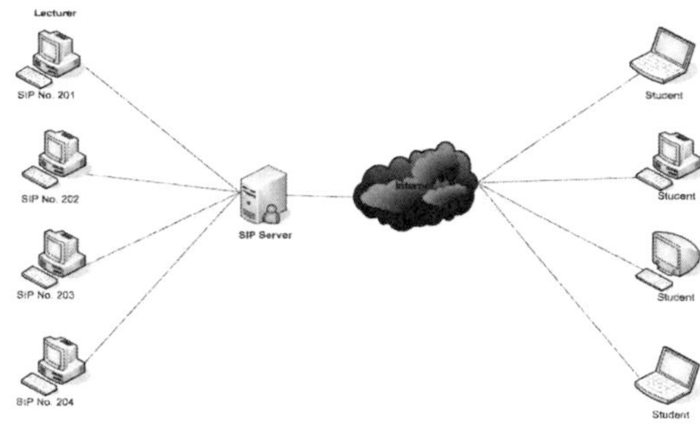

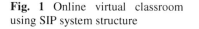

Fig. 1 Online virtual classroom using SIP system structure

The user agent consists of three main functions as follows:

Real-Time Voice Communication: is a user agent's function that connects to the SIP server (Asterisk) and broadcasting voice to all participants. The teacher and students are allowed to talk in the virtual classroom. The Soft-phone (User Agent) contains dial pad for entering the class room's SIP number. The call controlling buttons is consist of call button, hang-up button, pause button, automatic answering button, and disable call button respectively. The disable call button will disable student to make a voice call from their user agent. The voice controlling buttons is consist of voice button, microphone button and voice volume task bar.

Live Text Chat Module: is a user agent's function for receiving questions or messages from the students. If teacher received question from student then the teacher could answers that question via voice broadcasting or typing message.

Slide Control Module: is a function for controlling slides on the student's user agents. The teacher can click on the next button for the next slide or the back button for the previous slide or choose the slide number from the drop down menu to jump to the desired slide.

4 The Experiment

In the experiment, the virtual classroom using SIP system will be compared with a commercial web conference system called "DimDim"[11]. The bandwidth, jitter and delay of the online virtual classroom using SIP system and DimDim web conference system are measured and analyzed. However, the purposed online virtual classroom using SIP system transfers data by UDP and SIP protocols which theoretical more effective for the multimedia data than the TCP protocol. The experiment tests between the purposed system that use the user agent for voice broadcasting and the DimDim web conference system. Both systems were tested by one-to-one communication in the same LAN (Local Area Network) between teacher and student sites to measure bandwidth, delay, and jitter.

These scenarios do not include the case of animated PowerPoint slides because the web conference system does not support the animation feature of the PowerPoint. If user uploads animation PowerPoint slide then the web conference system allows this uploading but it changes animated slides to static slides and some slides are displayed incorrect format. The average bandwidth (Kbit/sec) of each case is measured.

The bandwidth that measure at the teacher system shows not significantly different between the two systems. The student does not need to download the slides file for the class. The slides' image will be shown on the browser during the class. The SIP system lets the teacher and student to share the PowerPoint file before the class start. The student must download the slide from the class's website manually. The slide control messages are sent from the teacher system to the student's system. No image download is required during the class. The proposed system shows significant improvement of the interactive between teacher and student during the class using SIP compare to the web conference system.

Both applications are delay and jitter sensitive. The jitter represents the continuous of the voice during the class or conference. The delay represents the response time of the counterparts.

The jitter experiment scenario is measured the voice data packet from two hops of the data transfer, which are jitter of the data transfer from the teacher's system to the server, and from the server to the student's system. The jitter experiment's result is shown in Fig. 2 and Fig. 3.

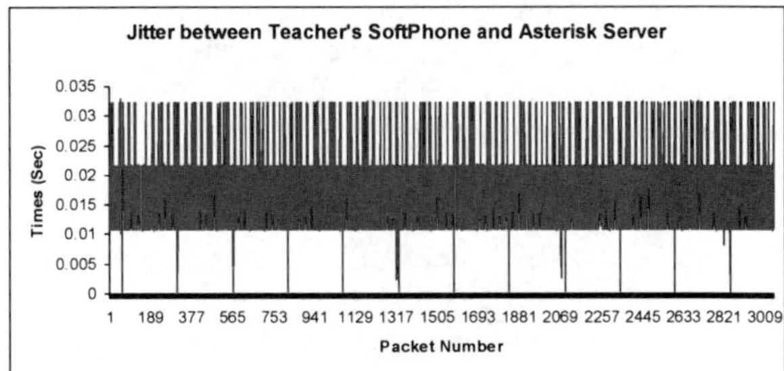

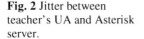

Fig. 2 Jitter between teacher's UA and Asterisk server.

The jitter graphs in Fig. 2 - Fig. 5 are plotted between discrepancy of the timestamp of current voice packet and previous voice packet and the packet numbers. The Fig. 2 shows the jitter of the SIP system. The packet source is the teacher system, and the destination is the Asterisk server. Fig. 3 shows the jitter of web conference system. The packet source is teacher's system, and the destination is the web conference server. From the both graphs, the jitter of the SIP system shows less average value than the web conference system. The average jitter value of SIP system is 25 millisecond, but the web conference system is 50 millisecond.

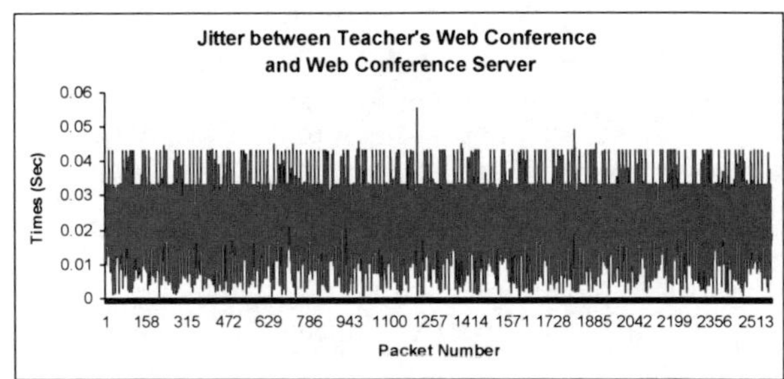

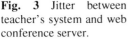

Fig. 3 Jitter between teacher's system and web conference server.

The Fig. 4 shows the jitter of the SIP system. The voice packet's source is the Asterisk server, and the destination of the voice packet is the student's system. Fig. 5 shows the jitter of web conference system. The packet source is the web conference server, and the destination is the student's system. From the both graphs, the average jitter of SIP system is less than the average jitter of the web conference system. The average jitter value of SIP system is 20 millisecond, but the web conference system is 50 millisecond.

The jitter of voice packets that are transmitted between Asterisk server and the student system are more stable than the jitter between the web conference server and the student system, as shown in Fig. 4 and Fig. 5. In the real situation, many students will connect to the server simultaneously. The less jitter implies to the smoother (continuous) sound.

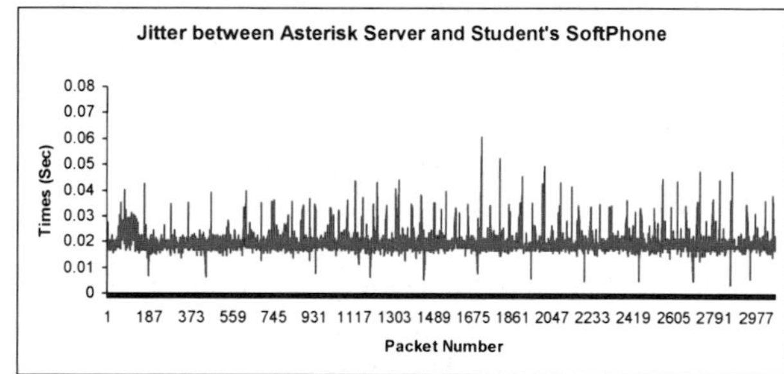

Fig. 4 Jitter between Asterisk server and student's Soft-phone.

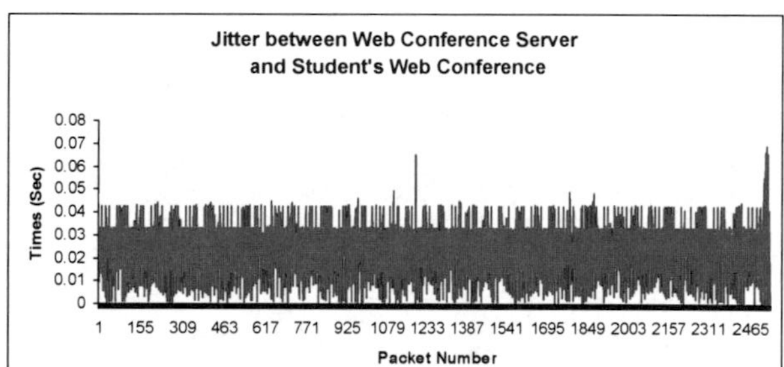

Fig. 5 Jitter between web conference server and student's system.

For delay experiment, the purposed virtual classroom using SIP and the web conference systems are compared in the scenario of voice broadcasting from the teacher to the student in one minute and is tested within a local area network.

The average end-to-end delay of the web conference system is 40.12 millisecond. The average end-to-end delay of the SIP system is 20.0 millisecond. The result shows that the web conference has two times higher average end-to-end delay than the purposed SIP system.

5 Conclusion

The new virtual classroom-using SIP is purposed. The system is designed to create an online virtual classroom using VoIP technology. A new user agent (Soft-phone) is developed. There are two types of user agents that are the user agent for the teacher and the user agent for the students. The user agent is designed to use with the unmodified SIP server (Asterisk). The key features of the new SIP user agent are the real-time voice conference, synchronization of PowerPoint slide between teacher's system and the students' systems, and live text chat.

The system uses UDP to transmit the voice over the Internet Protocol. The purposed system was tested and compared with a commercial web conference. The experiment's results show that the purposed system has better response time with less jitter and end-to-end delay. The real-time voice conference among all class attendances smoother than the web conference, which uses the TCP.

The web conference system is more convenient for all users because it does not require any installation on the client side. However, the TCP cause higher jitter and end-to-end delay than the SIP conference system which base on UDP. In the high congestion and capricious network condition as in the Internet, the virtual classroom-using SIP is more effective than the web conference.

The purposed system presents a new feature of the SIP conference, which is PowerPoint slide synchronization. The teacher and the student can share the PowerPoint file before the class start. The teacher can dominate the slides on the student's machine from his/her user agent. The technique reduces the slide's image download time during the class and makes the slide run smoother than the web conference approach.

Acknowledgments. This work is supported by NTC Telecommunication LAB (IP-Based Software), School of Information Technology, King Mongkut's University of Technology Thonburi.

References

1. Bian, L.: Research on E-learning Based on Network Technology. In: Proc. IEEE Conf. Networkingand Digital Society 2009, pp. 289–292. IEEE press (May 2009)
2. Sun, J., Fang, H., Wang, G., He, Z.: Information Push Technology and Its Application inNetwork Control system. In: Proc. IEEE Conf. Computer Science and Software Engineering 2008, pp. 198–201. IEEE press (December 2008)
3. Wang, Z., Xiong, X., Hou, Y.: The Study of Mobile Learning Based on Information PushTechnology. In: Proc. IEEE Conf. Education Technology and Computer Science, pp. 1140–1142. IEEE press (March 2009)
4. Handley, M.J., Jacobson, V.: SDP: Session Description Protocol. RFC 2327, IETF (1998)
5. Sinnreich, H., Johnston, A.B.: Internet Communications Using SIP. United States of America (2001)
6. Tan, K.K., Goh, H.L.: Session Initiation Protocol. In: Proc. IEEE Conf. Industrial Technology (ICIT 2002), vol. 2, pp. 1310–1314. IEEE press (December 2002)

7. Meggelen, J.V., Smith, J., Madsen, L.: Asterisk: The Future of Telephony, p. 408. O'Reilly, media (2005)
8. Ni, J., Luo, J.: Design of Multimedia Conference Control System based on SIP. In: Proc. IEEE Conf. International Conference on Software Engineering, Artificial Intelligence, Networking and Parallel/Distribution Computing (SNPD 2007), pp. 810–814. IEEE press (July 2007)
9. Wu, E.H.K., Hsieh, M., Wu, Z., Chiang, J.: Application layer multipoint conference using SIP. In: Proc. IEEE Symp. Personal, Indoor and Mobile Radio Communications 2004 (PIMRC 2004), pp. 2129–2133. IEEE press (September 2004)
10. Yanyan, Z., Yuan, Y.: SIP-based multimedia conference system design and implement. In: Proc. IEEE Conf. Computer Design and Applications 2010 (ICCDA 2010), vol. 3, pp. 607–610. IEEE press (June 2010)
11. DimDim (September 21 (2010), http://www.dimdim.com/

The Designing of the EPON System

Chen Guangjun

Abstract. Ethernet Passive Optical Network (EPON) is used in access network. , Optical Line Terminal (OLT) is connected with multiple Optical Network Unit (ONU) by the network connected by Optical Distribution Network （ODN） which consists of passive cable, optical splitters and optical combiners . Ethernet technology is by far the most successful and sophisticated Local Area Network (LAN) technology. EPON is the supplement of the existing agreement IEEE802.3, basically compatible. The compatibility of EPON and Ethernet is one of its greatest strengths. EPON's downstream channel is for the Fast/Gigabit of the broadcasting, while the upstream is for users to share the Fast/Gigabit channel. It offers symmetric uplink and downlink 1.25Gbit/s of bandwidth. In addition, EPON system can provide remote access within 20KM, with the characteristics of long distance and wide coverage.

1 Instruction

Today along with the rapid development of information technology, the end-user's demand has been extended from the initial invoice and internet access to video, data and other business , Then it will be further extended to High Definition Television (HDTV), video communication and other Multi-play business. Business development, especially the gradual promotion of the video category, is increasingly demanding for the stability, bandwidth and transmission distance of network[1,2,3,4].

In recent years, the rapid development of Fiber To The X (FTTx) technology and its application has become one of the international hot spots[2]. Fiber To The Home (FTTH) is fit for new residential and residential villas, which provides business solutions of voice, data and video communications package and builds high-quality information technology area[3]. FTTH needs one ONU device in each family to achieve various service access. FTTH needs high investment cost per household, so taking the acceptance of current actual consumer into account,

Chen Guangjun
Department of Computer and Commnunication Engineering Weifang University,
No. 5147, Dongfeng Street, Weifang City, China
e-mail: chengj_1211@163.com

F.L. Gaol (Ed.): Recent Progress in DEIT, Vol. 2, LNEE 157, pp. 395–400.
springerlink.com

Fiber To The Building (FTTB) is a generally applicable model. FTTB, as the prelude of FTTH, will be quickly transited to FTTH to meet the growing consumer demand of the users.

2 EPON'S Network Architecture

At present, the competition of our cable TV operators is mainly from the following four aspects[4] :

- The new media business as represented by IPTV has begun to threaten the basic cable TV service ;
- DBS(direct broadcasting by satellite) will launch and be operated, which will snatch the limited cable TV users of rural and urban-rural ;
- Development of family information terminal will shake the cable TV operators' monopoly on TV ;

One-way cable digital television can not meet users' needs; interactive television is the future direction of business.

Compared with other FTTx technology, EPON is a little economical by which we can reduce investment and long-term construction cost.

EPON technology can enable the network to achieve high-quality IPTV streaming media service, broadband data access service and VolP service, which can effictively support the service needs for the triple play[5].

The basic composition of EPON system:

- Optical Line Terminal（OLT）
- Optical Distribution Node（ODN, including optical splitter, optical fiber cable, optical fiber junction box, fiber cable cross-connecting cabinet and other passive components）
- Optical Network Unit（ONU）

EPON System's Network Connection Diagram as Fig 1.

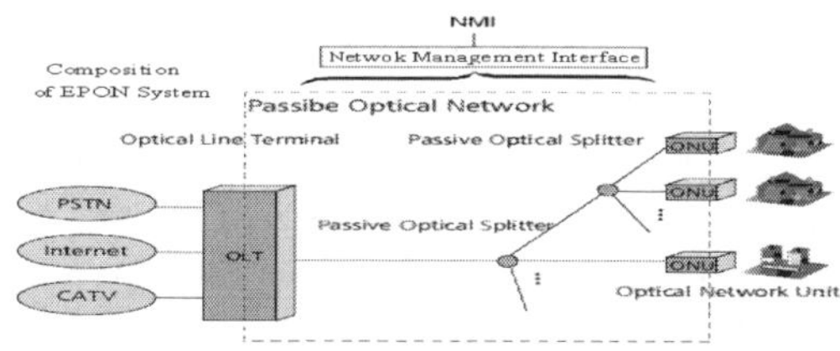

Fig. 1 EPON System's NetworkConnection Diagram

1. Signal Transmission

Signal transmission between OLT and ONU based on the Ethernet frame of IEEE 802.3 .

Line code is 8B/10B, data rate is symmetric 1Gbps uplink and downlink and line bit-rate of uplink and downlink is symmetric 1.25Gbps.

Based on MPCP (multi point control protocol) of MAC controling sublayer, MPCP control the topology of P2MP by message, state machine and timer. P2P emulation sub-layer is the key component of EPON/MPCP protocol.

2. Security of Downlink Data

For the feature of multicast of PON, all of the downlink data will be broadcasted to all ONU of PON system. If an anonymous user removes its ONU receiving function, then it can hear the downlink data of all users, which is called "Listening Threat" in PON system.

Another feature of PON network is that ONU in network can not hear other ONU's uplink data.

The solution to security in PON is to encrypt downlink message by OLT (including all data frames and OAM frames). The system should stir each LLID which has a separate key.

3. Encryption

The key questions of Encryption include:

Encryption algorithm, Generation and delivery of the key , Update and synchronization of the key.

Encryption algorithm:

- Advanced Encryption Standard (AES)-128: High security, but not meeting the national regulations for management of commercial key .
- Churning: Low security, simple.
- Triple-Churning: China Telecom's proprietary technology, improving the security of churning.

China Telecom uses the algorithm of Triple Churning.

The key is the 3-byte random extracted from uplink user's data from ONU, 3-byte random Exclusive OR (XOR) or the plus result.

The key is sent to OLT by the way of Operation Administration and Maintenance (OAM) message by ONU. OLT encrypts downlink data by the key according to specific encryption algorithm.

Key Update: The key must be periodically updated.

Key Synchronize: Because the key needs to be updated regularly, a synchronization mechanism is needed, making ONU know which key is currently used.

3 Design of OLT and ONU

1. OLT

Block Diagram of OLT as Fig2.

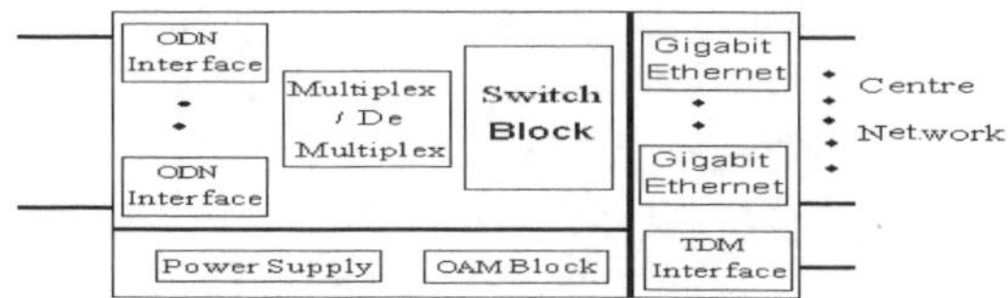

Fig. 2 Block Diagram of OLT

2. ONU
Block Diagram of ONU as Fig 3.

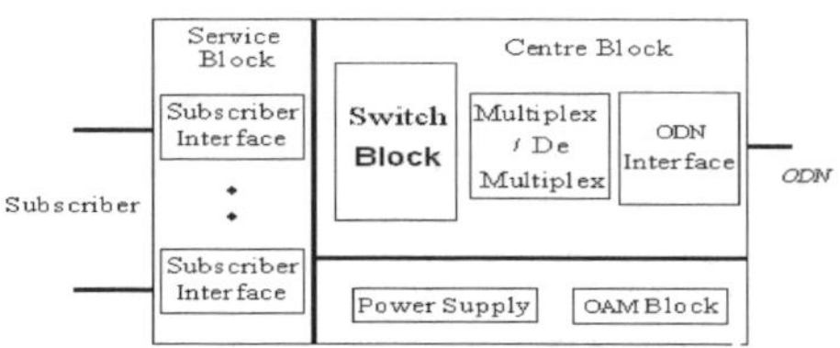

Fig. 3 Block Diagram of ONU

4 Full-Page Structure

The part of EPON chooses Cortina's CS8021 program. CS8021 Block Diagram as Fig 4.

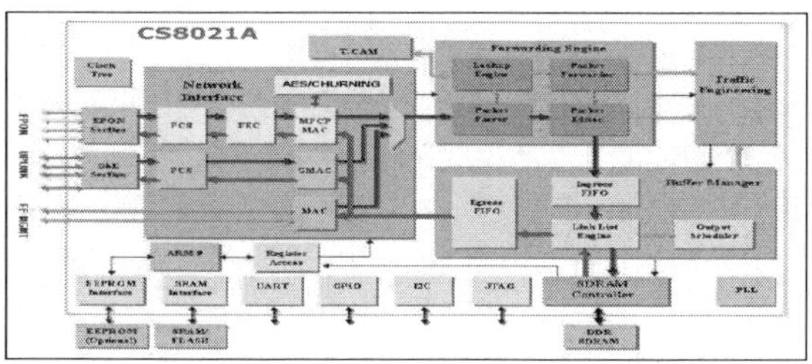

Fig. 4 CS8021 Block Diagram

From the figure we can see, for CS8021, 10/100/1000 MAC provides four SERDES interfaces or GMII interface, four EPON ports of SerDes and one FE management port.

BCM5464S Interface Method as Fig 5.

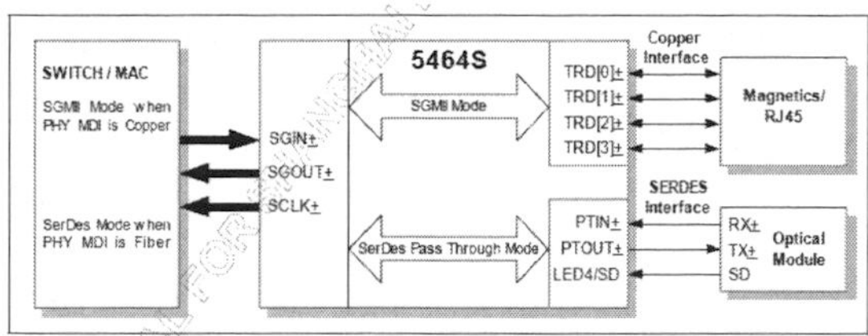

Fig. 5 BCM5464S Interface Method

1. PON Port

CS8021 provides four SerDes and TBI/GMII interfaces of 1.25Gbps, TBI/GMII for external SerDes, here we use built-in SerDes, so that we can plug light modules on it directly. PON Port as Fig 6.

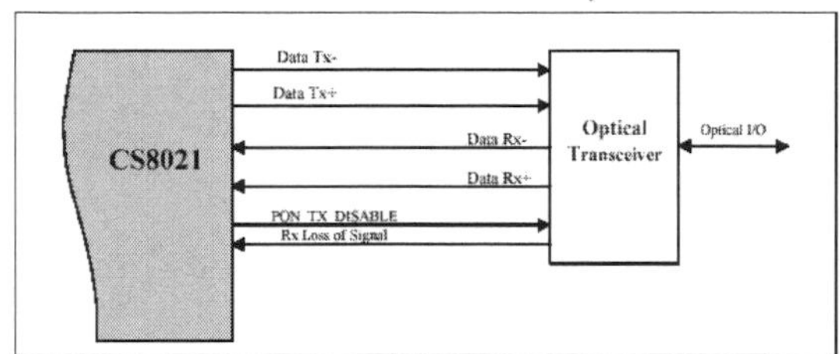

Fig. 6 PON Port

2. Management Interface

PPC405EP has two MII interfaces, one used for commission and management of the whole machine, the other used for the management of CS8021 and the download of OLT firmware, with MAC TO MAC way.

3. CPLD

CPLD needs to connect the following signals : twelve ID signal lines, LED lines of PON port 8+2=10, 15 line chips of data and address.

So that we can choose LC4064V-75T100C which we use most commonly.

4. Reset Watchdog Circuit

EPON OLT switch machine adopts hardware dog to achieve chip reset by combining software and hardware when putting electricity on a single board. DS1819 can generate reset signal and monitor voltage. When the voltage is lower than 2.85V, an effective and low reset signal is generated. ST pin must provide a clock lower than1.12s, otherwise it will produce reset signal. Those reset signals distributed from 74HC244 will reset CPU, OLT and PHY.

5. Hot Swap Design

It is mainly achieved by inseting card here, then responsed by software.

- For electrostatic discharge, it will conduct electricity on slideway.
- For inrush current, we adopt the existing FDS9926, making use of one way of them and adding RC circuit to control mosfet's opening time so as to achieve the slope control of voltage. It can play a role of rectifier and regulating voltage if adding rectifier ES2BA and 6V regulator on the front of FDS9926. There are some power control chips for hot-plug design on the market. Here, we have to set it aside.

5 Conclusion

Currently, EPON technology has matured, its system is stable and it is most suitable for the current market. Moreover, along with the gradual development of optical devices and technology, the difficulties of FTTx's cost and engineering has gone, fiber to the home becomes reality.

Optical fiber interface and FTTH take optical fiber as the transmission medium, with the advantages of high transmission capacity, high transmission quality, high reliability, long transmission distance and anti-electromagnetic interference, are the future development direction of fixed-access broadband.

References

1. Hajduczenia, M., da Silva, H.J.A., Monteiro, P.P.: EPON system Efficiency Evaluation with Extended GATE/REPORT MPCP Dus. In: Proceedings of the 11th IEEE Symposium on Computer and Commnunication, p. 45 (June 2006)
2. Sierra, A., Kartalopoulos, S.V.: Evaluation of two prevalent EPON Networks Using Simulation Methods. In: Proceedings of the Advanced Int'l Conference on Telecommuncations and Int'l Conference on Internet and Web Application and Services, fED 2006, p. 15 (February 2006)
3. Park, C.G., Lee, Y.: Performance analysis of DBA scheme with interleaved polling algorithm in an Ethernet PON. In: Proceedings of the Ninth International Symposium on Computer and Commnunication, ISCC 2004, vol. 2, p. 12 (June 2004)
4. Dhaini, A.R., Assi, C.M., Shami, A.: Quality of Service in TDM/WDM Ethernet Passive Optical Networks (EPONS). In: Proceedings of the 11th IEEE Symposium on Computer and Commnunication, p. 23 (June 2006)
5. Yang, Y.-M., Nho, J.-M., Ahn, B.-H.: An enhanced burst –polling based delta dynamic bandwidth allocation scheme for QOS over Epons. In: Proceedings of the 2004 ACM Worshop on Next –Generation Residential Broadband Challenges, p. 21 (October 2004)

A Case Study of a Successful mHealth Application: Cell-Life's EMIT System

Alfred Mukudu and Jean-Paul Van Belle

Abstract. There has been a massive investment in information system resources to aid in the fight against HIV/AIDS over the years. Recently, there has been an increased emphasis on mobile technologies as a key technology for developing countries in this fight. This paper describes a successful mobile health information system that is used by field workers in South Africa to gather and upload field data to a central database.

1 Introduction

In many parts of the world the HIV/AIDS pandemic continues to present grave challenges for healthcare providers and governments as they try to improve their populations' health. The joint United Nations Programme on HIV/AIDS (UNAIDS) and the World Health Organisation (WHO), estimates that there are 39.5 million people living with HIV at the end of 2006 [1; 2]. Over the years there has been an increasing mobilisation of civil societies, academia, private companies and other sectors of society to fight the spread of the pandemic and work towards eventually eradicating it. An essential component of this strategy is the effective deployment of health management information systems (HMIS).

This paper documents a successful mobile health information system used by field workers in South Africa to gather and upload field data to a central database. The system is called EMIT and is a data capture and analysis system developed by Cell-Life. The system was developed to allow data capture from different devices into a central database, from which analysis of data would be performed to produce operational information used by the program managers of the community-based organisations. Until now, only a high-level description of a pilot study was available to the research community [3]. This paper aims to disseminate a more

Alfred Mukudu · Jean-Paul Van Belle
University of Cape Town, Rondebosch, South Africa
e-mail: munyam@gmail.com, Jean-Paul.VanBelle@uct.ac.za

F.L. Gaol (Ed.): Recent Progress in DEIT, Vol. 2, LNEE 157, pp. 401–408.
springerlink.com

detailed system description along with initial implementation findings and user feedback of the larger scale roll-out and implementation.

2 Background on HIV/AIDS and Health Information Systems

Although accurate statistics are difficult to obtain, South Africa had an estimated total population of 48 282 000 in 2006, a HIV prevalence of 16 579 per 100 000 population in 2005, 21 % ARV coverage among people with advanced HIV infection in 2006, 675 deaths due to HIV/AIDS per 100 000 population per year in 2005 [4]. Dorrington *et al* estimated a slightly lower figure, with approximately 5.4 million South Africans living with HIV in 2006, including 294 000 children between the ages of 0 and 14 [5]. In 2006, an estimated 71% of all deaths in the 15-49 age groups were due to HIV [5]. Also, 66% of the 1.5 million orphans in the country in 2006 had been orphaned as a result of HIV and AIDS [6].

Over the years, government, public and private sector organizations have invested heavily in IT in healthcare in a bid to improve service delivery and performance of healthcare systems. In the South African health sector, 14.1% of government expenditure was on health in 2007 [7]. Introduction of information and communication technologies (ICT's) can radically affect healthcare organisation and healthcare delivery and outcomes, however it is estimated that up to 60-70% of all software projects fail [8]. The challenge is to not only provide tools for prevention of infection, monitoring and evaluation of efforts to combat the disease, but also to ensure these tools result in the increased effectiveness of the organisations and health care workers using them.

A health information system (HIS) is defined here as a system that integrates data collection, processing, reporting, and use of the information necessary for improving health service effectiveness and efficiency through better management at all levels of health services [9]. A health management information system (HMIS) thus becomes an information system specifically designed to assist in the management and planning health services (as opposed to delivery of care) [9].

An HMIS can be decomposed into two subsystems [10]. The first entity is the data collection level where data is collected and transformed into information. The second component is the conceptual level where analysis and feedback take place to transform the information into knowledge for decision making. The HMIS is thus more than a tool for collecting data as it is also an instrument to aid decision making by providing knowledge.

The use of mobile technologies in health care is a growing field that is now referred to as mHealth. It is a sub-segment of eHealth that utilises mobile communications for health services and information. The range of applications being designed is constantly expanding. Some key applications for mHealth in developing countries are: education and awareness, remote data collection, remote monitoring, communication and training for healthcare workers, disease and epidemic outbreak tracking, diagnostic and treatment support [11].

Some HMIS projects focus exclusively on using mobile technologies in the form of cellular phones i.e. they exclude PDA's and tablet computers. A South African example of such project is Project Masiluleke (meaning "hope and warm council in Zulu). The project began by sending one million text messages per day throughout South Africa that encourage people to be tested and treated for HIV/AIDS. By capitalizing on the ubiquity of mobile devices in even the most resource constrained areas, Project Masiluleke increases the reach of volunteer care workers and health educators. The project's efforts in increasing AIDS awareness among the community have led to nearly a 350% increase in call volume to the local South African HIV/AIDS helpline. The project has however been criticised for not being scalable due to the cost implications of using SMS, as such continuous funding is required to keep the service running and growing [11].

3 System Context and Sponsor

The following system description was inspired by one of the researchers' experiences during the implementation a mobile version of an existing HMIS by Cell-life, and the subsequent rollout of the mobile implementation to organisations countrywide. The case description builds on an initial study conducted by Cell-life on the implementation of the EMIT system for Community Media Trust (CMT), with a focus on the suitability of the mobile implementation for the organisational tasks across different organisations. At the time of writing, some information on the Cell-life's CMT case could be found on the EMIT website [3]. Additional background information for this case description was gathered from business analysts and program managers working on the implementation, as well as the researcher's own observations. A first hand observation of the implementation process, the organisations affected and the technical aspects of the EMIT system contributed to a greater understanding of the project dynamics of using mobile phones in health care systems.

Cell-Life is a non-profit organisation that started as a research project at the University of Cape Town in 2001. Cell-Life is based in Cape Town South Africa and aims at using technology-based applications to address different challenges of HIV/AIDS, such as: HIV/AIDS information awareness, ARV treatment monitoring and adherence enforcement in South Africa. More recently, the organisation has placed higher emphasis on mobile based technology solutions to approach the challenges of HIV/AIDS through their 'Cell Phones for HIV' project. The Cell Phone for HIV project aims at exploring the range of cell phone applications for information, communication and interactive services to support the HIV sector. Cell-Life is sponsored by the following organisations: Cape Peninsula University of Technology, the University of Cape Town, Vodacom, The Desmond Tutu HIV/AIDS Research Centre, Vodacom foundation and the Raith foundation.

4 Technical Description of the EMIT System

The EMIT system is a data capture and analysis system developed by Cell-Life. The system was developed to allow data capture from different devices into a central database, then from this database analysis of data would be performed to produce operational information used by the program managers of the community-based organisations.

The information flow in this HMIS is as follows. Data collection on the desktop system is performed through a survey application written in PHP that is running on a central server. The PHP application consists of a range of web forms that are used to capture user (fieldworker) data input. On the mobile system, the data collection forms are provided by a Java Mobile Edition (J2ME) application built on an open-source platform for data collection on mobile devices known as Javarosa [12]. The Javarosa platform uses the Xforms data collection web standard to define the way in which form data collected is transmitted across the internet and determines the formatting of questions in a web form. The application stores the forms on the device and, as such, the number of completed and uncompleted forms the fieldworkers can store is limited by the memory capabilities of the mobile phone. On the mobile EMIT system, the forms are transmitted to the database on the central server through GPRS.

The information can be retrieved in two different ways: through Internet access or through J2Me, a store and forward mode which allows data to be entered and saved on a mobile phone and later sent when there is GPRS network coverage. A core design objective of this system was to ensure that there would be a consistent look and feel between the desktop and mobile instances of the application. The reasoning was that the more consistent the look and feel, the easier it would be to train people who had been using the desktop system first. This would result in a higher take up and utilisation of the mobile system since it was familiar to the users. The mobile and desktop systems thus follow similar workflows and the basic functionality of the data capture system is common to both input types.

Another objective was that of ease of use, because of the varying levels of technical proficiency amongst the field workers and other potential users of the system. It is, however, interesting to note a comment made by a business analyst at Cell Life that, after distributing the cell phone application to some of the fieldworkers during the pilot stage, they took to the cell phone application quicker than they had to the desktop system earlier. This is because the fieldworkers interact with cell phone applications on a daily basis and are therefore familiar with navigation and general use of cell phone applications. This is not surprising in a country where more than 80% of the households own a cell phone [13]. The system's ease of use was also ensured by constant interaction with the users of the system during the design stage to ensure their requirements and needs were understood and incorporated into the final system.

The EMIT HMIS places emphasis on security and audit capabilities, as it has the capacity to capture and store sensitive information pertaining to HIV and other health information which is regarded as private. Thus, the systems employs password authentication on both the desktop and mobile system, with the passwords being stored on a central server as opposed to on the actual input devices. The ability to audit data captured is a requirement placed on the system by program coordinators and funders. The system provides a reference number for each record captured allowing data auditors to verify the reference number on the mobile device with the corresponding database reference.

The EMIT system follows the dual subsystem structure suggested by Fenenga and Jager [10]. The data collection system is used by fieldworkers (field agents) through either mobile devices or desktop computers. The desktop and mobile data collection devices transmit their data over the internet to the database, though it is important to note that the data collection is performed through a set of electronic forms. The forms differ between the desktop and mobile device interfaces, as the mobile forms are custom designed to suit the device constraints of the mobile device [14]. A key consideration is their reduced screen size or form factor. Also, considerations should be made to include measures to account for the limited quality of network connection (e.g., limited bandwidth requirements, indicators of network quality) which is a key differentiator between mobile systems and desktop systems. Other considerations are to be made regarding system elements such as processor and battery performance, storage capacity, bandwidth requirements, menu structures, setup requirements, system performance and user dialogue.

The second subsystem of the HMIS is that of data analysis. In the EMIT system, a server side application provides the data analysis capabilities of the system. The application allows users to securely login and access information regarding data collected such as descriptive statistics on raw data submissions, as well as more detailed analyses of responses to individual submissions.

5 Project Implementation

Experiences with the system were recorded for the implementation of the EMIT system for 20 organisations funded by Johns Hopkins Health Education South Africa (JHHESA). The system was piloted at Community Media Trust (CMT), a community based organisation in Cape Town South Africa. It is here that the mobile data capture devices were first piloted and introduced, in parallel to the desktop system. This allowed for the comparison of the effect of a mobile version of the system with the desktop version before rolling out the system to the rest of the organisations in the nationwide JHHESA network. Cell Life undertook to assess the individual organisations earmarked to use the EMIT system for different levels of technology and process maturity. Organisations with immature processes would be given assistance in process establishment and building up organisational capacity up to a point when they could be able to use the EMIT system. Organisations which were more process-mature would be provided with training on the EMIT system from the onset. The technology aptitude of the fieldworkers was determined through a brief questionnaire and if the field workers were adjudged to be proficient they would be moved onto the mobile version of the EMIT system. Upon completion of the roll out, the organisations utilising the EMIT system have users using a mix of the desktop and mobile system versions across South Africa.

The project has recorded significant success since its inception having resulted in 20025 forms being submitted over the first 9 months of the project. These submissions were from 312 web users and 70 mobile users. This is quite a significant number of submissions, as each form submitted represents a single training and/or literacy session delivered by a field worker on HIV/AIDS. The forms thus represent the number of messages delivered to various communities pertaining to HIV/AIDS treatment and prevention. The use of a mobile and web HMIS would have greatly enhanced the organisational capacity of the participating organisations, as they are able to reduce the turnaround time in processing information captured by the forms.

6 System Evaluation

In a survey, users of the mobile system were asked to make a comparison of the desktop system, which preceded the current mobile system and compare it to this system and communicate a preference. Although 12 (37%) of the 33 respondents did not express a preference either way, the majority of the others seemed to prefer the mobile implementation.

One fieldworker stated that "before the cell phones we all had to come to the office on Fridays and take turns to use the same computer to submit the forms and it was difficult because it is far for some of us." The reason for this evaluation could be that those users who were later moved onto the mobile system felt that using the mobile system was a form of reward for having done well with the desktop system. An interview with a fieldworker revealed that there was a sense of achievement amongst fieldworkers when they were asked to move from one input system to another. As one fieldworker explained it: "We were told that Cell Life was going to bring us a new system for submitting our reports and some of us had to wait before they were upgraded to the cell phones. But after we all got the phones, I think everyone likes it".

This stems from the way the implementation process was explained to the organisations and the staff. At the beginning of the project it was made clear that the implementation of the mobile data collection version of the EMIT system was dependent on the level of process maturity of the organisations. As such, being asked to move to the mobile system was an acknowledgement of more mature data collection processes and more capable staff. It is plausible that the users' perception of the system would change as they would be under the impression that, since the use of the mobile system was an advanced step, then the mobile system must be better, which in turn would influence their evaluations.

During an interview with a programme manager it emerged that the definition of the data the fieldworkers collect and the managers analyse is subject to regular change. This may have an impact on the meaning variable, and even the task characteristics. According to the manager "Every now and again we get new indicators to monitor which requires us to change the content of the information we are teaching in the community. As managers we always make sure the fieldworkers don't have to worry about that and they just have to know about the new content, not the indicators behind them."

7 Challenges and Benefits

As this is an ongoing project, the challenges and benefits from this system continue to emerge. The key challenges and benefits of the system that have been identified so far are listed below. These were gathered from observations, as well as comments from the business analysts from Cell Life and users at one of the implementing organisations based in Cape Town.

Some of the challenges experienced during this project are the following:

- The mobile version of the system is constrained by quality of cell phone coverage. A fieldworker from a Durban-based organisation commented: "I can't submit my forms when I am at the prison because the signal there is very poor, so I must wait until I get home at the end of the day to do that, you see."
- Accessing the web system over the web is dependent on the availability and quality of internet speeds in those areas. The organisations taking part in the implementation of EMIT are provided with internet access by the funding partner JHHESA. In some areas, this is slow and is a source of great frustration for the users as they try to submit data.
- There is a low usage of the data analysis capabilities of the system. Most usage is focused on the data collection and population of the HMIS.
- Field workers are at higher risk of theft as they have to carry certain (high value) cell phones into public areas while they conduct their literacy campaigns
- Training on the system is difficult where computer and cell phone (general IT) literacy is low. This has typically resulted in low usage of the system in some organisations. A Cell Life business analyst had this to say: "Utilisation of the system has been admittedly low at first in some of the organisations [...] probably due to a lack of understanding of what the system is for. When it comes to training though, we find it easier to train the cell phone version users as there is high literacy there. It's quite a leap to using the same system on a computer... "
- There is an added, minimal cost to organisations for submitting data across the mobile phone network via GPRS. Individual data submissions are observed to cost between 2c and 5c (US\$1 $\approx$ 700c) and, even though this is very low, it still requires organisations to ensure their field workers always have airtime.
- The mobile application is subject to memory issues. This means that on low-end mobile devices, memory is quickly filled up with submitted forms and these have to be deleted periodically. The application is also unable to run on the cheapest J2ME enabled cell phones, restricting the reach of the application.

The following key benefits were identified:

- A significant increase in data quality, as observed by comments made by one of the program managers at a Cape Town organization: "We are definitely spending less time cleaning our data for our monthly reports, and this is especially true for the data we get from mobile phone submissions."
- The turnover time between data collection and reporting on the data has reduced. This is due to the fact that the data is available for analysis in an electronic form upon submission by the field workers. This is as opposed to the paper-based system that was in use in most of the organisations prior to the implementation of EMIT.
- There is increased satisfaction amongst the fieldworkers using the mobile system, as they no longer have to travel distances in order to submit data collected during their field work. In the absence of a mobile version of the system, field workers would have to travel to the organisation's base to submit form data to a central internet-enabled computer.

8 Conclusion

This paper described a health information system designed to capture data from fieldworkers using mobile phones. The system replaced an earlier desktop-based system. From both a technical and user perspective, the system has been an unqualified success. However, a few practical challenges remain such as loss/theft of mobile device, form factor (memory), minimal costs and connectivity. It is hoped that the demonstration of a success mHealth application will lead other researchers to investigate the potential of mobile phones for data capture and feedback in health and other functional information systems.

References

1. UNAIDS, Report on the Global Aids Epidemic, UNAIDS/WHO (2008)
2. Department of Health, Republic of South Africa, Progress Report on Declaration of Commitment on HIV and AIDS, Reporting Period: January 2006-December 2007. Prepared for United Nations General Assembly Special Session on HIV and AIDS (2008)
3. EMITMobile, Mobile Monitoring and Evaluation with Community Media Trust, http://www.emitmobile.co.za/mobile-monitoring-and-evaluation-with-community-media-trust/ (last accessed February 25, 2010)
4. WHO, World Health Organisation's regions/countries indicators database, http://www.who.int/whosis/data/Search.jsp (last accessed December 17, 2009)
5. Dorrington, R., et al.: The Demographic Impact of HIV/AIDS in South Africa: National and Provincial Indicators for 2006. Centre for Actuarial Research, South African Medical Research Council and Actuarial Society of South Africa, Cape Town (2006)
6. Medical Research Council, Initial estimates from the South African National Burden of Disease Study. MRC Policy Brief, No. 1. MRC, Tygerberg (2003)
7. Health Systems Trust, Health Expenditure and Finance in South Africa (2009), http://www.hst.org.za/uploads/files/hstefsa.pdf (last accessed February 25, 2010)
8. Ammenwerth, E., Iller, C., Mahler, C.: IT-adoption and the interaction of task, technology and individuals: A fit framework and case study. BMC Medical Informatics and Decision Making (2006)
9. WHO, Developing health management information systems: a practical guide for developing countries (2004), http://www.wpro.who.int/NR/rdonlyres/Health_manage.pdf (accessed October 20, 2010)
10. Fenenga, C., de Jager, A.: Health Management Information Systems as a Tool for Organizational Development. The Electronic Journal of Information Systems in Developing Countries 31 (2007)
11. Vital Wave Consulting, mHealth for Development:The opportunity of mobile technology in the developing world, http://www.globalproblems-globalsolutions-files.org (accessed December 5, 2009)
12. Javarosa, What is Javarosa? (2009), http://code.javarosa.org (last accessed May 10, 2010)
13. Markettree Consulting, Cell phone usage in South Africa (2005), http://www.markettree.co.za (accessed March 15, 2010)
14. Gebauer, J., Shaw, M., Gribbins, M.: Success Factors and Impacts of Mobile Business Applications: Results from a Mobile e-Procurement Study. International Journal of Electronic Commerce 8(3), 19–41 (2004)

Combine DPC and TDM for MIMO Broadcast Channels in Circuit Data Scenarios

Yingbo Li, Da Wang, Guocheng Lv, Mingke Dong, and Ye Jin

Abstract. Dirty paper coding (DPC) is shown to achieve the capacity of multiple-input multiple-output (MIMO) Gaussian broadcast channels (BCs). Finding the optimal covariance matrices and order of users for maximizing user capacity requires high complexity computation. To deal with this problem, many researchers use fixed order of users e.g. minimum power first (MPF) and get subpar performance. Meanwhile, the complexity of DPC scheme grows linearly with number of users requiring to cancel known-interference. In this paper we present a scheme combining DPC and traditional Time Division Multiplexing (TDM) that aim to maximize user capacity and reduce complexity of implementation. Simulation results show that the proposed scheme achieves better performance than that of DPC MPF scheme.

1 Introduction

Multiple-input multiple-output (MIMO) techniques, by implementing multiple antennas at both the transmitter and receiver, have receive a lot of attention, since the heuristic research works by Telatar [1] and G. J. Foschini [2]. We consider downlink case of a cellular system, which is also referred as MIMO broadcast channel (MIMO BC). Dirty paper coding (DPC) is, to the best of our knowledge, the only way to achieve the capacity region of MIMO BCs. It is not convenient to study the whole capacity region of a broadcast system when multiple users exist. Many researchers turn to practical operating points of the capacity region. Most of published papers deal with sum rate capacity of MIMO BC [3].

The sum rate capacity is only a practical factor in packet data scenarios. However, in circuit data scenarios, services demanded by users are sensitive to delay.

Yingbo Li · Da Wang · Guocheng Lv · Mingke Dong · Ye Jin
School of Electronics Engineering & Computer Science, Peking University, Beijing, 100871, P.R. China
e-mail: {yingbo.li,wangda,lvguocheng,mingke.dong, jinye}@pku.edu.cn

F.L. Gaol (Ed.): Recent Progress in DEIT, Vol. 2, LNEE 157, pp. 409–414.
springerlink.com

Each user requires to get service from the base station at fixed rate. In this case, a practical performance evaluation factor is the user capacity. The user capacity is defined in [4] as the expected value of the number of users/receivers to which the transmitter can simultaneously transmit at an equal rate.

Although, DPC is capacity achieving in MIMO BC, finding optimal users' covariance matrices and encoding order for maximization of user capacity introduces high complexity. Some suboptimal schemes can be found in [5], where fixed order of users is used. Users are sorted by ascending order of the power desired to achieve the target rate, which is referred as minimum power first (MPF) scheme.

In this paper, we propose a better scheme combining DPC and traditional time division multiplexing (TDM). Users are divided into small groups with few users, e.g. 2 or 3, in the same group. We use TDM for different transmitting of different groups, and DPC for canceling known-interference within each group. Simulation results show that the proposed scheme achieves better performance than DPC MPF scheme with reduced implementation complexity.

2 System Model

Consider a MIMO BC with a base station (BS) employed with t antennas and K mobile users (MUs) with $r_1,\ldots,r_K$ antennas, respectively. The BS is required to send independent data streams to K MUs constraint to a total sum power limitation of P. The channel model of MIMO Gaussian BC at any time sample is represented as follows for each user:

$$\mathbf{y}_i^{r_i\times 1} = \mathbf{H}_i^{r_i\times t}\mathbf{x}^{t\times 1} + \mathbf{n}_i^{t\times 1}, i = 1,2,\ldots,K \tag{1}$$

In (1), $\mathbf{x} = \sum_{i=1}^{K}\mathbf{x}_i$, each $\mathbf{x}_i \in \mathbf{C}^{t\times 1}$ is signals intended for user i respectively. $\mathbf{x}_i$ is assumed to have independent circular symmetric stationary complex Gaussian distribution, with zero mean and covariance matrix $\Sigma_i = \mathbf{E}[\mathbf{x}_i\mathbf{x}_i^{\dagger}]$. The input signal is subject to an total power constraint P, $tr(\Sigma_x) \leq P$, where $\Sigma_x = \mathbf{E}[\mathbf{x}\mathbf{x}^{\dagger}] = \Sigma_1 + \Sigma_2 +, \ldots, +\Sigma_K$ is the covariance matrix of the input signal, assuming independent signals intended for different users.

$\mathbf{H}_i \in \mathbf{C}^{r_i\times t}$ is the complex channel gain matrix of user i. These matrix are considered to be perfectly known at both the transmitter and the receiver. We consider $\mathbf{H}_i$ to be fixed during every channel use, i.e. we have enough time to estimate channel side information. The length of each frame may be very long to achieve good performance of Dirty Paper Coding. Entries of $\mathbf{H}_i$ have independent and identical complex Gaussian distribution with zero mean and unit variance $\sigma_h^2 = 1$, i.e. $\mathscr{C}\mathscr{N}(0,1)$.

$\mathbf{n}_i \in \mathbf{C}^{r_i\times 1}$ is the complex additive white Gaussian noise at receiver side of user i. Entries of $\mathbf{n}_i$ are also independently and identically distributed with zero mean and unit variance.

The signal-to-noise ratio is defined as $SNR = \frac{\sigma_h^2}{\sigma_n^2}P = P$. Power limitation P is adjusted for different SNR consideration in the following sections.

3 User Capacity Evaluation

In a circuit data transmitting system, service provided by the base station such as voice communication and video conference, is very sensitive to time delay. Each user demands to get service at a fixed rate. Without loss of generality, we assume all the users in the system desire the same fixed rate, denoted as R_0. The user capacity of a system is defined as the number of users it can serve simultaneously at rate R_0 desired by each user.

3.1 MPF Scheme

In MPF scheme of [5], the users are encoded in increasing order of the power needed by them to achieve the desire rate R_0, seeing no interference from other users, i.e. single user case. The covariance matrices of each users are determined sequentially.

First sort the users in ascending power cost to achieve R_0, and choose the top N users. Assume users are already sorted as $1,\dots,K$. User 1 consumes least power, and is encoded last to make sure it sees no interference from other users, $\mathbf{y}_1 = \mathbf{H}_1\mathbf{x}_1 + \mathbf{n}_1$. The desired R_0 is achieved by the base station using as less power as possible. The covariance matrix Σ_1 of $\mathbf{x}_1$ is found by a optimization problem by minimizing $tr(\Sigma_1)$. This can be calculated either by a convex optimization algorithm or by standard water-filling algorithm. For user i, it is subject to interference from users that consume less power, whose covariance matrices already known as $\Sigma_j, j < i$.

$$\mathbf{y}_i = \mathbf{H}_i\mathbf{x}_i + \mathbf{H}_i\sum_{j<i}\mathbf{x}_j + \mathbf{n}_i \tag{2}$$

For calculating covariance matrix of user i, one can use the effective channel concept as in [6]. The effective channel gain matrix is defined as follows

$$\tilde{\mathbf{H}}_i = [\mathbf{I} + \mathbf{H}_i(\sum_{j<i}\Sigma_j)\mathbf{H}_i^{\dagger}]^{-1/2}\mathbf{H}_i \tag{3}$$

In this encoding order, the suboptimal user capacity is found to be

$$N_{DPC} = \max\{N : \sum_{i=1}^{N} tr(\Sigma_i) \leq P\} \tag{4}$$

4 Combining Scheme for User Capacity

DPC is the optimal scheme to serve as many users at fixed desired rate as possible, through simultaneously transmitting. When using DPC, only the message for the first-encoded user can be encoded by conventional channel coding, treating the interference from other users as Gaussian noise. Message intended for the following users have to be encoded by DPC canceling known interference at the transmitter. This brings high complexity in practical implementation.

On the other hand, TDM scheme can be easily implemented, although suboptimal. In TDM, only one user is served at a time. C_i is the rate achievable by user i when the base station uses all the power to serve it exclusively. Using a fraction $\frac{R_0}{C_i}$ of time for user i is enough to achieve the desirable rate. Thus, the total number of users simultaneously served is defined as follows.

$$N_{TDM} = \max\{N : \sum_{i=1}^{N} \frac{R_0}{C_i} \leq 1\} \tag{5}$$

The maximization can be done by offering time to users according to the descending order of their C_i.

We combine DPC and TDM for maximizing of user capacity. In TDM, only one user is served at any instant time. In MIMO, this is not an efficient way in using channel resource. We can add more users in each time slots, so that they will consume less time resource overall. In order to reduce the complexity, we divide users into m groups, $\mathcal{G}_1, \ldots, \mathcal{G}_m$. DPC is used in each group to cancel known interference while TDM is applied for transmitting of different groups. Thus, much less messages are encoded by DPC, and the complexity is greatly reduced. Within each group, we use whole power to achieve best performance. In this circumstance, we do not optimize the sum rate achieved by members in the group. That's because time slot assigned to some group of users is determined by the lowest rate within the group. As a matter of fact, we optimize the maximum equal rate. Then different time sharing are allocated to each group, just like the TDM scheme described previous.

For group $\mathcal{G}_j$, denote $\mathbf{H}_{\mathcal{G}_j}$ as all the channel matrices of users within this group. The optimization problem is

$$\begin{aligned} &\max_{\{r_1,\ldots,r_{|\mathcal{G}_j|}\} \in C_{BC}(P, \mathbf{H}_{\mathcal{G}_j})} \quad r_{equ} \\ &s.t. \quad r_1 = \ldots = r_{|\mathcal{G}_j|} = r_{equ} \end{aligned} \tag{6}$$

To find the optimal group pattern involves calculation of high complexity. We sort the users by simple order and members of each group are chosen one by one. Motivated by the TDM scheme, we sort users in descending order of single-user capacity. Then users are chosen to each group sequentially according to cardinality of each group. Through simulation we find that this order already works well.

Denote $C_{\mathcal{G}} = \max\{|\mathcal{G}_1|, \ldots, |\mathcal{G}_m|\}$. If $C_{\mathcal{G}}$ is chosen to be small, small portion of users have to be encoded by DPC. For example, when $C_{\mathcal{G}} = 2$, only half of the message have to be encoded by the complex DPC. This greatly reduce implementation complexity.

5 Performance Simulation Results

In this section we present simulation results to reveal performance of combining scheme for MIMO BC under criteria of user capacity. For different instance of channel matrices, the maximum number of users supported is a random variable. We will reveal average user capacity of combining scheme in Fig. 1-2 for different antenna configurations.

In Fig. 1 both the transmitter side and the receiver side are equipped with four antennas. For different number of users in the system, we calculate average maximum number of users supported by different schemes. For combining scheme, $C_{\mathscr{G}} = 2, 3$ are considered, and denoted as DPC TDM(2) and DPC TDM(3) respectively. The performance of TDM scheme is also revealed for comparison. DPC TDM(2) achieve nearly the same performance as DPC MPF scheme. And DPC TDM(3) outperforms DPC MPF scheme. And their complexity are nearly half and Two-thirds of DPC MPF scheme, respectively. Thus a complexity-performance trade-off can be made by adjusting parameter $C_{\mathscr{G}}$ in the combining scheme. Increasing $C_{\mathscr{G}}$ means put more users in one group, and the best performance can be achieved when $C_{\mathscr{G}}$ equals total number of users in system, where the combining scheme reduces to DPC scheme. When $C_{\mathscr{G}} = 1$, the combining scheme reduces to TDM scheme.

From Fig. 1 we can see that all the schemes considered have an increasing user capacity as total number of users enlarges. This comes from multi-user gain of the system, where opportunity communication can be carried out.

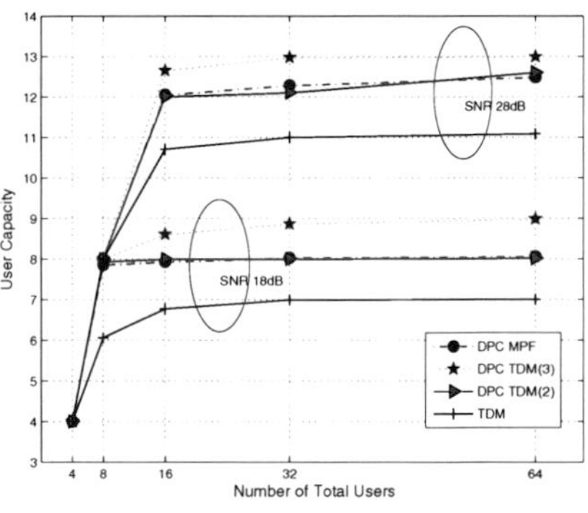

Fig. 1 User Capacity, support 3 bps/Hz for each user, antenna configuration t=4, r=4. The top four lines are user capacity at SNR 28dB, and the rest four 18dB.

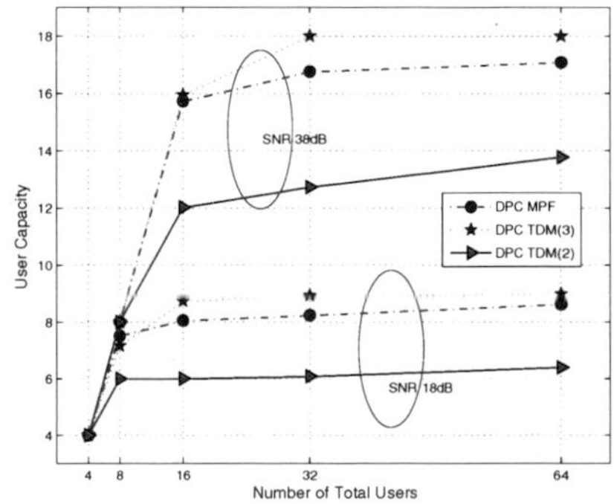

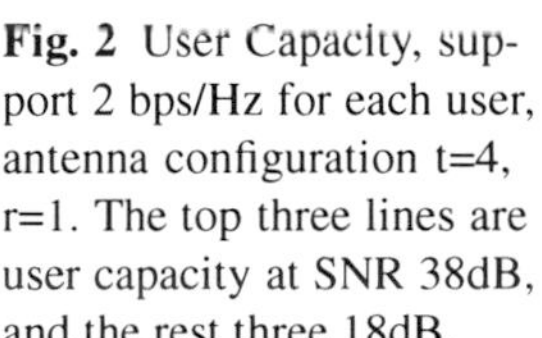
Fig. 2 User Capacity, support 2 bps/Hz for each user, antenna configuration t=4, r=1. The top three lines are user capacity at SNR 38dB, and the rest three 18dB.

In Fig . 1, both transmitter and receivers have equal number of antennas. However in practical environment, receivers are usually limited in power and scalability, which means less antennas will be used in the receiver side. In Fig. 2 the transmitter has four antennas, and each of the users have single antenna. Although DPC TDM(2) is not better than DPC MPF scheme, DPC TDM(3) scheme still outperforms it.

6 Conclusion

In this paper we propose scheme of combining DPC and TDM for MIMO Gaussian BC. Comparisons are made under user capacity criteria. Simulation results show that combining scheme has the advantage of achieving better user capacity than DPC MPF scheme by reduced complexity.

References

1. Telatar, I.: Capacity of multi-antenna gaussian channels. Bell Laboratories, Tech. Rep. (June 1995)
2. Foschini, G.J., Gans, M.J.: On limits of wireless communications in a fading environment when using multiple antennas. Wireless Personal Communications 6, 311–335 (1998)
3. Caire, G., Shamai, S.: On the achievable throughput of a multiantenna gaussian broadcast channel. IEEE Transactions on Information Theory 49(7), 1691–1706 (2003)
4. Boppana, S., Shea, J.: Downlink user capacity of cellular systems: Tdma vs dirty paper coding. In: Proc. IEEE Int. Symp. Inform. Theory, July 2006, pp. 754–758 (2006)
5. Viswanathan, H., Venkatesan, S., Huang, H.: Downlink capacity evaluation of cellular networks with known-interference cancellation. IEEE J. Sel. Areas Commun. 21(5), 802–811 (2003)
6. Vishwanath, S., Jindal, N., Goldsmith, A.: Duality, Achievable Rates and Sum-Rate Capacity of Gaussian MIMO Broadcast Channels. IEEE Transactions on Information Theory 49, 2658–2668 (2003)

Estimating Average Round-Trip Time from Bidirectional Flow Records in NetFlow

Chao Gao, Wei Ding, Yan Zhang, and Jian Gong

Abstract. Round-trip time (RTT) is significant to the network performance. Traditional RTT passive estimating methods are based on IP Traces, which need specialized packet capture equipment and mass storage space. This paper proposes a method to estimate average round-trip time between two hosts by analysis TCP steady traffic model, using specific bidirectional flow records in NetFlow. Test result shows that this method has acceptable error.

1 Introduction

Delay is an important metric of network performance, which may be applied to QoS Assurance, network planning, congestion detection, etc. Delay can be divided into two categories: one-way delay and round-trip delay. One-way delay is defined as the interval between when source host sends a packet and when destination host receives it. Round-trip delay, often called round-trip time (RTT), is the time between sending of a packet and the receipt of its acknowledgement. The detail definitions of one-way delay and round-trip delay are given in RFC 2679 and RFC 2681. User experience of network service is obviously dependent on the delay between sending a request and receiving the response, therefore, we mainly focus on the RTT of TCP flow.

There are many approaches to estimate the RTT, which can be classified in two main categories: active measurement and passive measurement. Active measurement approach estimate RTT by sending test packets to calculate the delay of link.

Chao Gao · Wei Ding · Yan Zhang
School of Computer Science and Engineering, Southeast University
e-mail: {chgao,wding}@njnet.edu.cn

Jian Gong
Jiangsu Province Key Laboratory of Computer Networking Technology,
Southeast University
e-mail: jgong@njnet.edu.cn

F.L. Gaol (Ed.): Recent Progress in DEIT, Vol. 2, LNEE 157, pp. 415–423.
springerlink.com

But active approach often brings additional overhead into traffic; it is not suitable for the monitor devices deployed in network. Passive approaches deploy monitor point in network, and analyze the traffic data to retrieve RTT of TCP flow. At present, many passive RTT estimate algorithms have been proposed: SYN-ACK method estimates RTT by monitoring SYN/SACK/ACK packets in TCP connection establish period[1]; SLOW-START method observes the round gap in TCP slow-start period to estimates RTT[1]; PRE method uses round gap caused by TCP congestion control to estimates RTT[2]; TIMESTAMP method calculate RTT by observing the timestamp field in TCP segments[3]; SELF-CLOCKING method uses autocorrelation to make RTT estimates, since bulk data stream exhibits certain self-clocking pattern[3]. RUNNING method maintains a finite-state-machine to track the congestion window and RTT[4]; SPECTRAL ANALYSIS method applies Lomb-Scargle periodogram to RTT estimating[5,6].

Estimating methods described above need IP trace captured by monitor devices. But in high speed network environment such as network management center, specialized equipment and mass storage space are required to recording detail information of every packet, such as address, arrival time, etc. Therefore, the application range of those methods is limited due to the cost.

NetFlow is a network traffic monitoring technique developed by CISCO in 1996. This technique is widely integrated in many commercial routers and switches provided by CISCO, Juniper and other major network device manufacturers. Because of the widely application of NetFlow, analyzing data set collected by NetFlow has an important significance on RTT estimation. Our objective in this paper is to estimate the end-to-end RTT of TCP flows in passive way by analyzing NetFlow traces. RTT discussed here is not a single measurement result at a particular time, but rather the average RTT value of a TCP flow in terms of statistics. In other word, RTT of a TCP NetFlow record is defined as the average RTT of source host and destination host in the duration of that record.

2 RTT Estimate Based on Netflow Records

A network flow is defined as a unidirectional sequence of packets from a source host to a destination host in RFC 2722 and RFC 3697. In traditional Cisco definition, packets in a flow share 7-tuple key, which are: source and destination IP address, source and destination port, IP protocol, IP Type of Service and ingress interface. The router examines 7-tuple key of the captured packets to determine whether it belongs to any recognized flows. If a match flow is found, its traffic information will be updated; otherwise a new flow record will be generated.

The router will send a NetFlow record to NetFlow data collector in four cases: a segment with FIN or RST flag is captured, which indicates that this TCP session is terminated; traffic has been stopped for a certain period, usually 15 seconds; the cache of router has been depleted; routers also output a flow record at a fixed interval (normally 30 minutes) even if the flow is still ongoing.

Latest version of NetFlow is v9, but the most common version is v5, which is available on many routers from different brand, but restricted to IPv4 flows. A NetFlow v5 record contains a wide variety of information about the traffic in a given flow, including timestamps for the flow start and end, number of bytes and packets in flow, source and destination IP addresses, source and destination IP addresses and port numbers, IP protocol, Type of Service (ToS) value and etc. In the case of TCP flows, it also has a tcp_flags field, which refers to the union of all TCP flags observed over the life of the flow.

Our RTT estimate methods are based on the data collected by NetFlow v5. TCP flows are differentiated into several categories according to TCP flags, port numbers and other fields in NetFlow records, and we propose different estimate methods for each category.

2.1 Bidirectional Flows Containing Starting and Ending Phase

TCP uses a three-way handshake to establish a connection. First, the client sends a SYN to the server. Second, the server replies with a SYN-ACK in response. Finally, the client sends an ACK back to the server. In most cases, the connection terminates with a four-way handshake. When a host wishes to stop its half of the connection, it sends a FIN packet, then the other host acknowledges with an ACK. Therefore, a typical tear-down requires a pair of FIN and ACK segments from each TCP endpoint. Each side of the connection should terminate independently. These two phases are shown in Fig. 1.

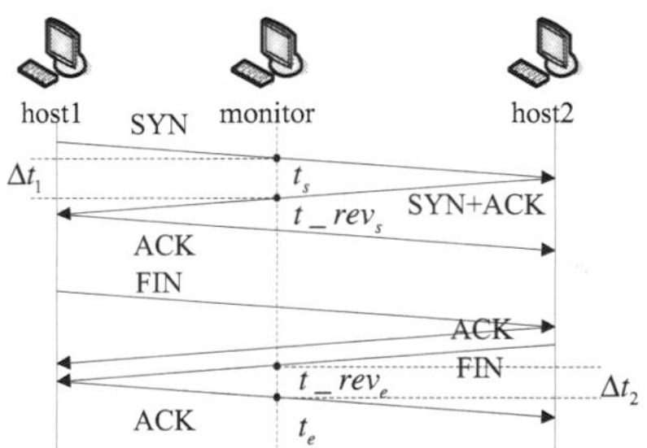

Fig. 1 Connection start and end

NetFlow only stores start and end time of flows instead of details information of each packet, we can not retrieve exact timestamps of three-way and four-way handshake packets. Thus, RTT could not be calculated directly from those timestamps. But if the monitor observes a flow and its reverse flow, let Δt_1 and Δt_2 represent the intervals between start and end times of two flows, then Δt_1 and Δt_2 can be added to obtain the RTT of this flow.

If a NetFlow record F contains connection establishment and termination phase (SYN and FIN bits are set in tcp_flags), let t_s and t_e be the start and end time of this flow. And if the reverse flow record F_rev exists, and also including two phases mentioned above, let t_rev_s and t_rev_e be the start and end time of the reverse flow. If $t_s < t\_rev_s$ and $t\_rev_e < t_e$, RTT can be calculated by (1).

$$RTT = (t\_rev_s - t_s) + (t_e - t\_rev_e) \quad (1)$$

2.2 Delayed ACK Flows

Delayed ACK technique described in RFC 1122 suggests that a host that is receiving a stream of TCP data segment should send less than one ACK segment per data segment received. Primary operation systems such as Windows usually send one ACK when it received two data segment. For a stable TCP data flow without packet loss, the ratio of packet number to its ACK flow packet number should be 2:1. In this case, the average intervals of ACK segments in ACK flow can be regarded as the mean RTT of this TCP session.

If a data flow record F and its reverse ACK flow record F_rev exists, let N and N_rev be the packet number of the data flow and the reverse flow, let t_rev_s and t_rev_e be the start and end time of the reverse flow, let S be the sampling rate. If $N : N\_rev = 2:1$, we can calculate mean RTT by (2).

$$RTT = \frac{t\_rev_e - t\_rev_s}{N\_rev \times S} \quad (2)$$

2.3 Bidirectional Flows without Congestion

In a wired network, random bit error rate may be negligibly small. Therefore, packets loss indicates the occurrence of congestion. Hua Wu represents a loss rate estimate method based on packet number ratio of bidirectional flow records in [7]. According to this method, bidirectional flows with round-trip packet ratio of 2:1 could be assumed to be congestion free.

In the case of congestion free, TCP flow begins with slow-start phase after connection establishment. When the threshold has been reached, TCP enters the congestion avoidance phase. At this phase, the sender's windows is linearly increased until the minim value of congestion window and receive windows is reached.

Let t_s and t_e be the start and end time of a flow, and let X be the round number has been sent during this flow. If we can estimate X from the NetFlow record, then RTT can be generated by (3).

$$RTT = \frac{t_e - t_s}{X} \quad (3)$$

Assume host sends i rounds in slow-start phase, and then it sends no more than ssthresh (usually 655535) bytes at $i-th$ round. Let p be the average packet length, then i is given as:

$$i = \left\lceil \log_2 \frac{ssthresh}{p} \right\rceil + 1 \quad (4)$$

After $i-th$ round, packets sent in each round increase linearly until it reaches maximum windows limit $W_{\max}$ (default value is 110KB). The packets be sent at $k-th$ $(k > i)$ round is $2^{i-1} + (k-i)$.

Assume N packets are sent in j rounds. We solve j by two cases:

- Sending Window does not Reaches Limitation

Packets number N increases exponentially in slow-start phase, and increases linearly in congestion-avoidance phase:

$$N = \sum_{k1}^{i} 2^{k1-1} + \sum_{k2=i+1}^{j} (2^{i-1} + k2 - i) \quad (5)$$

From (5), we can solve j as:

$$j = i + \sqrt{(\frac{1}{2} + 2^{i-1})^2 + 2(N - 2^i + 1)} - (2^{i-1} + \frac{1}{2}) \quad (6)$$

- Sending window reaches limitation

Assume host sending window reaches limit after t rounds at congestion avoidance phase. Packet number be sent at this round is $W_{\max}^{'} = W_{\max}/p$, and we have $2^{i-1} + t = W_{\max}^{'}$. Thus, total packet number sent before host reaches limit is given as:

$$N^{'} = \sum_{k1}^{i} 2^{k1-1} + \sum_{k2=i+1}^{W_{\max}^{'}} (2^{i-1} + k2 - i) = 2^i - 1 + \frac{(W_{\max}^{'} - 2^{i-1})(W_{\max}^{'} + 2^{i-1} + 1)}{2} \quad (7)$$

Host needs totally j rounds to send N packets:

$$j = i + t + \frac{N - N'}{W_{\max}^{'}} \quad (8)$$

In summary, assume a NetFlow record F represents a TCP bulk data stream, judging by its port number (21, 80 and etc.). Let N be its packet number, t_s and t_e be start time and end time, and S be the sampling rate. If its reverse flow record F_rev exists, let N_rev be the packet number of F_rev. If $N : N\_rev = 2:1$, we consider flow F has no congestion. Total packet number $N^{'}$ sent before host reaches limit can be calculated by (4) and (7). If total packet number in flow F is bigger than $N^{'}$, we use (8) to calculate rounds X, otherwise, (6) is used. Finally, put X in (3) to obtain RTT of this flow.

2.4 Bidirectional Flows with Congestion

If the round-trip packets ratio of bidirectional flows do not equals 2:1, which means packet loss rate is not 0, this flow can be considered as having congestion. According to [8], a TCP data stream is composed of many transmission periods. Each period consists three phases: Time-out, slow-start and congestion avoidance, as shown in Fig. 2. W^i is the windows size at $i-th$ period. Expected value of congestion windows is given in [9]:

$$E[W] = \frac{2+b}{3b} + \sqrt{\frac{8(1-p)}{3bp} + (\frac{2+b}{3b})^2} \tag{9}$$

In (9), p is the loss rate, b represents host send one ACK per b segments, usually b is 2.

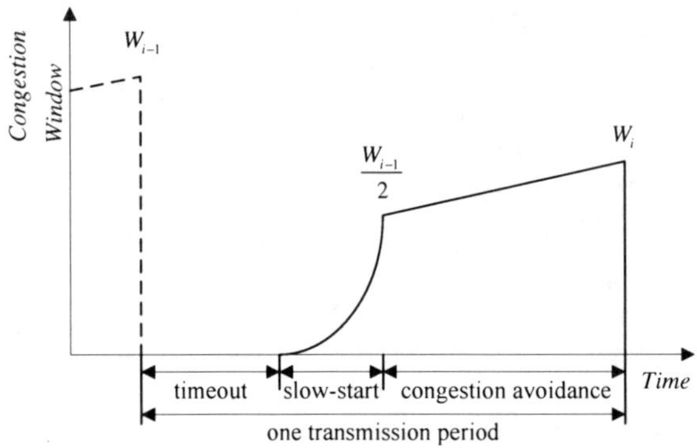

Fig. 2 One TCP transmission period

RTT estimating is discussed below according to whether sending window reaches limitation.

- Sending Window does not Reaches Limitation

$E[Y]$ is the expected number of packets been sent, and $E[A]$ is expected duration time of flow, which are given in (10) and (11):

$$E[Y] = \frac{1}{1-p} + \frac{b+1}{2}E[W] - b + (1+P_{TD})(\frac{1}{p}+1) \tag{10}$$

$$E[A] = T_0\frac{f(p)}{1-p} + RTT(\log_2\frac{E[W]}{2} + 1 + (1+P_{TD})\frac{b}{2}E[W]) \tag{11}$$

$$f(p) = 1 + p + 2p^2 + 4p^3 + 8p^4 + 16p^5 + 32p^6 \tag{12}$$

P_{TD} is the probability of packet loss caused by triple-duplicate ACK. T_0 is the initial retransmission timeout (RTO) estimated between the sender and receiver, we treat it as RTT roughly in this paper.

- Sending Window Reaches Limitation

Considering the effect of time-out, slow-start, congestion avoidance and receiving window limitation, [8] proposed a modified model of packet number and flow duration time:

$$E[Y]=\frac{1}{1-p}+\frac{b+1}{2}W_{\max}-b+(1+P_{TD})(\frac{1}{p}+1) \quad (13)$$

$$E[A]=T_0\frac{f(p)}{1-p}+RTT(\log_2\frac{W_{\max}}{2}+1+(1+P_{TD})\frac{b}{2}W_{\max}+\frac{1}{p\cdot W_{\max}}) \quad (14)$$

In summary, NetFlow record F is a TCP bulk data stream, let t_s and t_e be start time and end time, and S be the sampling rate. Assume its reverse flow record F_rev exists, let N_rev be the packet number of F_rev. If $N:N\_rev \neq 2:1$, loss rate p can be calculated by (15) described in [7]:

$$LossRate=\frac{3r^2}{4r+2+\sqrt{-11r^2+16r+4}} \quad (15)$$

In (15), $r=(2N-M)/M$, M and N are the packet number of a bidirectional flow, and $M>N$.

The throughput of a TCP sender can be regarded as a function of loss rate and average RTT. Further more, when throughput and loss rate are known, RTT can also be calculated. Packets number been sent in a flow can be obtained using $N\times S$, and the flow duration time is t_e-t_s. Thus using (10), (11), (13), (14), RTT_1 (RTT of a flow which not reaches window limitation) and RTT_2 (flow reaches limitation) can be solved. Because we cannot determine whether flow reaches limitation, thus RTT is estimated in (16), $P_{\lim}$ is the probability of flow reaches receiving window limitation, assumes to 0.5.

$$RTT=(1-P_{\lim})RTT_1+P_{\lim}RTT_2 \quad (16)$$

3 Evaluation

In order to collect testing data, we build an experiment environment as shown in Fig. 3. One client PC downloads file from a remote server, using FTP download tools. The NetFlow collector is a router deployed in northeast network center of CERNET. The file downloaded is big enough to assure the NetFlow collector stores enough data. And for comparison, client PC uses active measure to get the actual RTT, by ping the server while downloading.

The correlation curve between estimated RTT and actual RTT of the same host (202.112.23.35) is shown in Fig. 4. Experiment shows that the change tendency of estimated RTT is mostly same as actual RTT, although some errors still exist. We also noticed that packet number in NetFlow records also has affect on RTT estimating. Thus, we choose the flows with packet number bigger than 25 when estimating RTT.

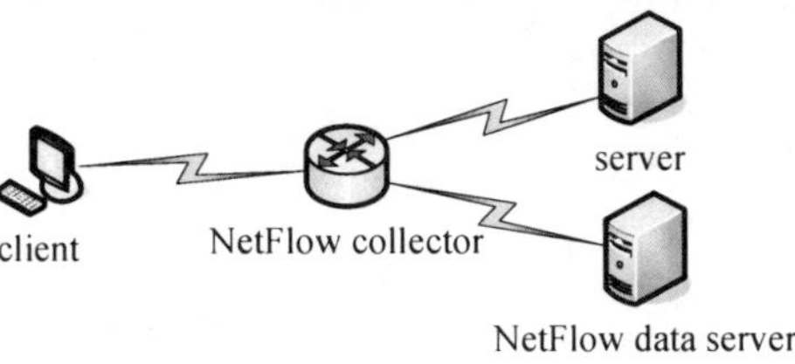

Fig. 3 Experiment environment

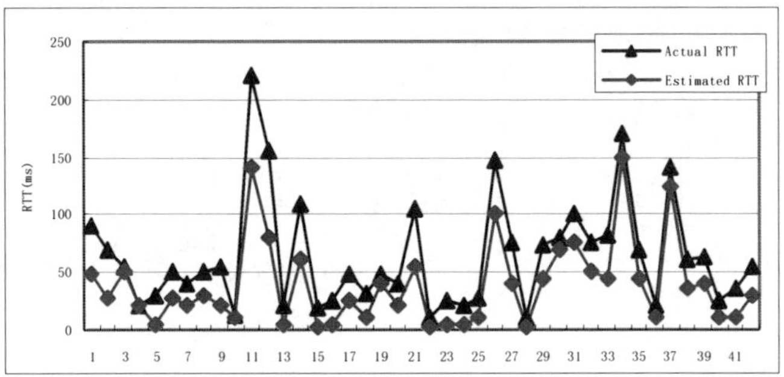

Fig. 4 Actual RTT and estimated RTT

4 Conclusion

Many passive measurement methods have been proposed for RTT estimating. But most of these methods are based on full record of IP traces. In this paper, we propose a method of estimating average RTT of TCP flows, using the NetFlow records collected by routers. This method has a precondition that the computed flow has a reverse flow record. And this method uses several fields in NetFlow record: start and end timestamp, packet number, TCP flags, and the packet number ratio of a flow and its reverse flow.

The basic idea is obtain the loss rate by packet number ratio of a flow and reverse flow, and analysis the TCP steady traffic model with given loss rate to calculate sending rounds in the flow duration time. Finally, the average RTT can be acquired from the flow duration divide by sending rounds. Furthermore, we put forward a simple approach for two special flow types: flows containing establishment and termination Phase, delayed ACK flows. Testing result shows that the estimated RTT has acceptable precision.

As future work, it would be of importance to investigate the traffic model of unidirectional flows, which is more popular in flow records.

Acknowledgments. This work was supported by the Major State Basic Research Development Program of China (973 Program) 2009CB320505 and National Key Technology R&D Program 2008BAH37B04.

References

1. Jiang, H., Dovrolis, C.: Passive estimation of TCP round-trip times. ACM SIGCOMM Computer Communication Review 32, 75–88 (2002)
2. Zhang, Y., Lei, Z.: Estimate round trip time of TCP in a passive way. In: Proc. of ICSP 2004, Beijing, China, pp. 1914–1917 (2004)
3. Veal, B., Li, K., Lowenthal, D.: New methods for passive estimation of TCP round-trip times. In: Proc. of Passive and Active Measurement, Boston, USA, pp. 121–134 (2005)
4. Jaiswal, S., Iannaccone, G., Diot, C., et al.: Inferring TCP connection characteristics through passive measurements. In: Proc. of INFOCOM 2004, Hong Kong, China, pp. 1582–1592 (2004)
5. Lance, R., Frommer, I.: Round-Trip Time inference via passive monitoring. In: Proc. of SIGMETRICS, pp. 32--38 (2005)
6. Carra, D., Avrachenkov, K., Alouf, S., et al.: Passive online RTT estimation for flow-aware routers using one-way traffic. Technical Report INRIA-00436444 (2009)
7. Wu, H., Gong, J.: Packet Loss Estimation of TCP Flows Based on the Delayed ACK Mechanism. In: Hong, C.S., Tonouchi, T., Ma, Y., Chao, C.-S. (eds.) APNOMS 2009. LNCS, vol. 5787, pp. 540–543. Springer, Heidelberg (2009)
8. Zhao, J., Zhang, S.-J., Zhou, Q.-G.: An adapted model for TCP steady state throughput and its performance analysis. Journal of China Institute of Communications 24(1), 52–59 (in Chinese)
9. Padhye, J., Firoiu, V., Towsley, D., et al.: Modeling TCP Reno performance: a simple model and its empirical validation. In: Proc. of SIGCOMM 1998, pp. 133–145 (1998)

PAR: Prioritized DTN Routing Protocol Based on Age Rumors

Luan Xu, Hongyu Wei, Zhengbao Li, and Haipeng Qu

Abstract. In this paper, we proposed a prioritized DTN Routing Protocol Based on Age Rumors (PAR). In order to reduce the waste of resources in message transference procedure and increase data delivery ratio, the protocol limits the total number of message copy through combining forwarding-based protocols and replication-based ones effectively. Through simulative analysis, the algorithm is found to be able to achieve high data delivery ratio and improve the network performance.

Keywords: delay tolerant networks (DTNs), routing protocol, age rumor.

1 Introduction

Delay tolerant networks (DTNs)[1], which are characterized by intermittent connection, high transfer delay and error rates, attract increasing interest these years. Initially, it is proposed for Inter Planetary Network (IPN), and now it covers many new networks, such as wildlife tracking networks, pocket switched networks and military networks.

The representative forwarding-based routing algorithms include Randomized Routing, Utility routing[2] and a utility-based algorithm named seek-and-focus[3]. The representative replication based routing algorithms include Epidemic routing[4], PRoPHET[5] and Spray-and-Wait/Focus[6]. In addition, a rumor routing algorithm based on age message was proposed by Jacquet[7]. The relatively new algorithms include EC (erasure-coding)[8] and EBEC (estimation-based erasure coding)[9].

Luan Xu
College of Information Science and Engineering, Ocean University of China, Qingdao
e-mail: xudipingxian@126.com

Hongyu Wei
Dept. of Network and System Development, National Marine Data & Information Service, Tianjin, China
e-mail: why@mail.nmdis.gov.cn

F.L. Gaol (Ed.): Recent Progress in DEIT, Vol. 2, LNEE 157, pp. 425–431.
springerlink.com

Conventional communication model of mobile ad hoc network (MANET), such as AODV, DSR, which require at least one path exists between source and destination nodes, do not work in delay tolerant networks. DTNs overcome the problems associated with intermittent connectivity by using store-carry-forward message switching. A message could be sent over an existing link, getting buffered at the next hop until the next link in the path comes up, and so on and so forth, until it reaches its destination or expires. So how to select optimal next hop becomes the key issue of designing efficient DTN routing algorithms.

In order to achieve certain transmission rate and reduce transfer delay in DTNs, multiple copies of message are usually transferred independently. Multiple copies will increase energy consumption. Therefore, controlling the number of message copies becomes the key issue. In this paper, a Prioritized DTN Routing Protocol Based on Age Rumors (PAR) is proposed to solve the problem of controlling the message copies. The basic idea of PAR is to limit the number of message copies by using spraying based on age rumor, thus reducing waste of resources and achieving better delivery ratio. Through simulative analysis, the algorithm is found to be able to achieve better delivery ratio and reduce waste of resources.

The rest of this paper is organized as follows. Our algorithm is described in detail in Section 2. And then the algorithm is verified through simulative analysis in Section 3. We conclude this paper and point out future research directions in Section 4.

2 Details of PAR

2.1 Network Model

Initially, N nodes are randomly distributed in a M*M square area, R is communication radius of node. In addition, the network has the following characteristics:

(1) All nodes move according to Random Waypoint model, the speed is constant (v), and all the nodes move independently. The Random Waypoint model can be described as follows: select destination D randomly, starting from source S to D in a straight line with speed v, and then treat D as the start node of next hop, so on and so forth.
(2) Before sending message, every node should explicit the destination node (represented by D).
(3) The algorithm has high requirements on timing, so a time synchronization mechanism should be executed after nodes deployment.

2.2 Algorithm Design and Implementation

2.2.1 Basic Idea

Destination *D* broadcast a very small age message, which is getting old over time. When a node receives more than one age message which is send by a same destination, it will keep the minimal one. The value of age message will serve as the criterion of forwarding decision making. A difference threshold is set to be turning point from forwarding to replication.

As shown in Fig.1, node *i* has minimal age in all neighbors of S, and the age difference between them is greater than the threshold, then we can consider that such transmitting shortens the distance with destination obviously, the decision is forwarding directly; in a similar way, the age difference between *i* and *j* is greater than the threshold too, then forwarding directly; the age differences between j and all of its neighbors are not significant, less than the threshold, then it can be treat as a turning point from forwarding to replication, transmitting three copies to three youngest neighbors, *k*, *m* and *n* (number of copies *L*=3). And then each copy is transmitted independently.

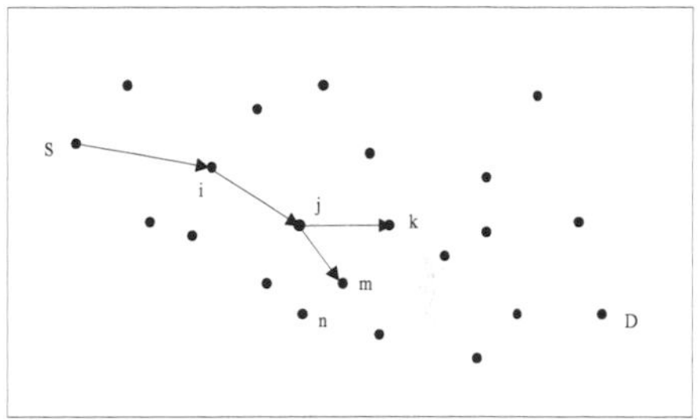

Fig. 1 Basic principles of the algorithm

2.2.2 Algorithm Parameters

The related parameters are described as follows:

(1) Age message: the time for transmitting a message from one node to another reflects the distance and the probability of communication between them to a great extent. The age message, which records the time information from destination to others, is set to 0 initially and getting old over time.

    Setting age of the message: node *i* may be neighbors of many nodes at the same time, which means simultaneously existing m nodes(m>1) j, k...which meet the conditions: $|ij|\leq R$, $|ik|\leq R$... Now *i* receives *m* age messages, then it should store *m* nodes and their age in buffer in the shape of 2-tuple $<a, b>$: *a* indicates neighbor nodes' ID, and *b* indicates the age from this neighbor to *i* (e.g. $<j, age_{ji}>$, $<k, age_{ki}>$). Finally, the age of i will be the minimal of *m* age messages, which is $age_i = \min\{age_{ji}, age_{ki}...\}$.

(2) The age difference threshold τ_{age} which is the turning point from forwarding to replication: the age difference between two nodes reflects their ability difference to transmit a same message to its destination. If the difference between a node and its next hop is not significant, it can be learnt the transmitting makes little sense to shorten the distance between the message and its destination. The threshold is predefined initially, whose value is decided according to the density of nodes.

(3) The number of copies: In Spray and Wait/Focus, the simplest way is determining *L* according to the expected delay of the application. *L* depends on *n* and number of nodes *N*. In our future work we will discuss the selection in detail.

2.2.3 Algorithm Description

(1) The source node *S* sends a message to *D*, firstly flooding a small message which records the destination node ID, in order that every node, including *D*, knows destination of this communication. And then, judging whether *D* is in communication radius of S, if $|SD| \leqslant R$, *S* can send message to *D* directly. Otherwise, go to (2);

(2) After *D* receives the small flooding message, it broadcasts an age message which is set to 0 initially. All of *D*'s neighbors can receive the message directly; all the neighbors broadcast the age message too, sending it to nodes in their communication radiuses. So on and so forth, until all the nodes in the network receive the age message.

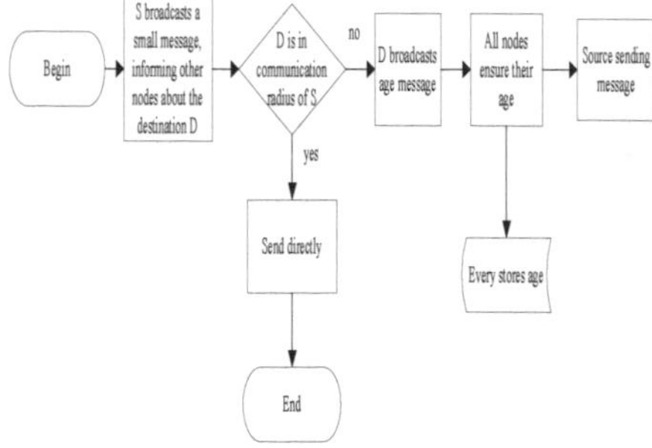

Fig. 2 Algorithm flow chart (Preparation stage)

(3) *S* sends message, at first it will select the neighbor whose age is minimal (assuming *S* has m neighbors, *1, 2, 3…m*), and then calculate the age difference Δ between *S* and the node, $\Delta = age_s\text{-}\min(age_1, age_2, age_3 \ldots age_m)$, if $\Delta < \tau_{age}$, go to (4); otherwise, go to (5);

(4) *S* sends *L* copies to *L* minimal age neighbors, and then the nodes which receive the copy should select the minimal age neighbor as its next hop; repeating the process until at least one of the copies arrives at destination or the message expires.

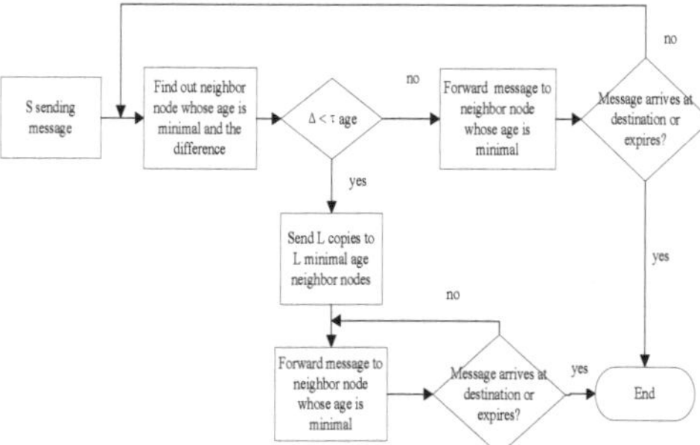

Fig. 3 Algorithm flow chart (transmitting stage)

(5) S sends the message to its neighbor whose age is minimal (assuming i), and then consider i as the source, go to (3), until transforming from forwarding to replication or destination receive the message or it expires.

2.3 Example

As shown in Fig.4, S is sending message to D, the dotted circle is the communication boundary of nodes, and R is communication radius, the number beside node is its age. The whole path is built according to age. For simplicity, τ_{age} is set to 10ms and L is 2.

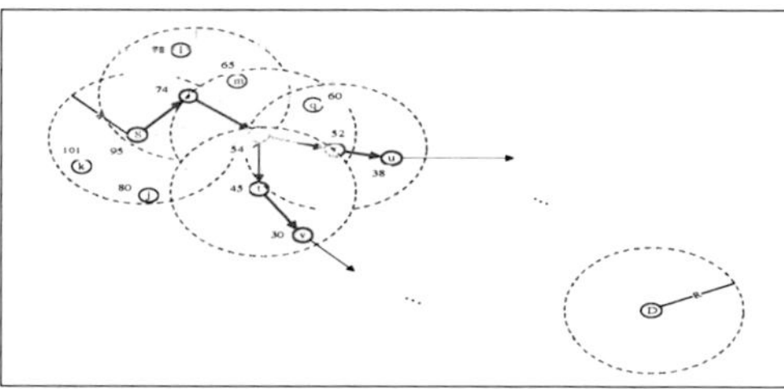

Fig. 4 Example

S(95) is sending message to D, it has three neighbors: i(74), j(80), k(101), now the age difference is Δ = 95-min{74, 80, 101}=21>τ_{age}, and then S sends message to its minimal age neighbor i; i(74)has three neighbors: l(78), m(65), n(54), now Δ = 74-min{78, 65, 54}=20>τ_{age}, i forwards it to n; n(54) has four neighbors: q(60), m(65), r(52), t(45), now Δ = 54-min{60, 65, 52, 45}=9<τ_{age}, it will be a turning point from forwarding to replication: n sends 2 copies to its two minimal age neighbors t and r; and then, the two copies will be transmitted independently, and the nodes which receive message copy will forward it to its minimal age neighbor, until one of the copies arrives at destination or the message expires.

3 Simulation

Table 1 Parameters

Parameters	Default values
Network size M * M	100*100
Number of Nodes N	20
Communication Radius R (m)	3
Speed of nodes V (m/s)	3
Threshold τ_{age} (s)	2
Number of copies L	3

We compare our algorithm with age rumors by simulation.

Delivery ratio: Comparing to age rumor, PAR considers much more about the dynamic topology. When network topology changes frequently, it will be probable that the message be sent to an isolated node in traditional age rumor and the packet is lost. But in PAR, the nodes selected are distributed wildly. The performance will be different according to different values of a parameter.

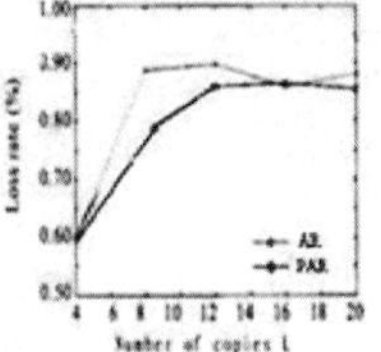
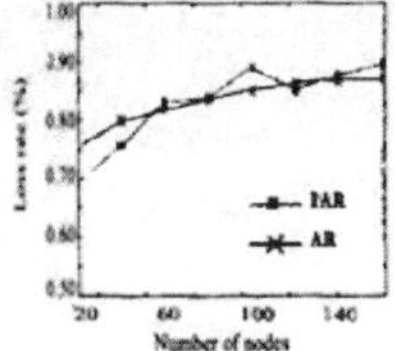

Fig. 5 Loss rate of different L and number of nodes

Average delay: Because an age message should be broadcast before communication, the average delay of PAR and age rumor is slightly higher than other protocols. However, better delivery ratio will be beneficial to the performance of the whole network.

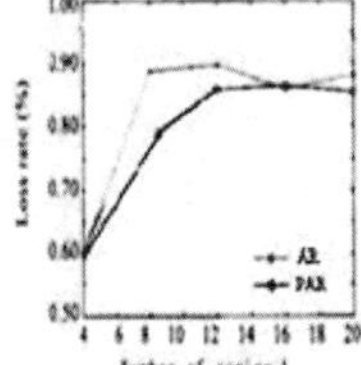
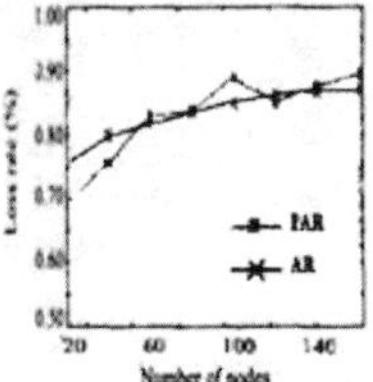

Fig. 6 Average delay of different L and number of nodes

4 Conclusion

Based on age rumors, we have proposed an effective routing algorithm by limiting the number of copies. Nodes forward or replicate the message according to the age difference between itself and its minimal age neighbor. Through simulative analysis, the algorithm is found to be able to achieve high data delivery ratio.

In future work, we intend to focus on the following problems: number of copies L, which is determined by specific application; time synchronization of nodes, it is necessary to design specific algorithms applicable to DTNs.

Acknowledgments. This work is supported by the National Natural Science Funds under grant No.60970129, No.60933011 and No.60873248.

References

[1] Fall, K.: A delay-tolerant network architecture for challenged Internets. In: Proc. of the 2003 Conf. on Applications, Technologies, Architectures, and Protocols for Computer Communications, pp. 27–34. ACM, Kalsruhe (2003)

[2] Lindgren, A., Doria, A., Schelen, O.: Probabilistic routing in intermittently connected networks. SIGMOBILE Mobile Comput. Commun. Rev. 7(3) (2003)

[3] Spyropoulos, T., Psounis, K., Raghavendra, C.S.: Efficient routing in intermittently connected mobile networks: The single-copy case. IEEE/ACM Trans. on Network 16(1), 63–76 (2008)

[4] Vahdat, A., Becker, D.: Epidemic routing for partially connected ad hoc networks. Technical Report, CS-2000-06, Duke University (2000)

[5] Lindgren, A., Doria, A., Schelén, O.: Probabilistic routing in intermittently connected networks. ACM SIGMOBILE Mobile Computing and Communications Review 7(3), 19–20 (2003)

[6] Spyropoulos, T., Psounis, K., Raghavendra, C.S.: Spray and focus: Efficient mobility-assisted routing for heterogeneous and correlated mobility. In: Proc. of the IEEE PerCom Workshop on Intermittently Connected Mobile Ad Hoc Networks (2007)

[7] Jacquet, P., Mans, B.: Routing in intermittently connected networks: Age rumors in connected components. In: Proc. of the 5th Annual IEEE Int'l Conf. on Pervasive Computing and Communications Workshops, PerCom Workshops, pp. 53–58 (2007)

[8] Wang, Y., Jain, S., Martonosi, M., Fall, K.: Erasure-Coding based routing for opportunistic networks. In: Proc. of the 2005 ACM SIGCOMM Workshop on Delay-Tolerant Networking, pp. 229–236. ACM, Philadelphia (2005)

[9] Liao, Y., Tan, K., Zhang, Z., Gao, L.: Estimation based erasure-coding routing in delay tolerant networks. In: Proc. of the 2006 Int'l Conf. on Wireless Communications and Mobile Computing. ACM, Vancouver (2006)

The Applications of GIS in Modern Community Management and Service

Haibin Gao, Bin Gao, and Yinghui Li

Abstract. GIS is a useful technology to manage and organize information effectively. It can combine and show objects' geography location and other characters directly, visually and entirely. GIS has a vast application future in community management and service. It is positive for the improvement of community management and residents' life quality. It will become an important part of community informationization. In this article, we introduced the study and practice of GIS using in modern community management and service.

1 Introduction

GIS is the abbreviated form of Geographic Information System, originating in the 1960s. It has been developing continuously with the progress of computer technology. GIS deals with geo-spatial data by the use of modern computer graphics and database technology, combining objects' geographic location and features organically then presenting them directly, visually and entirely by various display facilities. GIS is a useful technology to manage and organize information effectively, conforming to human's cognition habits. It has a vast application future and has gained general concern and rapid growth. It has been widely used in some important fields which effect the national economy and the people's livelihood such as resource survey, homeland management, city planning, telecommunication, traffic and transportation, military and public security, irrigation and hydroelectricity, public accommodation management and so on. With the unceasing progress of the information technology, GIS is ripping day by day and its

Haibin Gao
General Affairs Office Tsinghua University, Beijing China
e-mail: gaohaibin@tsinghua.edu.cn

Bin Gao · Yinghui Li
QingHuaYuan Sub-district Office, Beijing China
e-mail: zwbgb@tsinghua.edu.cn, daying_li@hotmail.com

F.L. Gaol (Ed.): Recent Progress in DEIT, Vol. 2, LNEE 157, pp. 433–438.
springerlink.com

implement cost is decreasing. The barriers for its application and development have been broken down one by one. Viewing from the perspective of historic development, GIS is sure to step down from the mysterious and expensive altar and enter to our common life.

Community, being the basic integral part of modern society, is the main place for residents' life and relaxation. The quality of residents' living largely depends on the level of community management and service. With the gradual acceleration of China's urbanization process, large communities are emerging constantly, covering larger and larger areas and supervising more and more residents. This process from quantitative change to qualitative change shows the characteristics of modern communities. It requires a higher level of community management and service than ever before. Modern community residents are generally highly educated. They not only require larger coverage and higher quality of community management and service, but also make new requests for the operation mode, efficiency and informationization degree of community management and service. In recent years, local governments make greater efforts to instruct and support the community construction. From the perspective that the progress of science and technology promotes community development, the Beijing Municipal Science and Technology Commission offers the challenge to institute "Demonstration District for the Applications of Science and Technology in Community Service" in some good conditional communities. QingHuaYuan community, which is the location of Tsinghua University, was selected as one of the first batch demonstration districts. An important part of QingHuaYuan demonstration district is using GIS technology in its community management and service, making this advanced technology serve residents' daily life practically.

2 GIS Manages Community Infrastructure Effectively and Raises the Working Efficiency

Community infrastructure includes diversiform equipments, facilities and pipelines of building, road, water/heating/electric/gas supply, sewage system, telecommunication and fire control. All of them, distributed widely in the community, are the basic supports of community operation and the important ensures for residents' normal life. Some of those infrastructures above the ground can be located easily, but the others are buried underground, which are hard to be traced. Therefore, the problems as to how to manage so many infrastructures effectively, how to keep them working normally and how to get rid of the accidents in time often puzzle the community managers. When these infrastructures are first established, most of them have detailed and complete documents such as blueprints, specifications and descriptions. But these paper documents are easily to be damaged by inappropriate safekeeping or to be lost by blockage in handover, which will cause a lot of trouble for the future management and maintenance. Especially for those infrastructures that have been established for a long time, that kind of problem is quite striking. The geographic information mainly depends on the memory of veteran workers, which means high maintenance costs and low efficiency. To get the statistical information of infrastructures is harder to be completed.

GIS technology can solve those problems well. Owing to the infrastructures' obvious geographical characteristics, all their information management, statistics and reporting can be done by GIS. GIS manages different information by layer conception, recording information of the same kind infrastructures on their correspondent layer, thus realizing digital storage and management. The manager can get all information by simply click the mouse. For example, selecting a water pipe, you will know its location, the depth of lay, caliber, material, service life, established time and higher level valve and so on. When the water pipe happens to be out of work, using that information, people can switch the higher level valve and find the exact location where the pipe has been laid, dig it out and repair it quickly. Comparing with the past way that digging and finding pipe according to memories, this way raises the work efficiency greatly. By using GIS technology, it becomes very easy for information gathering. People only need to set up the filter rules, and then the information such as the area of roads or the length of pipe lines can be reported through GIS automatically. This makes it easier for managers to grasp the overall situation. In the mean time, it provides important basis for related decisions.

According to the characteristics of infrastructure, we can do further development for GIS system, adding management criterions into daily operations. For example, fire-fighting equipments need to be checked and maintained regularly. GIS can automatically remind the relevant personnel to test equipment by adding an expiration reminder module, and generate a list of devices which require to be maintained in the near future. This mechanism forces the management criterions to be carried out in everyday work and eliminates potential security risks caused by human negligence.

3 Combining with GPS Technology, GIS Provides Object Position Tracing and Dispatch Service

GPS is the abbreviated form of Global Positioning System, locating objects by the signals from global positioning satellites in outer space. QingHuaYuan Community did a secondary development which combined with GPS technology to build an object position tracing and dispatch system. This system is based on GIS, GPS and GPRS technology, making it possible for authorized personnel to locate people or vehicles that are carrying GPS positioning terminals. GPS positioning terminal gathers the satellites' signals and figures out the geographical coordinates of its location and then uploads these information to GIS through GPRS communication network. The authorized personnel can get the information about the positioned people or vehicles directly and visually on an electronic map by any online computer, such as their locations, speeds and moving directions and so on. This position tracing and dispatch system has a promising future in community management and service. At present, it has been practically used in the following fields: security supervisory and instruction, aged people locating and position enquiry of community buses.

3.1 Security Supervisory and Instruction

By combining community supervisory system and GIS, the distribution of cameras can be presented on the electronic map through the graphical interface of GIS. Clicking the camera's icon, security can watch the real time images taken by that camera and supervise the on-the-spot situation directly. Equipping security personnel and vehicles with GPS positioning terminals, their real time position can also be presented on the electronic map, which makes the manager easier to know where his man is. Those two functions give technical supports to security supervisory and instruction system. When some emergent accident happens, on one hand, the manager can know the on-the-spot situation immediately through supervisory network. On the other hand, it is available to check the distribution of the security personnel around the spot and instruct the nearest ones to deal with the problems.

3.2 Aged People Locating System

In order to deal with the easily happens emergencies of aged people, such as sudden diseases or getting lost when they are out, QingHuaYuan community equipped GPS positioning terminals for those aged people who need locating service. When they are out, their families can enquiry the position of them on the electronic map through internet. So people no longer worry about those aged ones getting lost, both the young and the old can be at ease. In addition to that, the aged can also use the communication function on the terminals and talk with their families whenever they need to tell their conditions.

3.3 Position Enquiry of Community Buses

As a large community, QingHuaYuan equipped transfer buses to connect the living zone, commercial outlets, hospitals and leisure facilities. Because the buses' large intervals, sometimes residents have to wait for a long time. Based on the object position tracing and dispatch system, community buses are equipped with GPS positioning terminals. Therefore, residents can enquiry buses' real time position by internet or phone calls and choose an appropriate time to be at the bus stop, which decreases their waiting time effectively. This application brings the residents a lot of convenience, especially in frozen winter.

4 GIS Presents the Community Service Resources Visually, Making It More Convenient for Residents to Enquire and Get Service

With the fierce development of the tertiary industry, the service resources in communities become more and more abundant, including trade, catering, accommodation, property services, home services, maintenance, medical, logistics, banking, insurance and consulting and so on, almost covering every aspects of residents' daily life. Facing with such a large number of service resources, the residents are easily getting confused. Based on GIS technology, these service resources are categorized and presented on the electronic map where residents can directly search and compare. Information organization becomes more efficient and service finding becomes more convenient. This new way to organize and present information introduced by GIS is more conforming to people's cognition habits. Residents only need to choose the service pattern they want and then they can find the business sites' distribution of this kind of service on the electronic map, as well as information about their locations and distance. By clicking any site's icon, other information, such as name, business hours, phone number, address, website, pictures and profile, can be clearly seen. If necessary, residents can get more detailed information by logging on its website. By these functions provided by GIS, residents have a comprehensive grasp on the general condition of the community service resources, which gives them wider choices and more convenience for accessing services.

5 With New Organization Mode of Community Special Resources, Community Gets a New Impression

Different communities have different advantageous resources and these advantageous ones become the special resources of that community. To manage and present these unique resources effectively has significant meaning for community's culture accumulation and public advertisement.

QingHuaYuan community is not merely a residential one. The famous Tsinghua University is situated in. Tsinghua Science and Technology Park which consists of numerous prestigious enterprises is also part of it. What is more, it includes the picturesque relics of XiChunYuan royal garden. All these special resources demonstrate one aspect of QingHuaYuan community from their own side. Tsinghua campus covers an area of 392 hectares, with its construction floor space covering 1.98 million square meters, including a huge number of institutes and research centers. GIS can manage the distributing and using of all kinds of buildings and resources better, raising the level of resource management and the usage efficiency. Tsinghua University is making great efforts to establish "green campus". At present, there are over 1000 species of arbors and bush in the campus, with a green coverage of 54.8%, among which there are 240 ancient trees. Through GIS, one can enquiry the distribution of all kinds of plants, figuring out their occupying areas and get some ideas about the construction situation of the green campus from the development view. The protection and maintenance tracing of ancient trees can be better carried out as well. QingHuaYuan used to be part of the Xichunyuan royal garden. So there are a lot of natural and cultural scenery, such as JinChunYuan garden, Tsinghua Garden, Wen Pavilion, The Old Gate, Tsinghua Xue Tang and etc. GIS records not only the information about their locations, but also the detailed text and image data related to them. Even their historical transitions are included, which plays a positive role in community's culture accumulation and spread. GIS integrates all these different aspects into an electronic map through geographic coordinates to show a comprehensive and vivid picture of QingHuaYuan community, which now has a new name card.

6 Conclusion

All in all, GIS technology is a very powerful tool to organize and present information. It has great practice potential in modern community management and service. QingHuaYuan community has made preliminary attempts on it and gained positive effects. We will continue exploring and going deep into practice to make GIS technology serve residents' life better.

Acknowledgments. The authors would like to thank the Beijing Municipal Science and Technology Commission (grant No. Z09020600760902) for financial support.

References

1. Zhu, C., Gao, H.: Research and Practice on The Information Platform of Community Service and Mangemnet. Science and Technology Management Research 5, 490–491 (2009)
2. Zhang, F., Ji, J., Wang, Y.: Reform and Development of Tsinghua Logistic Service in New Period. China Higher Educiation Research 9, 20–22 (2005)
3. Zhu, C., Gao, H., Zhao, J.: The Application of E-Bussiness in University Logistic Service. University Logistics Research 2, 35–37 (2007)

Comp. Psy. 101: The Psychology behind High Performance Computing

Wendy Sharples, Louis Moresi, Katie Cooper, and Patrick Sunter

Abstract. Building software which can deliver high performance consistently, across a range of different computer clusters, is a challenging exercise for developers as clusters come with specialized architectures and differing queuing policies and costs. Given that optimal code configuration for a particular model on any machine is difficult for developers and end-users alike to predict, we have developed a test which can provide instructions for optimal code configuration, is instantly comprehensible and does not bombard the user with technical details. This test is in the form of a 'personality type' resonant with users' everyday experience of colleagues in the workplace. A given cluster is deemed suitable for either development and or production and small/composite models and or large/complex ones. To help users of our software to choose an efficient configuration of the code, we convert the personality assessment result into a series of optimization instructions based on their cluster's personality type.

1 Introduction

High performance computing is synonymous with efficient, scalable, parallel computation on thousands of cores, however building software which can deliver high performance consistently, is a challenging exercise for developers. Computer

Wendy Sharples
School of Mathematical Sciences, Monash University, Melbourne, Australia
e-mail: wendy.sharples@monash.edu

Louis Moresi
School of Mathematical Sciences and School of Geosciences, Monash University, Melbourne, Australia

Katie Cooper
School of Earth & Environmental Science, Washington State University, Pullman, WA, USA

Patrick Sunter
Victorian Partnership for Advanced Computing, Melbourne, Australia

F.L. Gaol (Ed.): Recent Progress in DEIT, Vol. 2, LNEE 157, pp. 439–452.
springerlink.com

clusters come with limited resources and availability, specialized architectures, queuing policies and costs and these factors can effect the efficiency of running any given software.

Several attempts have been made to provide consistency across architectures in other software such as PETSc whereby the option PETSC_DECIDE can reduce the run time by adapting to different hardware configurations [1]. Our first attempt, in the software Underworld, was to re-use the system testing suite to run specific model-based performance tests on a cluster. However, the arcane output of the timing suite (appendix B) is only legible to a subset of the developers of Underworld and given that a user's time is valuable, we wanted to bundle the test results into a format which is more meaningful, memorable and instantly comprehensible, without bombarding the user with technical details.

With accessibility and expediency in mind, we convert the result into a 'personality type' much like a Myers-Briggs personality type test [2]. Four Cluster personality types have been chosen, inspired by the current users of Underworld; development oriented, production oriented, complex proficient and ensemble proficient. Most users have two main stages to their research; development, in which a scientific model and usually code along with it, are built and production, in which a pre-built scientific model undergoes parameter sweeps at a resolution fit for publication, which are run with a static build of Underworld. The development stage usually consists of lower resolution models or a set of composite model components, whereas the production stage can entail high resolution or complex models.

Furthermore, we have develped personality traits which fall under the four types, based on typical examples from a case study of three different clusters. An assessment of which traits a given cluster possesses, determines whether that cluster is suitable for development or production or both and whether it can handle large/complex models and or small/composite models. To provide an optimal performance configuration of Underworld on any cluster, instructions based on the traits the cluster possesses can be given to the user after the personality test suite is completed (see appendix C).

2 Underworld

All the results from in this paper are obtained from running the software Underworld. Underworld is a mantle convection code whereby solutions for velocity and pressure are obtained using an FEM particle-in-cell method ([8], [9])

Underworld has two different system test suites; a three dimensional suite, whereby convection models are run and a two dimensional suite, whereby analytic models are run (for more details see appendix A) [3]. These tests suites are solved using a multi-grid method [4] as a pre-conditioner for the velocity solution, as it converges to a solution more quickly than a standard iterative solver. As the multi-grid name suggests, a problem domain is broken up into a hierarchy of grid resolutions-finest to coarsest, and a velocity solution for each grid resolution is obtained and then fed back in as the intitial starting velocity for the next grid level up or down.

The final preconditioned velocity solution is obtained when a full 'V cycle'; solving first from finest grid level to coarsest then back up to finest, is completed [4]. For each grid level, a user can specify a different set up for obtaining the velocity solution, for example one can specify that an iterative solver be used on the finest grid level and that a direct solver be used on the coarsest grid level. These options can be added to the command line arguments conveniently via an option file [1].

In our quest to find the multi-grid solving options which produce the fastest solve time, we have found the main bottlenecks (which are influenced by differing cluster architecture) to achieving a solution quickly are: the time taken to compute a dot product and sensitivity to over-decomposition at the coarsest grid level. To address those issues, we have subsequently added a basic multi-grid test to our system tests whereby a mid-resolution model from the test suite is run with two option files encompassing command line arguments which address each of those bottlenecks (one file with multi-grid options that lower the number of dot products computed and one file with multi-grid options that restrict the decomposition of the coarsest grid) and then also with default options. All three runs are then compared against each other to ascertain which options produces the fastest wall time (see appendix B).

Installation of Underworld and its dependencies on a cluster can be done by the user in the form of a custom build with a stable or up to date version of the code or by the system administrators of the cluster, where a stable release of the code plus dependencies are built in the form of modules [12]. Installation of Underworld and its dependencies on a cluster is not always straight forward, when compared to installation on a desktop, as a cluster's specialized architecture can be restrictive.

3 Case Studies

Underworld was run on three different clusters of differing sizes, architecture, components, queuing systems, charging rates and type of clientele. The clusters' features are summarized below.

Ranger is a Sun Constellation system which is intended for users with codes scalable to thousands of cores (1024 and above). Due to his large resources, applying for time on Ranger is an arduous process. A quick launching batch queue is available to help develop, test, and scale codes and provides up to 1024 compute CPUs however the wait is still longer than for Tako or Persephone. Four separate login nodes provide interactive connectivity to the system for compiling and interfacing with the batch queuing system. Ranger has a pay as you go system whereby user is charged per service system which is defined as one processor-hour of wallclock time. Ranger's clientele mainly run heavy resource jobs [11].

Tako is a single board system belonging to the geodynamics research group at Monash University, with a single image presented to the user. The single image machine makes compilation / installation / maintenance trivial. The user can launch jobs at no charge based on a good will system. Jobs are launched directly (and start running immediately) from mpi whereby the user has no control of the distribution across the CPUs / Node. This is handled by the linux scheduler. There is no distinct head-node. The configuration is out-of-the-box meaning that computational jobs compete with background processes / services. Tako's clientele run small to moderate resource jobs [10].

Persephone is a SiCortex computer belonging to Washington State University's Geodynamics Lab. Her set up contains one head node/login node for compilation and interactive connectivity and then the rest of the nodes are reserved for job runs. A queuing system is set up, however the wait from job launching to job running for Persephone's clientele is very short. Most of Persephone's clientele run parallel, moderate to large resource jobs [5].

Table 1 Summary of Ranger's, Tako's and Persephone's Technical Specifications

	Ranger	Tako	Persephone
CPU Type:	Operton, four socket, quad core	AMD	Proprietary, MIPS based
CPU Speed:	2.3 GHz	2.8 GHz	1.4 GHz
CPUs per Node:	16	2	6
Number of Nodes:	3936	8	108
Total Processors:	62976	16	648
Memory per Processor:	2 GB	16 GB	1.5 GB
Total Distributed Memory:	125952 GB	128 GB	432 GB
Total Teraflops:	579	$\ll 1$	1.2
Compilation Time:	over an hour	under 5 minutes	over 12 hours

Performance Reviews

Figure 1 shows a series of graphs demonstrating the effect of job size, communications between processors and variations in resources allocated on the wall time for the solution at the first time step for Ranger, Tako and Persephone. The scripted suite of tests were not run under ideal conditions to mimic the conditions under which the tests will be run by a user of the cluster. Because of this, the average percentage errors have been calculated for the three clusters based on running the same model multiple times. It was found that Tako contained the most variability in wall time (given the competition between background processes and running jobs, see figure 1 graph D) and Ranger the least. The percentage error has been incorporated into the data as error bars on the data points; Ranger 3%, Persephone 5% and Tako 34%.

On Persephone and Ranger, the 3D test suite was run and on Tako, the 2D test suite was run.

4 The Moresi-Cooper Cluster Personality Type Test

Loosely based on the Myers-Briggs personality test [2] and developed using the results from figure 1 and the technical specifications of the case studies above, we

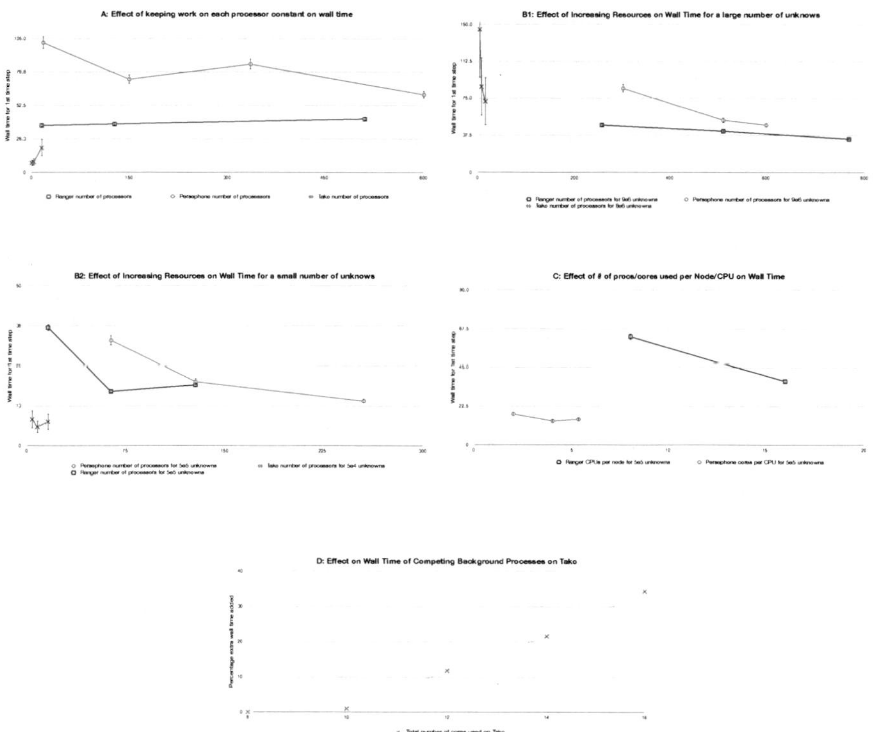

Fig. 1 Graphs A to C: showing the effect of keeping work on each processor constant (A), the effect of increasing resources for both a large ($\geq 10^8$) and small number of unknowns ($\leq 10^6$), (B1, B2) and the effect of the number of processors per CPU on wall time of the first time step whist keeping everything else constant (C). Graph D: shows the percentage extra time taken when running the same job multiple times with increasing processors in use on Tako, which illustrates the competition with background processes

have identified four distinct personality types based upon 'traits' which determine the usefulness of the cluster for each research stage and model type. They are *production oriented*, suitable for production purposes, *development oriented*, suitable for development purposes, *complex proficient*, handles large/complex models well and *ensemble proficient*, handles small/ensemble problems well. Not every cluster is going to fit into exactly one of these types, they are really just end member cases as most clusters are going to be some combination of each. The descriptions of each type are generic as computer cluster technical capabilities are not going to remain constant over time and different models may produce slightly different results. From Underworld's two test suites, small/composite models are defined as models with less than 10^6 unknowns and large/complex models are defined as models with greater than 10^9 unknowns. A summary of the traits belonging to each type is given in table 2.

Table 2 Summary of Personality Traits

	Production Oriented Personality Traits		**Development Oriented Personality Traits**
High maintenance	Compilation time is greater than a long lunch Software and dependencies are not straight forward to install when compared to installation on a desktop	Easy Going	Compilation time is less than a coffee break Software and dependencies are straight forward to install, equivalent to installation on a desktop
Consistent	A given problem encounters insignificant variation in the time taken when run multiple times Every processor is the same	Inconsistent	A given problem encounters significant variation in the time taken when run multiple times Not all processors are the same
Elitist	Difficult to obtain time on the cluster Wait time is two days or more in standard queue Costly to use	Accessible	Easy to obtain time on the cluster Wait time is less than a day in the standard queue Inexpensive to use
	Complex Proficient Personality Traits		**Ensemble Proficient Personality Traits**
Extensive	More than 100 times the number of processors in a standard desktop, available for use CPU speed greater than that of a standard desktop Speed to compute a dot product is not a bottleneck	Small Scale	Less than 100 times the number of processors in a standard desktop, available for use CPU speed less than that of a standard desktop Speed to compute a dot product is a bottleneck
Parallel	Efficient communication between processors Linear or better speed up when more resources are added Small/Composite models suffer from over decomposition when more resources are added Sensitive to over-decomposition at coarsest grid solve (Multi-grid)	Serial	Inefficient communication between processors Sub-optimal speed up when more resources are added Greater than 5 times a standard desktop's memory per processor for faster direct solve in serial Insensitive to over-decomposition at coarsest grid solve (Multi-grid)
Flat Out	Fastest wall time is achieved when all processors per node are in use	Under-driven	Fastest wall time is achieved when not all processors per node are in use

These traits are summarized in fig 2. into a map of all possible types and combinations, along with the assessment of where Ranger, Tako and Persephone fit. For example, DPE is a mixture of *development oriented* and *production oriented* and *ensemble proficient*.

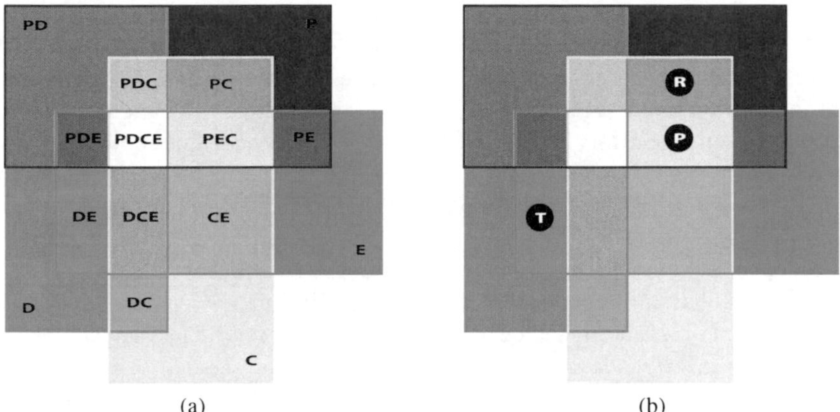

Fig. 2 Fig a: Cluster Personality Map where the individual types are shown (D, P, C, E) as well as all the possible combinations, Fig b: Cluster personality Map including dots with P (Persephone), R (Ranger) and T (Tako) to indicate which stage of research and model type each cluster is most suited to

5 Personality Assessment of Ranger, Tako and Persephone

This section contains more meaningful interpretations of the technical specifications (table 1) and the graphs in figure 1. The information presented will be interpreted in terms of the 'personality traits' developed in the previous section.

Ranger is a *production oriented* personality type as he is a high maintenance machine because Underworld compilation takes over an hour and the dependencies are not straight forward to install, he works consistently (there is less than 5% error in the same model being run) and is he is elitist because he is difficult to obtain time on and costly to use.

He is *complex proficient* as he handles large/complex models well because of the efficient communication between his processors (the wall time is fairly even when applying weak scaling, figure 1 graph A) and when more resources are added a linear speed up occurs (figure 1 graph B1). He is extensive because there are over 62000 processors for use on the cluster and each CPU has a speed of over 2 GHz. He works best when over-driven, as there is a distinct disadvantage to running a job on CPUs with less than the full 16 cores allocated as shown in figure 1 graph C. He does not handle small/composite jobs well (figure 1 graph B2) and has none of the traits of a *development oriented* type. Multigrid works best on Ranger if the coarsest grid is solved in parallel (non-redundant).

Tako is a *development oriented* personality type as he is an easy going machine because Underworld compilation takes less than five minutes and the dependencies are easy to install. He works inconsistently (there is more than 30% error in the same model being run, see figure 1 graph D) and he is very accessible because there is no wait time for jobs and he is free to use.

He is *ensemble proficient* as he handles small/composite models well because he has a large amount of memory per core to be able to solve directly (and therefore faster than an iterative solver) in serial. He has 16 processors for use and mainly resource light jobs run on this cluster so he is a small scale machine. Inefficient communications between processors is evident because the wall time is goes up significantly when weak scaling is applied (figure 1 graph A). He works best when under-driven, as there is a distinct disadvantage to running a job with all cores engaged (figure 1 graph D). Multigrid works best on Tako if the coarsest grid is solved redundantly.

Persephone is *production oriented* even though she is accessible, (it is easy to obtain time on her and users do not have a long wait time for jobs to run) because she is a high maintenance machine; compilation of Underworld takes over 12 hours and the software dependencies are not straight forward to install. In addition, she works consistently as there is approximately 5% error in the time taken for multiple runs of the same model and all processors are the same.

She is both *complex proficient* and *ensemble proficient* as she handles both large/complex models and small/composite models well. She can deliver a fast wall time for both (figure 1 graphs B1, B2) although she has a slow CPU speed. She is extensive as she has over 600 processors available for use and mainly resource heavy jobs are in the queue. However she is not entirely parallel as she is an inefficient communicator, shown in figure 1 graph A, because the wall time fluctuates with the number of processes, when the work on each processor is kept constant. She works best slightly under-driven, as illustrated in figure 1 graph C although there is not a significant difference to her being used flat out. When using multigrid, the best result is given when the coarsest grid is not solved redundantly.

Assessment of Type

The three clusters studied above are a good mix of different characteristics. To make a comparison between three, we have put them on the cluster personality map (figure 2 a) which takes into account the four personality types: *development orientated*, *production orientated*, *complex proficient* and *ensemble proficient*. In figure 2 b, the dots marked P for Persephone, R for Ranger and T for Tako show what personality type combinations the clusters have, where a cluster is said to have a given type when it is deemed to have 50% or more of that type's hardware and software traits. An example of what the output to screen could look like, in terms of the Cluster personality types, is given in appendix C.

In summary, Persephone is a PEC type, she is suited to production and is both *complex proficient* and *ensemble proficient* because she can handle both large/complex models and small/composite models well. Ranger is a PC type suitable for production and is *complex proficient*; suitable for large/complex models only. Tako is a DE suited to development and is *ensemble proficient* as he handles simple/composite models well.

Optimization Based On Personality Assessment

Based on their resources and personality type, the following tables are a series of optimization suggestions (the output from appendix C in tabular form). A custom build is suggested (for either Underworld or its dependencies) when it is relatively easy for the user to do themselves, this does not mean that the system administrators can not also install a stable build on the cluster.

Table 3 Suggestions For PC Type

Suggestions For Ranger's Performance Optimization
Build: Stable
Dependencies: Pre-Built
MG option file: add -option_file mg_underdecomp.opt to command line
Serial: Not recommended
Saturation: Reliable scaling at 16 CPUs per node
Large number of unknowns ($\geq 10^9$) use 512 or more processors
Small number of unknowns ($\leq 10^6$) use 64 - 128 processors

Table 4 Suggestions For DE Type

Suggestions For Tako's Performance Optimization
Build: Custom
Dependencies: Custom
MG option file: add -option_file mg_lowdotprodcount.opt to command line
Serial Solve: add -option_file options-lu.opt to command line
Saturation: Reliable scaling using up to 8 cores
Large number of unknowns ($\geq 10^9$) do not run
Small number of unknowns ($\leq 10^6$) use up to 16 processors

Table 5 Suggestions For PEC Type

Suggestions For Persephone's Performance Optimization
Build: Stable
Dependencies: Pre-Built
MG option file: add -option_file mg_underdecomp.opt to command line
Serial: Not recommended
Saturation: Reliable Scaling up to 5 cores per CPU
Large number of unknowns ($\geq 10^9$) use 480 or more processors
Small number of unknowns ($\leq 10^6$) use up to 480 processors

6 Generalized Optimization of Underworld for Any Cluster

In order to streamline the process above for Underworld model runs on any cluster, some automation is required. This means that the process for determining the traits which lead to type and specialized optimization, needs to be scripted. As well as re-using the system tests, the development team need to record: the compilation time of Underworld, the CPU speed, memory, total number of CPUs/Nodes of the cluster, Cores per CPU or CPUs per Node, the cost of use and the type of clientele, plus a post-processing script where the output is presented in the form of the graphs from figure 1, the cluster personality map (figure 2) and the output to screen in appendix C.

Born out of the need to automate the scripts above as well as other applications, such as automated system testing and ensuring Underworld's continuing reliability via scientific benchmarking, work on a analysis toolkit for Underworld has recently commenced [3]. This toolkit is implemented in Python which provides a succinct but powerful interface to specify 'suites' of model runs that need to be performed to allow this key profiling and performance information to be much more effectively gathered. The analysis toolkit is designed to operate as both an interactive scripting tool for scientist-driven exploration of numerical and computational performance and also for regular, automated performance gathering as it is able to coordinate, launch, post-process, analyze and publish the results of Underworld model runs. The toolkit will provide the functionality to produce the graphs, cluster personality map and optimization suggestions, to assess the 'personality types' of both existing and new high-performance systems that become available to a research group using Underworld. In the future we hope to investigate the implications and benefit to users of this type of assessment of grid computing, when more cluster personality types are recorded and compared.

7 Conclusion

In conclusion, clusters are analogous to people, they have commonalities but they are not all alike. A cluster's 'personality' can affect the speed and cost of running any software, thus it would be beneficial for users to get a feel for what type of models and stage of research the cluster is suitable for, via the personality test and automatically obtain different optimization specifications, such as we have described in this paper. The one size fits all mentality is becoming increasingly irrelevant when time and cost of use are important constraints for scientists using high performance computer clusters.

Acknowledgements. The authors would like to thank the Underworld development and documentation team and for funding from Auscope.

Appendix

A: More about Underworld

Underworld is a mantle convection code whereby a velocity and pressure solution is obtained as a result of solving stokes equation (equation 1) for a very low reynolds number on a cartesian grid.

$$f_i = \tau_{ij,j} - p_{,i} \tag{1}$$

Material properties such as density and viscosity are advected with the particles, whereas field variables such as velocity and pressure are solved at the grid points.

Underworld's two different system test suites are comprised of; a three dimensional suite where a 2x2x1 convection model is run which has a periodic boundary and an embedded buoyant viscous block ($\Delta\eta = 10^3$) representing a continent [7] and a two dimensional suite where a 1x1 analytic model is run which has density driven flow where density is given by $\rho = -\sigma \sin(2\pi z) \cos(2\pi x)$. The analytic model contains a viscosity jump of $\Delta\eta = 10^{15}$ in the middle of the box in the horizontal direction and the boundary conditions are free slip everywhere on the surfaces of the system box [6].

B: Verbatim Output from Scripted Tests

```
login4.ranger.tacc.utexas.edu x86_64 GNU/Linux: 3D - timing suite (Convection)

run with mg_default.opt
run with mg_lowdotprodcount.opt
run with mg_underdecomp.opt

best result with -option_file mg_underdecomp.opt,
2nd: mg_default.opt
3rd: mg_lowdotprodcount.opt

results with -option_file mg_underdecomp.opt

|Grid size  |Unknowns  |  Cores      |MG levels|Solver t(s)|Wall t (s)|
|64x64x32   | 5e5      |  8way16     |    5    |    63     |  (-)     |
|64x64x32   | 5e5      | 16way16     |    5    |    37     |  (-)     |
|64x64x32   | 5e5      |  8way32     |    5    |    35     |  (-)     |
|64x64x32   | 5e5      |  8way32     |    4    |    35     |  (-)     |
|64x64x32   | 5e5      | 16way32     |    5    |    23     |  2184    |
|64x64x32   | 5e5      | 16way64     |    5    |    17     |  1328    |
|64x64x32   | 5e5      | 16way128    |    5    |    19     |  1400    |
|64x64x32   | 5e5      | 16way256    |    4    |    20     |  1300    |
|128x128x64| 4e6       |  8way32     |    5    |    284    |  25000   |
|128x128x64| 4e6       | 16way32     |    5    |    (*)    |          |
|128x128x64| 4e6       |  8way64     |    5    |    101    |  12000   |
|128x128x64| 4e6       | 16way64     |    5    |    64     |  7076    |
|128x128x64| 4e6       | 16way128    |    5    |    38     |  4444    |
|128x128x64| 4e6       | 16way128    |    4    |    34     |  4100    |
|128x128x64| 4e6       | 16way128    |    3    |    35     |  4200    |
|128x128x64| 4e6       | 16way256    |    5    |    35     |  3210    |
|128x128x64| 4e6       | 16way512    |    5    |    35     |  3200    |
|192x192x64| 9e6       | 16way256    |    5    |    48     |  5040    |
|192x192x64| 9e6       | 16way512    |    5    |    42     |  3910    |
|192x192x64| 9e6       | 16way768    |    5    |    34     |  3400    |
|256x256x64| 1.6e7     | 16way512    |    5    |    42     |  6889    |
|256x256x64| 1.6e7     | 16way768    |    4    |    38     |  6030    |
|256x256x64| 1.6e7     | 16way1024   |    4    |    (#)    |          |

(-) -- job failed: solver diverged
(*) -- job failed: not enough memory
(#) -- job failed: unknown cause
```

```
tako.maths.monash.edu.au x86_64 GNU/Linux: 2D - timing suite (Analytic)

running MG config test .....

run with mg_default.opt
run with mg_lowdotprodcount.opt
run with mg_underdecomp.opt

best result with -option_file mg_lowdotprodcount.opt
2nd: mg_default.opt
3rd: mg_underdecomp.opt

results with -option_file mg_default.opt

|Grid size   | Unknowns  |  Cores     |MG levels|Solver t(s)|Problem|
|   64x64    | 1.2e4     |  -np 1     |    4    |    7.6    | Cx    |
|   64x64    | 1.2e4     |  -np 4     |    4    |    2.1    | Cx    |
|  128x128   | 5e4       |  -np 4     |    5    |    8.3    | Cx    |
|  256x256   | 2e6       |  -np 4     |    6    |    35.2   | Cx    |
|  512x512   | 8e6       |  -np 4     |    7    |    145    | Cx    |
| 1024x1024  | 1.6e7     |  -np 4     |    8    |    524    | Cx    |
|   64x64    | 1.2e4     |  -np 4     |    4    |    0.4    | Kz    |
```

```
|  128x128   | 5e4      |   -np 4     |    5     |    2.0     | Kz     |
|  256x256   | 2e6      |   -np 4     |    6     |    7.8     | Kz     |
|  512x512   | 8e6      |   -np 4     |    7     |    27.1    | Kz     |
| 1024x1024  | 1.6e7    |   -np 4     |    8     |    151     | Kz     |
|   64x64    | 1.2e4    |   -np 8     |    4     |    3.2     | Cx     |
|  128x128   | 5e4      |   -np 8     |    5     |    5.9     | Cx     |
|  256x256   | 2e6      |   -np 8     |    6     |    20.6    | Cx     |
|  512x512   | 8e6      |   -np 8     |    7     |    86.6    | Cx     |
| 1024x1024  | 1.6e7    |   -np 8     |    8     |    392     | Cx     |
|   64x64    | 1.2e4    |   -np 8     |    4     |    0.6     | Kz     |
|  128x128   | 5e4      |   -np 8     |    5     |    1.1     | Kz     |
|  256x256   | 2e6      |   -np 8     |    6     |    4.8     | Kz     |
|  512x512   | 8e6      |   -np 8     |    7     |    21.9    | Kz     |
| 1024x1024  | 1.6e7    |   -np 8     |    8     |    86.7    | Kz     |
|   64x64    | 1.2e4    |   -np 16    |    4     |    6.2     | Cx     |
|  128x128   | 5e4      |   -np 16    |    5     |    7.5     | Cx     |
|  256x256   | 2e6      |   -np 16    |    6     |    19.4    | Cx     |
|  512x512   | 8e6      |   -np 16    |    7     |    71.9    | Cx     |
| 1024x1024  | 1.6e7    |   -np 16    |    8     |    266     | Cx     |
|   64x64    | 1.2e4    |   -np 16    |    4     |    1.2     | Kz     |
|  128x128   | 5e4      |   -np 16    |    5     |    1.6     | Kz     |
|  256x256   | 2e6      |   -np 16    |    6     |    4.3     | Kz     |
|  512x512   | 8e6      |   -np 16    |    7     |    18.1    | Kz     |
| 1024x1024  | 1.6e7    |   -np 16    |    8     |    87.1    | Kz     |

(-) -- job failed: solver diverged
(*) -- job failed: not enough memory
(#) -- job failed: unknown cause



persephone.geol.wsu.edu mips64 GNU/Linux: 3D - timing suite (Convection)

running MG config test .....

run with mg_default.opt
run with mg_lowdotprodcount.opt
run with mg_underdecomp.opt

best result with -option_file mg_underdecomp.opt
2nd: mg_default.opt
3rd: mg_lowdotprodcount.opt

results with -option_file mg_lowdotprodcount.opt

|Grid size |Unknowns |  Cores        |MG levels|Solver t(s)|Wall t (s)|
| 72x72x24 | 5e5     | -n600 -N100 |   4     |     10     | 1163      |
| 72x72x24 | 5e5     | -n300 -N50  |   4     |     11     | 1469      |
| 72x72x24 | 5e5     | -n256 -N64  |   4     |     10     | 1294      |
| 72x72x24 | 5e5     | -n256 -N48  |   4     |     11     | 1387      |
| 72x72x24 | 5e5     | -n128 -N32  |   4     |     14     | 2069      |
| 64x64x32 | 5e5     | -n256 -N64  |   4     |     14     | 1506      |
| 64x64x32 | 5e5     | -n256 -N48  |   4     |     15     | 1603      |
| 64x64x32 | 5e5     | -n128 -N64  |   4     |     18     | 2281      |
| 64x64x32 | 5e5     | -n128 -N48  |   4     |     19     | 2357      |
| 64x64x32 | 5e5     | -n128 -N32  |   4     |     20     | 2413      |
| 64x64x32 | 5e5     | -n64  -N32  |   4     |     32     | 4163      |
| 64x64x32 | 5e5     | -n64  -N16  |   4     |     33     | 4300      |
| 64x64x32 | 5e5     | -n18  -N3   |   4     |     102    | 17000     |
|128x128x64| 4e6     | -n600 -N100 |   6     |     (#)    |           |
|128x128x64| 4e6     | -n600 -N100 |   5     |     48     | 10470     |
|128x128x64| 4e6     | -n600 -N100 |   4     |     45     | 10550     |
|128x128x64| 4e6     | -n600 -N100 |   3     |     48     | 10300     |
|128x128x64| 4e6     | -n512 -N96  |   5     |     26     | 5500      |
|128x128x64| 4e6     | -n384 -N96  |   5     |     31     | 6270      |
|128x128x64| 4e6     | -n256 -N96  |   5     |     41     | 8240      |
|128x128x64| 4e6     | -n384 -N64  |   5     |     31     | 8000      |
|128x128x64| 4e6     | -n256 -N64  |   5     |     44     | 9050      |
|128x128x64| 4e6     | -n300 -N50  |   5     |     39     | 8136      |
|128x128x64| 4e6     | -n256 -N48  |   5     |     43     | 8840      |
|128x128x64| 4e6     | -n150 -N25  |   5     |     73     | 19100     |
|128x128x64| 4e6     | -n128 -N48  |   5     |     77     | 15200     |
|192x192x64| 9e6     | -n600 -N100 |   5     |     48     | 10700     |
|192x192x64| 9e6     | -n512 -N96  |   5     |     53     | 11273     |
|192x192x64| 9e6     | -n336 -N56  |   5     |     (*)    |           |
|192x192x64| 9e6     | -n300 -N50  |   5     |     85     | 17920     |
|256x256x64| 1.6e7   | -n600 -N100 |   5     |     61     | 18562     |
|256x256x64| 1.6e7   | -n400 -N100 |   5     |     84     | 22820     |
|256x256x64| 1.6e7   | -n300 -N50  |   5     |     (*)    |           |

(-) -- job failed: solver diverged
(*) -- job failed: not enough memory
(#) -- job failed: unknown cause
```

C: Example of Output to Screen from Cluster Personality Assessment

```
login4.ranger.tacc.utexas.edu x86_64 GNU/Linux: 3D - timing suite (Convection)

Information supplied by the developers at configuration/build time:

    UW compilation time: 01:34
    Dependencies: Installed by Sys Admin
    Clientele: Heavy resource
    Cost: $$$$$
    CPU Speed - 2.3 GHz
    CPUs per Node - 16
    Number of Nodes - 3936
    Queueing system - PBS

running 3D suite .......

analyzing graphs ........

personality diagnosis ........

*P*roduction Orientated
*C*omplex

optimization suggestions:

MG option file: add -option_file mg_underdecomp.opt to command line
Serial Solve: not recommended
Saturation: Reliable scaling at 16 cores per node
Large number of unknowns (> 10^9) use 512 or more cores
Small number of unknowns (< 10^6) use less than 64 - 128 cores



tako.maths.monash.edu.au x86_64 GNU/Linux: 2D - timing suite (Analytic)

Information supplied by the developers at configuration/build time:

    UW compilation time: 00:04
    Dependencies: custom
    Clientele: Light resource
    Cost: $
    CPU Speed - 2.8 GHz
    Cores per CPU - 4
    Number of  CPUs - 4
    Queuing - None

running 3D suite .......

analyzing graphs ........

personality diagnosis ........

*D*evelopment Orientated
*E*nsemble

optimization suggestions:

MG option file: add -option_file mg_lowdotprodcount.opt to command line
Serial Solve: add -option_file options-lu.opt to command line
Saturation: Reliable Scaling using up to 8 cores
Large number of unknowns (> 10^9) do not run
Small number of unknowns (< 10^6) can use up to 16 cores



persephone.geol.wsu.edu mips64 GNU/Linux: 3D - timing suite (Convection)

Information supplied by the developers at configuration/build time:

    UW compilation time: 13:43
    Dependencies: Installed by Sys Admin
    Clientele: Heavy resource
    Cost: $$
    CPU Speed - 1.4 ~GHz
    Cores per Node - 6
    Number of CPUS - 108
    Queueing system: SLURM

running 2D suite .......

analyzing graphs ........
```

```
personality diagnosis ........

*P*roduction Orientated
*C*omplex
*E*nsemble

optimization suggestions:

MG option file: add -option_file mg_underdecomp.opt to command line
Serial Solve: not recommended
Saturation: Reliable scaling up to 5 cores per node
Large number of unknowns (> 10^9) use 480 or more cores
Small number of unknowns (< 10^6) can use up to 480 cores
```

References

1. Balay, S., Buschelman, K., Eiukhout, V., Gropp, W., Kaushik, D., Knepley, M., McInnes, L.C., Smith, B., Zhang, H.: PETSc Users Manual (2007), http://www.mcs.anl.gov/petsc/petsc-as/snapshots/petsc-2.3.3/docs/manual.pdf
2. Briggs, K.C., Myers, I.B.: Myers-Briggs Type Indicator. Consulting Psychologists Press, Palo Alto (1943)
3. Credo Documentation, http://auscope.monash.edu.au/credo-doc/
4. Hackbusch, W.: Multi-grid methods and applications. Springer Series in Computational Mathematics, vol. 4, 391 pages. Springer, Berlin (1985)
5. HPCwire: Washington State University Uses SiCortex Systems to Study Earthquakes (2009), http://www.hpcwire.com/offthewire/Washington-State-University-Uses-SiCortex-Systems-to-Study-Earthquakes-43773062.html
6. Humble, K.: Underworld Documentation (2010), https://csd.vpac.org/doxygen-v1.4.1/dd/dbe/Analytic_solCx_8c.html
7. Moresi, L., Cooper, C., Mansour, J.: Size and Scaling of small-scale instabilities beneath continental lithosphere. AGU Poster 90(52) (2009)
8. Moresi, L., Dufour, F., Muhlhaus, H.-B.: A Lagrangian integration point finite element method for large deformation modeling of viscoelastic geomaterials. Journal of Computational Physics 184, 476–497 (2003)
9. Moresi, L.M., Quenette, S., Lemiale, V., Meriaux, C., Appelbe, B., Muhlhaus, H.-.B.: Computational approaches to studying non-linear dynamics of the crust and mantle. Earth and Planetary Science Letters 305, 149–168 (2007)
10. Spencer, J.: Auscope Simulation and Modelling Newsletter, http://www.auscope.org.au/userfiles/file/Communications/AuScopeNewsletter_Feb20081.pdf
11. Texas Advanced Computing Center: HPC, http://www.tacc.utexas.edu/resources/hpc/
12. Underworld Installation Guide, http://www.underworldproject.org/documentation/InstallGuide.html

Quality Assessment for MPEG-2 Video Streams

Caihong Wang, Xiuhua Jiang, Yuxia Wang, and Fang Meng

Abstract. Depending on different contents of video sequences, the compression settings also differ. So many parameters can be obtained from video steams. We extract and analyze a large set of parameters from different layer. Based on the correlation between features and the subjective perceived quality, the most important objective parameters are picked out. A low complexity objective quality assessment metric is obtained by a linear calculation on the selected parameters. The presented method can perform continuous objective quality assessment in unit of GOP (Group of Pictures). The experimental results show that our model can achieve good performance for video quality prediction. In addition, our model does not require the source, or the decoded picture, it is suitable for real-time applications. And continuous quality assessment can provide an automatic warning without delay when picture quality problems occur.

1 Introduction

Compressed video streams are widely used in digital television, multimedia network and mobile communication. How to maximize the perceived quality of video streaming services to a given cost is an essential problem for service providers. Quality assessment plays a crucial role in video processing and application.

Subjective evaluation, which is recognized as the most reliable method, asks human viewers to score the quality of the scenes. However, the application is limited because subjective experiment is a time-consuming, laborious and complex operation. So an increasing interest is placed in objective evaluation, which can obtain the quality of test image automatically by using mathematical models. However, objective evaluation can replace subjective evaluation only when the objective result has a good consistency with the one gotten by subjective experiment [1].

It's a natural method of making a comparison between the processed image and the unprocessed source to measure the effect on picture quality of a compression processing [2-4]. This kind of method is known as double-ended measurement.

Caihong Wang · Xiuhua Jiang · Yuxia Wang · Fang Meng
School of Information Engineering, Communication University of China, Beijing, China

F.L. Gaol (Ed.): Recent Progress in DEIT, Vol. 2, LNEE 157, pp. 453–458.
springerlink.com

Double-ended measurement is not well suited to monitoring applications where the source isn't available. While single-ended approach, which can also be called no-reference (NR) method and doesn't need to access to the source image, is appropriate for real-time applications.

There are many kinds of NR methods. Most of the methods focus on evaluating video quality in fully decoded image [5] [6]. Compared with the methods performed on decoded image, the metrics based on compressed video streams [7-10] have a great advantage. Firstly, compression settings can be extracted from bit streams, for example bit rate. Bit rate can be looked as a benchmark of video quality and can't be known by the decoded picture. In addition, the process only needs partly decoding, so the method in the compressed domain is appropriate for real-time monitoring.

So many parameters can be gotten from bit steams. However most of metrics only discuss one or several parameters. The literature [7] obtains quality metric from frame rate and bit rate. An exponential curve is approximated between subjective perceived quality and bit rate. In [8], temporal features of the video sequence are mainly analysed, including movement speed, direction and so on. The metric is generated by a series of calculations on motion vector. An artificial neural network is utilized to estimate the video quality in [9] [10]. Compare with these two metrics, our method applies only a linear calculation on picked-out features. So our method is easier and can also perform a good assessment.

An extensive survey is done to various objective parameters extracted from video steams in this paper, including compression settings and content features. Based on the correlation between these parameters and mean opinion score (MOS) gotten by subjective experiment, a set of the most important objective parameters are picked out. A low complexity metric is obtained by a linear calculation on the picked-out parameters. The idea of our method can be applied for the streams coded with any conventional motion compensated video codec, although the specific application in this paper is MPEG-2 standard definition streams. The experiments have shown the good predictive performance on continuous quality assessment when objective score is gotten for every GOP (Group of Pictures) in video sequence.

2 No-Reference Quality Assessment

2.1 Feature Definition and Extraction

Images will undergo different impairments during compression due to different movement and complexity features. The more critical video with rapid movement and rich details, the more serious flaw will be shown. Depending on different contents of video sequences, the compression settings also differ. So we can obtain an accurate evaluation of image impairment by analyzing these parameters.

2.1.1 Sequence Layer

Compared with the method performed in fully decoded images, the method in compressed domain can know bit rate from the streams. It is a remarkable advantage. The bit rate can determine the image quality in a large extent.

2.1.2 Macroblock Layer

We can calculate the percentage of a kind of macroblock in a P (Predicated) or B (Bidirectional) frame. For example, intra MB in a P frame can be described as below.

$$MB_{intra} = \frac{1}{N}\sum_{k=1}^{N} mb_k \qquad (1)$$
$$\begin{cases} mb_k = 1, & \text{the } kth \text{ MB is intra MB} \\ mb_k = 0, & \text{the } kth \text{ MB is not intra MB} \end{cases}$$

Where N is the number of macroblocks in a P frame. MBs in P (Predicated) frame and B (Bidirectional) frame will take the intra encoding mode if the motion is so rapid that the MBs can not search out the matching blocks. Therefore, the number of intra MBs can reflect movement degree of the video.

Quantization is the source of image distortion in compression processing. Many types of image impairments are caused by the quantization, such as blocking artifacts, image blurring, ringing artifacts and etc.

The feature indicating the quantization degree can be extracted from the streams. We calculate the average value of quantiser-scale factors over a frame, as shown below:

$$QS = \frac{1}{N}\sum_{k=1}^{N} qs_k \qquad (2)$$

Where qs_k is the quantiser-scale factor in the *kth* macroblock and N is the number of macroblocks in a frame.

To express the motion degree of the picture, mean magnitude of the vectors averaged over a frame are used, which is calculated as follows:

$$\mu_{mv} = \frac{1}{N}\sum_{k=1}^{N} |MV_k| \qquad (3)$$

Where MV_k is the MV of the *kth* MB and N is the number of MBs with non-zero MV in a frame.

2.1.3 Block Layer

A statistics of skipped block in a frame can be carried out when getting CBP parameter from steams. A statistics similar to the percentage of macroblock coding type is used in our analysis on skipped luminance block and skipped chroma Block.

AC energy is defined to be the square sum of the DCT coefficients except for the DC coefficient.

$$ac_k = \sum_{m=0}^{7}\sum_{n=0}^{7} C(m,n)^2 - C(0,0)^2 \tag{4}$$

$$AC = \frac{1}{N}\sum_{k=1}^{N} ac_k \tag{5}$$

Where $C(m,n)$ denotes the DCT coefficients and N is the number of blocks in a frame. We calculate AC energy of each I frame in the sequence. The complexity of image has a great deal with image quality. The more details the image has, the more bits will be produced by encoding. The image will be damaged more seriously than the one with lower complexity. AC energy can reflect the image complexity.

2.2 Feature Selection

The relationship between every the above feature and the corresponding subjective scores is analysed. Three features with the best correlation are selected in our method: quantiser-scale factor, bit rate and intra macroblock. The correlation coefficient between quantiser-scale factor and MOS in our experiments is 84.69%. For bit rate and intra macroblock, the correlation coefficient is respectively -72.41% and 61.02%.

2.3 No-Reference Quality Assessment Metric

In order to derive the final metric, three features are first normalized to the interval of (0.1~0.9). Further, the least square method is applied to fit the features and the corresponding subjective MOS. 8 sequences encoded at 6 kinds of bit rates is used in the fitting stage. And finally, the objective assessment metric in this paper is as follows.

$$\begin{aligned} SCORE_{obj} &= 95.28 - \overline{QS} * 14.89 + \overline{R} * 8.05 \\ &- \overline{MB}_{intra} * 4.70 \end{aligned} \tag{6}$$

For each feature, we try to get a value in units of GOP. That is, we calculate the average bit rate over all the frames in a GOP, the average quantiser-scale factor over all the frames in a GOP and the intra macroblock percentage over all the P frames in a GOP. Then we feed the three values to the metric to get the objective assessment quality of a GOP. The objective score of entire sequence can be obtained by an average over all the GOP scores.

3 Experiment Results

A wide variety of source materials including 16 frame-coded sequences are used in the experiments. Each one has more than 10s duration and the video resolution is 720*576. All the sequences are encoded at bit rates of 3Mbps, 4Mbps, 5Mbps,

6Mbps, 8Mbps and 10Mbps by a MPEG-2 video encoder. Sequences have been carefully selected with a wide range of spatial and temporary complexity.

Subjective evaluation scores are from the subjective experiments with DSCQM (Double-Stimulus Continuous Quality-scale Method). To obtain MOS, we worked with 25 paid test persons. The chosen group ranged different ages (between 17 and 40), sex, education and experience with image processing. For each video, observers have to give a quality score between 0 and 100 (100 means the best quality and 0 the worst). The statistical analysis is performed on all the scores to obtain an average value as the final MOS for each video.

In our experiment, 16 sequences are divided into two sets. Each set has 8 sequences. One set is used for training to get the coefficient of three features in our objective assessment metric. Another set with 8 sequences is used to testify the performance of our algorithm. Experiment results show the accuracy of this method. The quality ratings by human assessors with the corresponding results of the model for 48 samples (8 test sequences encoded at 6 kinds of bit rates) are compared. And the correlation coefficient is 91.29%. The correlation between objective assessment scores and subjective perceived scores for all 96 samples takes the value 91.96%.

Fig.1 shows the continuous objective quality scores in GOPs of the sequence *habor* and *spring festival*. The corresponding subjective MOS can be seen in Tab. 1.Perceived subjective quality of the sequence *habor* is better than the sequence *spring festival*. A dance scene with fast movement of objects is in *spring festival*, while only a slow camera movement is in *habor*. The objective results correspond with subjective perception.

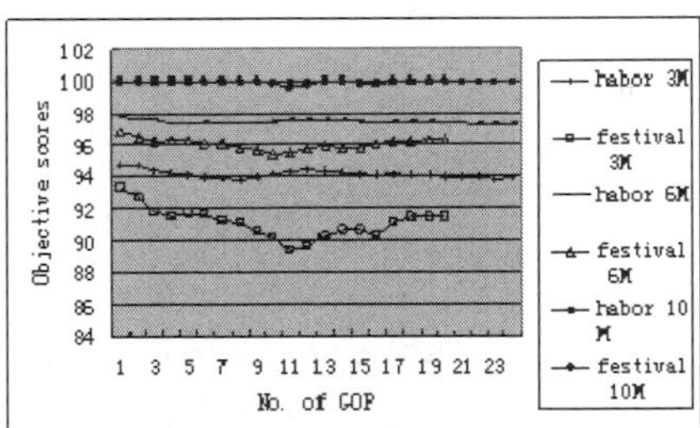

Fig. 1 Continuous objective assessment of *habor* and *spring festiva* in unit of GOP

Table 1 Subjective quality scores of *habor*

Bitrate (Mbps)	3	4	5	6	8	10
habor	91.7	93.3	95.7	96	96.8	97.6
spring festival	80.7	85.3	89.4	94	97	97.2

4 Conclusions

By analysis, research and selection on various kinds of features in the video stream, a simple and effective no-reference image quality assessment algorithm of MPEG-2 is presented.

Three kinds of features is employed in the method, which are bit rate, intra macroblock and quantiser-scale factor. These three features are extracted from streams directly, so our method does not require access to the source picture, nor to the decoded picture.

Quality assessment in unit of GOP shows the continuing validity of our algorithm. Once some quality problems occur, the corresponding score of the GOP will become low. A warning can be provided. Therefore, the method is appropriate for real-time monitoring and in-service quality evaluation.

References

[1] Winkler, S., Mohandas, P.: The Evolution of Video Quality Measurement: From PSNR to Hybrid Metrics. IEEE Trans. on Broadcasting 54, 660–668 (2008)

[2] Lubin, J.: The use of psychophysical data and models in the analysis of display system performance. In: Waston, A.B. (ed.) Digital Images and Human Vision, pp. 163–178. MIT press (1993)

[3] Pinson, M.H., Wolf, S.: A New Standardized Method for Objectively Measuring Video Quality. IEEE Trans. on Broadcasting, 312–322 (2004)

[4] Wang, Z., Lu, L., Bovik, A.C.: Video Quality Assessment Based on Structural Distortion Measurement. Signal Processing: Image Communication 19(2), 121–132 (2004)

[5] Wang, X., Tian, B., Liang, C., et al.: Blind Image Quality Assessment for Measuring Image Blur. IEEE Congress on Image and Signal Processing, 467–470 (2008)

[6] Farias, M.C.Q., Mitra, S.K.: No-reference video quality metric based on artifact measurements. In: IEEE International Conference on Image Processing, September 2005, vol. 3, pp. 141–144 (2005)

[7] Koumaras, H., Kourtis, A., Martakos, D.: Evaluation of Video Quality Based on Objectively Estimated Metric (2005)

[8] Ries, M., Nemethova, O., Rupp, M.: Motion Based Reference-Free Quality Estimation for H.264/AVC Video Streaming. In: International Symposium on Wireless Pervasive Computing (February 2007)

[9] Gastaldo, P., Rovetta, S., Zunino, R.: Objective quality assessment of MPEG-2 video streams by using CBP neural networks. IEEE Transactions on Neural Networks, 939–947 (July 2002)

[10] Jiang, X., Wang, X., Wang, C.: No-reference video quality assessment for MPEG-2 video streams using BP neural networks. In: International Conference on Interaction Sciences: Information Technology, Culture and Human (2009)

Portuguese Education Going Mobile with M-Learning

Rosa Reis and Paula Escudeiro

Abstract. M-learning intend to endorse new forms of learning encouraging the use of mobile technologies, but this purpose is not been fully implemented yet in institutions Portuguese. In this paper, we describe an environment addresses the mobile learning environment. Before we design the environment was investigated existing solutions using mobile technologies to support learning process and analyzed practical implications of design and education when they come into virtual learning environments.

1 Introduction

In this paper we present an ongoing mobile learning environment to improve the acquisition of knowledge on abstract concepts by software engineering. To stimulate the mobile learning we argue that a new thinking and behavioral changes should be incorporated into the traditional teaching and e-learning, encouraging discussion and research by professionals and educators.

The use of Mobile Technologies opens new possibilities for the teaching and learning processes, taking us to this new reality. With the Mobile learning, the teachers can do reviews of small units, update the information, and send messages to students. Likewise, the students can access to m-learning environment, at anytime and anywhere. Thus, the environment is characterized as a facilitator for the distribution of teaching materials because it offers access through mobile devices such as notebooks, mobile phones and PDAs.

The planning of a course that uses these resources, "should consider this new student profile, the content, the pedagogical framework that gives the teaching-learning process a collaborative environment, participation and teacher-student interactions"[7].

Rosa Reis · Paula Escudeiro
GILT- Graphics, Interaction and Learning Technologies – ISEP, Porto-Portugal
e-mail: {rmr,pmo}@isep.ipp.pt

F.L. Gaol (Ed.): Recent Progress in DEIT, Vol. 2, LNEE 157, pp. 459–464.
springerlink.com

Based on these ideas, the paper was organized as following: the first part (section 2) we describe some works developed about m-learning in Portugal. In the second part (section 3) we describe our environment, the architecture and design phases. Finally in the last section we present some problems encountered in design and direction for future work.

2 M-Learning and E-Learning

With the rapid growth of information and communication technologies (ICTs) it becomes possible the emergence of new forms of education. These forms provide new means to combat the shortcomings of traditional teaching.

E-Learning offers new methods of distance education based on computers and network technologies. Furthermore, the m-learning is part of e-learning.

The meaning of m-learning is defined as the access to learning through the use of mobile devices with wireless communication in a transparent manner and with high degree of mobility. "This new paradigm emerges advantage of the availability of mobile devices and considering the need for specific education and training"[7]. The use of m-learning as a teaching resource, enables new thinking and behavioral changes that are being incorporated into traditional teaching and e-learning process teaching-learning distance, encouraging discussion and research by professionals and educational technology. For this, the planning of a course that uses these resources, "should be considered the profile of this new student, and tailor strategies, content and a pedagogical framework that gives the teaching-learning process a collaborative environment, participation and interaction teachers and students."[8].

The literature has documented, broadly, the multiple m-learning effects with e-learning. (see table 1)

Table 1 M-learning effect with e-learning.

Features
Self-Study- M-learning allows greater flexibility for participants to learn at their own pace in anytime and anywhere. [8]
Evaluation and Feedback - To monitor student's progress and produce reports we can include into eLearning packages e m-learning assessment tools. The teacher can give feedback to their students. [8]
Access of Online Repository- Through m-learning system, the online materials offer a place for the students and teachers to interact frequently. [8]
Engagement- The students work with different dimensions of interactivity and respond with immediacy; they have more active participation[7,8]

3 M-Learning in Portugal

In Portugal m-learning is given the first steps. The work described hereby is a result of an exploratory research carried out on the Internet. From a consistent pursuit, it was found the following research studies that demonstrate the advantages and disadvantages of m-learning in educational contexts.

The University Fernando Pessoa, in 2007, built one system, which allows the construction of forms (tests and surveys), that are available in class [5]. The purpose of this study was observing the potential of m-learning in higher education. It was developed a prototype consisting of two modules. The first module provides the resources the workspaces to users of the platform of collaborative learning. The second module allows the teacher to build quizzes and make them available in the context of the classroom. With this study, it was concluded that the m-learning has great potential to be used in higher education, because the system was well accepted by students and teachers. Moura and Carvalho, in 2009, conducted two studies: " one, using pod casts for study of Portuguese Literature and another one in Portuguese Language learning focused on the use mobile phones as a productive tool: text, audio, image, video and as information repository"[4]. First study was applied in a public school and it concluded that the mobiles phones contribute to the motivation of school activities, collaborative work and quick access to learning content. The second study aimed at a vocational school. The aims of project were: - create opportunity for students to use mobile phones and create their own content; build knowledge and respect the learning time of each student. They planned a set of learning activities in order to assess the influence of mobile learning in engaging students in activities. The results led to conclude that: the use of mobile phones in the classroom, enables a better understand of concepts which increase their comprehension skills; this makes it the access to information and, the students can take notes faster.

The present authors intend to identify problems or benefits from the incorporation of m learning in our current educational system. Our academic community can incorporate the m-learning in its educational context and the students can develop applications that use this type of technology, ensuring that education be accessible to all.

4 Mobile Learning Environment

The Department of Computer Science of ISEP is taking the first steps in terms of mobile learning. It is mainly in the area of Graphics Systems and Multimedia, in the subject of learning in Multimedia Systems (SIMU), which students have been developing small curricular work where they have been adapting educational software (Educational Games) to mobile technology.

In order to study the impact of m-learning in education, we decided to develop a prototype of a mobile learning environment, which addresses a specific content of software engineering. The prototype is under development and evaluation is ongoing. Our aim is to observe if the use of mobile technology can contributes to facilitate the knowledge acquisition of abstract concepts. So, we intend to identify which activities may be fit into a learning of abstract concepts. In first stage, it will be used the application in the real classroom and also observe the students' feedback and outcomes. The students should be able to follow and solve specific questions of content and keep on their social interactions through their PDA's without being affected by time and place. We want to encourage student's active participation in the learning process. It attempts to engage them in constructivist learning through social and intellectual interactions.

The environment allows teachers and students interact with course material either a personal computer or form mobile device. Students have a user login while teacher have an dministrative login for configuration and monitoring the contents and the activities learning. After login, the environment delivers the unit content through the web browser.

The prototype design was divided into several phases, as shown in Fig. 1

Fig. 1 Process of Prototype Design.

Content Definition – The content selected was the use case diagram, which was divided it into small units of information: the concept of use case, actors, and type of relationships. The description of each unit reports the basic concepts, resorting to different types of media such as text, videos and images. Each unit has a set of questions.

Scenarios Design – This step was designed to describe the structure of the presentation's content (storyboard). A special attention was paid on the type of device with where the application will work to enhance training delivery in our content field

Learning Activities Definition and Evaluation– Different ways were chosen to perform the learning activities: the SMS because the teenagers frequently use SMS as communication medium. "The use of SMS in M-learning has been identified as an effective tool to enhance both students' learning experience and tutors' instructing experience, yet the requirements of the two parts were quite different in terms of their previous experience of SMS" [7]. The activities will consist in the presentation of small case studies and it will be carried out by group's teams, encouraging collaborative learning. Students should outline the use case diagram corresponding to the situation presented using any graphical tool. Finally, the resolution should send via MMS to the teacher. In order to clarify doubts that may arise, the students communicate verbally with the teacher. ISEP has a protocol with a Portuguese Mobile Operator.

Prototype Development – Design prototype based on the scenarios. It will be tested and evaluated before being available to students who will attend for the first time the subject of software engineering. The aim is to ensure that it performs correctly in all situations. It will be used a small group of individuals to evaluate the performance of our strategy across a range of devices and learning styles.

Dimension: Functionality	
Factor	Criteria
Easy of use	Does the student use the m-learning application without having to read the manuals exhaustively?
	An on-line system exists to help the user overcome the difficulties?
Content's quality	Is the information well structured and does it adequately
	Is the content related with situations and problems of student's interest?
	Are examples, simulations and graph part of the system?
Dimension: Efficiency	
Factor	Criteria
Audiovisual quality technical and static elements	Is there no excess of information?
	Has it a rigorous scenario design.
Navigation and interaction	Does the application have a good program structure that allows easy access to content and activities
	Is the navigation system transparent, allowing the user to control action?
	Has the system a good communication between the users and the program?
Dimension: Adaptability	
Factor	Criteria
Pedagogical Aspects	Does it allow for activities that keep the curiosity and the interest of the students in the content, without provoking anxiety?
Didactical resources	Does it provide different activity types, concerning the knowledge acquisition, that allow for different forms of using the system?
	Does it provide help for students as tutoring actions, guiding activities and reinforcements?
Simulates the initiative and self learning	Does it allow for student's decisions concerning the tasks to carry through, the choice of study of subject matter?
Cognitive effort of the activities	Does it allow for easy memorization, interpretation, syntheses and experimentation?

Fig. 2 Evaluation Matrix by Escudeiro [3]

5 Conclusion

In Considering that our intention is to evaluate the implications of use of mobile learning in teaching and learning of abstract concepts, we design this environment by giving a special attention to objectives of m-learning: improve the capabilities of preparation tasks; have access to educational content at anytime and anywhere; increase the accessibility of content; promote formal and non-formal learning; expand the boundaries of the classroom and enable the development of innovative methods of teaching using the technological resources [7]. Although the environment is still under development, one set of problems arose from technical limitations of mobile devices, the small memory capacity and the devices do not operate with a mouse- they rely on touch. However we emphasize:

- Design of content - The content shouldn't be designed ah doc. We need to develop the contents adaptively to a wide range of devices, due limitation of memory, resources reduced and phone display. The content must be small so that the student not be discouraged. The complex content, with small font and displayed in several pages can lead students to abandon what he is doing;
- Potential synchronous communication – If used as part of learning system, the interactions can be more efficient and can bring a high value to the learning process.

In future, we will designing new courses to curriculum of computer engineering; create hands-on, whose purpose is to increase awareness of the use of m-learning in the classroom and "investigate and test strategies to capture and keep students' attention by involving them in the learning activity" [7]."Because the use of m-learning is new in education, it is important for educators, researchers, and practitioners to share what works and what does not work in m-mobile learning can be implement in a more timely and effective manner" [1].

References

1. Ally, M.: Mobile Learning: Transforming the Delivery of Education and Training. Issues in Distance Education series. Au Press, Athabasca University (2009) ISBN- 978-1-897425-44-2
2. Escudeiro, P.: X-Tec model and QEF model: A case study. In: Bastiaens, T., Carliner, S. (eds.) Proceedings of World Conference on E-Learning in Corporate, Government, Healthcare, and Higher Education, pp. 258–265. AACE, Chesapeake (2007)
3. Moura, A., Carvalho, A.: Mobile Learning: Teaching and Learning with Mobile Phones and Podcasts. In: Eighth IEEE International Conference on Ad-vanced Learning Technologies, ICALT (2008) 05ISBN: 978-0-7695-3167-0
4. Rodrigues, J.L.S.: Mysynaps: Uso de m-learnnig no ensino uperior. Tese de mestrado, Universidade Aveiro. Aveiro (2007), http://ww2.uft.pt/~lmg/monografias/jrodrigues_ms_v07.pt (accessed in September, 2010)
5. Valentim, H.: Para uma Compreensão do Mobile Learning. Reflexão sobre a utilidade das tecnologias móveis na aprendizagem informal e para a construção de ambientes pessoais de aprendizagem. Tese de mestrado em Gestão de Sistemas de e-Learning, Universidade Nova de Lisboa, Lisboa, Recuperado de (2009), http://hugovalentim.com
6. Nyíri.: Towards a Philosophy of M- Learning. Presented at the IEEE International Workshop on Wireless and Mobile Technologies in Education (2002)
7. Rosman, P.: M-learning-as a paradigm of new forms in education. E+M, Ekonomie a Management 1, 119–125 (2008)
8. Campbell, J.: Mobile Learning Evaluation tools. In: Mobile Learning Evaluation Tools | eHow.com, http://www.ehow.com/list_7322519_mobile-learning-evaluationtools.html#ixzz1AYrqOs4Z

Measurement and Improvement of the Delay of SCTP Multihoming Path Switch

Kuang-Yuan Tung, Richard Chun-Hung Lin, Ying Chih Lin, and Yu-Hsiu Huang

Abstract. SCTP multihoming feature has been widely exploited to develop schemes for mobile host handoff, e.g., mSCTP. However, there is no research work on the measurement of path switch delay in practical SCTP implementations. In this paper, we consider the possible case of SCTP path switch delay on common Linux system testbed with SCTP implementation. Our experiment demonstrates that many packets are blocked in the duration of path switch, where the delay usually results from the timeout/retransmission trial process. And we propose an approach to intelligently bypass retransmission phase without modifying the SCTP protocol. The experiment results show that the path switch delay can be significantly reduced and has no negative impact on the whole performance.

1 Introduction

The type of mobile host handoff can be classified according to the handling responsibility of the communication protocol layers. Mobile host does not need to go through IP re-configuration during the data link layer handoff. Hence, just the handoff process of the data link layer has to be involved. Mobile IP is a well known method to support host mobility at the network layer. And Its mechanism results in longer handoff delay and wasting more bandwidth, and cannot meet the QoS requirement of real-time application. In contrast, Mobile SCTP (mSCTP) [1, 2, 3] has proposed to deal with host mobility based on the Stream Control Transmission

Kuang-Yuan Tung · Richard Chun-Hung Lin · Ying Chih Lin · Yu-Hsiu Huang
Department of Computer Science and Engineering, National Sun Yat-Sen University, Kaohsiung, 80424, Taiwan, R.O.C
e-mail: {beck,lin,yclin,yhhuang}@cse.nsysu.edu.tw

Yu-Hsiu Huang
Department of Computer Science and Information Engineering, Cheng Shiu University, Kaohsiung, 83347, Taiwan, R.O.C
e-mail: yhhuang@cse.nsysu.edu.tw

F.L. Gaol (Ed.): Recent Progress in DEIT, Vol. 2, LNEE 157, pp. 465–470.
springerlink.com

Protocol (SCTP) [4], which is a message-oriented and reliable protocol with good features of UDP and TCP. SCTP provides two core features, multihoming and multistreaming mechanism. Furthermore, mSCTP extends the base SCTP to facilitate mobility in the Internet at the transport layer [5]. The dynamic address reconfiguration process [6] of mSCTP allows two SCTP end-points to add new IP addresses, and subsequently, reset primary IP addresses for the association after the deletion of IP addresses from an active association.

2 Handoff Handling of SCTP

An SCTP association, on the other hand, supports multihoming service [7]. The sending and receiving hosts can define multiple IP addresses in each end for an association. Once one path fails, another interfaces can be used for data delivery without interruption.

A large amount of packets are blocked and lost in the duration of path switch because SCTP uses retransmission to make sure if the original path is broken. The excessive path switch delay makes SCTP unsuitable to design mobile network handoff schemes, especially for delay-sensitive applications like VoIP. We will demonstrate the SCTP path switch delay with using Linux kernel SCTP library [8, 9] in our experiments. Furthermore, we propose an approach which intelligently bypasses retransmission phase without modifying the SCTP protocol.

3 Experiments and Performance Analysis

3.1 Experimental Setup

The client and the server experiment hosts were running Ubuntu Linux 8.10, and both machines installed Linux kernel SCTP library. The client host is equipped with two network interface cards (NICs), Ethernet and WiFi, and the server is with Ethernet only. Ethernet and WiFi NICs in the client have different IP addresses. The client uses the multihoming feature of SCTP to setup two paths through Ethernet and WiFi interfaces individually to connect to the server. We let SCTP bulk data flow from server to client which is first coming client host through the Ethernet NIC. To emulate the handoff, we disconnect the cable and force the client to make the path switch. Then we observe the incoming data throughput fluctuation during the time period of path switch. Note that the server has no idea about when the path is switched in client. It just follows SCTP path switch scheme to gradually switch the path (i.e., timeout/retransmission, path switch, and then enabling slow start scheme), and keeps transmitting data to client host.

We let Wireshark [10] network analyzer running in client side to capture the incoming packets received from the server to measure the data throughput. Next, we do the same experiment, but changing the RTO (Retransmission TimeOut) parameter to measure the throughput fluctuation. Finally, we do the experiment by using our approach to conspicuously improve the SCTP path switch performance.

3.2 Performance Analysis

In the following section, we will show the throughput fluctuation in the above three experiments.

3.2.1 Experiment *A*

Both IP addresses in client host are known at the time of SCTP association establishment. As we know, a SCTP association supports multihoming service. In this experiment, we will measure SCTP multihoming path switch performance. The switch delay from the old path to new path is our most concern. The time period of delay we are measuring here is the time of the throughput from going down to coming back and stable again. This is the handoff period in the mobile network. The smaller the handoff period is, the more seamless the transmission is. This is very important for real-time application, e.g., VoIP. The SCTP association is not interrupted during the handoff.

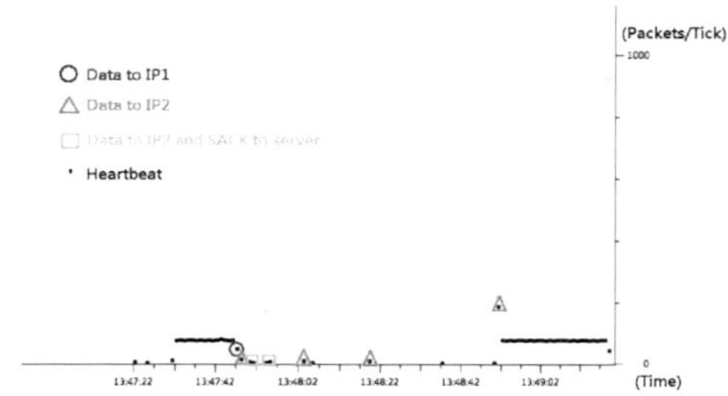

Fig. 1 The IO Graphs of Experiment. *A*

The packets captured by Wireshark show that when the path switches form Ethernet (binding IP1) to WiFi (binding IP2), the client will immediately sends SACK for those data coming from IP1 through IP2. Then the server starts to send small data flow to IP2, and in the meantime the server sends HEARTBEAT chunk (HB) to IP1 to test if the path to IP1 is still active. Server keeps trying to send data to IP1. Obviously, the timeout occurs and exponential backoff is performed to retransmit. Linux kernel SCTP is set 5 times retrying as the default. Thus, we observe that a serial of 1, 2, 4, 8, 16, and 32 seconds timeout delay of retransmission procedure occur in our experiment. Finally, the data is still lost, and then server gives up IP1 and switches to IP2. Totally it takes 63 seconds before giving up the original path.

In Figure 1, we illustrate an IO Graph of throughput in client host from Wireshark. After five consecutive timeout, backoff and retransmission cycles, finally the server decide to give up the original path and to switch to the new path. Then the slow-start phase is enabled and gradually the throughput can reach the stable state. The total time spent is too large to apply SCTP multihoming feature in many applications, especially the delay-sensitive ones, e.g., VoIP. Those research works for applying SCTP in mobile network handoff are never measuring the practical path switch delay. The performance is definitely not suitable for VoIP in mobile network handoff. In the following sections, we will show the methods to improve this delay. Obviously, adjusting the RTO seems to be a good direction.

3.2.2 Experiment *B*

The related parameters of RTO (Retransmission TimeOut) are adjustable in Linux kernel SCTP library. In this experiment, we change the values of RTO parameters to check the variation of total delay time spent on the path switch. Therefore, we set them as following: the minimal value of RTO (RTOmin) = 100ms, the maximal value of RTO (RTOmax) = 500ms, and the maximal number of retransmissions (MPR) = 5 (default value). We repeat the same procedure of experiment *A* and the connection experiences a shorter path switch delay. It occurs 5 retransmission timeouts and we get 6 timeout intervals: 100ms, 200ms, 400ms, 500ms, 500ms, and 500ms. Thus, the total delay of path switch is greatly reduced to 2.2 seconds (it still needs extra time to make the throughput become stable again).

In Figure 2, an IO Graph is depicted for the throughput from Wireshark. We could see HB probing packets generated in 1.5 seconds, and then the path switch completes. The result for VoIP looks fine, but the time for throughput to become stable again will take 7 seconds. It may affect the QoS and may have some packet loss. In our experiments, we use the default MPR 5. That means, there must be 5 retransmission trials before path switch. It contributes the main delay of path switch. However, retransmission procedure is necessary in SCTP and cannot be disabled to avoid incorrect path switch decision. In the next experiment, we will propose our approach which makes the path switch immediately and correctly without doing retransmission trials.

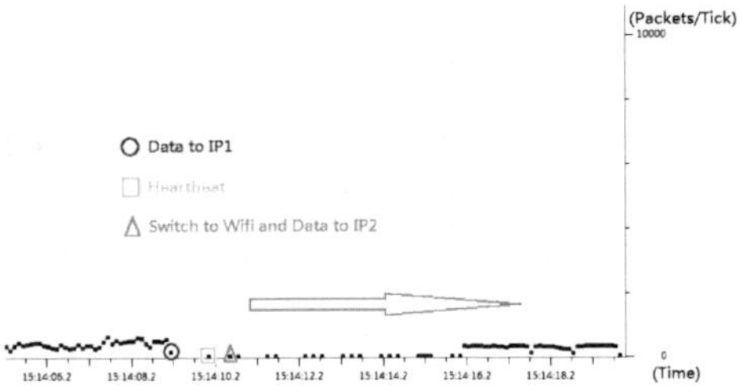

Fig. 2 The IO Graphs of Experiment. *B*

3.2.3 Experiment *C*

From the above two experiment results, we can figure out the main delay to be taken in the detection of path broken. Consider the mobile network applications. The mobile client handoff will make the path broken. So in most cases, the client can detect the path failure immediately. However, the server side cannot get this information and must use timeout and retransmission scheme to make sure the path failure. It is time-wasting. So we can exploit the characteristic of client site's immediate path failure detection to make server switch path earlier without enabling time-wasting retransmission. In our approach, when the client finds the path broken, it immediate sends a short data packet (the payload can be empty or just one byte dummy data) to server from the new path (i.e., new NIC's IP address). The server receives this dummy data packet from client through new path. It must send a SACK to client for acknowledgement through new client NIC. Like TCP, SCTP is always piggy-backing data over SACK packet to save the transmission cost. Thus, the data flow from server to client will be unconsciously directed into the new path to client. In this experiment, the server no longer enables the retransmission procedure to make data flow path switch decision. But it is a client triggered path switch. This is shown in Figure 3. When the client finds the IP1 NIC disconnected, it immediately sends 1-byte dummy data packet to server to direct data flow of server to client into IP2 NIC. The following experiment results will show our approach to be very useful and efficient, and will completely avoid retransmission delay.

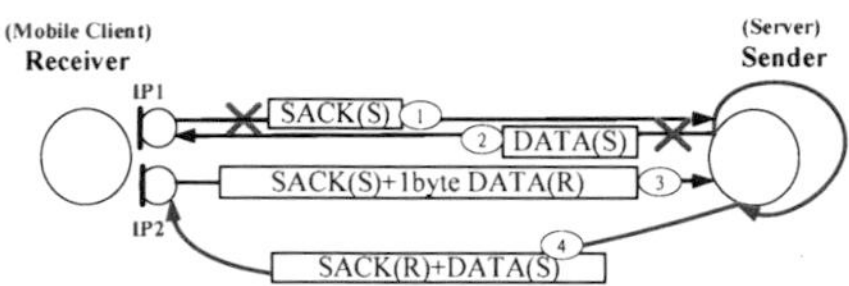

Fig. 3 1 byte pseudo data from receiver to sender.

In our experiment, we let IP1 be Ethernet NIC and IP2 be WiFi NIC. The server has only Ethernet NIC. The data flow is coming from the server to the client. SACK(S) and SACK(R) mean the acknowledgements for server and client, respectively. DATA(S) is the regular data flow from server to client, and DATA(R) is 1-byte dummy packet to speedup server path switch. From the output of Wireshark network analyzer, the data flow is completely matching our expectation to switch to the new path without enabling retransmission. So the path switch delay is much reduced.

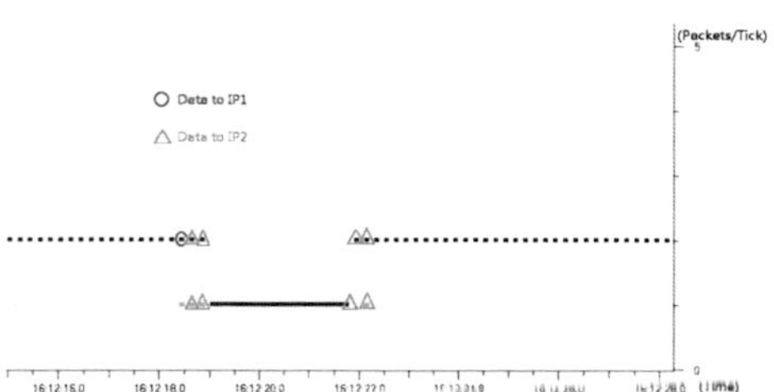

Fig. 4 The IO Graphs of Experiment. *C*

Figure 4 is the Wireshark IO Graph output. We could see old NIC replaced by new NIC very quickly. It is almost seamless (no gap existing) during the path switching. The result is absolutely accepted by mobile network application, e.g.,

VoIP. The duration is lasting for 2.5 seconds until data flow into new NIC is stable. Also, the throughput is almost keeping the same during the path switch (Observe that the time ticks become twice faster than the original flow and the number of data flow received becomes half). So we never feel any interruption during the path switching. When sender receives the 1 byte data, it piggybacks its data to SACK to client IP2, and no longer sends data to client IP1. This is why we do not feel any interruption.

4 Conclusion

SCTP's retransmission-based switching decision policy results in excessive path switch delay. We demonstrate the poor performance of the current SCTP implementation (Linux kernel SCTP library), and subsequently, propose an approach to intelligently bypass the retransmission-based path switching policy. The experiment results show that the path switch delay can be significantly reduced and has no negative impact on the performance.

References

1. Koh, S., Xie, Q., Park, S.: Mobile sctp (msctp) for ip handover support. Internet draft (2005), http://tools.ietf.org/html/draft-sjkoh-msctp-01
2. Ma, L., Yu, F., Leung, V., Randhawa, T.: A new method to support utms/wlan vertical handover using sctp, vol. 11(4), pp. 44–51 (2004)
3. Riegel, M., Tuexen, M.: Mobile sctp. Internet draft (2005), http://tools.ietf.org/html/draft-riegel-tuexen-mobile-sctp-05
4. Hakkinen, A.: Sctp - stream control transmission protocol. Seminar on Transport of Multimedia Streams in Wireless Internet, Dep. Comp. Sci., Univ. Helsinki (October 2003)
5. Aydin, I., Seok, W., Shen, C.-C.: Cellular sctp: A transport-layer approach to internet mobility. In: Proc. ICCCN 2003, Dallas, Texas, October 2003, pp. 285–290 (2003)
6. Stewart, R., Xie, Q., Tuexen, M., Kozuka, M.: Stream control transmission protocol (sctp) dynamic address reconfiguration. RFC 5061 (September 2007)
7. Iyengar, J., Shah, K., Amer, P., Stewart, R.: Concurrent multipath transfer using sctp multihoming. In: SPECTS 2004, San Jose, California (2004)
8. Linux kernel stream control transmission protocol (lksctp) project, http://lksctp.sourceforge.net/
9. Sctplib version 1.3.1., http://tdrwww.exp-math.uniessen.de/inhalt/forschung/sctp_fb/sctp_links.html
10. A network protocol analyzer: Wireshark, http://www.wireshark.org/

A Holistic Game Inspired Defense Architecture

Sajjan Shiva, Harkeerat Singh Bedi, Chris B. Simmons,
Marc Fisher II, and Ramya Dharam

Abstract. Ad-hoc security mechanisms are effective in solving the particular problems they are designed for, however, they generally fail to respond appropriately under dynamically changing real world scenarios. We discuss a novel holistic security approach which aims at providing security using a quantitative decision making framework inspired by game theory. We consider the interaction between the attacks and the defense mechanisms as a game played between the attacker and the defender. We discuss one implementation of our holistic approach, namely, game inspired defense architecture in which a game decision model decides the best defense strategies for the other components in the system.

1 Introduction

The research and practicing community have been paying attention to the Internet and data security problems for more than two decades. However, these problems are far from being completely solved. The main limitation of the current security practice is that the approach to security is largely heuristic, is increasingly cumbersome, and is struggling to keep pace with rapidly evolving threats. The core security breaches occur in terms of confidentiality, integrity, and availability.

To overcome these problems, four types of security endeavors have been employed in the past. (a) Implementation of Secure Communication Infrastructure where, cryptographic algorithms are used to build secure networking protocols such as Internet Protocol Security (IPSEC) or Transport Layer Security (TLS). However if one endpoint is compromised, the crypto becomes helpless. (b) Utilizing Monitoring and Response Systems such as firewalls, intrusion detection systems (IDS) and antivirus programs. Further, with the advent of the virtualization technology, researchers are advocating to host applications on a virtual machine

Sajjan Shiva · Harkeerat Bedi · Chris Simmons · Marc Fisher II · Ramya Dharam
University of Memphis, Memphis, Tennessee, USA
e-mail: {sshiva,hsbedi,cbsmmons,mfisher4,
rdharam}@memphis.edu

F.L. Gaol (Ed.): Recent Progress in DEIT, Vol. 2, LNEE 157, pp. 471–476.
springerlink.com

X, so that all activities in X could be observed by a monitor application residing outside X. Nevertheless, a perfectly safe monitor is yet to be designed. (c) In the Built-In Approaches for System Development, security features are designed up front and form part of the system development.

However the Bolt-On approaches compensate for the errors made earlier in the development cycle and emergent errors introduced after the system is deployed. Bolt-on approach is the only solution for deployed (legacy) systems. (d) Code Instrumentation Tools and Self Checking Modules provide for enforcement of data and control flow integrity of a software component to provide security. Such techniques compute a flow graph using static or dynamic analysis, and instrument the program to check if the execution at runtime conforms to the flow graph. However, these techniques are not generally effective against polymorphic exploits.

Despite the past considerable effort to protect and secure software and data, it can be observed that the goal of securing the same is far from being accomplished.

Data security is a data engineering problem, which we aim to address in our proposed solution. We propose a holistic security approach [4] which provides a framework that encompasses a whole system in a layered and organized manner. Most strategies to implementing security mainly focus on one specific area at a time. To help advance the cyber and data security community, our approach collectively uses monitoring tools, knowledge management systems, and game inspired decision models for achieving an optimal level of security.

The following Section 2 discusses our proposed holistic security scheme. Section 3 discusses an implementation of our holistic approach. Section 4 provides the concluding remarks and future work with respect to our holistic approach.

2 A Holistic Security Scheme

We envision a 4-layer holistic security scheme as illustrated in Figure 1. At the innermost layer are the core hardware and software components. We envision each of these components having a provision of being wrapped with a self-checking module (with inspiration from the traditional BIST architecture). At the second layer reside the Secure Applications which are designed with Built-In or Bolt-On security approaches utilizing self-checking concepts and components. At the third layer lies the traditional Security Infrastructure that is built using firewalls, anti-virus software, etc. At the fourth layer, we envision a Game Inspired Decision Module (GIDM) which is responsible for choosing the best security strategy for all the inner layers.

We visualize this fourth layer as one placed directly above the three previously defined layers, emphasized by the pie formation in Figure 1. This placement stresses the fact that GIDM can obtain input from any of the layers and can recommend probable defense actions for the same. The solid arrows in Figure 1 represent the progression of information pertinent to attack detection to the outer layer. The dotted line represents the flow of corresponding corrective action strategy as decided by the outer layer for the inner layers.

We observe that in the past, majority of the security efforts have only been in the second and the third layers. Traditional intrusion detection systems can be considered as residing in the third layer, which can be made more effective by the use of game inspired decision techniques, which resides in the outermost layer. Note that our layered view is an operational one and does not have any direct relationship with the traditional ring-oriented privilege separation principle or the OSI network stack.

We now define the layers contributing to our holistic security approach and characterize their purpose and interaction with the other layers to form a cohesive secure solution.

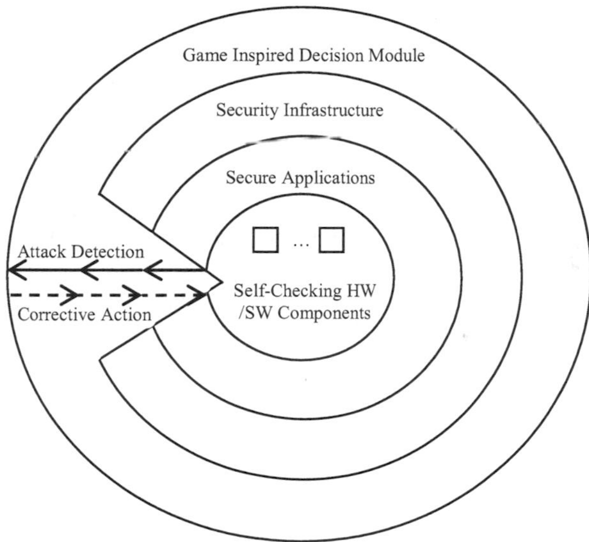

Fig. 1 The Holistic Security Approach

Self-Checking HW/SW Components: The innermost layer consists of self-checking hardware and software components. Monitoring allows observing the behavior of the software to determine whether it complies with its intended behavior. We intend to use the BIST methodology [2] for monitoring hardware components. For monitoring software components we intend to use run time monitoring tools [1] to specify the software properties that need to be monitored. When these monitors recognize a deviation of the software behavior we intend to use protocols for communicating the said information with the Game Inspired Decision Module (GIDM), which is explained later. GIDM will use a knowledge management system and a honey pot to effectively communicate the flow of information. This information is used to take preventative action against software executing malicious activity.

Secure Applications: The second layer in our holistic security model contains security applications, which offer a built-in or bolt-on approach. The built-in approach uses pre-release security components in a software application which ensure agreement with secure software specifications. The bolt-on approach use post-release patches and software updates to achieve the optimal level of security.

In addition to built-in and bolt-on approaches we intend to perform monitoring in this layer by enabling the developers to develop monitors for the existing secure applications. The monitors can notify violations of specified properties to GIDM via protocols and perform recovery actions recommended by GIDM to avoid further deviation of the software.

Security Infrastructure: The third layer is the security infrastructure within GIDA, which primarily focuses on the use of tools such as intrusion detection system (IDS) and firewalls capturing raw input. The third layer also provides protocols for communicating with the outermost layer which is the Game Inspired Decision Module.

Game Inspired Defense Model (GIDM): This outermost layer functions as the brain behind our holistic approach whose main purpose is to evaluate these attack vectors using game theoretic analysis and decide the best action strategy for the inner layers to counter committed or probable attacks. GIDM and a knowledge management system (KMS) accept input from the inner three layers, which contain the information related to malicious activity. The KMS classifies the nature of an attack using attack vectors and associated defense measures, which are propagated as input to GIDM. This decision module can also interact with honeypots which are used to gain knowledge of attack activity in a mendacious manner without the attacker being aware. This information is filtered into GIDM for selecting the optimal decision for defense. Information regarding the handling of defense recovery based on the attack is propagated back to the inner layers.

3 Game Inspired Defense Architecture

Game Inspired Defense Architecture (GIDA) is an implementation of the above described holistic security approach. It focuses on the concept of offering defense strategies against probable and committed attacks by modeling such situations as multi-player game scenarios.

Figure 2 shows our Game Inspired Defense Architecture. We illustrate its flow of execution using a network based attack scenario. An attacker on the Internet aims at exploiting a vulnerability in one of components cx on the Target Application which includes self-monitoring modules mx. The modules monitor the individual components and contain the provision to send their findings to either the KMS or the Game Inspired Decision Module. These modules also include the ability to stop the execution of such components if deemed necessary. The Game Inspired Decision Module acts as the brain of this architecture and computes defense strategies based on feedback from the monitoring modules, KMS, the Honeypot and the Intrusion Detection System (IDS). Execution of such strategies aims to protect and defend the Target Application against adversaries.

We now discuss the several components used by GIDA and explain their functionality and purpose for accomplishing the aim of providing robust security for target systems.

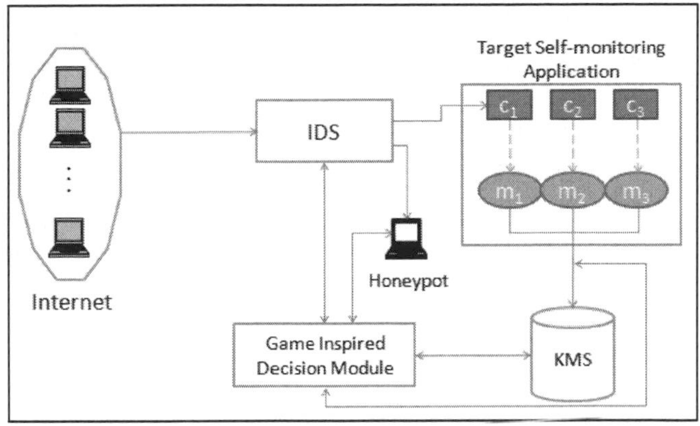

Fig. 2 Game Inspired Defense Architecture

Software Monitoring: Monitors can either be embedded along with the application code or can be separated and placed away from the application code. Monitors developed using monitoring tools will provide users flexibility to specify software properties to be monitored using logical formalisms. Monitors generated from formal specifications are then used to verify the execution of the program behavior.

Monitors thus present in the innermost two layers will be responsible for notifying GIDM about the behavioral changes of the software that could indicate an attack. Once GIDM performs a decision analysis on the probability of attack, it informs the monitor of the appropriate action to take to minimize the damage to the application software. This information is also transferred to the KMS for appropriate attack classification. This approach to self-checking software enables verification of control flow towards an applicable output for defense measures.

Knowledge Management System: In [5], we proposed a cyber-attack taxonomy called AVOIDIT that classifies attack vectors to assist defenders with disseminating defense strategies. Five major classifiers are used to characterize the nature of an attack, which are classification by attack vector, classification by attack target, classification by operational impact, classification by informational impact, and classification by defense. It is presented in a tree-like structure to neatly classify attack vectors and common vulnerabilities used to launch cyber-attacks.

The proposed cyber-attack taxonomy is used as a repository schema for a knowledge management system (KMS). The KMS is a component within the game inspired decision model that captures monitoring related data from the inner layers in an attempt to classify an attack and output to GIDM for decision analysis.

The KMS is used for regenerating the consummate path to an attack for propagating appropriate defenses. Notification is sent to GIDM, which investigates the applicability of determining the action space of the defender and attacker. Integrating attack information into the GIDM allows game agents to locate data easier for the most relevant defense method. This approach towards attack classification and defense dissemination provides seamless transfer of knowledge for our holistic security approach.

4 Conclusion and Future Work

In this paper we presented a holistic security approach which provides a multilayer framework to achieve an optimal level of security. GIDA focuses on use of security monitoring tools to observe deviations in the function of software and hardware, a KMS for classifying attacks, and a game decision module for deciding the best defense strategy based on captured information.

Our experimentation [3, 6] included modeling and simulation of game theory-based solutions against DoS and DDoS attacks. In future work, we aim to extend the functionality of such previously proposed game models by incorporating them in architectures like GIDA for addressing a broader array of cyber and data engineering problems. We also aim to use attack test-beds like DETERLAB to further investigate the efficiency of our proposed holistic security approach.

References

1. Delgado, N., Gates, Q., Roach, S.: A Taxonomy and Catalog of Runtime Software-Fault Monitoring Tools. IEEE Transactions on Software Engineering 30(12) (December 2004)
2. Kranitis, N., Paschalis, A., Gizopoulos, D., Xenoulis, G.: Software Based Self-Testing of Embedded Processors. IEEE Transactions on Computers 54(4) (April 2005)
3. Shiva, S., Roy, S., Bedi, H., Dasgupta, D., Wu, Q.: An Imperfect Information Stochastic Game Model for Cyber Security. In: The 5th International Conference on i-Warfare and Security (2010)
4. Shiva, S., Roy, S., Dasgupta, D.: Game theory for cyber security. In: Proceedings of the Sixth Annual Workshop on Cyber Security and Information Intelligence Research (2010)
5. Simmons, C., Shiva, S., Dasgupta, D., Wu, Q.: AVOIDIT: A cyber attack taxonomy. Technical Report: CS-09-003. University of Memphis (August 2009)
6. Wu, Q., Shiva, S., Roy, S., Ellis, C., Datla, V.: On Modeling and Simulation of Game Theory-based Defense Mechanisms against DoS and DDoS Attacks. In: Part of the 2010 Spring Simulation Multi Conference on 43rd Annual Simulation Symposium (ANSS 2010), April 11-15 (2010)

Performance of VoIP on IPv4 and IPv6to4 Tunnel Using Windows Vista and Windows 7

Samad Salehi Kolahi, Mohib Ali Shah, and Justin Joseph

Abstract. In this paper, the performance VoIP is investigated for IPv4, IPv6to4 tunneling mechanism, for various VoIP codecs, namely the G.711.1, G.711.2, G.723.1, G.729.2 and G.729.3 codecs using Windows Vista and Windows 7 operating systems. Parameters studied are jitter, RTT (round trip time) and throughput. The results indicate that IPv6to4 tunneling generates extra RTT and jitter compared to the IPv4 networks. For the system studied, the highest RTT was at 0.8msec for G711.2 using Pv6to4 tunneling. For the system studied, the G.711.1 codec provides the highest bandwidth of 688Kbps while G.723.1 has the lowest bandwidth of 77.4 Kbps of all codecs studied.

1 Introduction

Due to the overwhelming number of desktop, laptop and other computing machines that require IP addresses for accessing the internet and networking, all available IP addresses in IPv4 will be exhausted in the near future. IPv6 was designed and introduced by the Internet Engineering Task Force (IETF) to be the successor of IPv4. The most important character of IPv6 is it supports large numbers of address, 2^{128} IP addresses (128-bit address field in IPv6 packet).

VoIP (Voice over Internet Protocol) was designed to provide voice communications over a packet switched network (internet). VoIP network system uses a range of voice protocols and audio codecs to manage communication. These protocols look after the connection, set-up and tear-down of calls, while the codecs encode the speech so that it can be transported over the internet. The main reason VoIP has become so popular over the last few years is because of the reduced cost associated with using VoIP compared to the PSTN (Public Switched Telephone Network). As VoIP becomes more and more popular and IPv6 is being introduced to society, some technologies such as IPv6to4 tunneling mechanism allow both IPv4 and IPv6 networks to exist in the same network infrastructure.

Samad Salehi Kolahi · Mohib Ali Shah · Justin Joseph
Unitec Institute of Technology, New Zealand
e-mail: skolahi@unitec.ac.nz

F.L. Gaol (Ed.): Recent Progress in DEIT, Vol. 2, LNEE 157, pp. 477–483.
springerlink.com

Windows 7 is newer version of Microsoft Operating Systems and was introduced to replace Windows XP and Windows Vista. There is not much work in literature on VoIP performance using Windows 7.

The IPv6to4 is a tunneling mechanism which allows IPv6 users to communicate with various IPv6 networks through IPv4 network. This mechanism was designed to connect IPv6 networks through an IPv4 cloud network. The ability of IPv6to4 tunnel is as it carries IPv6 packets and encapsulates them into IPv4 header and sends them via IPv4 network. It de-capsulates the packets at the other end and delivers to its destination.

As VoIP gains greater popularity and soft phones are being introduced to provide more flexibility to the users; the evaluation of parameters affecting VoIP performance is becoming very important. The authors, therefore, carried out experiments to identify the impact of different operating systems (Windows 7 and XP), VoIP codecs, Pv4 and IPv6 to4 tunnelling, on VoIP networks.

The organization of this paper is as follows: next section is related work and contribution of the paper, section 3 contains network set-up and section 4 mentions the VoIP traffic generating tool. Section 5 outlines the outcome of the experiments conducted and last sections contain conclusion and acknowledgments followed by the references.

2 Related Work

In [1], the performance comparison of VoIP network using SIP (Session Initiation Protocol) over both IPv4 and IPv6 networks were carried out. The tests involved IPv4 and IPv6 as well as IPv6to4 tunnels and Teredo tunnels. The results conclude that the native IPv4 had slightly less delay than the native IPv6 and IP6to4 tunneling had much less delay than the Teredo tunneling. In [2], the authors discussed the connection of IPv6 domains using the current IPv4 network without setting up an explicit tunnel between the two domains. The report explains the use of the IP6to4 pseudo-interface, which is when the IPv6 packet is encapsulated in an IPv4 packet at one end, and is then sent over the IPv4 cloud. When the packet reaches the other end, it is then unpacked. The authors in [3] compared various aspects of VoIP using IPv4 and IPv6. The parameters involved in this study were jitter, delay, packet loss and throughput on various network traffic loads ranging from 0-200 Mbps. The authors indicated that for windows XP, the average delay for IPv4 and IPv6 is almost exactly the same, except at 100Mbps when IPv6 delay is approximately 0.002ms more than IPv4. From 0 – 50Mbps of traffic, packet loss was equal (0 packet loss) between the two IP versions but for 100, 150 and 200Mbps, IPv6 had 4, 13, and 17 packet losses respectively, whereas IPv4 packet loss was 0, 12 and 17 respectively. The average jitter reduced fairly consistently up to 100Mbps, but from 100Mbps to 200Mbps, IPv4 showed less jitter than IPv6 by 0.05ms. Overall, IPv4 had better performance across the tests compared to IPv6. In [4], researchers discussed the effect of speech and audio compression on speech recognition performance. The investigation involved GSM (Global System for Mobile Communications) full rate, G.711, G.723.1 and MPEG (Moving Picture Experts Group) coders and stated that MPEG transcoding degrades the speech recognition performance for low bitrates and remains the performance of specialized speech coders like GSM or G.711. In [5], experiments were carried out to identify the performance of VoIP with IPv4, IPv6, IP6to4 tunnel and NAT (Network Address Transition) mechanism. The results were obtained for the jitter, packet loss, and throughput. Their outcome was that the quality of VoIP is impacted due to transition mechanism and voice quality decreases as network traffic capacity is exceeded. In [6], the authors discussed that "VoIP technology is fundamentally changing telephony, enabling not just cheaper calls but also richer and more flexible services." The authors pointed out that VoIP still has some challenges in business enterprise communication system. The major challenges in VoIP technologies mentioned by the authors are security and NAT (Network Address Transition); however SIP (Session Initiation Protocol) based VoIP network has improved many of the challenges and it also has replaced PBX (Private Branch Exchange) network system.

To the authors knowledge no one has compared the performance of VoIP codecs on IPv4 and IPv6to4 tunneling mechanism using Microsoft Windows Vista and Windows 7 operating systems. The contribution of this paper is to identify the performance of five different VoIP codecs with IPv4 and IPv6to4 tunnel using Windows Vista and Windows 7 operating systems and compare their results. This performance evaluation of new technologies mentioned above will be helpful for the IT managers to make efficient decision on parameters involved in network upgrades.

3 Network Setup

The network test-bed was configured based on two different network setups involving the IPv4 based network setup and IPv6to4 tunneling. The IPv4 based network was configured, where two IPv4 based networks were connected via our campus network (figure 1 below). The IPv6to4 based network setup configuration includes two IPv6 networks were connected through the campus's IPv4 Network, using 6to4 tunneling mechanism (figure 2 below). In Figure 1, in both workstations, one of the operating system was installed (Windows Vista or Windows 7) and then was configured based on IPv4 addresses. The workstations were wired to a router's node (Cisco 2811) using IPv4 address. The second node of the router was wired to the Campus's IPv4 network.

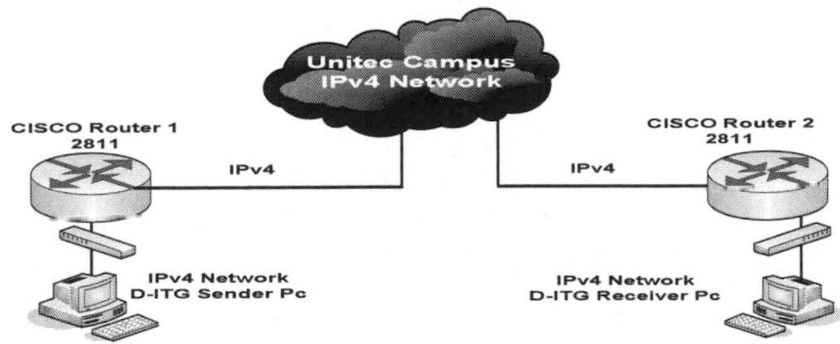

Fig. 1 Network test bed based on IPv4 via Campus's IPv4 Network

In figure 2 below, IPv6 based computers were linked via IPv4 based Network. The computer with IPv6 was linked to a router node (Cisco 2811) using IPv6 addresses and RIPng (Routing Information Protocol Next Generation) routing protocol. The router was linked to Campus's IPv4 network. The IPv6to4 tunneling mechanism was configured in both routers as illustrated in the figure 2.

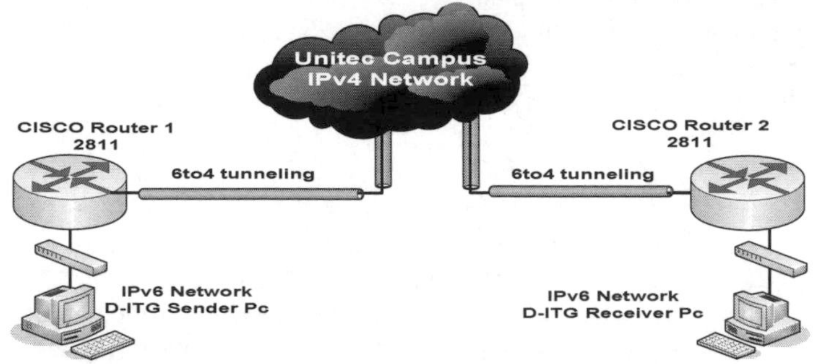

Fig. 2 Network test bed based on IPv6to4 tunnel via Campus's IPv4 Network

The experiments were then carried out in order to analyze the performance of VoIP codecs on IPv4 network and IPv6to4 network for Windows Vista and Windows 7 operating systems. The parameters measured in this experiment were delay, jitter and throughput and their results were compared.

The hardware & software involved in this experiment are Intel® Core™ 2 Duo 6300 1.87 GHz processor with 2.00 GB RAM for Windows Vista and Windows 7 OS, an Intel Pro/100 S Desktop Adapter NIC and a Western Digital Caviar SE 160 GB hard-drive on each workstation. In order to conduct comparison of the results, we used identical hardware for all our tests. To assess the elements of the workstations, a tool known as CPU-Z was used to ensure all components in both computers were identical. To connect the computers to the network, two switches and cat5e fast Ethernet cables were also used in this experiment (figures 1 and 2).

4 Data Generation and Traffic Measurements

D-ITG (Distributed Internet Traffic Generator) [7] is a tool that was chosen to generate and evaluate the VoIP packets. This tool is capable of supporting both IPv4 and IPv6 traffic and various operating systems such as Linux and Windows. D-ITG tool was designed to generate fixed frame size and packets per seconds for each VoIP codec as mentioned in the table 1 below:

Table 1 D-ITG Codecs for VoIP traffic [7]

Codecs	Samples	Frame-size	Packets (per sec)
G.711.1	1	80	100
G.711.2	2	80	50
G.729.2	2	10	50
G.729.3	3	10	33
G.723.1	1	30	26

D-ITG has multiple features and generates various kind of traffic such as data and voice. It has two different modes such as GUI mode and command mode. In this experiment, a command mode version was used to send and receive VoIP packets across the networks. D-ITG sender was installed on a workstation and D-ITG receiver was installed on another workstation. The experiments includes of performing 10 flows concurrently (a flow is equal to one VoIP call) with IPv4 and IPv6. Each flow contains 1000 packets of a codec, sending from a workstation to its destination. In order to generate 10 simultaneously flows, a script was used and these flows were repeated more than 10 times to obtain average results. Each codec has its own standard packet size (table 1).

5 Results

The RTT (Round Trip Time) delay, jitter and throughput were evaluated for G.711.1, G.711.2, G.723.1, G.729.2, and G.729.3 codecs with IPv4 and IPv6to4 using Windows Vista and Windows 7 operating systems over a fast Ethernet networks as seen in the network test bed diagram (figures 1 and 2) above. Figure 3 below shows that for IPv4, the RTT measured for all codecs on Windows Vista and Windows 7, were very close and had only less than 0.01 milliseconds difference between them. The best RTT performance of a codec was calculated on the G.723.1 codec, where the RTT was approximately 0.45 milliseconds and it was for IPv4. The highest amount of RTT was measured on G.711.2 codec using IPv6to4 tunnel at approximately 0.8 milliseconds. Comparison between IPv4 and IPv6to4 for each codec indicates that IPv6to4 tunnel added approximately 0.1 milliseconds extra RTT on each codec performance. This is due to the added delay for creating the IPv6to4 tunnels.

Figure 4 shows for IPv4, Windows Vista had higher jitter for each codec except for G.711.1 codec whereas Windows 7 performed better than Windows Vista. For IPv4, the least amount of jitter was calculated on G.729.3 codec using Windows 7 at approximately 0.075 milliseconds and highest jitter was calculated using Windows 7 on G.711.1 at approximately 0.165 milliseconds. For IPv6to4, the comparison between Windows Vista and Windows 7 shows that IPv6to4 performed better with Windows 7 for all codecs except for G.711.1.

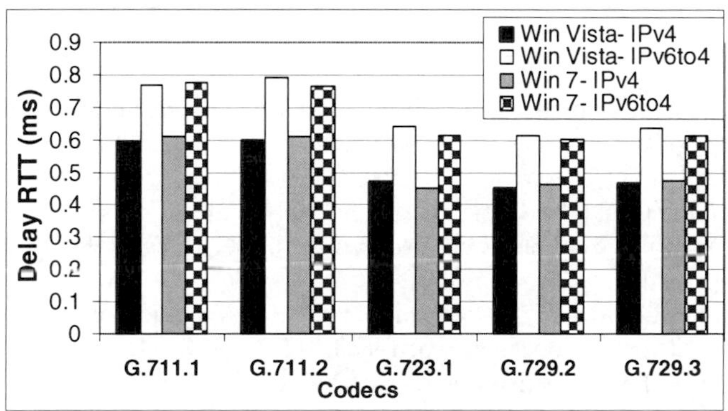

Fig. 3 RTT Comparison for IPv4 and IPv6to4 tunnel on Windows Vista and Windows 7

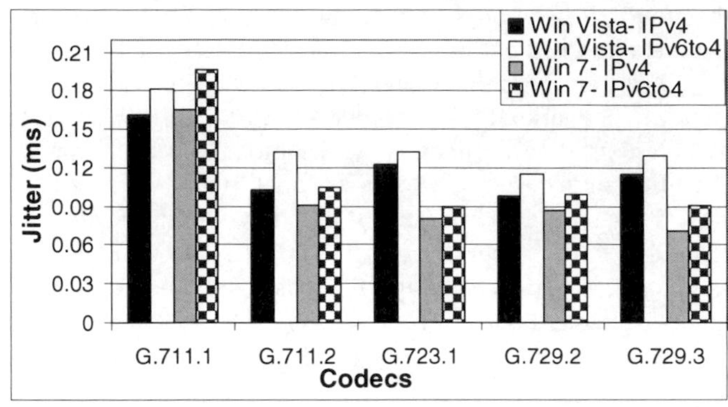

Fig. 4 Jitter Comparison for IPv4 and IPv6to4 tunnel on Windows Vista and Windows 7

The highest amount of jitter was measured on G.711.1 codec with IPv6to4 and Windows 7 at approximately 0.195 milliseconds and the least amount of jitter was measured for G.729.3 with IPv4 and Windows 7 at approximately 0.075 milliseconds. It could be due to G.729.3 frame size whereas other codecs have bigger frame size than G.729.3. The results indicate that networks using IPv6to4 tunneling degraded the performance of jitter and delay compared to all IPv4 networks (figures 3 and 4 above). This is mainly due to the need for IPv6to4 tunneling to encapsulate at the sending side and de-capsulate at the receiving side.

Table 2 Throughput for IPv4 and IPv6to4 on Windows Vista and Windows 7 (Kbps)

	Throughput IPv4		Throughput IPv6to4	
Codec	Win Vista	Win 7	Win Vista	Win 7
G.711.1	688.3	687.6	688.0	687.4
G.711.2	656.5	656.4	655.9	658.0
G.723.1	77.4	77.6	77.4	77.7
G.729.2	109.4	109.3	109.8	109.5
G.729.3	98.1	98.3	98.2	98.4

For the system studied and amount of calls generated, the throughput for the G.711.1 and G.711.2 codecs are 655-688Kbps that is much higher the other three codecs (G.723.1, G.729.2 and G.729.3) at 77-110Kbps. The highest throughput was seen on the G.711.1 codec, using Windows Vista at approximately 688.3kbps, while the lowest throughput was observed on the G.723.1 codec using Windows Vista at approximately 77.4kbps.

Windows 7 almost had the same amount of throughput as Windows Vista with IPv6to4 tunnel for all codecs except for G.711.2 codec where 2.1kbps difference was observed. The comparison between IPv4 and IPv6to4 tunnel shows that IPv4 had performed approximately the same as IPv6to4 using all codecs on both operating systems.

6 Conclusions

Based on the results the use of IPv6to4 tunneling mechanism, no matter which operating system it was working on, added extra RTT compared to its IPv4 counterpart. This is mainly due to the need to encapsulate at the sending side and de-capsulate at the receiving side. For the system studied, the overall jitter performance was marginally better on Windows 7 than on Windows Vista. For IPv6to4, Windows 7 performed better than Windows Vista for all codecs except for G.711.1. It might be the ability of Windows 7 to provide more efficient performance with IPv6 or IPv6to4 traffic whereas it was not so efficient in Windows Vista. Among all codecs investigated, the best jitter performance was measured on the G.729.3 codec using IPv4, where the jitter was approximately 0.075 milliseconds, for the system studied. It could be duc to its frame size whereas other codecs have bigger frame size than G.729.3. The G.711.1 codec provided the highest bandwidth while G.723.1 has the lowest bandwidth of all codecs studied.

References

1. Hoeher, T., Petraschek, M., Tomic, S., Hirschbichler, M.: Evaluating Performance Characteristics of SIP over IPv6. IEEE Journal of Networks 2(4), 10 (2007)
2. Carpenter, B., Moore, K.: Connection of IPv6 Domains via IPv4 Clouds. Internet proposed standard RFC 3056 (2001)
3. Yasinovskyy, R., Wijesinha, A.L., Karne, R.K., Khaksari, G.: A Comparison of VoIP Performance on IPv6 and IPv4 Networks. In: IEEE/ACS International Conference on Computer Systems and Applications, pp. 603–609 (2009)
4. Besacier, L., Bergamini, C., Vaufreydaz, D., Castelli, E.: The effect of speech and audio compression on speech recognition performance. In: IEEE Fourth Workshop on Multimedia Signal Processing 2001, pp. 301–306 (August 2002)
5. Yasinovskyy, R., Wijesinha, A.L., Karne, R.: Impact of IPSec and 6to4 on VoIP Quality over IPv6. In: IEEE International Conference on 10th, Telecommunications, ConTEL, pp. 235–242 (2009)
6. Zourzouvillys, T., Rescorla, E.: An Introduction to Standards-Based VoIP. IEEE Journal on Internet Computing 14(2), 69–73 (2010)
7. Alessio, B., Alberto, D., Antonio, P.: Multi-Protocol and Multi-Platform Traffic Generation and Measurement (2009), `http://www.grid.unina.it/software/ITG/`
8. Khaksari, G.H., Wijesinha, A.L., Karne, R.K., He, L., Girumala, S.: A Peer-to-Peer Bare PC VoIP Application. In: IEEE Proc. CCNC 2007, January 2007, pp. 803–807 (2007)
9. Schulzrinne, H., Casner, S., Frederick, R., Jacobson, V.: RTP: A Transport Protocol for Real-Time Applications. RFC 3550 (July 2003)

A PageRank-Based Heuristic Algorithm for Influence Maximization in the Social Network

Zhi-Lin Luo, Wan-Dong Cai, Yong-Jun Li, and Dong Peng

Abstract. The influence maximization is the problem of how to find a small subset of nodes (seed nodes) that could maximize the spread of influence in social network.However,it proved to be NP-hard.We propose a new heuristic algorithm, the High-PageRank greedy algorithm (HPR_Greedy),which searches the seed nodes in a small portion containing only the high-PageRank nodes, based on the power-law influence distribution in non-uniform networks. The experimental results showed that, compared with classical algorithms, the HPR_Greedy algorithm reduced search time and achieved better scalability without losing influence.

1 Introduction

With the development of WEB2.0, online social networking has become a popular way to share and disseminate information over Internet. According to the Nielsen Company's reports[1], up to two-thirds of the global online population visited social networks and blogs. Social network has become a fundamental part of the online society. Some large-scale social net-working sites are very popular, e.g. Facebook has become the largest website in the US, surpassing Google. Social networking sites provide huge potential business opportunities for companies to market their products. How to efficiently find a small subset of nodes (seed nodes) with the greatest influence in large-scale social networks is the focus of this study.

How to find the seed nodes with the greatest influence is the influence maximization problem, which is of great interest to many companies or people who want to market their products or service through Internet . When studying the viral-style marketing, Domingos and Richardson et al. [3,4]are the first to discuss it.Kempe et al. [5] formally proposed the influence maximization problem as the discrete optimization and proved that it is NP hard. They presented the original greedy approximation algorithm.Leskovec et al.[6] presented the greedy with CELF

Zhi-Lin Luo · Wan-Dong Cai · Yong-Jun Li · Dong Peng
School of Compute Science Northwestern Polytechnic University, Xi'an China
e-mail: {lzluo007,justastriver}@gmail.com,
{caiwd,lyj}@nwpu.edu.cn

F.L. Gaol (Ed.): Recent Progress in DEIT, Vol. 2, LNEE 157, pp. 485–490.
springerlink.com

(Cost-Effective Lazy Forward selection) optimization algorithm based on the submodularity function theory. Their results showed that the influence of the Greedy with CELF was nearly close to those of the original Greedy , while achieved as much as 700 times speed.Chen Wei et al.[8] proposed the NewGreedy and the MixGreedy algorithms. MixGreedy improves the efficiency.But both algorithms' scalability is poor in Linear threshold model.

2 Information Dissemination Models

There are three basic information dissemination models.

Independent cascade model (IC)

The network is modeled as an undirected graph.Each node has two states: active or inactive.nodes are active if they have accepted information, inactive otherwise. Nodes only can switch from being inactive to being active. Active nodes activate inactive nodes with a constant influence factor p ($0<p<1$).

Weighted cascade model (WC)

WC is very similar to IC. The difference is that a social network is modeled as an directed graph in WC model, with asymmetric influence factors, The influence factor of a node v is *1/dv* , where *dv* is the degree of the node v.

Linear threshold model (LT)

LT is different from the two models above. A node v is randomly assigned a threshold and influenced by each neighbor w with a weight.The condition for the node v to be activated is that the total weight of its active neighbors is greater than threshold

3 The Algorithms for Influence Maximization

In this section, we will first introduce the classical greedy with CELF optimization, and then propose a new heuristic algorithm. Before describing the algorithms, let's define $\sigma(\bullet)$,S, and U as the influence function, the set of the seed nodes and the set of all nodes, respectively.

Theorem 1. *THEOREM 1 (Nemhauser et al.1978)*

$$\forall A \subseteq B \subseteq N, \forall j \in N \backslash B, if\ \mathrm{f}(A+j)-\mathrm{f}(A) \geq \mathrm{f}(B+j)-\mathrm{f}(B) \tag{1}$$

f *is submodular function.*

So submodular function is non-negative and monotone.

Theorem 2. *(Kempe et al.2005)For an arbitrary instance of the IC or WC or LC model, the resulting influence function* $\sigma(\bullet)$ *is submodular.*

Based on theorems 2, It can be deduced that the influence function $\sigma(\bullet)$ of each node in the set U gradually weakens with the number of the set S increasing.

The classical greedy with CELF can be divided into two different rounds: in the first round it compute the influence of each node in the set U and select the node with the largest influence as the first seed node and remove it from U; in the second round it select remaining seed nodes,when selecting seed node each time, it compute the influence of the small number of nodes in the set U based on the submodularity of the influence function and choose the node with largest influence in U.

The social networks is non-uniform and information dissemination is asymmetric, that indicate a large number of low-influence nodes and a small number of large-influence nodes. In influence maximization problem the seed nodes are those with larger influence, it is very low possibility for a large number of low-influence nodes to be selected as seed nodes, we can consider to rank nodes' influence based on the topological dependence and choose the seed nodes in the small set of large-influence nodes. This problem of ranking nodes' influence based on the topological dependence in a network is similar to ranking web pages based on its connectivity. Google uses the PageRank algorithm to rank web pages in their search results [13].The key idea behind PageRank is to allow propagation of influence along the network of web pages, instead of just counting the number of other web pages pointing at the web page. In this section we rank nodes' influence by the PageRank algorithm, but there is a problem that the PageRank is not fit in the undirected social network. In order to extend PageRank algorithm in undirected social network, we can regard it as bipartite network.

Based on the analysis above, we present a new heuristic greedy algorithm, that is the High-PageRank greedy algorithm (HPR_Greedy), with a basic idea to search for the seed nodes in the small part of the high-PageRank nodes.The pseudo code is as follow.

```
Algorithm HPR_Greedy
Initialize: S=∅,M=50,σ(•)=NULL
Input:r
Step1:
    if (G is Gu)G⟶Gb; //Gu is undirected network, Gb is bipartite network.
    Ranking all nodes' influence by PageRank;
    Choosing top r% nodes into U;
    Computing and saving σ(w) of each node w in U;
    Choosing the most influence node n into S and removing it from U;
Step2:
    for(i=1;i++;i<M-1){
        Choosing the most influence node j in U
        Computing σ(j)=σ(S+j)-σ(S);
       Set flag=σ(j),sn=j;
       for(Every node in U){
         if(flag<σ(z){///node z in U
            Computing σ(z)=σ(S+z)-σ(S);
            If(flag<σ(z))
               Set flag =σ(S+z)-σ(S),sn=z;}
         Set sn⟶S and remove sn from U;}
    }
```

4 The Experimental Results and Discussion

Two real social network dataset which are derived from paper-sharing sites arXiv (www.arXiv.org). In the networks a node represents a scholar and an edge represents the collaboration relationship between two scholars. The first scholars collaboration network comes from the "high-energy physics theory" section with papers from 1991 to 2003 and is denoted as NetHEPT that contains 15,233 nodes and 58,891 edges. The second scholars collaboration network is from the "physics" section and is denoted as NetPHY that contains 37,154 nodes and 231,584 edges.

All experiments are executed on the same PC, with a CPU of 2.53GH Intel Core Duo E7200 and memory of 2 GB. Because HPR_Greedy performs best with r% set to 1%, we will compare HPR_Greedy_01 with other classical algorithms in IC,WC and LT model ,respectively, with the influence factor p in the IC model is consistent at 0.01. Other classical algorithms include the geedy with CELF(Greedy), the New-Greedy, and the MixGreedy algorithms. The number of seed nodes is 50. Each node is calculated 20,000 times to ensure the accuracy of the influence.

Fig1 and Fig2 show the influence from the IC model , on NetHEPT and NetPHY. The growth curves of the influence from HPR_Greedy_01 and other three greedy algorithms almost overlap and the influence difference among them is less than 1%.

Fig3 and Fig4 give the runtime. Compared with Greedy ,HPR_Greedy_01 is reduced by 75% on NetHEPT and by 64% on NetPHY, It is also significantly faster than MixGreedy and NewGreedy.Both NewGreedy and MixGreedy are instable, e.g.,on NetHEPT NewGreedy is nearly as fast as Greedy; however, in NetPHY it is 2.5 time faster than Greedy. MixGreedy has the same problem.

Fg5 and Fig6 give the influence in WC model. The influnce of HPR_Greedy_01 almost overlap those of other greedy algorithms, and the difference is less than 1.5%.

Fig7 and Fig8 report the runtime of different greedy algorithms. HPR_Greedy_01 runs faster than Greedy, with 55.6% and 48.5% savings in runtime on NetHEPT and NetPHY, respectively. The performance of NewGreedy is poor, slower than Greedy, with 244% and 275% cost in runtime on NetHEPT and NetPHY, respectively.

As NewGreedy and MixGreedy are proposed based on the influence factor in an IC or a WC model, they doesn't fit in LT model. In order to extend the MixGreedy into the LT model, Chen Wei(2009) first use MixGreedy to obtain the seed nodes in an IC or a WC model and then calculate the influence of the seed nodes in a LT model. So in the following figures MixGreedyIc represents obtaining the seed nodes in an IC model and then calculating the influence of the seed nodes in a LT model.

Figures 9 and 10 give the influence results of different algorithms. HPR_Greedy_01 perform as well as Greedy on both NetHEPT and NetPHY,. the MixGreedyWc also gets the same influence of Greedy. It is worth mentioning that the influence gap between MixGreedyIc and Greedy is very small in NetHEPT , however, it becomes large in NetPHY, and MixGreedyIc only gets 76.1% influence of Greedy. This indicates that MixGreedy runs not very well in the LT model.

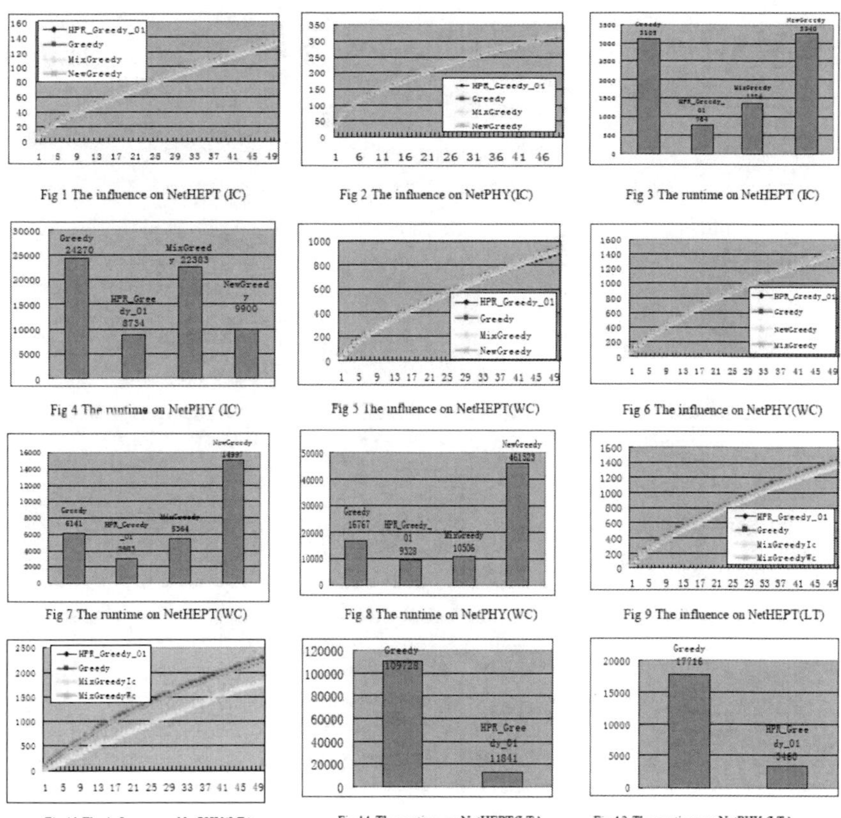

Fig 1 The influence on NetHEPT (IC)

Fig 2 The influence on NetPHY(IC)

Fig 3 The runtime on NetHEPT (IC)

Fig 4 The runtime on NetPHY (IC)

Fig 5 The influence on NetHEPT(WC)

Fig 6 The influence on NetPHY(WC)

Fig 7 The runtime on NetHEPT(WC)

Fig 8 The runtime on NetPHY(WC)

Fig 9 The influence on NetHEPT(LT)

Fig 10 The influence on NetPHY(LT)

Fig 11 The runtime on NetHEPT(LT)

Fig 12 The runtime on NetPHY (LT)

Figure 11 and Figure 12 show the runtime in the LT model, from which we can see that HPR_Greedy_01 is 5 times and 9 times faster than Greedy on NetHEPT and NetPHY, respectively.

5 Conclusion

In this paper,we proposed the HPR_Greedy algorithm, based on the power-law influence distribution in non-uniform networks. The experimental results show that HPR_Greedy is more efficient without the loss of the influence, especially in large social networks, compared with other greedy algorithms.Furthe rmore, HPR_Greedy has a good scalability in different models.

Further efforts are needed to explore HPR_Greedy more. One is how to determine the ratio r to make HPR_Greedy achieve the best performance. When r is too small the influence of HPR_Greedy will be lost, however, if r is too large the efficiency of HPR_Greedy will be poor, and in the extreme case when r%= 100% ,it degenerates into the Greedy ,and it is associated with some parameters, such as the number of nodes,the number of seed nodes ,the influence factor of model and so on. We will focus on these issues in our future studies.

Acknowledgements. This work was supported by the State High Technology Research and Development Program (863)Fund in china (2009AA01Z424), by the Special Fund for Fast Sharing of Science Paper in Net Era by CSTD(20096102410001.)

References

1. Nielsen Online Report: Social networks blogs now 4th most popular online activity (2009)
2. Goldenberg, J., Libai, B., Muller, E.: Talk of the Network: A Complex Systems Look at the Underlying Process of Word-of-Mouth. Marketing Letters 12(3), 211–223 (2001)
3. Domingos, P., Richardson, M.: Mining the network value of customers. In: Proceedings of the 7th ACM SIGKDD Conference on Knowledge Discovery and Data Mining, pp. 57–66 (2001)
4. Richardson, M., Domingos, P.: Mining knowledge-sharing sites for viral marketing. In: Proceedings of the 8th ACM SIGKDD Conference on Knowledge Discovery and Data Mining, pp. 61–70 (2002)
5. Kempe, D., Kleinberg, J.M., Tardos, E.: Maximizing the spread of influence through a social network. In: Proceedings of the 9th ACM SIGKDD Conference on Knowledge Discovery and Data Mining, pp. 137–146 (2003)
6. Leskovec, J., Krause, A., Guestrin, C., Faloutsos, C., vanBriesen, J., Glance, N.S.: Cost-effective outbreak detection in networks. In: Proceedings of the 13th ACM SIGKDD Conference on Knowledge Discovery and Data Mining, pp. 420–429 (2007)
7. Nemhauser, G., Wolsey, L., Fisher, M.: An analysis of the approximations for maximizing submodular set functions. Mathematical Programming 14, 265–294 (1978)
8. Wei, C., Yajun, W., Siyu, Y.: Efficient influence maximization in social networks. In: Proceedings of the 15th ACM SIGKDD International Conference on Knowledge Discovery and Data Mining, pp. 199–208 (2009)
9. Newman, M.E.J.: The Structure and Function of Complex Networks. SIAM Review 45(2), 167–256 (2003)
10. Mossel, E., Roch, S.: On the submodularity of influence in social networks. In: Proceedings of the 39th ACM Symposiumon Theory of Computing, pp. 128–134 (2007)
11. Kempe, D., Kleinberg, J.M., Tardos, E.: Influential nodes in a diffusion model for social networks. In: Proceedings of the 32nd International Conference on Automata, Languages, and Programming, pp. 1127–1138 (2005)
12. Kimura, M., Saito, K.: Tractable models for information diffusion in social networks. In: Proceedings of the 10th European Conference on Principles and Practice of Knowledge Discovery in Databases, pp. 259–271 (2006)
13. Page, L., Brin, S., Motwani, R., Winograd, T.: The pagerank citation ranking:Bringing order to the web. Technical Report 1999-66, Stanford InfoLab (November 1999)

Examining Learning Attitude toward ICT in Digital Photography

Sin-Ho Chin, Jung-Hui Lien, and Yi-Xin Huang

Abstract. Fast developed Internet transmission and digital environment has been an impact to on-line learning and E-learning. Researchers however hold diverse aspects toward ICT learning, positive such as self-direct learning and online group learning; negative, unwilling change and loneliness. In order to clarify students' learning attitude toward web learning environment, one semester experimental course digital photography is held. 5 factors, awareness, communication, usefulness, learning motivation, and satisfaction, are examined. The research result indicates that the satisfaction factor is significant correlated with achievement performance. However, other applications and perceptions of ICT may not bring to high achievement in this research.

1 Introduction

The digital learning environment is advocated by educational authorities in many countries. The Information and Communication Technology (ICT) blended into learning and curriculum becomes major tendency in modern education. Many researchers found ICT blended into teaching and curriculum can motivate student to learn the way of learning, independent learning, information accumulation, group learning, and interaction with teachers and pupils (Lofstrom and Nevgi, 2006; Sabry & Backer, 2009).

More importantly, immersed in the digital learning environment not only can build up what Yelland, Cope, & Kalantzis (2008) maintained the most natural and effective learning, "life learning"(p. 199), but also accommodate the full digital work and life environment in the future. Some researchers echoed the digital learning environment has turned the traditional learning into constructive learning (Passerini & Granger, 2000), because of the authorial learning and setting the learning goal and objectives.

Sin-Ho Chin · Jung-Hui Lien · Yi-Xin Huang
Tatung University, Taipei 10452, Taiwan
e-mail: {shchin,lien,g9904026}@ttu.edu.tw

F.L. Gaol (Ed.): Recent Progress in DEIT, Vol. 2, LNEE 157, pp. 491–497.
springerlink.com

However, the transformation to digital learning environment grows gradually, and not synchronized in all school. School has own aspects and steps in building digital learning environment. That the heyday of online learning and digital learning environment can only bring to positive and high level of learning effectiveness is doubtful (Wellington, 2005). For example, Lin (2008) dimmed the ICT blended into teaching as the result of government's advocacy or endorsement, and can not fit all curricula and schools. It can create more work loads and time loads for teachers and becomes pressure and burden. Such unwilling change of digitalization in teaching and learning objective sometimes can move one to feel depressed and lost (Deryakulu et. al, 2008). Mukama & Ansersson (2008) promulgated that school authorities should have students and teachers prepared to change psychologically and physically rather than simply introduce people into an unfamiliar learning environment. In fact, ICT blended into learning environment has its technical limitation, such as hardware and running time, also the connecting problem may ask student spend more time in homework searching (Fallshaw & McNaught, 2005). Lofstrom & Nevgi (2007) found the main learning obstacle laid in loneliness that student along faced self-directed learning in digital learning environment. Hirschheim (2005) criticized further that digitalization in education did not bring up the learning quality, in some cases; the personality and humanity are stifled by uniformed digital learning environment.

Digital image device evolves rapidly. Even cell-phone can produce high quality image. Image transmission and sharing becomes a social life style in communication. Thus, students in digital photography are working with digital image by up-loading and down-loading for sharing to friends or turning the assignment. They are so to said more comfortable in the digital learning environment than other students.

As many researchers stated, the more students are acquainted with ICT learning environment the higher acceptance in ICT learning and achievement, and more positive attitude student holds toward Internet and ICT (Helmi, 2002; Palmer, 2002). However, those researches are testing attitude toward internet and ICT learning environment in general. They were not focusing on specific course and not tracing in a long term. The digital photography course demands large amount of information collection and transmission. Thus, the question of whether the digital photography students who are more acquainted with digital image and learning environment embed positive attitude toward ICT blended learning environment, also high performance in achievement test or assignment. This research aims at analyzing the relation between high acquainted rate in ICT learning environment and the attitude, and achievement performance in photography, by using one-semester digital photography course as vehicle. This research also exams the factors that enhance effective learning.

2 Related Works

2.1 Information and Communication Technology

The web connection and transmission in WWW builds a two-way interactive communication and is a powerful way in ICT communicating (Chen & Yen, 2004).

According to the statistics of Internet World Stats (2009), there are over 1.5 billion users in internet surfing. The digital hardware develops fast and expands the capacity to 200% in every two years, in 2010 the web transmission speed via light-fiber can soar to one billion bits, it implies the speed and quantum of information transmission bursts out to a mass that never seen before (Glumbert, 2008).

Nevertheless, whether the bursting mass of information satisfies people cognitive desire, solves people problem, or helps student learning, is questionable. Barker & Finnie (2004) defined the problem as the mass of information is not full exploited, because the computer technology leaps tremendous so that people cannot find information easily. Some researchers dimmed the problem of not exploited information as the results of user inactive attitude toward web communication (e.g. Liaw, 2002). Islam (2009) added the factors not providing useful information, such as paid information in an online library that cannot be viewed before checking out and even unsearchable by keywords. These factors impede the satisfaction of user with the mass of online information.

2.2 Constructivist Learning with ICT

Modern digitalization learning environment has brought education into a new era, in which students can be satisfied with the knowledge, information, and learning method for self constructive learning (Chuang & Tsai, 2005). The core concept of constructive learning lies at the knowledge constructed by student not conveyed by the teacher. Teacher acts as coordinator or facilitator to help student search and build own knowledge; hence, student is active to learn, not passive (Mater, 1999).

Blended learning seems to be the idea way combining the strength of traditional and digital learning. Nevertheless, some researchers are cautious dealing with ICT blended learning (Wellington, 2005). Hirschheim (2005) criticized the uniformity and mass production of ICT learning and environment can rub down people's individuality, and personality, and impede the critical thinking. Lin (2008) argued the uniformity of ICT learning curriculum can only produce uniformed student. Thus, Selwyn (2007) suggested focusing on the reflexivity and reflectivity of teaching activities as well as promoting student's creativity and instinct, and to prevent the uniformity of thinking and learning pattern in the uniformed ICT learning environment.

3 Method

This research explores the relationship of attitude and satisfaction toward ICT learning in photography course. Observe objectives are samples who attended one semester digital photography class, and n=47, male= 17, female=30. Also, the time spent in internet surfing per day is another category, >5 hours=7, <5 hours=40. The activities in the digital photography include pictures upload via web-hard-disc, lecturing and discussion via E-classroom and face-to-face classroom, shooting activities, and photographs searching and analysis. Students are expected spending large amount of time in immersing in ICT learning environment. Achievement performance is collected by assignments, mid term works, and final works. Scoring criteria encompasses originality, aesthetics, and media execution.

Three variables in this research are two attitude and perception instruments and an achievement performance in the semester end. The first instrument is Internet Perception Scale (IPS), focusing on the preferences and perception about the web learning environment; and is modified from Web Attitude Scale (Liaw, 2002), and Student Course Experience Questionnaire (Ginns & Ellis, 2009). The second instrument is Satisfaction Scale to ICT Photography Learning (SSICTPL) investigating the satisfaction and learning willingness toward ICT learning is modified from Questionnaires of Attitude and Involvement to E-Learning by Milic, Martinovic, & Fercec (2009). 5 factors elicited from IPS and SSICTPL are examined with achievement performance. They are awareness, communication, usefulness, learning motivation, and satisfaction. The IPS has the Cronbach α=.82, and the SSICTPL has a correlation coefficient of .90 that indicates higher test reliability. Data from questionnaires are collected at the end of semester.

4 Result

Samples in two groups of spending more than 5 hours in internet surfing per day and less than 5 hours are 7 to 40. Samples in male and female are 17 to 30. The mean score equaling 70 indicates student hold average satisfaction in ICT learning environment. The >5 hours group shows average high satisfaction rate by SD=2.878 over the <5 hours group. Female group shows higher score in satisfaction than male by mean = 37.4 and SD = 6.495 (see Table 1).

Table 1 Mean and sd for time and gender

	N	Mean	SD
<5hours/day Surfing	40	66.70	6.358
>5hours/day Surfing	7	69.57	2.878
Male	17	71.52	7.102
Female	30	74.80	6.495

Two factors satisfaction and usefulness gain higher score than other three (see Table 2), while factor communication gains less score than others. And in the Pearson correlation analysis, satisfaction is significant correlative with student's achievement performance by r = .317 (p=0.05, two tails, also see Table 3). However, other 4 factors do not show significantly correlative with achievement performance.

Table 2 Mean and sd for ict learning factors and achievement in photography

	N	Mean	SD
Achievement	47	41.59	6.357
Satisfaction	47	23.15	6.653
Awareness	47	15.96	2.331
Communication	47	11.06	2.497
Usefulness	47	19.74	2.714
Motivation	47	14.74	2.260

Table 3 Correlations for ict learning factors and achievement in photography

95% Confidence interval of the difference.(2-tailed)

	Achievement	Satisfaction	Cognition	Communicate	Useful	Motivate
Achieve.	1	.317*	.134	-.182	-.273	.156
Satisfac.	.317*	1	.007	-.079	-.013	.101
Aware.	.134	.007	1	-.093	.472*	.266
Commu.	-.182	-.079	-.093	1	.092	-.463**
Useful	-.273	-.013	.472**	.092	1	.251
Motivat.	.156	.101	.266	-.463**	.251	1

*. Significant at P=0.05 (two tails) **. Significant at P=0.01 (two tails)

5 Summary and Discussion

As test results showed, digital photography students who spent a lot time immersing in the ICT learning environment are comfortable and satisfied with ICT learning and its learning environment. Data analysis also shows that the more time spent on internet navigating the higher the satisfaction rate one possesses. Female students tend to be more satisfaction than males. The high involvement in ICT learning environment does significantly correlated with high achievement performance of photography class. Especially, the satisfaction factor which is highly correlated with achievement score. However, other 4 factors of web learning gaining a less correlation rate with the achievement score indicates that there may be some other factors that are related with achievement performance. Further researches are suggested to delve deeper with factors that bring the learning effectiveness and the achievement performance. The photographic students are supposed to be acquainted with digital learning environment, but the realities show they did not get higher achievement as expected. The reason may lies at distraction or inertia. In the web communication world, people communicate whenever, and wherever, people are used to access the Internet in anytime and anywhere. The wireless internet promotes people in readiness to surf and communicate. The question of whether student can concentrate in learning and not being influenced by allures of asking to react to the online chatting is unsolved. In sum, the ICT blended learning may be concerned about merging ethical topics into curriculum to bring out high quality of learning.

References

1. Barker, J.R., Finnie, G.: A model for global material management using dynamic information. In: Proceedings of the Americas Conference on Information Systems (2004), AIS Electronic Library, http://aisel.aisnet.org/amcis2004/493 (retrieved September 2009)
2. Chen, T.: Recommendations for crating and maintaining effective networked learning communities. International Journal of Instructional Media 30, 35–44 (2003)
3. Chuang, S.C., Tsai, C.C.: Preferences toward the constructivist Internet-based learning environments among high school students in Taiwan. Computers in Human Behavior 21, 255–272 (2005)
4. Deryakulu, D., Buyukozturk, S., Karadeniz, S., Olkum, S.: Satisfying and frustrating aspects of ICT teaching: A comparison based on self-efficacy. PWASET 36, 481–484 (2008)
5. Fallshaw, E., McNaught, C.: Quality assurance issues and processes to ICT-based learning. In: Follows, S., Bhanot, R. (eds.) Quality Issues in ICT-Based Higher Education, pp. 23–36. Routledge Falmer, Oxon (2005)
6. Glumbert. Shift happens (2008), http://glumbert.com/media/shift2 (retrieved September 2009)
7. Helmi, A.: An analysis on the impetus of online education. The Internet and Higher Education 4, 243–253 (2002)
8. Hirschheim, R.: Look before you leap. Communications of the ACM 48, 97–101 (2005)
9. Hirumi, A.: Student-centered, technology-rich learning environments: Operationalizing constructivist approaches to teaching and learning. Journal of Technology and Teacher Education 10, 497–537 (2002)
10. Internet World Stats (2007), 1,574 Million Internet Users! http://www.internetworldstats.com/blog.htm (retrieved September 2009)
11. Karal, H., Aydin, Y., Ursavas, O.F.: Struggles for Integration of the Technologies into Learning Environment in Turkey. International Journal of Social Sciences 4, 102–111 (2009)
12. Liaw, S.S.: An internet survey for perceptions of computers and the World Wide Web: Relationship, prediction, and difference. Computers in Human Behavior 18, 17–35 (2002)
13. Lin, M.C.: ICT education: To integrate or not integrate? British Journal of Education Technology 39(6), 1121–1123 (2008)
14. Lofstrom, Nevgi: From strategic planning to meaningful learning: Diverse perspectives on the development of web-based teaching and learning in higher education. British Journal of Educational Technology 38, 312–324 (2007)
15. Milic, D.C., Martinovic, D., Fercec, I.: E-learning: Situation and perspectives. Technical Gazette 16, 31–36 (2009)
16. Mukama, E., Ansersson, S.B.: Coping with change in ICT-based learning environments: Newly qualified Rwandan teachers' reflections. Journal of Computer Assisted Learning 24, 156–166 (2008)
17. Passerini, K., Granger, M.J.: A development model for distance learning using the Internet. Computers and Education 34, 1–15 (2000)
18. Salmon, G.: E-Moderating: The key to online teaching and learning, 2nd edn. Taylor & Francis, London (2004)

19. Selwyn, N.: The use of computer technology in university teaching and learning: A critical perspective. Journal of Computer Assisted Learning 23, 83–94 (2006)
20. Wellington, J.: Has ICT come of age? Recurring debates on the role in education, 1982-2004. Research in Science & Technological Education 23, 25–39 (2005)
21. Yelland, N., Cope, B., Kalantzis, M.: Learning by design: creating pedagogical frameworks for knowledge building in the twenty-first century. Asia-Pacific Journal of Teacher Education 36(3), 197–213 (2008)
22. Zhang, P., von Dran, G.M.: Satisfiers and dissatisfiers: A two-factor model for website design and evaluation. Journal of the American Society for Information Science 51, 1253–1268 (2000)

Study on the Information Technology-Based Lean Construction Supply Chain Management Model

Chen Yan and Xu Zhangong

Abstract. Lean supply chain management, originally generated from manufacturing, has been proved to be an effective supply chain management model mainly by eliminating all forms of wastes in the entire supply chain. This paper first analyzes the status quo and major problems of our construction supply chain. Then it introduces the idea of lean construction supply chain management. Based on this, it puts forward the information technology-based lean construction supply chain management (IT-based LCSCM) model and elaborates its structure. Finally, the paper fully discusses the main issues of the implementation of this IT-based LCSCM model, including constructing lean supply chain network node enterprises and establishing lean supply chain linkages among construction supply chain parties.

1 Introduction

Nowadays business competition has shifted from competition among individual enterprises to that among supply chains. In order to win the increasingly fierce market competition, enterprise in any sector has to focus on the effectiveness and competence of its supply chain. Construction Supply Chain (CSC) is the network of upstream and downstream organizations that are involved in a construction project, including the general contractor, subcontractors, raw material suppliers, equipment suppliers, the designer and the owner, etc. Currently, in terms of property right, member enterprises in our CSC are completely independent. Their independent and isolated operations have greatly affected the sharing of information as well as the supply chain integration, which results in the high cost, weak per formance and low competence of our CSC. Based on the successful application of lean thinking in manufacturing and good practices of LCSCM in the developed

Chen Yan · Xu Zhangong
College of Management, Qingdao Technological University, Qingdao City, P.R. China
e-mail: chenyan0714@126.com, qdxzg@163.com

F.L. Gaol (Ed.): Recent Progress in DEIT, Vol. 2, LNEE 157, pp. 499–505.
springerlink.com

countries, we attempt to introduce lean thinking to our construction sector and propose building the information technology-based lean construction supply chain management model to solve the current problems of our CSC.

2 Major Problems of Our Construction Supply Chain

Under traditional construction management model, most member enterprises in our CSC typically operate independently and have their own strategies and plans on purchasing, inventory, etc. The lack of collaborative operations among CSC parties has led to the high cost and low competence of our CSC. To be specific, the major problems of our CSC are as follows:

(1) High supply chain procurement cost caused by immature procurement model

Currently, most member enterprises in our CSC still adopt traditional separate procurement model and consequently the supply chain procurement cost has been greatly increased. For example, in a typical EPC project, the general contractor often authorizes procurement right to its subcontractors. Each subcontractor has its own procurement department and makes individual procurement plan. This immature procurement model results in the establishment of duplicate procurement departments as well as the increase of relevant human and material resources in the entire CSC. And the small lot size of separate procurement also makes it difficult to get the favorable purchasing price and achieve the economies of scale in transportation.

(2) High supply chain inventory cost caused by separate inventory management practices

There are many unpredictable factors in the process of construction, including design modifications, climate changes, untimely replenishment of materials, etc. In order to avoid the suspension or postponement of the project because of insufficient inventories, a certain amount of inventory has to be held at the construction site. As discussed above, most member enterprises in our CSC used to operate independently and make individual inventory strategies and plans. Consequently, their separate inventory management practices have not only led to a large amount of inventory at the site, but also increased the number of personnel and field equipments required by inventory management. These all contribute to high inventory cost of our CSC.

(3) Ineffective supply chain information exchange caused by the lack of coordinative information platform

Nowadays there is no integrated and coordinative information management platform in our CSC and each supply chain party acts as an "Information Island". In most cases, the general contractor acts as an information intermediary. Most member enterprises in our CSC only exchange information with the general contractor and do not directly communicate with each other. This approach has increased the layers of information transmission and caused the delays and distortions of information transmission.

3 Concept of Lean Construction Supply Chain Management

The main reason for the above discussed problems of our CSC can be attributed to the lack of a highly integrated and wholly coordinative SCM model. Lean thinking is originated from manufacturing and its philosophy has now been widely applied in various sectors, such as healthcare, service and insurance, etc. In 1992, Lauri Koskela submitted a report named "Application of the New Production Philosophy to Construction", in which lean thinking was introduced to the construction sector for the first time. So far, over 15 countries including the UK, the U.S.A., Finland, Singapore, Australia, Brazil, etc., have already applied lean thinking to their construction sectors. Furthermore, it is proved that construction companies implementing LCSCM have achieved significant benefits, such as the shortened construction time, the decreased engineering changes and claims, and the low project cost, etc. Therefore, LCSCM provides us a good idea in solving the current problems of our CSC. The idea of LCSCM combines two advanced ideas, i.e., lean thinking and SCM. The ultimate purpose of LCSCM is to achieve cost-effective collaboration among all CSC parties (including tendering, design, consulting, purchasing, supplying, construction, final completion and inspection, the delivery to the owner, etc.), therefore best satisfying customer, getting rid of all the wastes, and strengthening the overall competence of the CSC.

4 Structure of IT-Based Lean Construction Supply Chain Management Model

In order to achieve the sharing of information and real time interactive communication among CSC parties, we propose building an IT-based LCSCM model. The key of this model is the establishment of a CSC network information system based on Internet/Intranet. The network should be built around an information management center (IMC) which is constructed and managed by the focal/core company (mainly the general contractor), using three-tier browser/server (B/S) database model.

The focal/core company needs to build a project-based information management platform acting as the IMC of the IT-based LCSCM model. With the aid of this platform, the focal company can not only use internal Intranet to implement internal lean management, but also take responsibilities for the design and the construction of the LCSCM system, daily data collection and processing as well as real-time monitoring of the operations of the IMC. Through the inter-company Intranet, the sharing of information and real-time interactive communications between the core company and its partners can be ensured. And through Internet, the communications between the owner and all CSC parties can be achieved (see Figure 1). The Information Management Platform (Center) is composed of four sub-platforms. They are as follows: internal lean management sub-platform; strategic partnership evaluation sub-platform; online project management sub-platform; procurement and inventory management sub-platform.

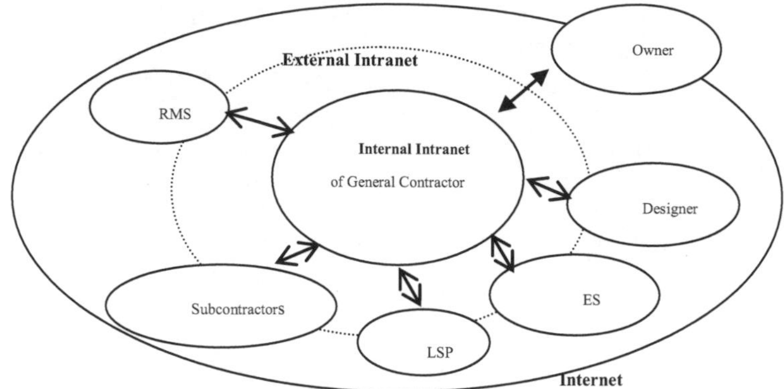

(Note: RMS-Raw Material Supplier, LSP-Logistics Service Provider, ES-Equipment Supplier)

Fig. 1 Structure of the IT-based LCSCM Model

5 Implementation of IT-Based Lean Construction Supply Chain Management Model

From the structure viewpoint, since any network is composed of the nodes and linkages among the nodes, building a lean supply chain means obtaining lean nodes as well as lean linkages among the nodes. Specifically speaking, building the LCSCM model can be carried out from the following two major aspects (see Figure 2):

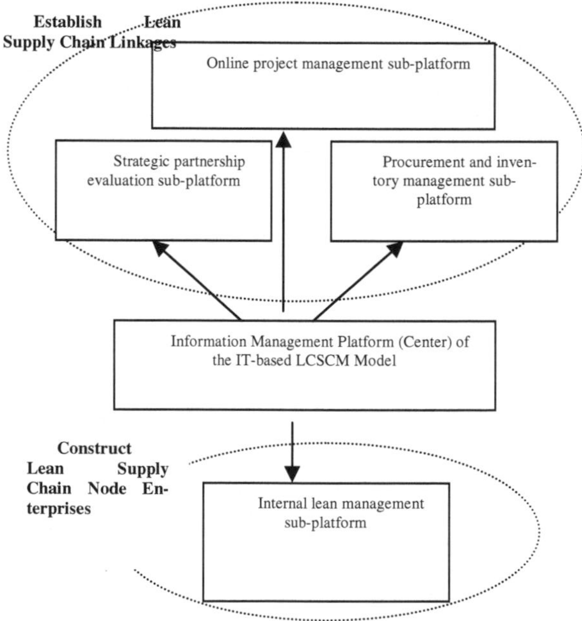

Fig. 2 Implementation of the IT-based LCSCM Model

A. Constructing Lean Construction Supply Chain Network Node Enterprises (Lean CSC Parties)

In terms of property right, our CSC parties are independent businesses and have their own interests. Therefore, it is inevitable that there would be conflicts of individual enterprise interests. In order to make LCSC operate as a whole, the idea of LCSCM requires the core company to play the role of the coordinator, conductor and manager of the entire CSC. First of all, the core company should implement internal lean management by using internal lean management sub-platform. This sub-platform enables it to conduct process-focused, standard and lean management within the enterprise including lean design, lean plan, lean supply, lean construction, lean marketing, etc. Secondly, the core company should guide and assist other CSC parties' internal lean management with the aid of inter-company intranet. Finally, through the Internet and CRM, timely communication with the owner can be ensured, which will greatly help to raise the owner's satisfaction.

B. Establishing Lean Construction Supply Chain Linkages by Forming Good Quality Inter-company Relationship among CSC Parties

Under traditional management model, the relationship among our CSC parties has always been short-term and hostile. Nowadays, good quality inter-company relationship is often cited as a highly important ingredient of effective supply chain. Therefore, another key aspect of building LCSC is to establish lean supply chain linkages by forming good quality inter-company relationship among all CSC parties with the aid of strategic partnership evaluation sub-platform, online project management sub-platform, procurement and inventory management sub-platform. There are three key points:

The first is to establish good quality inter-company relationship among CSC parties by using strategic partnership evaluation sub-platform. To simplify and optimize the CSC network, the core company should select fewer but best partners to establish strategic partnership. The main function of this sub-platform is to help construction enterprise make careful selection and correct evaluation of strategic partners. Its main tasks include: using decision support system to make initial evaluation of potential partners; selecting best strategic partners based on initial evaluation; establishing strategic partnership with selected partners; monitoring and maintaining strategic partnership; evaluating real performance of strategic partners at fixed time period, etc.

Secondly, online project management sub-platform based on inter-company Intranet is set up to ensure the sharing of information and real-time interactive communication between the core company and its CSC partners. Its key functions include: online encrypted electronic file transfer; online time processing of various engineering programs; online information updates about the progress of subcontractors, quality inspection, funds usage, etc.; online coordination to solve the uncertain events including design changes, engineering disputes, etc.

Finally, procurement and inventory management sub-platform is built to adopt innovative procurement models and integrated inventory management technologies to reduce or eliminate various forms of wastes across the entire CSC. With the aid of this sub-platform, centralized procurement and Just-in-time delivery by a dedicated Third Party Logistics (3PL) company can be adopted to greatly cut down procurement cost as well as transportation cost. Centralized procurement can help achieve economies of scale in procurement and transportation. In order to ensure timely and reliable supply of raw materials to the construction site, the delivery is performed by a dedicated 3PL company instead of subcontractors or suppliers. As for inventory management, integrated inventory management technologies including VMI and JMI can be applied to strategically plan and position inventory in the entire CSC to meet customers' requirements at the lowest inventory investment. The major functions of this sub-platform are as follows: subcontractors timely publish their inventory information online; the core company makes centralized procurement and issues purchasing orders; suppliers obtain and process purchasing orders online; the 3PL company makes distribution plan for materials according to procurement plans, etc.

6 Conclusion

The status quo and major problems of our CSC indicate that traditional management ideas and models can not adapt to the new requirements of today's fierce market competition. Therefore, this paper introduces lean thinking to our construction sector, puts forward an IT-based LCSCM model and fully elaborates its structure and implementation. With the aid of this model, not only lean CSC parties(including lean core company and lean CSC member enterprises) can be obtained, but also CSC network can be simplified and optimized by establishing lean linkages and good quality inter-company relationship among all CSC parties. The key of this IT-based LCSCM model is to build an information management platform based on Internet/Intranet which is composed of four sub-platforms: internal lean management sub-platform, strategic partnership evaluation sub-platform, online project management sub-platform and procurement and inventory management sub-platform.

References

[1] Vrijhoef, R., Koskela, L.: The four roles of supply chain management in construction. European Journal of Purchasing & Supply Chain (6), 169–178 (2000)

[2] Akintola, A., George, M., Eamon, F.: A survey of supply chain collaboration and management in the UK construction industry. European Journal of Purchasing & Supply Chain 6(3-4), 159–168 (2000)

[3] Love, P.E.D., Irani, Z., Edwards, D.J.: A seamless supply chain management model for construction. Supply Chain Management: An International Journal 9(1), 43–56 (2004)

[4] Ma, S., Lin, Y., Chen, Z.: Supply Chain Management. Mechanical Press (2003)

[5] Hao, S., Qiao, F.: Application of Supply Chain Management in EPC. Logistics Technology (6), 43–46 (2006)

[6] Wang, Y., Xue, X.: Supply chain management application research in the construction industry. Journal of Civil Engineering 37(9), 86–89 (2004)
[7] Wang, T., Xie, J.: Study on the Application of Construction Supply Chain Management Model. Construction Management Modernization (2), 5–8 (2005)
[8] He, Y., Lu, H.: Construction Supply Chain Management Based on Lean Construction. Construction and Architecture (18), 30–32, 36 (2009)

Design and Implementation of Value-Added Services Based on Parlay/OSA API*

Chunyue Zhang and Youqing Guan

Abstract. Development of value-added services in traditional telecommunication network has some problems, such as high costs and long period. API which hides the protocol details of the underlying network is used in NGN nowadays to speed up service creation and deployment. This paper describes a key development technology of value-added services—Parlay, analyses the architecture of Parlay/OSA API, and then presents a detailed design of "SMS Filter" service and "Positioning and Calling" service, so as to resolve problems of spam messages and user location respectively. At the end of this paper, the two services are implemented in the Ericsson NRG simulator.

Keywords: Parlay/OSA API, SMS Filter, Positioning and Calling, Ericsson NRG.

1 Introduction

The traditional service framework is not an open platform for service creation, because development is related to specific communication protocols and signaling[1]. So, open service access and integration with existing network resources and technologies for service creation and deployment have been gaining attention recently. NGN (Next Generation Network) [2] is a business-driven network, with the characteristic of separation of service layer, call control and underlying bearer. It is a packet-based network that can provide voice, data, multimedia and other services. The major feature of NGN services is the use of object-oriented API (Application Programming Interface) [3], which allows third party application developers to have access to network resources to create all

Chunyue Zhang · Youqing Guan
College of the Internet of Things, Nanjing University of Posts and Telecommunications, Nanjing, P.R. China
e-mail: zhangcy_19860112@163.com, guanyouq@njupt.edu.cn

* **Foundation item:** This work is supported by the Natural Science Foundation of Jiangsu Provincial Educational Commission of China (No. 05KJD520146).

F.L. Gaol (Ed.): Recent Progress in DEIT, Vol. 2, LNEE 157, pp. 507–514.
springerlink.com

kinds of value-added services, thus accelerating the development of NGN services.

The conception of service creation using API is to abstract the underlying network capabilities to a range of standard software interfaces that the application developers can use, so that it is not necessary for the developers to know the details of network technologies and protocols. It is only necessary for them to focus on service logic. Thus, service layer is separated from basic call control layer. Parlay Group[4][5] is a forum of many important organizations both in telecommunications and computer industry. The objective of the Parlay Group is to define network API to support creation of services. In this paper two new value-added services based on Parlay/OSA (Open Service Access) API are designed and implemented in Ericsson NRG (Network Resource Gateway) platform.

2 Architecture of Parlay/OSA API

The architecture of Parlay/OSA API is composed of applications, application server, Framework, Service Capability Server (SCS) and core network element, as shown in Fig 1.

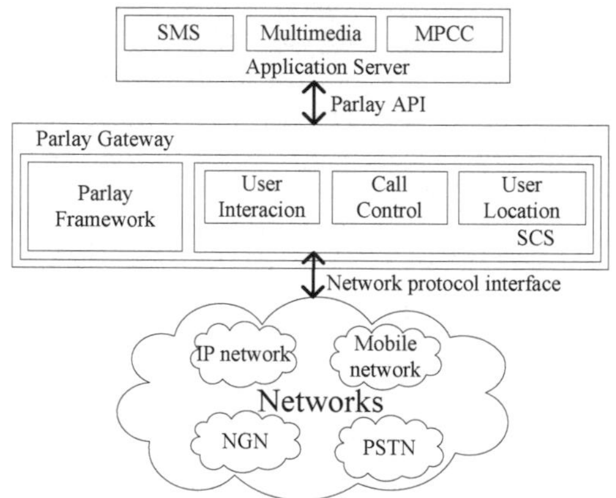

Fig. 1 Parlay Architecture

Applications: Applications refers to some specific services, like short message service, multimedia service, multi-party call control service.

Application Server: Application server is where service logic runs on, and it gains access to the underlying network by means of standard API provided by Parlay gateway.

Framework[6]: Authentication occurs between applications and Framework and also between SCS and Framework. The Framework is used to establish the identity of each entity involved. Once the mutual authentication is successful, the

Framework provides an interface which allows the application to discover what kind of SCF (Service Capability Feature) it wants. After performing service discovery, the Framework and the application will mutually sign service level agreement covering the interface of SCF that the application can use. A service instance is then created for the application, which will return the callback interface to the application [7].

SCS: SCS provides applications with the capability of the underlying network, such as Call Control, User Interaction and User Location.

Core Network Element: It includes PSTN (Public Switched Telephone Network), mobile network, IP network and NGN, etc.

Framework and SCS constitute the Parlay gateway. In general, the network operators have the standard API of Parlay gateway exposed to third party application developers, enabling the latter to create services that can utilize the underlying network. The Parlay gateway is seen as the abstraction of the underlying network from the application perspective.

3 Design and Implementation of Value-Added Services

The Ericsson NRG [8] acts as a gateway between applications and the underlying network technologies. The NRG simulates the underlying network resources for developers without real telecommunication networks. The NRG provides API with such service capabilities as Framework, Multi-Party Call Control, User Interaction, User Location and User Status for applications[9]. The NRG SDK (Software Development Kit) enables developers to speed up development of their own services using pure JAVA API easily.

3.1 Service Description

Value-added service system includes "SMS Filter" service and "Position and Calling"service.

SMS (Short Message Service) is the text communication service that comes with the development of digital mobile communication system. SMS has become the most widely used service in the world though it is a non real-time and un-voiced application. With the popularity of SMS, it brings problem of spam messages. Therefore, "SMS Filter" service is designed. By using "SMS Filter", the application intercepts all SMS messages based on the sender who is not listed as a friend of the subscriber. And if the sender is specified in the configuration as a friend, the messages will be received successfully, which avoids interference of spam messages.

Location-based service can provide reliable location of mobile end users by using some specific positioning technology. It has a wide range of applications, for example, GPS navigation, location-based information services, target location tracking. This service is used to track the location of the subscriber, and then forward to the reserved numbers of the subscriber in turn. The service comes into effect in times of emergency.

3.2 Design of the New Value-Added Services

Value-added services are created in the Ericsson NRG simulator, and the system for these new services is composed of the following functional classes: Main, ServiceMngSysCtrl, ServiceMngSysView, DescriptionPanel, SMSFilterPanel, UserLocationPanel, FWproxy, Logic, SMSProcessor, SecLogic, Map, LocationProcessor, CallLegProcessor, CallProcessor, MPCCProcessor, which are depicted in Fig 2.

The key functions of each class are as follows:

Main: loads configuration files; calls ServiceMngSysCtrl class; starts application

ServiceMngSysCtrl: adds action listener for each button of the two value-added services; makes corresponding responses when different buttons are clicked; calls ServiceMngSysView class to show different views of each panel

ServiceMngSysView: sets title for the application; initializes the menu; initializes the panel of the services; sets layout of the panel

DescriptionPanel: displays detailed function description of the services the system provides

SMSFilterPanel: provides function description of "SMS Filter" service; displays the buttons

UserLocationPanel: provides function description of "Positioning and Calling" service; displays the buttons

FWproxy: interacts with the Framework

Logic: implements service logic of "SMS Filter"; uses an FWproxy class to obtain the User Interaction service manager IpHosaUIManager from the NRG

SMSProcessor: interacts with the NRG; finishes sending and receiving messages

SecLogic: implements service logic of "Positioning and Calling"; uses an FWproxy class to obtain the User Location service manager IpUserLacation and Multi-Party Call Control service manager IpMultiPartyCallControlManager from the NRG

Map: adds label of user; moves label location; removes the label

LocationProcessor: requests the NRG for the location of the subscriber; reports the response of the event

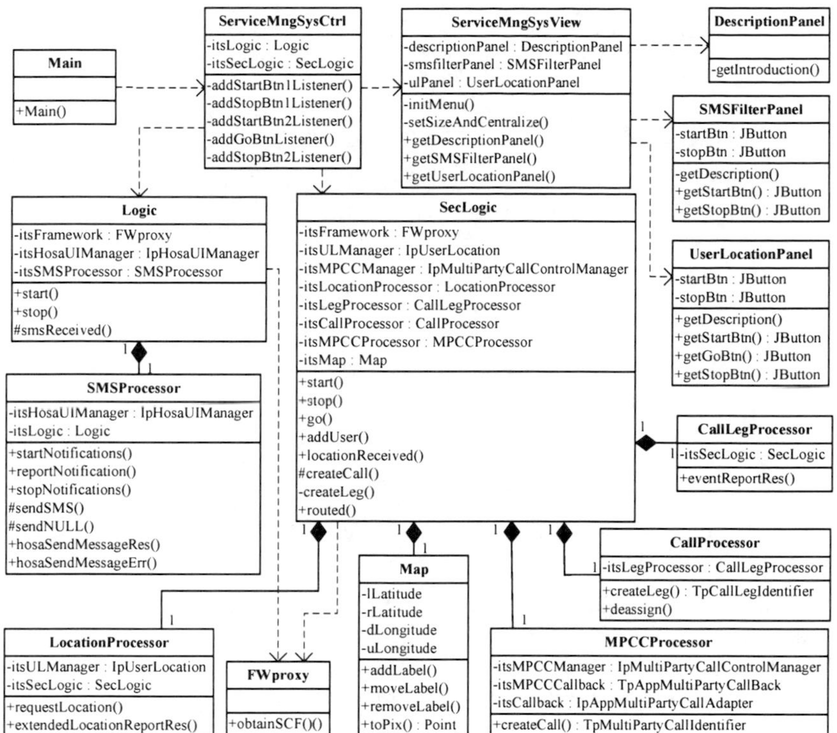

Fig. 2 Functional Classes

MPCCProcessor: creates a new multi-party call object
CallProcessor: creates and routes new legs for the reserved numbers in turn
CallLegProcessor: reports the reserved number whether it has been routed

3.3 Sequence Diagram of Value-Added Services

The sequence diagram of "SMS Filter" service is shown in Fig 3.

The addStartBtn1Listener() method of the ServiceMngSysCtrl class adds action listener for "start" button. Once the "start" button is clicked, the start() method is invoked. The application gains the User Interaction service manager IpHosaUIManager through calling the obtainSCF() method of the FWproxy class. The service manager provides many methods for the application to use. The createNotification() method is used to create specific notification criteria requesting the NRG to inform the application when the end users send SMS, the NRG then reports to the application by means of the reportEventNotification() method, and the smsReceived() method of the Logic class is called by the SMSProcessor class to judge. If the sender is a friend specified in the configuration file, the sendSMS() method is invoked to send SMS.

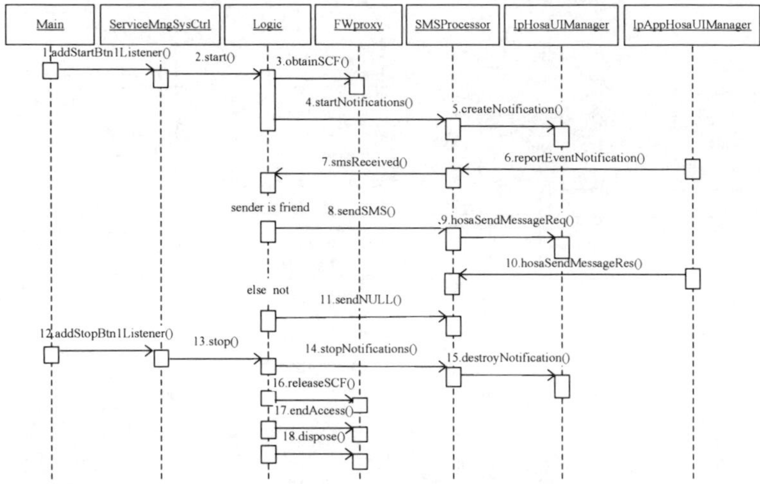

Fig. 3 SMS Filter Sequence Diagram

The hosaSendMessageReq() method is used to request the NRG to send SMS to the subscriber, while the hosaSendMessageRes() method is invoked when the SMS is received successfully. If the sender is not a friend, the sendNull() method will be called. The addStopBtn1Listener() method adds action listener for "stop" button. Once the "stop" button is clicked, the stop() method is invoked. When the application needs to be terminated, the destroyNotification() method can be invoked to destroy notification the application requests. The releaseSCF() method is invoked to release the service manager by terminating its service level agreement. The endAccess() method is called to release resources occupied by the Framework. The dispose() method is invoked to dispose of all resources allocated by the application.

3.4 Implementation of Value-Added Services

To implement the two services mentioned above it is first necessary to add end users in the NRG simulator. Select "Run" command in the "Application" menu, and value-added services window will be displayed. The main interface is composed of Description panel, "SMS Filter" panel and "Positioning and Calling" panel.

"SMS Filter" service: Click "start" button to connect to the NRG and request the service that is required for the application. In order to implement "SMS Filter" service, 10456 is added as a friend user and 10999 as a non-friend user, as is shown in Fig 4. User 10456 and user 10999 both send SMS to the subscriber 10123, but the subscriber will only receive SMS from the friend user, the spam SMS has been filtered successfully.

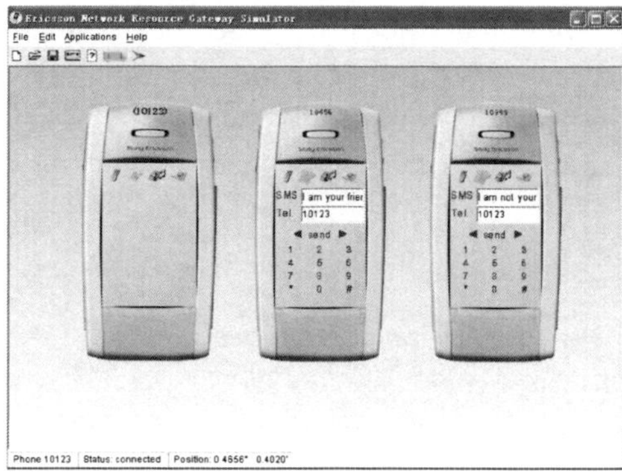

Fig. 4 SMS Sending

"Positioning and Calling" service: This service provides location of the subscriber and forwards to the reserved numbers automatically. The subscriber is 10123, and the reserved numbers are 10789 and 10456. Click "start" button to connect to the NRG and then Click "go" button to request the service that is required for the application. Once the location of the subscriber is positioned, it appears on the map, as is shown in Fig 5. At the same time, it forwards to the users 10789 and 10456.

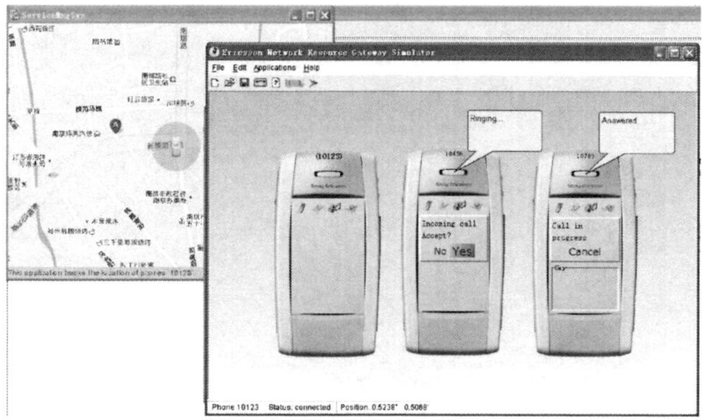

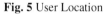

Fig. 5 User Location

4 Conclusions

Open API is a key feature of NGN and Parlay/OSA API has become factual interface standard oriented to NGN. Therefore, it is significant to research Parlay/OSA API for the development of NGN and the creation and deployment of new value-added services. Parlay/OSA API allows application development by third parties in order to speed up new service creation and deployment. A third party service provider, by using API, supplies telecommunications value-added services that can run with no dependence on the underlying network technology. The development of the new services may help achieve a further understanding of the Parlay/OSA specification, and contribute to development of applications based on Parlay/OSA API.

References

1. Wan, X., Yao, P.: The technologies of Next Generation Network Services, pp. 10–14, 76–77. Beijing University of Posts and Telecommunications Press, Beijing (2005) (in Chinese)
2. Zhang, Y., Zhang, Z., Liu, Y.: Next Generation Network open technologies, pp. 4–6. Electronic industry Press, Beijing (2004) (in Chinese)
3. Mi, Z.: Open Service Architecture and API Technologies. ZTE Technology (Z1), 33–36 (2002) (in Chinese)
4. Unmehopa, M., Vemuri, K., Bennett, A.: Parlay/OSA From Standards to Reality, pp. 49–126. John Wiley&Sons, Ltd, England (2006)
5. Wang, H., Zou, H., Yang, F.: A Parlay Based Service Creation Architecture for Next Generation Network. Journal of Beijing University of Posts and Telecommunications 23(9), 15–20 (2004) (in Chinese with English abstract)
6. ETSI ES 202 196-3. Open Service AccessApplication Programming Interface Part3: Framework
7. Liu, B., Zhou, W., Song, J.: Research and Implementation of 3G Service Simulation Based on Parlay. Computer Applications 23(9), 221–223 (2006) (in Chinese with English abstract)
8. Ericsson. Ericsson Network Resource Gateway User Guide, pp. 27-92. Ericsson Press (2005)
9. van Eck, E.: Ericsson NRG SDK for Parlay/OSA(EB/OL) (November 1, 2010), http://www.wirelessdevnet.com/articles/ericsson/index.html

A Design of Integrated Management System for Building Constructions Based on USNs

Byoung-Kug Kim, Won-Gil Hong, and Doo-Seop Eom

Abstract. This paper designs an Integrated Management System for building constructions based on USNs, which consist of diverse sensors, actuators and RFID attached devices so that the system can monitor the status the construction environments via the USNs and analyze the status information of the construction sites. Through the designed system, we expect the construction management to be carried out efficiently and rapidly for managing and monitoring the resources of workers, materials and equipment. Finally, the system can reduce both the construction cost and its building period.

1 Introduction

Construction industry has increased steadily. Contrastingly, in the case of resource management, including workers, materials and equipment for building constructions, an efficient management system for the construction is still required.

These days, many accurate sensor devices have emerged and some of them have the ability to transmit data with their RF devices and run simultaneously with some batteries. Therefore, the sensors can be deployed in any-place and enable the networked node to run.

The advantages of the ubiquitous sensor networks can facilitate a more rapid growth of the construction industry and enable to manage building constructions in almost real time. To establish a management system for the building construction based on USNs, the sensor nodes and RFID [1] tags can be mounted

---

Byoung-Kug Kim · Doo Seop Eom
Dept. of Electronics and Computer Engineering, Korea University, South Korea
e-mail: dearbk@final.korea.ac.kr, eomds@final.korea.ac.kr

Won-Gil Hong
Sensorway Co., Ltd., South Korea
e-mail: hong@sensorway.co.kr

F.L. Gaol (Ed.): Recent Progress in DEIT, Vol. 2, LNEE 157, pp. 515–523.
springerlink.com

on construction workers, materials and equipment. The RFID technology is generally used to identify the existences of materials and equipment. Accordingly, it is appropriate to take the RFID technology into account for progress checking, cost estimating and inventory managing. [2, 3, 4].

In the case of the building constructions, the methods of BIM (Building Information Modelling) [5, 6, 7] and PMIS (Project Management Information System)[8] are still mostly used.

Firstly, the BIM is the process of generating and managing building information. In general, this supports either three- or four- dimensional modelling systems depending on time information. Through the BIM, the building geometry, spatial relationships, geographical information, quantities and properties for building materials can be simulated.

Secondly, the PMIS is a part of MISs (Management Information Systems) and the project management is practically centralised. PMIS provides the management of all stakeholders in, which the people, who work in the same project, participate and accomplish the steps for its completion. Furthermore, through the PMIS, a manager can also reach a compromise amongst the workforce.

Both BIM and PMIS can be useful for building construction. However, these tools cannot be used to monitor the progress of the building construction and can't support managing the project in real time either. To the best of our knowledge, through using state of the art variable sensors with building construction materials, we could inspect the construction sites and achieve a better understanding of their status. Then finally, they could maximize efficiency of real time monitoring and managing for the building site.

This paper designs a system for building constructions named Integrated Management System using the techniques of diverse sensors and RFIDs (RF Identifies) based on USNs (Ubiquitous Sensor Networks). Through the designed system, the work for the building construction could be done in a shorter period with a lower cost.

The sections of this paper are organized as follows. Section II: An introduction on related researches for constructions with USNs supporting a real time service. Next, we design an integrated management system for building constructions in Section III. Finally, we make some results and conclude that the designed system could be suitable for reducing the period and the cost for the building constructions in the remaining sections.

2 Related Works

Every device, which has capabilities of sensing its environments and transmitting its sensed data to a root node (called a sink node) via WSNs (Wireless Sensor Networks), can become a member of the USNs. A sensor node typically adapts diverse sensors and actuators on it to be aware of its surroundings and to react by controlling its actuators from sensor managers of a root node through its WSNs.

The core technologies for the WSNs can be divided into sensors, processors, communication, a tiny operating system etc. In order to organize a sensor network and to connect to the Internet, each WSN requires sensor nodes, router nodes, a sink node and optionally a gateway, unless the sink node has an Ethernet functionality to connect to the Internet as shown in Fig.1.

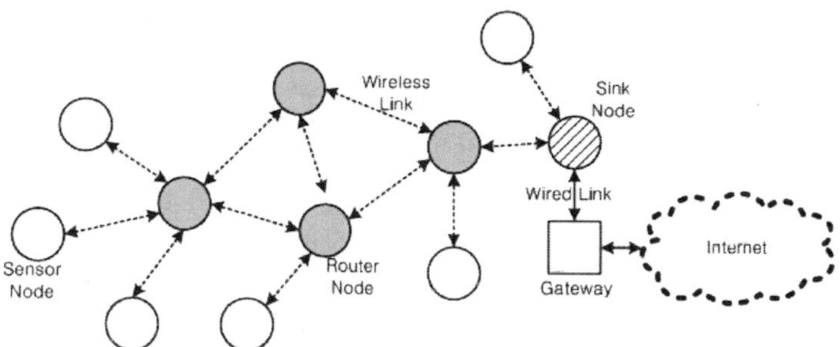

Fig. 1 The Construction of nodes for WSNs.

Sensor Nodes have diverse sensors to measure the situation around their environments and some of them also have actuators. Every sensor node is operated with its own batteries so it has a restriction of running time. In order to reduce the power consumption in the sensor nodes, they can co-operate with each other in the same sensing area so that every sensor node can save its energy and finally the WSN can be maintained longer.

Route Nodes generally have all of the functionalities of a sensor node. In addition, they provide networking amongst sensor nodes and a sink node to pass data, which a sensor node has put in, to the sink node and vice-versa.

The Sink Node has the same capabilities of both Sensor Nodes and Route Nodes. It also has an external interface for either gateway or Internet connection. The Sink Node can form its own network, called WSN, and manage all of its WSN and networked members. Depending on the capacity of the Internet connectivity, a gateway device can be required.

In terms of *Zigbee* technology, the type of Sensor Nodes are referred to as *RFD (Reduced Function Device)* and the type of Route Nodes are named *FFD (Full Function Device)* in the same manner as the Sink Node, which is called the *Coordinator Device.*

WSNs are suitable for sensing environments. However, in the case of controlling the influx of building materials, the RFID technology can be more appropriate. The RFID is also a part of the USNs. To check existing materials, we can attach some RFID tags on the materials. Then, by using the RFID reader, we can check, which kinds of materials come in. Moreover, the information on the RFID tag could be transmitted to a server through the WSNs. Therefore, by using both WSNs and RFID technology, construction managers can finally process their work efficiently and precisely in real time.

Table 1 Device Functionalities

USN equipment	Functionalities
Sensor Nodes	The nodes measure their environments and periodically report their data to a sink node.
Router Nodes	The nodes do their work as a sensor node does and they forward data like an intermediate node does, as well forwarding RFID reader's data to a sink node.
A Sink Node	It forms a network topology called WSN and manages the network. It gathers data from all sensor nodes and RFID readers via its WSN as well.
A Gateway	It relays data between a sink node and an Internet server.
RFID tags	It identifies the material and its ID is broadcasted to an RFID reader when the reader requests to send it back.
RFID Readers	It reads RFID tags and informs the data to Router Nodes for WSNs.

3 System Design

In the past, MEMS (Micro Electro Mechanical Systems) technology has developed hugely. As the MEMSs have developed, the sensors and RFID for USNs have developed too. Fig.2 shows how to employ USN devices and diverse communication devices such as GPS satellites, Smart Phones with either 3 or 4G in order to use them in building construction. Through networked backbones, which involve Internet based on wireless telecommunication systems, WSNs, existing LANs etc., the information for the building constructions could be delivered rapidly and be re-manufactured so that all groups of construction workers would improve their co-operation.

The workers can share their jobs if they have both reporting and monitoring systems. To do this efficiently, using a smart phone would be appropriate for those workers. In general, most 3G smart phones have capabilities to stream multimedia data and voice information. Additionally, construction managers can use a management application, which has been developed for building construction, to monitor the construction workers and construction materials. Even if the construction workers are out-doors, they can still alternatively use a mobile office APP. However, it is uncountable for measurement of the construction materials. Therefore, we adapt RFID tags on sampled materials. (e.g. one tag on every 100 materials)

Fig.3 designs the diagram of Integrated Management System for building constructions using USNs. We classified the system modules into the following functional criteria; Material Manager, Mobile Node Monitor, Data Pre-processor, Query Engine, Mobile Office Server and Common Manager. The DBs for both the User Management and the Construction Resource were built. To support all the aforementioned, the Common Manager was centralized and it scheduled the process of each module.

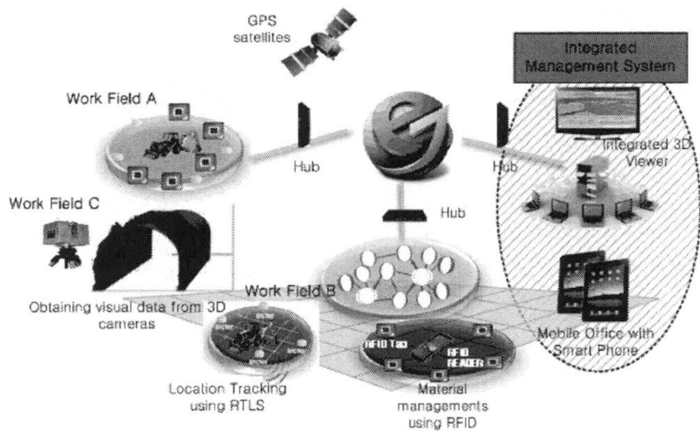

Fig. 2 Overview of Integrated Management System for Building Constructions

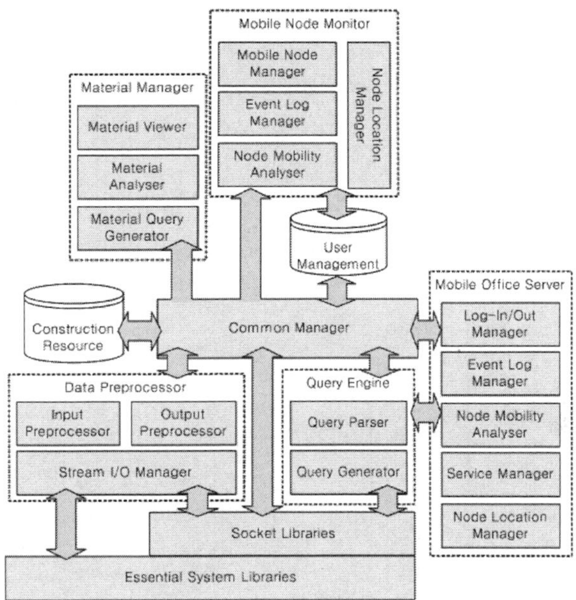

Fig. 3 A Block Diagram for Integrated Management System for Building Construction with WSNs

The Materials Manager's responsibility is to manage construction materials. This manager provides a GUI interface related with material statement so that construction workers can notify, either what kinds of construction materials were there, or, when they came in at a certain time. This way they can estimate the progress of the building construction. The data of construction materials is stored in the Construction Resource DB and the Material Analyzer uses the DB.

The Data Pre-processor gets the data of construction materials, labourers and construction equipment etc., parses the data and generates new data for the DB. Finally, the pre-processor stores it into the Construction Resource DB.

For this system, every construction worker should put on a suit, on which a WSN supporting sensor node could be attached. Through the sensor node, the Integrated Management System can become aware of the worker's status, i.e. worker's location, working time, precaution and health condition.

The modules of the Material Manager and the Mobile Node Monitor were especially designed for PC-like systems. It is imperative that the new application, which supports mobility, should be required for working outside. Therefore, we designed an application running as an APP in smart phones (called Mobile Office Client) for mobile managers as shown in Fig.4. To implement this, we designed the Mobile Office System on the Client/Server model.

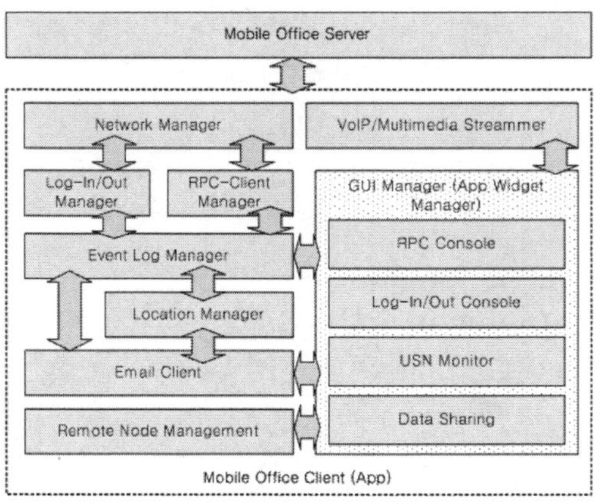

Fig. 4 A Block Diagram of Mobile Office Client

4 Results

Based on the design, we have implemented the Integrated Management System with UI for building constructions. (The Fig.5 is an example of the UIs via a Management Server and a smart phone.) Through this, construction managers could monitor the progress of building status and workers' behaviors, on an individual basis, and would analyze how much work has been done in real-time.

Fig.6 shows the expected effectiveness of managing construction resources using the integrated management system. Traditionally, construction managers can only control workers in the worksite and certain managers carry out most material and equipment management manually. Therefore, certain managers who can manage construction materials and equipment should be required.

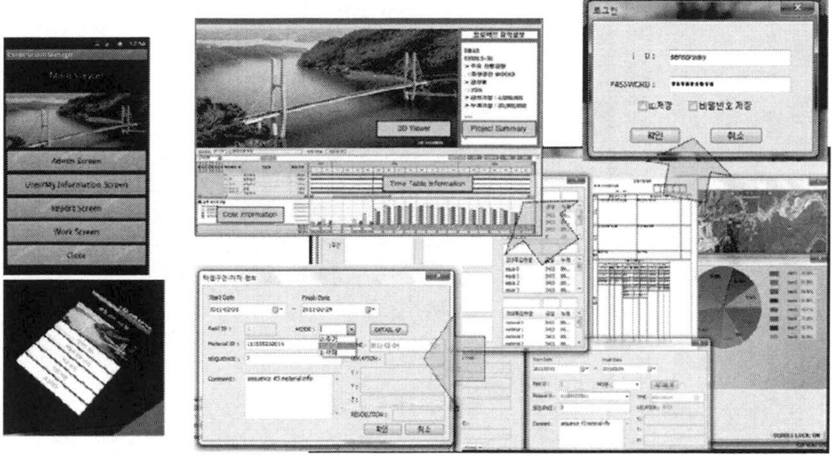

Fig. 5 Working Examples Integrated Management System via a Management Server and a Smart Phone

In order to cope with this problem, the integrated management system, commitment and the active mass of each worker are automatically checked and measured. Additionally, checking the existences of construction materials and equipment can be done via RFID. Furthermore, all of the above resources' locations can be logged into the system and used to analyse their tracking.

Consequently, we expect that integrated management system can increase labour productivity and also reduce the number of employees, who work for a certain construction site. In addition, it can also reduce the total cost for managing both materials and equipment for the building construction. By monitoring construction resources in functionalities through the system, this can be further employed as an antitheft system.

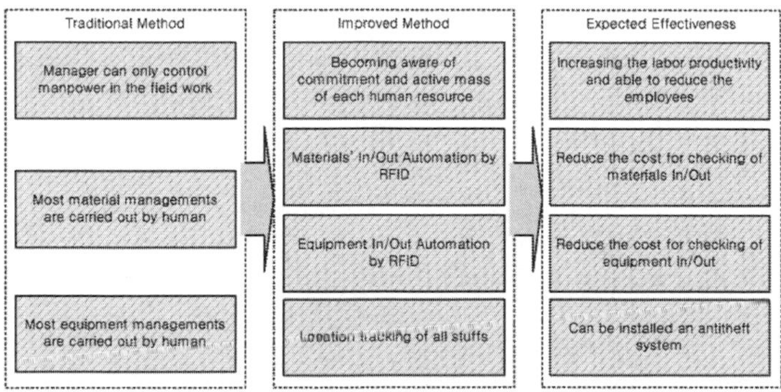

Fig. 6 Major Alterations in the Management of Construction Resources

In the case of risk management, people who work on site as a construction manager have to manage the risk and monitor both the status and the quality of a building construction in the traditional method as shown in Fig.7. The integrated management system can overcome the lack of risk management.

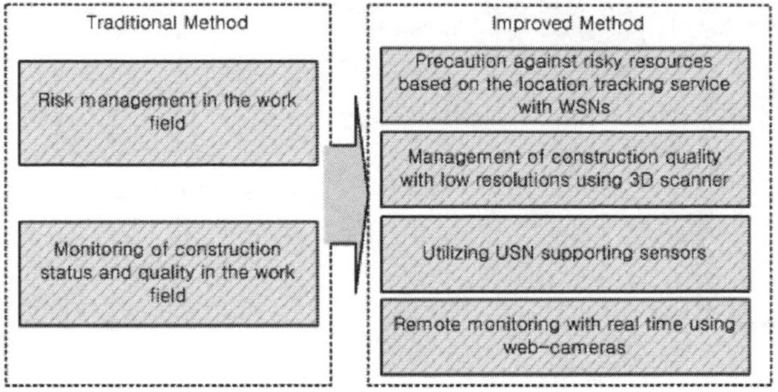

Fig. 7 Major Alterations in the Risk Management

5 Conclusions

The methods of BIM and PMIS are commonly used in construction fields. However, the BIM is mostly used for modelling and simulating of building constructors but the PMIS is used for real work fields. Therefore, the construction fields require a different system for managing and controlling of the real work fields. The Integration Management System based on the USNs can be useful for this.

By incorporating USN technology into the construction building industry, it will allow for a more efficient and precise management of resources in real time. In order to facilitate the improvement of throughput in building construction, we designed and implemented an Integration Management System based on USNs, and we outlined some expected benefits, which could be used by the construction industry.

Through the systemized building construction management, we expect that the cost and the period for the building construction can be reduced. Additionally, our designed management system is expected to improve the working capability of an organization, no matter what the size.

In the future, we aim to realize its implementation into the building construction industry. We plan to show its suitability, potential and measure its throughput. Finally, we intend to apply this system to both bridge and tunnel constructions.

Acknowledgments. This work was supported by the Korea Institute of Construction & Transportation Technology Evaluation and Planning (KICTEP) (Program No.: 09 Technology Innovation-E05).

References

1. Want, R.: An introduction to RFID technology. IEEE Pervasive Computing 5(1), 25–33 (2006)
2. Bae, K.-S., Lee, S.-B.: Status of practical use of RFID/USN in construction industry. In: Conference on Architectural Institute of Korea, pp. 423–426 (October 2007)
3. Lee, S.C., Kim, C.S.: Design and Implementation of Remote Monitoring System for Supporting Safe Subways Based on USN. SCI, vol. 142, pp. 321–330 (2008)
4. Kim, K.-T.: A Study on the Implementation of USN Technologies for Safety Management Mornitoring of Architectural Construction Sites. In: The Conference on the Korea Institute of Building Construction, vol. 9-4, pp. 103–107 (August 2009)
5. Won, J.-S., Lee, J.-J., Lee, G.: A Case Study On BIM Collaboration and Information Management Methods. The Journal of Architectural Institute of Korea 24(8), 3–330 (2008)
6. Moon, H.-J., Choi, M.-S., Ryu, S.-H., Park, J.-W.: Building Performance Analysis Interface based on BIM. The Journal of Architectural Institute of Korea 25(10), 271–278 (2009)
7. Eastman, C., Teicholz, P., Sacks, R., Liston, K.: BIM Handbook: A Guide to Building Information Modeling for Owners, Managers, Designers (March 2008) ISBN 978-0-470-18528-5
8. Raymond, L., Bergeron, F.: Project management information systems: An empirical study of their impact on project managers and project success. International Journal of Project Management 26(2), 213–220 (2008)

Design of Public Practical Training Service Platform Based on Cloud Computing for Students Majoring in Software

Jing Tang and Wen-ai Song

Abstract. For some problems in Chinese colleges and universities such as misdistributions and low share of teaching resources, the good and bad were intermingled in graduates majoring in software. This paper introduces a public practical training service platform. This platform is based on cloud computing which will realize virtual and network training. Firstly presents the characters about cloud computing. These characters can gather all kinds of practical training recourses, so it will happen that practical training is regarded as a public service. Then designs the architecture for public practical training service platform and introduces how students practice with this platform; provides major technologies. At last expatiates the advantages of the platform.

1 Introduction

Recently Chinese software industry is developing at full speed; global outsourcing industry has also transferred to China. To satisfy increasing demand for IT talent, Ministry of education of PRC began the plan to construct schools of software from 2001[1], and then some provinces and cities constructed schools or institutes of software. But for some problems such as misdistributions and low share of teaching resources[2], there was a huge difference in practical teaching in Chinese colleges and universities. A few graduates studied at their own expense in software technology training organizations such as APTECH, Tarena technology, which virtually made waste of teaching resources.

Jing Tang · Wen-ai Song
Software School of North University of China
e-mail: tj_ma@hotmail.com, songwenai@gmail.com

F.L. Gaol (Ed.): Recent Progress in DEIT, Vol. 2, LNEE 157, pp. 525–532.
springerlink.com

Will it happen that practical training is regarded as a public service combined with excellent teachers in famous universities, IT cases from enterprises, updated equipments and instruments and so on. It was too difficult to realize. Now with the development of network, the appearance of cloud computing provides a great probability of solution. This paper introduces a new practical training mode based on cloud computing: virtual and network training, researches a popular, efficient, standardized public practical training service platform for students majoring in software.

2 Characters of Cloud Computing

Jian-xun Zhang[3] concluded the characters of cloud computing after analyzed and summarized current cloud computing systems. As one of characters, virtualization breaks down the physical barriers. Because there are three layers where clouds are used: Infrastructure as a Service (IaaS), Platform as a Service (PaaS), and Software as a Service (SaaS)[4]. Infrastructure, application platform and software are regarded as services to users. So virtualized equipments and instruments, development tools in public practical training service platform can update in time. Cloud computing manager collocates and manages these recourses. QoS can provide standardized services to users and guarantee high quality practical training service. Scalability and self-management offers the automatic resizing of virtualized hardware recourses such as server and storage, so reduces manual workload of data center and improves efficiency of deployment, running and maintenance, strengths usability and flexibility. Finally reduces running cost according to strategy and target.

Recently researchers have used the concept of cloud computing into higher education. Yang Wang[5] designed a cloud computing education service platform based on P2P combined with the educational practice of computer curriculums in college; Ze-ang Zhang[6] studied an education platform based on cloud computing and analyzed the operation of advantages using this platform to carry out teacher training.

3 Architecture of Public Practical Training Service Platform

The public practical training service platform is based on cloud computing resource management platform, combines all kinds of hardware and software. According to a web portal, the platform provides students with open source, development tools and different practical training to satisfy different demands. The architecture reads as follow.

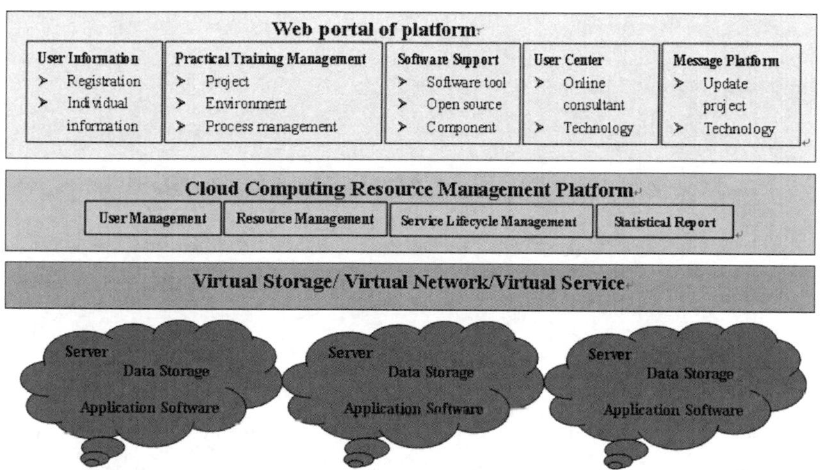

Fig. 1 Architecture of public practical training service platform

3.1 Web portal of platform

The functional modules as follows:

3.1.1 User Information

User information includes practical training student's registration, login, individual information and report form. Report form contains training schedule, process, and occupation of recourses.

3.1.2 Practical Training Management

Practical training project in this module is that which project student chooses; according to chosen project, then selects operation system, development tools data-base etc. in **practical training environment**. **Process management** supervises the training process with CMM and guarantees high quality. **Project testing need** relies on another cloud testing platform that isn't introduced in this paper.

3.1.3 Software Support

Namely application software such as software tools, open sources, component architecture etc.

3.1.4 User Center

If student has questions, he can get online-consultant at any moment; or log in technology forum to discuss with others.

3.1.5 Message Platform

This module will update new project from IT enterprises, and release updated technology messages.

3.2 *Cloud Computing Resource Management Platform*

The functions in service platform are based on cloud computing resource management platform.

As follows:

3.2.1 User Management

As soon as training student logins in portal website, cloud computing platform will provide computing recourse management: whether this user can use computing recourse, the amount of computing recourses used and the relationship of application projects. Above all need perfect user management system in data center. Cloud user management module is based on LDAP which is easy to extend and assemble with other systems. At the same time user management system should run with server, storage and net recourses management.

In our case, there are three kinds of users: cloud computing platform manager, practical training instructor and student.

Cloud computing platform manager can examine and approve all projects in the platform, add or reduce the amount of resource, change project duration, end or cancel project.

Practical training instructor can apply resource distribution according to project demand, so he can manage the project.

Practical training student can do some self-service operations.

3.2.2 Resource Management

According to cloud centralized resource management, cloud computing manager can manage all computing resource in clouds including server, web, storage and software etc.By using this module, cloud computing manager can add, cancel, mend and deploy. Centralized resource management module offer Web access interface, background is composed with resource data-base, middleware module and resource interface. Via centralized resource management, cloud computing manager can do:

- Physical equipments collocate and manage
- System platform collocate and manage
- Application software collocate and manage (including operation system, middleware, data-base etc.)
- Web resource collocate and manage

Convenient web access: cloud centralized resource management module uses B/S architecture, user uses browser to get uniform management. Web access protocol adopts HTTPS Security encryption to protect efficiently the safety of user's core computing message.

Resource data-base: resource data-base stores all computing resource messages which can realize classifying, sortable computing resource management.

Resource management access: now cloud computing is able to manage thousands of different manufactures, different kinds of physical equipments, operation system and application software. To every computing resource, resource management access does manage.

3.2.3 Service Lifecycle Management

Service lifecycle management module can provide the whole service from application, reservation, examination and approval, supply, mend to free. Just clicking mouse can finish all these service flows.

3.2.4 Statistical Report

Cloud management platform provides two kinds of statistical reports:

- The statistical report of usage of the whole platform resource, each resource pool and correlative services
- The statistical report of details of every user's usage including duration, type and amount.

4 Practical Training Process

After student chooses practical training project and environment in public practical training service platform, cloud computing resource management platform begins to build system circumstance via demand (Figure 2).

- Confirm required machine type, amount, operation system and software;
- Confirm beginning time and ending time of project;
- Choose required computing resource from available resource (server, storage, web);
- Choose required operation system;
- Choose required application software;
- Submit application;
- Develop project after application approval.

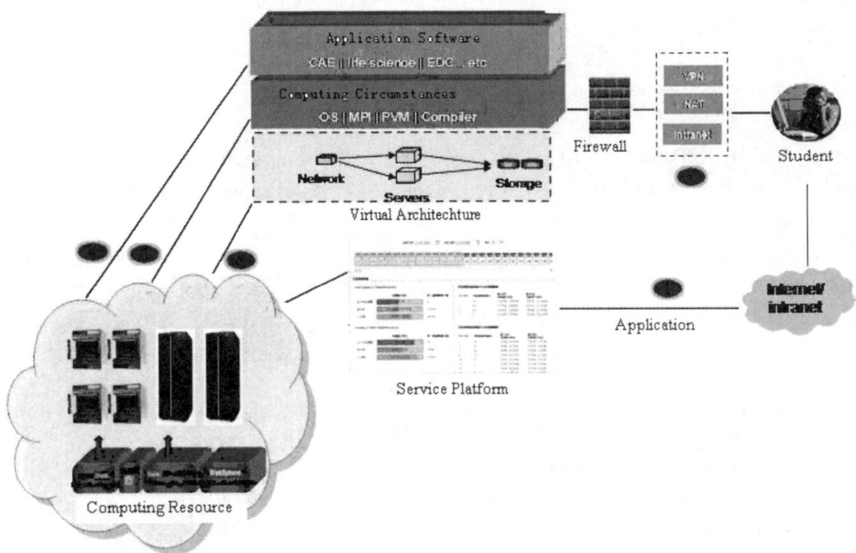

Fig. 2 Service flow

For example: for software engineering curriculum, we can divide students into a few project groups. In clouds each project group will get a project development environment to simulate a true software development project.

For each project, we can distribute a few virtual machines to each student in cloud computing platform. They have their own work desktop. Access method may be chosen from graphic desktops such as long-distance desktop, XWindow etc. By using this access method, does not install or only install lightweight software in clients. Next, customizes running environment and group work space in each virtual machine according to project demand. For example constructs Oracle data-base and Tomcat server for project 1, constructs DB2 data-base and WAS server for project 2, and installs correlative software such as demand management, architecture design, group development, quality management, code testing and so on.

So this virtual, network practical training makes students experience to do project from business needs to design requirements, and then develop, testing to last application. In such a whole process, everyone also simulates each role with responsibility taken. It is realized that theory into practice.

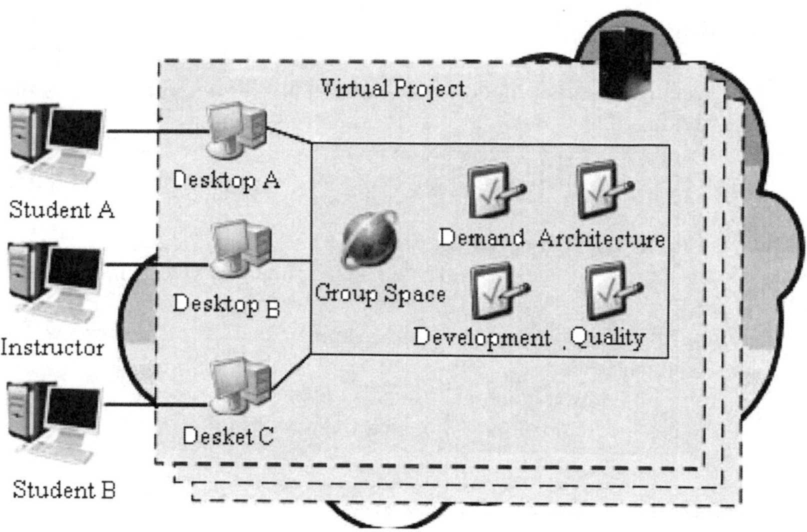

Fig. 3 Sketch map of virtual, network practical training based on cloud computing

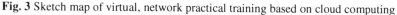

5 Main Technologies

Client-cloud Computing

In client-cloud computing, information is permanently stored in datacenters on the Internet and cached temporarily on clients that range from single chip sensors, handhelds, notebooks, desktops, and entertainment centers to huge data centers. Client-cloud computing will provide new capabilities including the following[7]:

- Maintaining the privacy of client information by storing it on datacenters encrypted so that it can be decrypted only by using the client's private key.
- Allowing greater integration of information obtained from datacenters of competing integrators.
- Allowing clients to provide convenient ways to share their information.

Virtualization

Virtualization is the key technology in cloud computing[8], virtualization can partition hardware and thus provides flexible and scalable computing platforms.

Web service and SOA[9]

Cloud services are normally exposed as Web services which follow the industry standards like WSDL, SOAP and UDDI . The service organization and orchestration inside clouds could be managed in a Service Oriented Architecture (SOA). A set of Cloud services furthermore could be organized in a SOA, make themselves available on various distributed platforms and could thus be accessed across networks.

6 Advantage and Conclusion

Traditional practical training mode for students majoring in software was limited by training field, students must arrive at practical training base; or limited by equipment and personnel. The actual training time is very limited. The platform mentioned in this paper help realize public service function and improve training efficiency.

Network training environment makes areas with Internet or LAN training location. Apply teaching resources via Internet, share public services in platform. This training mode also reduces training cost of school. Because cloud computing provides automatization and standardization management for IT resources: centralized resource management.

Virtual environment help students use all kinds of resources incessantly, which virtually extends the actual training time and improve training efficiency.

References

1. MOE of PRC, Inform of constructing software school, MOE, (2001), http://www.moe.gov.cn/publicfiles/business/htmlfiles/moe/s3862/201010/xxgk_109642.html
2. Zhang, Y.-Y., Zhao, G.-Y., et al.: Cloud computing and conformity of university teaching resources. Modern Science, 23 (April 2009)
3. Zhang, J.-X., Gu, Z.-M.: Survey of research progress on cloud computing. Application Research of Computers (2), 430–433 (2010)
4. Vaquero, L.M., Rodero-Merino, L., et al.: A Break in the Clouds: Towards a Cloud Definition. ACM SIGCOMM Computer Communication Review (39), 50–55 (2009)
5. Wang, Y., Yan, Y.-T.: Design and implementation of cloud computing educational service platform based on P2P. Computer Education (16), 147–150 (2010)
6. Zhang, Z.-A., Wu, J.-W.: A study on cloud-computing-based education platform. Distance Education in China, 66–69 (June 2010)
7. Hewitt, C.: historical perspective on developing foundations for iInfo^{TM} information systems: iConsult^{TM} and iEntertain^{TM} apps using iInfo^{TM} information integration for iOrgs^{TM} information systems, October 4 (2010), http://arxiv.org/abs/0901.4934
8. Jones, M.T.: Cloud computing with Linux cloud computing platforms and applications, Septembr 10 (2008), http://www.ibm.com/developerworks/library/l-cloud-computing/
9. Wang, L., Tao, J., et al.: The Cumulus Project: Build a Scientific Cloud for a Data Center, http://www.chinacloud.cn/download/research/Cumulus.pdf

Author Index